Mathematica By Example

Third Edition

Mathematica By Example

Third Edition

Martha L. Abell and James P. Braselton

ELSEVIER
ACADEMIC
PRESS

Amsterdam   Boston   Heidelberg   London   New York   Oxford   Paris
San Diego   San Francisco   Singapore   Sydney   Tokyo

Senior Acquisition Editor:	*Barbara Holland*
Project Manager:	*Brandy Palacios*
Associate Editor:	*Tom Singer*
Cover Design:	*Eric Decicco*
Interior Design:	*Julio Esperas*
Composition:	*Integra*
Printer:	*Maple Vail Press*

Elsevier Academic Press
200 Wheeler Road, Burlington, MA 01803, USA
525 B Street, Suite 1900, San Diego, California 92101-4495, USA
84 Theobald's Road, London WC1X 8RR, UK

This book is printed on acid-free paper. ∞

Library of Congress Cataloging-in-Publication Data
Abell, Martha L., 1962-
   Mathematica by example / Martha L. Abell, James P. Braselton. – 3rd ed.
     p. cm.
   Includes index.
   ISBN 0-12-041563-1
     1. Mathematica (Computer file) 2. Mathematics–Data processing. I. Braselton, James P., 1965- II. Title.

QA76.95.A214 1997
510'.285536–dc22

2003061665

British Library Cataloguing in Publication Data
A catalogue record for this book is available from the British Library

ISBN: 0-12-041563-1

For all information on all Academic Press publications
visit our website at www.academicpressbooks.com

PRINTED IN THE UNITED STATES OF AMERICA
03 04 05 06 07 08 9 8 7 6 5 4 3 2 1

Contents

Preface

Mathematica By Example bridges the gap that exists between the very elementary handbooks available on Mathematica and those reference books written for the advanced Mathematica users. *Mathematica By Example* is an appropriate reference for all users of Mathematica and, in particular, for beginning users like students, instructors, engineers, business people, and other professionals first learning to use Mathematica. *Mathematica By Example* introduces the very basic commands and includes typical examples of applications of these commands. In addition, the text also includes commands useful in areas such as calculus, linear algebra, business mathematics, ordinary and partial differential equations, and graphics. In all cases, however, examples follow the introduction of new commands. Readers from the most elementary to advanced levels will find that the range of topics covered addresses their needs.

Taking advantage of Version 5 of Mathematica, *Mathematica By Example*, Third Edition, introduces the fundamental concepts of Mathematica to solve typical problems of interest to students, instructors, and scientists. Other features to help make *Mathematica By Example*, Third Edition, as easy to use and as useful as possible include the following.

1. **Version 5 Compatibility.** All examples illustrated in *Mathematica By Example*, Third Edition, were completed using Version 5 of Mathematica. Although most computations can continue to be carried out with earlier versions of Mathematica, like Versions 2, 3, and 4, we have taken advantage of the new features in Version 5 as much as possible.

2. **Applications.** New applications, many of which are documented by references, from a variety of fields, especially biology, physics, and engineering, are included throughout the text.

3. **Detailed Table of Contents.** The table of contents includes all chapter, section, and subsection headings. Along with the comprehensive index, we hope that users will be able to locate information quickly and easily.

4. **Additional Examples.** We have considerably expanded the topics in Chapters 1 through 6. The results should be more useful to instructors, students, business people, engineers, and other professionals using Mathematica on a variety of platforms. In addition, several sections have been added to help make locating information easier for the user.

5. **Comprehensive Index.** In the index, mathematical examples and applications are listed by topic, or name, as well as commands along with frequently used options: particular mathematical examples as well as examples illustrating how to use frequently used commands are easy to locate. In addition, commands in the index are cross-referenced with frequently used options. Functions available in the various packages are cross-referenced both by package and alphabetically.

6. **Included CD.** All Mathematica input that appears in *Mathematica By Example*, Third Edition, is included on the CD packaged with the text.

We began *Mathematica By Example* in 1990 and the first edition was published in 1991. Back then, we were on top of the world using Macintosh IIcx's with 8 megs of RAM and 40 meg hard drives. We tried to choose examples that we thought would be relevant to beginning users – typically in the context of mathematics encountered in the undergraduate curriculum. Those examples could also be carried out by Mathematica in a timely manner on a computer as powerful as a Macintosh IIcx.

Now, we are on top of the world with Power Macintosh G4's with 768 megs of RAM and 50 gig hard drives, which will almost certainly be obsolete by the time you are reading this. The examples presented in *Mathematica By Example* continue to be the ones that we think are most similar to the problems encountered by beginning users and are presented in the context of someone familiar with mathematics typically encountered by undergraduates. However, for this third edition of *Mathematica By Example* we have taken the opportunity to expand on several of our favorite examples because the machines now have the speed and power to explore them in greater detail.

Other improvements to the third edition include:

1. Throughout the text, we have attempted to eliminate redundant examples and added several interesting ones. The following changes are especially worth noting.

(a) In Chapter 2, we have increased the number of parametric and polar plots in two and three-dimensions. For a sample, see Examples 2.3.8, 2.3.9, 2.3.10, 2.3.11, 2.3.17, and 2.3.18.

(b) In Chapter 3, Calculus, we have added examples dealing with parametric and polar coordinates to every section. Examples 3.2.9, 3.3.9, and 3.3.10 are new examples worth noting.

(c) Chapter 4, Introduction to Lists and Tables, contains several new examples illustrating various techniques of how to quickly create plots of bifurcation diagrams, Julia sets, and the Mandelbrot set. See Examples 4.1.7, 4.2.5, 4.2.7, 4.4.6, 4.4.7, 4.4.8, 4.4.9, 4.4.10, 4.4.11, 4.4.12, and 4.4.13.

(d) Several examples illustrating how to graphically determine if a surface is nonorientable have been added to Chapter 5, Matrices and Vectors: Topics from Linear Algebra and Vector Calculus. See Examples 5.5.8 and 5.5.9.

(e) Chapter 6, Applications Related to Ordinary and Partial Differential Equations, has been completely reorganized. More basic–and more difficult–examples have been added throughout.

2. We have included references that we find particularly interesting in the **Bibliography**, even if they are not specific Mathematica-related texts. A comprehensive list of Mathematica-related publications can be found at the Wolfram website.

```
http://store.wolfram.com/catalog/books/
```

Finally, we must express our appreciation to those who assisted in this project. We would like to express appreciation to our editors, Tom Singer, who deserves special recognition for the thoughtful attention he gave to this third edition, and Barbara Holland, and our production editor, Brandy Palacios, at Academic Press for providing a pleasant environment in which to work. The following reviewers should be acknowledged: William Emerson, Metropolitan State University; Mariusz Jankowski, University of Southern Maine; Brain Higgins, University of California, Davis; Alan Shuchat, Wellesley College; Rebecca Hill, Rochester Institute of Technology; Fred Szabo, Concordia University; Joaquin Carbonara, Buffalo State University. We would also like to thank Keyword Publishing and Typesetting Services for their work on this project. In addition, Wolfram Research, especially Misty Mosely, have been most helpful in providing us up-to-date information about Mathematica. Finally, we thank those close to us, especially Imogene Abell, Lori Braselton, Ada Braselton, and Mattie Braselton for enduring with us the pressures of meeting a deadline and for graciously accepting our demanding work schedules. We certainly could not have completed this task without their care and understanding.

Martha Abell   (E-Mail: `somatla@gsvms2.cc.gasou.edu`)
James Braselton   (E-Mail: `jimbras@gsvms2.cc.gasou.edu`)

Statesboro, Georgia
June, 2003

Getting Started 1

1.1 Introduction to Mathematica

Mathematica, first released in 1988 by Wolfram Research, Inc.,

```
http://www.wolfram.com/,
```

is a system for doing mathematics on a computer. Mathematica combines symbolic manipulation, numerical mathematics, outstanding graphics, and a sophisticated programming language. Because of its versatility, Mathematica has established itself as the computer algebra system of choice for many computer users. Among the over 1,000,000 users of Mathematica, 28% are engineers, 21% are computer scientists, 20% are physical scientists, 12% are mathematical scientists, and 12% are business, social, and life scientists. Two-thirds of the users are in industry and government with a small (8%) but growing number of student users. However, due to its special nature and sophistication, beginning users need to be aware of the special syntax required to make Mathematica perform in the way intended. You will find that calculations and sequences of calculations most frequently used by beginning users are discussed in detail along with many typical examples. In addition, the comprehensive index not only lists a variety of topics but also cross-references commands with frequently used options. *Mathematica By Example* serves as a valuable tool and reference to the beginning user of Mathematica as well as to the more sophisticated user, with specialized needs.

For information, including purchasing information, about Mathematica contact:

Corporate Headquarters:
Wolfram Research, Inc.
100 Trade Center Drive
Champaign, IL 61820
USA
telephone: 217-398-0700
fax: 217-398-0747
email: info@wolfram.com
web: http://www.wolfram.com

Europe:
Wolfram Research Europe Ltd.
10 Blenheim Office Park
Lower Road, Long Hanborough
Oxfordshire OX8 8LN
UNITED KINGDOM
telephone: +44-(0) 1993-883400
fax: +44-(0) 1993-883800
email: info-europe@wolfram.com

Asia:
Wolfram Research Asia Ltd.
Izumi Building 8F
3-2-15 Misaki-cho
Chiyoda-ku, Tokyo 101
JAPAN
telephone: +81-(0)3-5276-0506
fax: +81-(0)3-5276-0509
email: info-asia@wolfram.com

For information, including purchasing information, about *The Mathematica Book* [22] contact:

Wolfram Media, Inc.
100 Trade Center Drive
Champaign, IL 61820,
USA
email: info@wolfram-media.com
web: http://www.wolfram-media.com

A Note Regarding Different Versions of Mathematica

With the release of Version 5 of Mathematica, many new functions and features have been added to Mathematica. We encourage users of earlier versions of Mathematica to update to Version 5 as soon as they can. All examples in *Mathematica By Example*, Third Edition, were completed with Version 5. In most cases, the same results will be obtained if you are using Version 4.0 or later, although the appearance of your results will almost certainly differ from that presented here. Occasionally, however, particular features of Version 5 are used and in those cases, of course, these features are not available in earlier versions. If you are using an earlier or later version of Mathematica, your results may not appear in a form identical to those found in this book: some commands found in Version 5 are not available in earlier versions of Mathematica; in later versions some commands will certainly be changed, new commands added, and obsolete commands removed. For details regarding these changes, please see *The Mathematica Book* [22]. You can determine the version of Mathematica you are using during a given Mathematica session by entering either the command $Version or the command $VersionNumber. In this text, we assume that Mathematica has been correctly installed on the computer you are using. If you need to install Mathematica on your computer, please refer to the documentation that came with the Mathematica software package.

On-line help for upgrading older versions of Mathematica and installing new versions of Mathematica is available at the Wolfram Research, Inc. website:

$$\texttt{http://www.wolfram.com/.}$$

1.1.1 Getting Started with Mathematica

We begin by introducing the essentials of Mathematica. The examples presented are taken from algebra, trigonometry, and calculus topics that you are familiar with to assist you in becoming acquainted with the Mathematica computer algebra system.

We assume that Mathematica has been correctly installed on the computer you are using. If you need to install Mathematica on your computer, please refer to the documentation that came with the Mathematica software package.

Start Mathematica on your computer system. Using Windows or Macintosh mouse or keyboard commands, activate the Mathematica program by selecting the Mathematica icon or an existing Mathematica document (or notebook), and then clicking or double-clicking on the icon.

If you start Mathematica by selecting the Mathematica icon, a blank untitled notebook is opened, as illustrated in the following screen shot.

When you start typing, the thin black horizontal line near the top of the window is replaced by what you type.

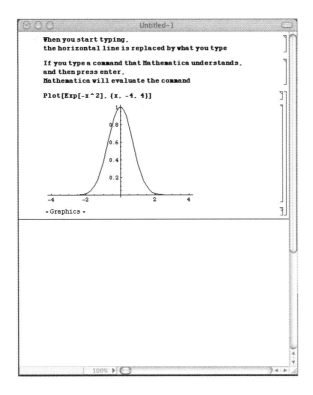

With some operating systems, **Enter** evaluates commands and **Return** yields a new line
The **Basic Input** palette:

Once Mathematica has been started, computations can be carried out immediately. Mathematica commands are typed and the black horizontal line is replaced by the command, which is then evaluated by pressing **Enter**. Note that pressing **Enter** or **Return** evaluates commands and pressing **Shift-Return** yields a new line. Output is displayed below input. We illustrate some of the typical steps involved in working with Mathematica in the calculations that follow. In each case, we type the command and press **Enter**. Mathematica evaluates the command, displays the result, and inserts a new horizontal line after the result. For example, typing N[, then pressing the π key on the **Basic Input** palette, followed by typing ,50] and pressing the enter key

```
In[1]:= N[π,50]
Out[1]= 3.1415926535897932384626433832795028841971693993751062.09749446
```

returns a 50-digit approximation of π. Note that both π and Pi represent the mathematical constant π so entering N[Pi,50] returns the same result.

The next calculation can then be typed and entered in the same manner as the first. For example, entering

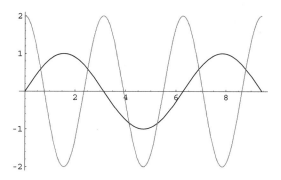

Figure 1-1   A two-dimensional plot

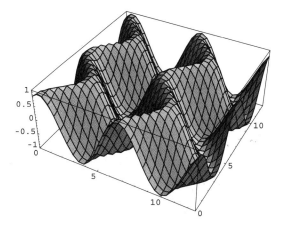

Figure 1-2   A three-dimensional plot

Notice that every Mathematica command begins with capital letters and the argument is enclosed by square brackets [. . .].

$In[2] :=$ **Plot[{Sin[x], 2 Cos[2x]}, {x, 0, 3π},**
                 PlotStyle- > {GrayLevel[0], GrayLevel[0.5]}]

graphs the functions $y = \sin x$ and $y = 2 \cos 2x$ on the interval $[0, 3\pi]$ shown in Figure 1-1. Similarly, entering

$In[3] :=$ **Plot3D[Sin[x + Cos[y]], {x, 0, 4π}, {y, 0, 4π},**
                 PlotPoints- > {30, 30}]

graphs the function $z = \sin(x + \cos y)$ for $0 \leq x \leq 4\pi$ and $0 \leq y \leq 4\pi$ shown in Figure 1-2.

Notice that all three of the following commands

To type x^3 in Mathematica, press the ⬚ on the **Basic Input** palette, type x in the base position, and then click (or tab to) the exponent position and type 3.

$In[4] :=$ **Solve[x^3 - 2x + 1 == 0]**

$Out[4] =$ $\left\{ \{x \to 1\}, \left\{x \to \frac{1}{2}\left(-1-\sqrt{5}\right)\right\}, \left\{x \to \frac{1}{2}\left(-1+\sqrt{5}\right)\right\} \right\}$

```
In[5] := Solve[x^3 - 2 * x + 1 == 0]
```

$$Out[5] = \{\{x\ \text{->}\ 1\},\ \{x\ \text{->}\ \frac{1}{2}\ (-1\ -\ \text{Sqrt}[5])\},\ \{x\ \text{->}\ \frac{1}{2}\ (-1\ +\ \text{Sqrt}[5])\}\}$$

```
In[6] := Solve[x^3 - 2 x + 1 == 0]
```

$$Out[6] = \{\{x \to 1\},\ \{x \to \tfrac{1}{2}\left(-1-\sqrt{5}\right)\},\ \{x \to \tfrac{1}{2}\left(-1+\sqrt{5}\right)\}\}$$

solve the equation $x^3 - 2x + 1 = 0$ for x.

In the first case, the input and output are in **StandardForm**, in the second case, the input and output are in **InputForm**, and in the third case, the input and output are in **TraditionalForm**. Move the cursor to the Mathematica menu,

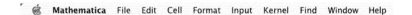

select **Cell**, and then **ConvertTo**, as illustrated in the following screen shot.

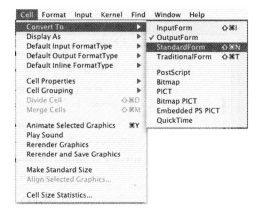

You can change how input and output appear by using **ConvertTo** or by changing the default settings. Moreover, you can determine the form of input/output by looking at the cell bracket that contains the input/output. For example, even though all three of the following commands look different, all three evaluate $\int_0^{2\pi} x^3 \sin x \, dx$.

```
Integrate[x^3 * Sin[x], {x, 0, 2 * Pi}]
```

$$\int_0^{2\pi} x^3\ \text{Sin}[x]\ dx$$

$$\int_0^{2\pi} x^3\ \sin(x)\ dx$$

A cell bracket like this ⌋ means the input is in **InputForm**; the output is in **OutputForm**. A cell bracket like this ⌉ means the contents of the cell are in **StandardForm**. A cell bracket like this ⌇ means the contents of the cell are in **TraditionalForm**. Throughout *Mathematica By Example*, Third Edition, we display input and output using **InputForm** or **StandardForm**, unless otherwise stated.

To enter code in **StandardForm**, we often take advantage of the **BasicTypesetting** palette, which is accessed by going to **File** under the Mathematica menu and then selecting **Palettes**

followed by **BasicTypesetting**.

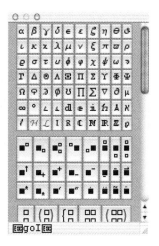

Use the buttons to create templates and enter special characters. Alternatively, you can find a complete list of typesetting shortcuts in *The Mathematica Book*, Appendix 12, Listing of Named Characters [22].

Mathematica sessions are terminated by entering `Quit[]` or by selecting **Quit** from the **File** menu, or by using a keyboard shortcut, like **command-Q**, as with other applications. They can be saved by referring to **Save** from the **File** menu.

Mathematica allows you to save notebooks (as well as combinations of cells) in a variety of formats, in addition to the standard Mathematica format.

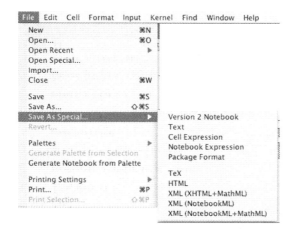

Remark. Input and text regions in notebooks can be edited. Editing input can create a notebook in which the mathematical output does not make sense in the sequence it appears. It is also possible to simply go into a notebook and alter input without doing any recalculation. This also creates misleading notebooks. Hence, common sense and caution should be used when editing the input regions of notebooks. Recalculating all commands in the notebook will clarify any confusion.

Preview

In order for the Mathematica user to take full advantage of this powerful software, an understanding of its syntax is imperative. The goal of *Mathematica By Example* is to introduce the reader to the Mathematica commands and sequences of commands most frequently used by beginning users. Although all of the rules of Mathematica syntax are far too numerous to list here, knowledge of the following five rules equips the beginner with the necessary tools to start using the Mathematica program with little trouble.

Five Basic Rules of Mathematica Syntax

1. The arguments of *all* functions (both built-in ones and ones that you define)are given in brackets [...]. Parentheses (...) are used for grouping operations; vectors, matrices, and lists are given in braces {...}; and double square brackets [[...]] are used for indexing lists and tables.

2. Every word of a built-in Mathematica function begins with a capital letter.

3. Multiplication is represented by * or a space between characters. Enter 2*x*y or 2x y to evaluate $2xy$ *not* 2xy.

4. Powers are denoted by ^. Enter (8*x^3)^(1/3) to evaluate $(8x^3)^{1/3} = 8^{1/3}(x^3)^{1/3} = 2x$ instead of 8x^1/3, which returns 8x/3.

5. Mathematica follows the order of operations *exactly*. Thus, entering (1+x)^1/x returns $\frac{(1+x)^1}{x}$ while (1+x)^(1/x) returns $(1+x)^{1/x}$. Similarly, entering x^3x returns $x^3 \cdot x = x^4$ while entering x^(3x) returns x^{3x}.

Remark. If you get no response or an incorrect response, you may have entered or executed the command incorrectly. In some cases, the amount of memory allocated to Mathematica can cause a crash. Like people, Mathematica is not perfect and errors can occur.

1.2 Loading Packages

Although Mathematica contains many built-in functions, some other functions are contained in **packages** that must be loaded separately. A tremendous number of additional commands are available in various packages that are shipped with each version of Mathematica. Experienced users can create their own packages; other packages are available from user groups and MathSource, which electronically distributes Mathematica-related products. For information about MathSource, visit

> http://library.wolfram.com/infocenter/MathSource/

or send the message "help" to mathsource@wri.com. If desired, you can purchase MathSource on a CD directly from Wolfram Research, Inc. or you can access MathSource from the Wolfram Research World Wide Web site

> http://www.wri.com or http://www.wolfram.com.

Descriptions of the various packages shipped with Mathematica are found in the **Help Browser**. From the Mathematica menu, select **Help** followed by **Add-Ons...**

to see a list of the standard packages.

Information regarding the packages in each category is obtained by selecting the category from the **Help Browser**'s menu.

Packages are loaded by entering the command

<<directory`packagename`

where **directory** is the location of the package **packagename**. Entering the command <<directory`Master` makes all the functions contained in all the packages in **directory** available. In this case, each package need not be loaded individually. For example, to load the package **Shapes** contained in the **Graphics** folder (or directory), we enter <<Graphics`Shapes`.

In[7]:= << Graphics`Shapes`

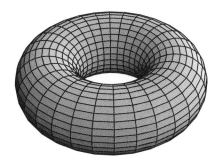

Figure 1-3   A torus created with Torus

Figure 1-4   A Möbius strip and a sphere

After the **Shapes** package has been loaded, entering

> *In[8]:=* **Show[Graphics3D[Torus[1, 0.5, 30, 30]], Boxed → False]**

generates the graph of a torus shown in Figure 1-3. Next, we generate a Möbius strip and a sphere and display the two side-by-side using GraphicsArray in Figure 1-4.

> *In[9]:=* **mstrip = Graphics3D[MoebiusStrip[1, 0.5, 40], Boxed → False];**
>
> **sph = Graphics3D[Sphere[1, 25, 25], Boxed → False];**
>
> **Show[GraphicsArray[{mstrip, sph}]]**

The **Shapes** package contains definitions of familiar three-dimensional shapes including the cone, cylinder, helix, and double helix. In addition, it allows us to perform transformations like rotations and translations on three-dimensional graphics.

A Word of Caution

When users take advantage of packages frequently, they often encounter error messages. One error message that occurs frequently is when a command is entered before the package is loaded. For example, the command `GramSchmidt [{v1,v2,...,vn}]` returns an orthonormal set of vectors with the same span as the vectors $\mathbf{v}_1$, $\mathbf{v}_2$, ..., $\mathbf{v}_n$. Here, we attempt to use the command `GramSchmidt`, which is contained in the **Orthogonalization** package located in the **LinearAlgebra** folder before the package has been loaded. Mathematica does not yet know the meaning of `GramSchmidt` so our input is returned.

```
In[10]:= GramSchmidt[{{1, 1, 0}, {0, 2, 1}, {1, 0, 3}}]
Out[10]= GramSchmidt[{{1, 1, 0}, {0, 2, 1}, {1, 0, 3}}]
```

At this point, we load the **Orthogonalization** package, which contains the `GramSchmidt` command, located in the **LinearAlgebra** folder. Several error messages result.

```
In[11]:= << LinearAlgebra`Orthogonalization`
GramSchmidt :: shdw : Symbol GramSchmidt appears in multiple contexts
{LinearAlgebra`Orthogonalization`, Global`};
definitions in context LinearAlgebra`Orthogonalization`
may shadow or be shadowed by other definitions.
```

In fact, when we reenter the command, we obtain the same result as that obtained previously.

```
In[12]:= GramSchmidt[{{1, 1, 0}, {0, 2, 1}, {1, 0, 3}}]
Out[12]= GramSchmidt[{{1, 1, 0}, {0, 2, 1}, {1, 0, 3}}]
```

However, after using the command `Remove`, the command `GramSchmidt` works as expected. Alternatively, we can quit Mathematica, restart, load the package, and then execute the command.

```
In[13]:= Remove[GramSchmidt]
```

```
In[14]:= GramSchmidt[{{1, 1, 0}, {0, 2, 1}, {1, 0, 3}}]
```

$$Out[14]= \left\{ \left\{ \frac{1}{\sqrt{2}}, \frac{1}{\sqrt{2}}, 0 \right\}, \left\{ -\frac{1}{\sqrt{3}}, \frac{1}{\sqrt{3}}, \frac{1}{\sqrt{3}} \right\}, \left\{ \frac{1}{\sqrt{6}}, -\frac{1}{\sqrt{6}}, \sqrt{\frac{2}{3}} \right\} \right\}$$

Similarly, we can take advantage of other commands contained in the **Orthogonalization** package like `Normalize` which normalizes a given vector.

```
In[15]:= Normalize[{1, 2, 3}]
```

$$Out[15]= \left\{ \frac{1}{\sqrt{14}}, \sqrt{\frac{2}{7}}, \frac{3}{\sqrt{14}} \right\}$$

1.3 Getting Help from Mathematica

Becoming competent with Mathematica can take a serious investment of time. Hopefully, messages that result from syntax errors are viewed lightheartedly. Ideally, instead of becoming frustrated, beginning Mathematica users will find it challenging and fun to locate the source of errors. Frequently, Mathematica's error messages indicate where the error(s) has (have) occurred. In this process, it is natural that you will become more proficient with Mathematica. In addition to Mathematica's extensive help facilities, which are described next, a tremendous amount of information is available for all Mathematica users at the Wolfram Research website

```
http://www.wolfram.com/.
```

One way to obtain information about commands and functions, including user-defined functions, is the command ?. ?object gives a basic description and syntax information of the Mathematica object object. ??object yields detailed information regarding syntax and options for the object object.

EXAMPLE 1.3.1: Use ? and ?? to obtain information about the command Plot.

SOLUTION: ?Plot uses basic information about the Plot function

```
?Plot

Plot[f, {x, xmin, xmax}] generates a plot of f as a
    function of x from xmin to xmax. Plot[{f1, f2, ... },
    {x, xmin, xmax}] plots several functions fi. More...
```

while ??Plot includes basic information as well as a list of options and their default values.

```
?? Plot

Plot[f, {x, xmin, xmax}] generates a plot of f as a
   function of x from xmin to xmax. Plot[{f1, f2, ... },
   {x, xmin, xmax}] plots several functions fi. More...

Attributes[Plot] = {HoldAll, Protected}

Options[Plot] = {AspectRatio → 1/GoldenRatio,
   Axes → Automatic, AxesLabel → None, AxesOrigin → Automatic,
   AxesStyle → Automatic, Background → Automatic,
   ColorOutput → Automatic, Compiled → True,
   DefaultColor → Automatic, DefaultFont → $DefaultFont,
   DisplayFunction → $DisplayFunction, Epilog → {},
   FormatType → $FormatType, Frame → False, FrameLabel → None,
   FrameStyle → Automatic, FrameTicks → Automatic,
   GridLines → None, ImageSize → Automatic, MaxBend → 10.,
   PlotDivision → 30., PlotLabel → None, PlotPoints → 25,
   PlotRange → Automatic, PlotRegion → Automatic,
   PlotStyle → Automatic, Prolog → {}, RotateLabel → True,
   TextStyle → $TextStyle, Ticks → Automatic}
```

Options[object] returns a list of the available options associated with object along with their current settings. This is quite useful when working with a Mathematica command such as ParametricPlot which has many options. Notice that the default value (the value automatically assumed by Mathematica) for each option is given in the output.

EXAMPLE 1.3.2: Use Options to obtain a list of the options and their current settings for the command ParametricPlot.

SOLUTION: The command Options[ParametricPlot] lists all the options and their current settings for the command ParametricPlot.

```
Options[ParametricPlot]

{AspectRatio → 1/GoldenRatio, Axes → Automatic,
   AxesLabel → None, AxesOrigin → Automatic,
   AxesStyle → Automatic, Background → Automatic,
   ColorOutput → Automatic, Compiled → True,
   DefaultColor → Automatic, DefaultFont → $DefaultFont,
   DisplayFunction :→ $DisplayFunction, Epilog → {},
   FormatType :→ $FormatType, Frame → False, FrameLabel → None,
   FrameStyle → Automatic, FrameTicks → Automatic,
   GridLines → None, ImageSize → Automatic, MaxBend → 10.,
   PlotDivision → 30., PlotLabel → None, PlotPoints → 25,
   PlotRange → Automatic, PlotRegion → Automatic,
   PlotStyle → Automatic, Prolog → {}, RotateLabel → True,
   TextStyle → $TextStyle, Ticks → Automatic}
```

As indicated above, ??object or, equivalently, Information[object] yields the information on the Mathematica object object returned by both ?object and `Options[object]` in addition to a list of attributes of `object`. Note that `object` may be either a user-defined object or a built-in Mathematica object.

EXAMPLE 1.3.3: Use `??` to obtain information about the commands `Solve` and `Map`. Use `Information` to obtain information about the command `PolynomialLCM`.

SOLUTION: We use `??` to obtain information about the commands `Solve` and `Map` including a list of options and their current settings.

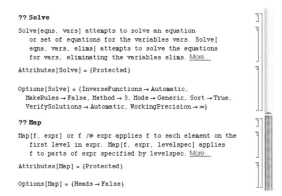

Similarly, we use `Information` to obtain information about the command `PolynomialLCM` including a list of options and their current settings.

The command `Names["form"]` lists all objects that match the pattern defined in `form`. For example, `Names["Plot"]` returns `Plot`, `Names["*Plot"]` returns all objects that end with the string `Plot`, `Names["Plot*"]` lists all objects that begin with the string Plot, and Names["*Plot*"] lists all objects that contain the string Plot. Names["form",SpellingCorrection->True] finds those symbols that match the pattern defined in form after a spelling correction.

EXAMPLE 1.3.4: Create a list of all built-in functions beginning with the string Plot.

SOLUTION: We use Names to find all objects that match the pattern Plot.

> *In[16]:=* **Names["Plot"]**
> *Out[16]=* {Plot}

Next, we use Names to create a list of all built-in functions beginning with the string Plot.

> *In[17]:=* **Names["Plot * "]**
> *Out[17]=* {Plot, Plot3D, Plot3Matrix, PlotDivision, PlotJoined,
>   PlotLabel, PlotPoints, PlotRange, PlotRegion,
>   PlotStyle}

■

As indicated above, the ? function can be used in many ways. Entering ?letters* gives all Mathematica objects that begin with the string letters; ?*letters* gives all Mathematica objects that contain the string letters; and ?*letters gives all Mathematica commands that end in the string letters.

EXAMPLE 1.3.5: What are the Mathematica functions that (a) end in the string Cos; (b) contain the string Sin; and (c) begin with the string Polynomial?

SOLUTION: Entering

returns all functions ending with the string Cos, entering

returns all functions containing the string `Sin`, and entering

returns all functions that begin with the string `Polynomial`.

■

Mathematica Help

Additional help features are accessed from the Mathematica menu under **Help**. For basic information about Mathematica, go to **Help** and select **Help Browser...**

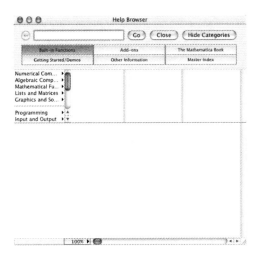

If you are a beginning Mathematica user, you may choose to select **Welcome Screen...**

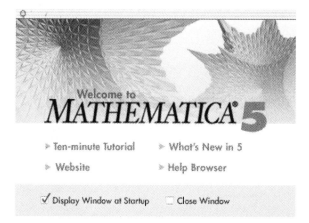

and then select **Ten-Minute Tutorial**

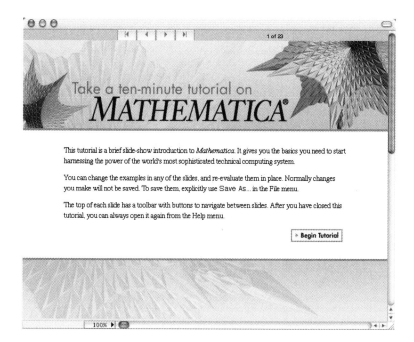

or **Help Browser**.

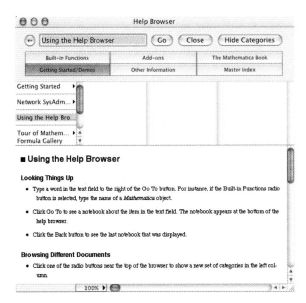

To obtain information about a particular Mathematica object or function, open the
Help Browser, type the name of the object, function, or topic and press the **Go**
button. Alternatively, you can type the name of a function that you wish to obtain
help about, select it, go to **Help**, and then select **Find in Help...** as we do here with
the DSolve function.

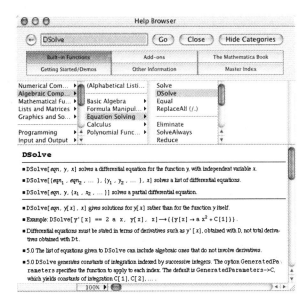

A typical help window not only contains a detailed description of the command and its options but also several examples that illustrate the command as well as hyperlinked cross-references to related commands and *The Mathematica Book* [22], which can be accessed by clicking on the appropriate links.

You can also use the **Help Browser** to access the on-line version of *The Mathematica Book* [22]. Here is a portion of Section 3.6.3, Operations on Power Series.

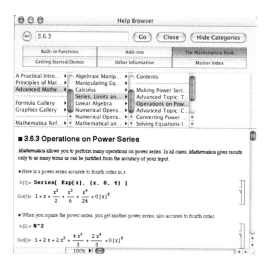

The **Master Index** contains hyperlinks to all portions of Mathematica help.

The Mathematica Menu

Basic Operations on Numbers, Expressions, and Functions

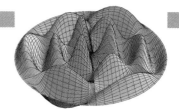

2

Chapter 2 introduces the essential commands of Mathematica. Basic operations on numbers, expressions, and functions are introduced and discussed.

2.1 Numerical Calculations and Built-In Functions

2.1.1 Numerical Calculations

The basic arithmetic operations (addition, subtraction, multiplication, division, and exponentiation) are performed in the natural way with Mathematica. Whenever possible, Mathematica gives an exact answer and reduces fractions.

1. "*a* plus *b*," $a + b$, is entered as a+b;
2. "*a* minus *b*," $a - b$, is entered as a-b;
3. "*a* times *b*," ab, is entered as either a*b or a b (note the space between the symbols a and b);
4. "*a* divided by *b*," a/b, is entered as a/b. Executing the command a/b results in a fraction reduced to lowest terms; and
5. "*a* raised to the *b*th power," a^b, is entered as a^b.

EXAMPLE 2.1.1: Calculate (a) $121 + 542$; (b) $3231 - 9876$; (c) $(-23)(76)$; (d) $(22341)(832748)(387281)$; (e) $\dfrac{467}{31}$; and (f) $\dfrac{12315}{35}$.

SOLUTION: These calculations are carried out in the following screen shot. In (f), Mathematica simplifies the quotient because the numerator and denominator have a common factor of 5. In each case, the input is typed and then evaluated by pressing **Enter**.

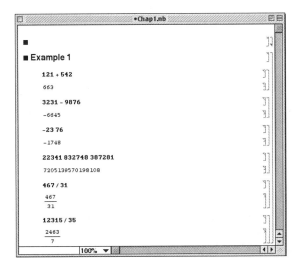

The term $a^{n/m} = \sqrt[m]{a^n} = \left(\sqrt[m]{a}\right)^n$ is entered as a^(n/m). For $n/m = 1/2$, the command Sqrt[a] can be used instead. Usually, the result is returned in unevaluated form but N can be used to obtain numerical approximations to virtually any degree of accuracy. With N[expr,n], Mathematica yields a numerical approximation of expr to n digits of precision, if possible. At other times, Simplify can be used to produce the expected results.

Remark. If the expression b in a^b contains more than one symbol, be sure that the exponent is included in parentheses. Entering a^n/m computes $a^n/m = \frac{1}{m}a^n$ while entering a^(n/m) computes $a^{n/m}$.

EXAMPLE 2.1.2: Compute (a) $\sqrt{27}$ and (b) $\sqrt[3]{8^2} = 8^{2/3}$.

SOLUTION: (a) Mathematica automatically simplifies $\sqrt{27} = 3\sqrt{3}$.

> $In[18] :=$ **Sqrt[27]**
>
> $Out[18] = 3\ \sqrt{3}$

We use N to obtain an approximation of $\sqrt{27}$.

> $In[19] :=$ **N[Sqrt[27]]**
>
> $Out[19] = 5.19615$

N[number] and number//N return numerical approximations of number.

(b) Mathematica automatically simplifies $8^{2/3}$.

> $In[20] :=$ **8^(2/3)**
>
> $Out[20] = 4$

∎

When computing odd roots of negative numbers, Mathematica's results are surprising to the novice. Namely, Mathematica returns a complex number. We will see that this has important consequences when graphing certain functions.

EXAMPLE 2.1.3: Calculate (a) $\dfrac{1}{3}\left(-\dfrac{27}{64}\right)^2$ and (b) $\left(-\dfrac{27}{64}\right)^{2/3}$.

SOLUTION: (a) Because Mathematica follows the order of operations, `(-27/64)^2/3` first computes $(-27/64)^2$ and then divides the result by 3.

> $In[21] :=$ **(-27/64)^2/3**
>
> $Out[21] = \dfrac{243}{4096}$

(b) On the other hand, `(-27/64)^(2/3)` raises $-27/64$ to the 2/3 power. Mathematica does not automatically simplify $\left(-\frac{27}{64}\right)^{2/3}$.

> $In[22] :=$ **(-27/64)^(2/3)**
>
> $Out[22] = \dfrac{9}{16}\ (-1)^{2/3}$

However, when we use N, Mathematica returns the numerical version of the principal root of $\left(-\frac{27}{64}\right)^{2/3}$.

> $In[23] :=$ **N[(-27/64)^(2/3)]**
>
> $Out[23] = -0.28125 + 0.487139\ i$

To obtain the result

$$\left(-\frac{27}{64}\right)^{2/3} = \left(\sqrt[3]{\frac{-27}{64}}\right)^2 = \left(-\frac{3}{4}\right)^2 = \frac{9}{16},$$

which would be expected by most algebra and calculus students, we load the **RealOnly** package that is contained in the **Miscellaneous** directory. Then,

> *In[24]:=* << **Miscellaneous`RealOnly`**

> *In[25]:=* **(-27/64)^(2/3)**
> *Out[25]=* $\frac{9}{16}$

returns the result 9/16.

∎

2.1.2 Built-In Constants

Mathematica has built-in definitions of many commonly used constants. In particular, $e \approx 2.71828$ is denoted by E, $\pi \approx 3.14159$ is denoted by Pi, and $i = \sqrt{-1}$ is denoted by I. Usually, Mathematica performs complex arithmetic automatically.

Other built-in constants include ∞, denoted by Infinity, Euler's constant, $\gamma \approx 0.577216$, denoted by EulerGamma, Catalan's constant, approximately 0.915966, denoted by Catalan, and the golden ratio, $\frac{1}{2}\left(1 + \sqrt{5}\right) \approx 1.61803$, denoted by GoldenRatio.

EXAMPLE 2.1.4: Entering

In[26]:= **N[e,50]**
Out[26]= 2.7182818284590452353602874713526624977757247093699959
                7.49669676000000000000000000000000000000000000000000 10^8

returns a 50-digit approximation of *e*. Entering

> *In[27]:=* **N[π,25]**
> *Out[27]=* 3.141592653589793238462643

returns a 25-digit approximation of π. Entering

> *In[28]:=* **(3 + i)/(4 - i)**
> *Out[28]=* $\frac{11}{17} + \frac{7\,i}{17}$

performs the division $(3 + i)/(4 - i)$ and writes the result in standard form.

2.1.3 Built-In Functions

Mathematica contains numerous mathematical functions.

Functions frequently encountered by beginning users include the exponential function, `Exp[x]`; the natural logarithm, `Log[x]`; the absolute value function, `Abs[x]`; the trigonometric functions `Sin[x]`, `Cos[x]`, `Tan[x]`, `Sec[x]`, `Csc[x]`, and `Cot[x]`; the inverse trigonometric functions `ArcSin[x]`, `ArcCos[x]`, `ArcTan[x]`, `ArcSec[x]`, `ArcCsc[x]`, and `ArcCot[x]`; the hyperbolic trigonometric functions `Sinh[x]`, `Cosh[x]`, and `Tanh[x]`; and their inverses `ArcSinh[x]`, `ArcCosh[x]`, and `ArcTanh[x]`. Generally, Mathematica tries to return an exact value unless otherwise specified with `N`.

Several examples of the natural logarithm and the exponential functions are given next. Mathematica often recognizes the properties associated with these functions and simplifies expressions accordingly.

EXAMPLE 2.1.5: Entering

$In[29]:=$ **N[Exp[-5]]**

$Out[29]=$ `0.00673795`

returns an approximation of $e^{-5} = 1/e^5$. Entering

$In[30]:=$ **Log[Exp[3]]**

$Out[30]=$ `3`

computes $\ln e^3 = 3$. Entering

$In[31]:=$ **Exp[Log[4]]**

$Out[31]=$ `4`

computes $e^{\ln 4} = 4$. Entering

$In[32]:=$ **Abs[-π]**

$Out[32]=$ π

computes $|-\pi| = \pi$. Entering

$In[33]:=$ **Abs[(3 + 2i)/(2 - 9i)]**

$Out[33]=$ $\sqrt{\dfrac{13}{85}}$

computes $|(3 + 2i)/(2 - 9i)|$. Entering

$In[34]:=$ **Sin[π/12]**

$Out[34]=$ $\dfrac{-1 + \sqrt{3}}{2\ \sqrt{2}}$

N[number] or number//N return approximations of number. Exp[x] computes e^x. Enter E to compute $e \approx 2.718$.

Log[x] computes $\ln x$. $\ln x$ and e^x are inverse functions ($\ln e^x = x$ and $e^{\ln x} = x$) and Mathematica uses these properties when simplifying expressions involving these functions.

Abs[x] returns the absolute value of x, $|x|$.

N[number] and
number//N return
approximations of number.

computes the exact value of sin($\pi/12$). Although Mathematica cannot compute the exact value of tan 1000, entering

```
In[35]:= N[Tan[1000]]

Out[35]= 1.47032
```

returns an approximation of tan 1000. Similarly, entering

```
In[36]:= N[ArcSin[1/3]]

Out[36]= 0.339837
```

returns an approximation of $\sin^{-1}(1/3)$ and entering

```
In[37]:= ArcCos[2/3]//N

Out[37]= 0.841069
```

returns an approximation of $\cos^{-1}(2/3)$.

Mathematica is able to apply many identities that relate the trigonometric and exponential functions using the functions TrigExpand, TrigFactor, TrigReduce, TrigToExp, and ExpToTrig.

```
In[38]:= ?TrigExpand

"TrigExpand[expr]expandsouttrigonometric
    functionsinexpr."

In[39]:= ?TrigFactor

"TrigFactor[expr]factorstrigonometricfunctions
    inexpr."

In[40]:= ?TrigReduce

"TrigReduce[expr]rewritesproductsandpowers
    oftrigonometricfunctionsinexprinterms
    oftrigonometricfunctionswithcombinedarguments."

In[41]:= ?TrigToExp

"TrigToExp[expr]convertstrigonometricfunctions
    inexprtoexponentials."

In[42]:= ?ExpToTrig

"ExpToTrig[expr]convertsexponentialsinexpr
    totrigonometricfunctions."
```

EXAMPLE 2.1.6: Mathematica does not automatically apply the identity $\sin^2 x + \cos^2 x = 1$.

> *In[43]:=* **Cos[x]^2 + Sin[x]^2**
>
> *Out[43]=* $\text{Cos}[x]^2 + \text{Sin}[x]^2$

To apply the identity, we use `Simplify`. Generally, `Simplify[expression]` attempts to simplify `expression`.

> *In[44]:=* **Simplify[Cos[x]^2 + Sin[x]^2]**
>
> *Out[44]=* 1

Use `TrigExpand` to multiply expressions or to rewrite trigonometric functions. In this case, entering

> *In[45]:=* **TrigExpand[Cos[3x]]**
>
> *Out[45]=* $\text{Cos}[x]^3 - 3 \ \text{Cos}[x] \ \text{Sin}[x]^2$

writes $\cos 3x$ in terms of trigonometric functions with argument x. We use the `TrigReduce` function to convert products to sums.

> *In[46]:=* **TrigReduce[Sin[3x] Cos[4x]]**
>
> *Out[46]=* $\dfrac{1}{2} \ (-\text{Sin}[x] + \text{Sin}[7 \ x])$

We use `TrigExpand` to write

> *In[47]:=* **TrigExpand[Cos[2x]]**
>
> *Out[47]=* $\text{Cos}[x]^2 - \text{Sin}[x]^2$

in terms of trigonometric functions with argument x. We use `ExpToTrig` to convert exponential expressions to trigonometric expressions.

> *In[48]:=* **ExpToTrig[1/2 (Exp[x] + Exp[-x])]**
>
> *Out[48]=* $\text{Cosh}[x]$

Similarly, we use `TrigToExp` to convert trigonometric expressions to exponential expressions.

> *In[49]:=* **TrigToExp[Sin[x]]**
>
> *Out[49]=* $\dfrac{1}{2} \ i \ \left(e^{-i \ x} - e^{i \ x}\right)$

Usually, you can use `Simplify` to apply elementary identities.

> *In[50]:=* **Simplify[Tan[x]^2 + 1]**
>
> *Out[50]=* $\text{Sec}[x]^2$

A Word of Caution

Remember that there are certain ambiguities in traditional mathematical notation. For example, the expression $\sin^2(\pi/6)$ is usually interpreted to mean "compute $\sin(\pi/6)$ and square the result." That is, $\sin^2(\pi/6) = [\sin(\pi/6)]^2$. The symbol sin is not being squared; the number $\sin(\pi/6)$ *is* squared. With Mathematica, we must be especially careful and follow the standard order of operations exactly, especially when using **InputForm**. We see that entering

$$In[51]:= \textbf{Sin[}\pi\textbf{/6]}\,\textbf{\textasciicircum 2}$$
$$Out[51]= \frac{1}{4}$$

computes $\sin^2(\pi/6) = [\sin(\pi/6)]^2$ while

$$In[52]:= \textbf{Sin\textasciicircum 2[}\pi\textbf{/6]}$$
$$Out[52]= \mathrm{Sin}^2\left[\tfrac{\pi}{6}\right]$$

raises the symbol Sin to the power $2\left[\frac{\pi}{6}\right]$. Mathematica interprets

$$In[53]:= \textbf{sin\textasciicircum 2 (}\pi\textbf{/6)}$$
$$Out[53]= \frac{\pi \sin^2}{6}$$

to be the product of the symbols $\sin^2$ and $\frac{\pi}{6}$. However, using **TraditionalForm** we are able to evaluate $\sin^2(\pi/6) = [\sin(\pi/6)]^2$ with Mathematica using conventional mathematical notation.

$$In[54]:= \textbf{Sin}^2\left(\frac{\pi}{6}\right)$$
$$Out[54]= \frac{1}{4}$$

Be aware, however, that traditional mathematical notation does contain certain ambiguities and Mathematica may not return the result you expect if you enter input using **TraditionalForm** unless you are especially careful to follow the standard order of operations, as the following warning message indicates.

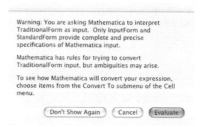

2.2 Expressions and Functions: Elementary Algebra

2.2.1 Basic Algebraic Operations on Expressions

Expressions involving unknowns are entered in the same way as numbers. Mathematica performs standard algebraic operations on mathematical expressions. For example, the commands

1. Factor[expression] factors expression;
2. Expand[expression] multiplies expression;
3. Together[expression] writes expression as a single fraction; and
4. Simplify[expression] performs basic algebraic manipulations on expression and returns the simplest form it finds.

For basic information about any of these commands (or any other) enter ?command as we do here for Factor.

or access the **Help Browser** as we do here for Simplify.

When entering expressions, be sure to include a space or * between variables to denote multiplication.

EXAMPLE 2.2.1: (a) Factor the polynomial $12x^2 + 27xy - 84y^2$. (b) Expand the expression $(x + y)^2(3x - y)^3$. (c) Write the sum $\dfrac{2}{x^2} - \dfrac{x^2}{2}$ as a single fraction.

SOLUTION: The result obtained with Factor indicates that $12x^2 + 27xy - 84y^2 = 3(4x - 7y)(x + 4y)$. When typing the command, be sure to include a space, or *, between the x and y terms to denote multiplication. xy represents an expression while x y or x*y denotes x multiplied by y.

> $In[55]:=$ **Factor$\left[12x^2 + 27xy - 84y^2\right]$**
>
> $Out[55]=$ 3 (4 x - 7 y) (x + 4 y)

We use Expand to compute the product $(x + y)^2(3x - y)^3$ and Together to express $\dfrac{2}{x^2} - \dfrac{x^2}{2}$ as a single fraction.

> $In[56]:=$ **Expand$\left[(x + y)^2 (3x - y)^3\right]$**
>
> $Out[56]=$ 27 x^5 + 27 x^4 y - 18 x^3 y^2 - 10 x^2 y^3 + 7 x y^4 - y^5
>
> $In[57]:=$ **Together$\left[\dfrac{2}{x^2} - \dfrac{x^2}{2}\right]$**
>
> $Out[57]=$ $\dfrac{4 - x^4}{2 x^2}$

■

Factor[x^2-3] returns $x^2 - 3$.

To factor an expression like $x^2 - 3 = x^2 - \left(\sqrt{3}\right)^2 = \left(x - \sqrt{3}\right)\left(x + \sqrt{3}\right)$, use Factor with the Extension option.

> $In[58]:=$ **Factor[x^2 - 3, Extension → {Sqrt[3]}]**
>
> $Out[58]=$ $-\left(\sqrt{3} - x\right)\left(\sqrt{3} + x\right)$

Similarly, use Factor with the Extension option to factor expressions like $x^2 + 1 = x^2 - i^2 = (x + i)(x - i)$.

> $In[59]:=$ **Factor[x^2 + 1]**
>
> $Out[59]=$ 1 + x^2
>
> $In[60]:=$ **Factor[x^2 + 1, Extension → {I}]**
>
> $Out[60]=$ (-i + x) (i + x)

Mathematica does not automatically simplify $\sqrt{x^2}$ to the expression x

```
In[61]:= Sqrt[x^2]
Out[61]= √x²
```

because without restrictions on x, $\sqrt{x^2} = |x|$. The command `PowerExpand[expression]` simplifies `expression` assuming that all variables are positive.

```
In[62]:= PowerExpand[Sqrt[x^2]]
Out[62]= x
```

Thus, entering

```
In[63]:= Simplify[Sqrt[a^2 b^4]]
Out[63]= √a² b⁴
```

returns $\sqrt{a^2 b^4}$ but entering

```
In[64]:= PowerExpand[Sqrt[a^2 b^4]]
Out[64]= a b²
```

returns ab^2.

In general, a space is not needed between a number and a symbol to denote multiplication when a symbol follows a number. That is, `3dog` means 3 times variable `dog`; `dog3` is a variable with name `dog3`. Mathematica interprets `3 dog`, `dog*3`, and `dog 3` as 3 times variable `dog`. However, when multiplying two variables, either include a space or `*` between the variables.

1. `cat dog` means "variable `cat` times variable `dog`."
2. `cat*dog` means "variable `cat` times variable `dog`."
3. But, `catdog` is interpreted as a variable `catdog`.

The command `Apart[expression]` computes the partial fraction decomposition of `expression`; `Cancel[expression]` factors the numerator and denominator of `expression` then reduces `expression` to lowest terms.

EXAMPLE 2.2.2: (a) Determine the partial fraction decomposition of $\dfrac{1}{(x-3)(x-1)}$. (b) Simplify $\dfrac{x^2-1}{x^2-2x+1}$.

SOLUTION: `Apart` is used to see that $\dfrac{1}{(x-3)(x-1)} = \dfrac{1}{2(x-3)} - \dfrac{1}{2(x-1)}$.

Then, `Cancel` is used to find that $\dfrac{x^2-1}{x^2-2x+1} = \dfrac{(x-1)(x+1)}{(x-1)^2} = \dfrac{x+1}{x-1}$.

In this calculation, we have assumed that $x \neq 1$, an assumption made by Cancel but not by Simplify.

$In[65] := $ **Apart** $\left[\dfrac{1}{(x - 3)\ (x - 1)}\right]$

$Out[65] = \dfrac{1}{2\ (-3 + x)} - \dfrac{1}{2\ (-1 + x)}$

$In[66] := $ **Cancel** $\left[\dfrac{x^2 - 1}{x^2 - 2x + 1}\right]$

$Out[66] = \dfrac{1 + x}{-1 + x}$

■

In addition, Mathematica has several built-in functions for manipulating parts of fractions.

1. Numerator[fraction] yields the numerator of fraction.
2. ExpandNumerator[fraction] expands the numerator of fraction.
3. Denominator[fraction] yields the denominator of fraction.
4. ExpandDenominator[fraction] expands the denominator of fraction.

EXAMPLE 2.2.3: Given $\dfrac{x^3 + 2x^2 - x - 2}{x^3 + x^2 - 4x - 4}$, (a) factor both the numerator and denominator; (b) reduce $\dfrac{x^3 + 2x^2 - x - 2}{x^3 + x^2 - 4x - 4}$ to lowest terms; and (c) find the partial fraction decomposition of $\dfrac{x^3 + 2x^2 - x - 2}{x^3 + x^2 - 4x - 4}$.

SOLUTION: The numerator of $\dfrac{x^3 + 2x^2 - x - 2}{x^3 + x^2 - 4x - 4}$ is extracted with Numerator. We then use Factor together with %, which is used to refer to the most recent output, to factor the result of executing the Numerator command.

$In[67] := $ **Numerator** $\left[\dfrac{x^3 + 2x^2 - x - 2}{x^3 + x^2 - 4x - 4}\right]$

$Out[67] = -2 - x + 2\,x^2 + x^3$

$In[68] := $ **Factor[%]**

$Out[68] = (-1 + x)\ (1 + x)\ (2 + x)$

Similarly, we use Denominator to extract the denominator of the fraction. Again, Factor together with % is used to factor the previous result, which corresponds to the denominator of the fraction.

$In[69] :=$ **Denominator** $\left[\dfrac{x^3 + 2x^2 - x - 2}{x^3 + x^2 - 4x - 4}\right]$

$Out[69] = -4 - 4\,x + x^2 + x^3$

$In[70] :=$ **Factor[%]**

$Out[70] = (-2 + x)\,(1 + x)\,(2 + x)$

Cancel is used to reduce the fraction to lowest terms.

$In[71] :=$ **Cancel** $\left[\dfrac{x^3 + 2x^2 - x - 2}{x^3 + x^2 - 4x - 4}\right]$

$Out[71] = \dfrac{-1 + x}{-2 + x}$

Finally, Apart is used to find its partial fraction decomposition.

$In[72] :=$ **Apart** $\left[\dfrac{x^3 + 2x^2 - x - 2}{x^3 + x^2 - 4x - 4}\right]$

$Out[72] = 1 + \dfrac{1}{-2 + x}$

■

You can also take advantage of the **AlgebraicManipulation** palette, which is accessed by going to **File** under the Mathematica menu, followed by **Palettes**, and then **AlgebraicManipulation**, to evaluate expressions.

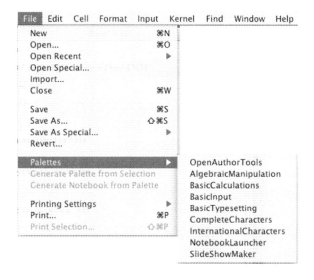

Expand [■]
Factor [■]
Together [■]
Apart [■]
Cancel [■]
Simplify [■]
FullSimplify [■]
TrigExpand [■]
TrigFactor [■]
TrigReduce [■]
ExpToTrig [■]
TrigToExp [■]
PowerExpand [■]
ComplexExpand [■]

EXAMPLE 2.2.4: Simplify $\dfrac{2(x-3)^2(x+1)}{3(x+1)^{4/3}} + 2(x-3)(x+1)^{2/3}$.

SOLUTION: First, we type the expression.

$$\frac{2\,(x-3)^2\,(x+1)}{3\,(x+1)^{4/3}} + 2\,(x-3)\,(x+1)^{2/3}$$

Then, select the expression.

$$\frac{2\,(x-3)^2\,(x+1)}{3\,(x+1)^{4/3}} + 2\,(x-3)\,(x+1)^{2/3}$$

Move the cursor to the palette and click on Simplify. Mathematica simplifies the expression.

$$\frac{8\,(-3+x)\,x}{3\,(1+x)^{1/3}}$$

■

2.2.2 Naming and Evaluating Expressions

In Mathematica, objects can be named. Naming objects is convenient: we can avoid typing the same mathematical expression repeatedly (as we did in Example 2.2.3) and named expressions can be referenced throughout a notebook or Mathematica session. Every Mathematica object can be named – expressions, functions, graphics, and so on can be named with Mathematica. Objects are named by using a single equals sign (=).

Because every built-in Mathematica function begins with a capital letter, we adopt the convention that *every* mathematical object we name in this text will begin with a *lowercase* letter. Consequently, we will be certain to avoid any possible ambiguity with any built-in Mathematica objects.

Expressions are easily evaluated using `ReplaceAll`, which is abbreviated with `/.` and obtained by typing a backslash (/) followed by a period (.), together with `Rule`, which is abbreviated with `->` and obtained by typing a forward slash (/) followed by a greater than sign (>). For example, entering the command

$$x^2 \ /. \ x->3$$

returns the value of the expression x^2 if $x = 3$. Note, however, this does not assign the symbol x the value 3: entering x=3 assigns x the value 3.

EXAMPLE 2.2.5: Evaluate $\dfrac{x^3 + 2x^2 - x - 2}{x^3 + x^2 - 4x - 4}$ if $x = 4$, $x = -3$, and $x = 2$.

SOLUTION: To avoid retyping $\dfrac{x^3 + 2x^2 - x - 2}{x^3 + x^2 - 4x - 4}$, we define `fraction` to be $\dfrac{x^3 + 2x^2 - x - 2}{x^3 + x^2 - 4x - 4}$.

Of course, you can simply copy and paste this expression if you neither want to name it nor retype it.

$In[73]:=$ **fraction** $= \dfrac{x^3 + 2x^2 - x - 2}{x^3 + x^2 - 4x - 4}$

$Out[73]= \dfrac{-2 - x + 2x^2 + x^3}{-4 - 4x + x^2 + x^3}$

`/.` is used to evaluate `fraction` if $x = 4$ and then if $x = -3$.

If you include a semi-colon (;) at the end of the command, the resulting output is suppressed.

$In[74]:=$ **fraction /. x- > 4**

$Out[74]= \dfrac{3}{2}$

$In[75]:=$ **fraction /. x- > -3**

$Out[75]= \dfrac{4}{5}$

When we try to replace each x in fraction by 2, we see that the result is undefined: division by 0 is always undefined.

```
In[76]:= fraction/.x → -2
Power :: infy : Infinite expression 1/0 encountered.
∞ :: indet :
Indeterminate expression 0 ComplexInfinity encountered.
Out[76]= Indeterminate
```

However, when we use Cancel to first simplify and then use ReplaceAll to evaluate,

```
In[77]:= fraction2 = Cancel[fraction]
Out[77]= (-1 + x)/(-2 + x)

In[78]:= fraction2/.x → -2
Out[78]= 3/4
```

we see that the result is 3/4. The result indicates that $\lim_{x\to-2} \frac{x^3+2x^2-x-2}{x^3+x^2-4x-4} = \frac{3}{4}$. We confirm this result with Limit.

```
In[79]:= Limit[fraction, x → -2]
Out[79]= 3/4
```

Generally, Limit[f[x],x->a] attempts to compute $\lim_{x\to a} f(x)$. The Limit function is discussed in more detail in the next chapter.

■

Two Words of Caution

Be aware that Mathematica *does not* remember anything defined in a previous Mathematica session. That is, if you define certain symbols during a Mathematica session, quit the Mathematica session, and then continue later, the previous symbols must be redefined to be used. When you assign a name to an object that is similar to a previously defined or built-in function, Mathematica issues an error message.

```
function = x^2

General::spell1 :
    Possible spelling error: new symbol name "function" is
    similar to existing symbol "Function".

x²
```

We have adopted the convention that every user-defined object begins with a lowercase letter so we know that we have not made an error and the message can be ignored. Sometimes, however, the message can occur frequently and become annoying. If desired, the message

`General::spell1:`

can be suppressed by entering

`Off[General::spell1].`

Generally, `Off[s::tag]` switches off the message `s::tag` so that it is not printed. `On` is used to switch on warning or error messages. Specific messages may be permanently turned off by inserting the desired `Off` commands in the **init.m** file, which is contained in the **Packages** folder (or directory).

2.2.3 Defining and Evaluating Functions

It is important to remember that functions, expressions, and graphics can be named anything that is not the name of a built-in Mathematica function or command. As previously indicated, every built-in Mathematica object begins with a capital letter so every user-defined function, expression, or other object in this text will be assigned a name using lowercase letters, exclusively. This way, the possibility of conflicting with a built-in Mathematica command or function is completely eliminated. Because definitions of functions and names of objects are frequently modified, we introduce the command `Clear`. `Clear[expression]` clears all definitions of expression, if any. You can see if a particular symbol has a definition by entering `?symbol`.

In Mathematica, an elementary function of a single variable, $y = f(x) = expression$ *in x*, is typically defined using the form

`f[x_]=expression in x` or `f[x_]:=expression in x.`

Notice that when you first define a function, you must always enclose the argument in square brackets (`[...]`) and place an underline (or blank) "_" after the argument on the left-hand side of the equals sign in the definition of the function.

EXAMPLE 2.2.6: Entering

In[80]:= **f[x_] = x/(x^2 + 1)**

Out[80]= $\dfrac{x}{1 + x^2}$

defines and computes $f(x) = x/(x^2 + 1)$. Entering

$$In[81] := \mathbf{f[3]}$$
$$Out[81] = \frac{3}{10}$$

computes $f(3) = 3/(3^2 + 1) = 3/10$. Entering

$$In[82] := \mathbf{f[a]}$$
$$Out[82] = \frac{a}{1 + a^2}$$

computes $f(a) = a/(a^2 + 1)$. Entering

$$In[83] := \mathbf{f[3 + h]}$$
$$Out[83] = \frac{3 + h}{1 + (3 + h)^2}$$

computes $f(3 + h) = (3 + h)/((3 + h)^2 + 1)$. Entering

$$In[84] := \mathbf{n1 = Simplify[(f[3 + h] - f[3])/h]}$$
$$Out[84] = -\frac{8 + 3\,h}{10\,(10 + 6\,h + h^2)}$$

computes and simplifies $\dfrac{f(3 + h) - f(3)}{h}$ and names the result n1. Entering

$$In[85] := \mathbf{n1/.h \rightarrow 0}$$
$$Out[85] = -\frac{2}{25}$$

evaluates n1 if $h = 0$. Entering

$$In[86] := \mathbf{n2 = Together[(f[a + h] - f[a])/h]}$$
$$Out[86] = \frac{1 - a^2 - a\,h}{(1 + a^2)\,(1 + a^2 + 2\,a\,h + h^2)}$$

computes and simplifies $\dfrac{f(a + h) - f(a)}{h}$ and names the result n2. Entering

$$In[87] := \mathbf{n2/.h \rightarrow 0}$$
$$Out[87] = \frac{1 - a^2}{(1 + a^2)^2}$$

evaluates n2 if $h = 0$.

Often, you will need to evaluate a function for the values in a **list**,

$$\texttt{list} = \{a_1, a_2, a_3, \ldots, a_n\}.$$

Once $f(x)$ has been defined, Map[f,list] returns the list

$$\{f(a_1), f(a_2), f(a_3), \ldots, f(a_n)\}$$

Also,

The Table function will be discussed in more detail as needed.

1. Table[f[n],{n,n1,n2}] returns the list

$$\{f(n_1), f(n_1 + 1), f(n_1 + 2), \ldots, f(n_2)\}$$

2. Table[(n,f[n]),{n,n1,n2}] returns the list of ordered pairs

$$\{(n_1, f(n_1)), (n_1 + 1, f(n_1 + 1)), (n_1 + 2, f(n_1 + 2)), \ldots, (n_2, f(n_2))\}$$

EXAMPLE 2.2.7: Entering

```
In[88]:= Clear[h]

        h[t_] = (1 + t)^(1/t);

In[89]:= h[1]
Out[89]= 2
```

defines $h(t) = (1 + t)^{1/t}$ and then computes $h(1) = 2$. Because division by 0 is always undefined, $h(0)$ is undefined.

```
In[90]:= h[0]
Power :: infy : Infinite expression 1/0 encountered.
∞ :: indet : Indeterminate expression 1^ComplexInfinity encountered.
Out[90]= Indeterminate
```

However, $h(t)$ is defined for all $t > 0$. In the following, we use Random together with Table to generate 6 random numbers "close" to 0 and name the resulting list t1. Because we are using Random, your results will almost certainly differ from those here.

Random[Real,{a,b}] returns a random real number between a and b.

```
In[91]:= t1 = Table[Random[Real, {0, 10^(-n)}], {n, 0, 5}]
Out[91]= {0.786833, 0.0937732, 0.00653261,
          0.000949186, 1.86913 × 10^-6, 2.32266 × 10^-7}
```

We then use Map to compute $h(t)$ for each of the values in the list t1.

```
In[92]:= Map[h, t1]
Out[92]= {2.09112, 2.60089, 2.70946, 2.71699, 2.71828,
          2.71828}
```

In each of these cases, do not forget to include the blank (or underline) (_) on the left-hand side of the equals sign in the definition of each function. Remember to always include arguments of functions in square brackets.

Including a semi-colon at the
end of a command
suppresses the resulting
output.

EXAMPLE 2.2.8: Entering

> $In[93]:=$ **Clear[f]**
>
> **f[0] = 1;**
>
> **f[1] = 1;**
>
> **f[n_] := f[n - 1] + f[n - 2]**

defines the recursively-defined function defined by $f(0) = 1$, $f(1) = 1$, and $f(n) = f(n-1) + f(n-2)$. For example, $f(2) = f(1) + f(0) = 1 + 1 = 2$; $f(3) = f(2) + f(1) = 2 + 1 = 3$. We use Table to create a list of ordered pairs $(n, f(n))$ for $n = 0, 1, \ldots, 10$.

> $In[94]:=$ **Table[{n, f[n]}, {n, 0, 10}]**
>
> $Out[94]=$ {{0, 1}, {1, 1}, {2, 2}, {3, 3}, {4, 5},
>   {5, 8}, {6, 13}, {7, 21}, {8, 34}, {9, 55}, {10, 89}}

In the preceding examples, the functions were defined using each of the forms f[x_]:=... and f[x_]=.... As a practical matter, when defining "routine" functions with domains consisting of sets of real numbers and ranges consisting of sets of real numbers, either form can be used. Defining a function using the form f[x_]=... instructs Mathematica to define f and then compute and return f[x] (**immediate assignment**); defining a function using the form f[x_]:=... instructs Mathematica to define f. In this case, f[x] is not computed and, thus, Mathematica returns no output (**delayed assignment**). The form f[x_]:=... should be used when Mathematica cannot evaluate f[x] unless x is a particular value, as with recursively-defined functions or piecewise-defined functions which we will discuss shortly.

Generally, if attempting to define a function using the form f[x_]=... produces one or more error messages, use the form f[x_]:=... instead.

To define piecewise-defined functions, we use Condition (/;).

EXAMPLE 2.2.9: Entering

> $In[95]:=$ **Clear[f]**
>
> **f[t_] := Sin[1/t] /; t > 0**

defines $f(t) = \sin\dfrac{1}{t}$ for $t > 0$. Entering

```
In[96]:= f[1/(10 π)]
Out[96]= 0
```

is evaluated because $1/(10\,\pi) > 0$. However, both of the following commands are returned unevaluated. In the first case, -1 is not greater than 0. In the second case, Mathematica does not know the value of a so it cannot determine if it is or is not greater than 0.

```
In[97]:= f[-1]
Out[97]= f[-1]
```

```
In[98]:= f[a]
Out[98]= f[a]
```

Entering

```
In[99]:= f[t_] := -t/; t ≤ 0
```

defines $f(t) = -t$ for $t \le 0$. Now, the domain of $f(t)$ is all real numbers. That is, we have defined the piecewise-defined function

$$f(t) = \begin{cases} \sin\dfrac{1}{t}, & t > 0 \\ -t, & t \le 0 \end{cases}$$

We can now evaluate $f(t)$ for any real number t.

```
In[100]:= f[2/(5 π)]
Out[100]= 1
```

```
In[101]:= f[0]
Out[101]= 0
```

```
In[102]:= f[-10]
Out[102]= 10
```

However, $f(a)$ still returns unevaluated because Mathematica does not know if $a \le 0$ or if $a > 0$.

```
In[103]:= f[a]
Out[103]= f[a]
```

Recursively-defined functions are handled in the same way. The following example shows how to define a periodic function.

EXAMPLE 2.2.10: Entering

$In[104]:=$ **Clear[g]**

g[x_] := x/; 0 ≤ x < 1

g[x_] := 1/; 1 ≤ x < 2

g[x_] := 3 - x/; 2 ≤ x < 3

g[x_] := g[x - 3]/; x ≥ 3

defines the recursively-defined function $g(x)$. For $0 \le x < 3$, $g(x)$ is defined by

$$g(x) = \begin{cases} x, & 0 \le x < 1 \\ 1, & 1 \le x < 2 \\ 3 - x, & 2 \le x < 3. \end{cases}$$

For $x \ge 3$, $g(x) = g(x - 3)$. Entering

$In[105]:=$ **g[7]**

$Out[105]=$ 1

computes $g(7) = g(4) = g(1) = 1$. We use **Table** to create a list of ordered pairs $(x, g(x))$ for 25 equally spaced values of x between 0 and 6.

$In[106]:=$ **Table[{x, g[x]}, {x, 0, 6, 6/24}]**

$Out[106]= \left\{ \{0,0\}, \left\{\frac{1}{4}, \frac{1}{4}\right\}, \left\{\frac{1}{2}, \frac{1}{2}\right\}, \left\{\frac{3}{4}, \frac{3}{4}\right\}, \{1,1\}, \left\{\frac{5}{4}, 1\right\}, \left\{\frac{3}{2}, 1\right\}, \right.$

$\left\{\frac{7}{4}, 1\right\}, \{2,1\}, \left\{\frac{9}{4}, \frac{3}{4}\right\}, \left\{\frac{5}{2}, \frac{1}{2}\right\}, \left\{\frac{11}{4}, \frac{1}{4}\right\}, \{3,0\},$

$\left\{\frac{13}{4}, \frac{1}{4}\right\}, \left\{\frac{7}{2}, \frac{1}{2}\right\}, \left\{\frac{15}{4}, \frac{3}{4}\right\}, \{4,1\}, \left\{\frac{17}{4}, 1\right\}, \left\{\frac{9}{2}, 1\right\},$

$\left. \left\{\frac{19}{4}, 1\right\}, \{5,1\}, \left\{\frac{21}{4}, \frac{3}{4}\right\}, \left\{\frac{11}{2}, \frac{1}{2}\right\}, \left\{\frac{23}{4}, \frac{1}{4}\right\}, \{6,0\} \right\}$

We will discuss additional ways to define, manipulate, and evaluate functions as needed. However, Mathematica's extensive programming language allows a great deal of flexibility in defining functions, many of which are beyond the scope of this text. These powerful techniques are discussed in detail in texts like Gaylord, Kamin, and Wellin's *Introduction to Programming with Mathematica* [9], Gray's *Mastering Mathematica: Programming Methods and Applications* [12], and Maeder's *The Mathematica Programmer II* and *Programming in Mathematica* [15, 16].

2.3 Graphing Functions, Expressions, and Equations

One of the best features of Mathematica is its graphics capabilities. In this section, we discuss methods of graphing functions, expressions, and equations and several of the options available to help graph functions.

2.3.1 Functions of a Single Variable

The command

$$\texttt{Plot[f[x],\{x,a,b\}]}$$

graphs the function $y = f(x)$ on the interval $[a, b]$. Mathematica returns information about the basic syntax of the `Plot` command with `?Plot` or use the **Help Browser** to obtain detailed information regarding `Plot`.

Remember that every Mathematica object can be assigned a name, including graphics. `Show[p1,p2, . . . ,pn]` displays the graphics p1, p2, ..., pn together.

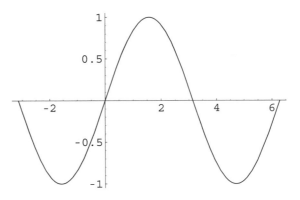

Figure 2-1 $y = \sin x$ for $-\pi \le x \le 2\pi$

EXAMPLE 2.3.1: Graph $y = \sin x$ for $-\pi \le x \le 2\pi$.

SOLUTION: Entering

$$In[107] := \mathbf{p1 = Plot[Sin[x], \{x, -\pi, 2\pi\}]}$$

graphs $y = \sin x$ for $-\pi \le x \le 2\pi$ and names the result p1. The plot is shown in Figure 2-1.

■

EXAMPLE 2.3.2: Graph $s(t)$ for $0 \le t \le 5$ where $s(t) = 1$ for $0 \le t < 1$ and $s(t) = 1 + s(t - 1)$ for $t \ge 1$.

SOLUTION: After defining $s(t)$,

$$In[108] := \mathbf{s[t\_] := 1 \,/; \; 0 \le t < 1}$$

$$\mathbf{s[t\_] := 1 + s[t - 1] \,/; \, t \ge 1}$$

we use Plot to graph $s(t)$ for $0 \le t \le 5$ in Figure 2-2.

$$In[109] := \mathbf{Plot[s[t], \{t, 0, 5\}, AspectRatio \to Automatic]}$$

Of course, Figure 2-2 is not completely precise: vertical lines are never the graphs of functions. In this case, discontinuities occur at $t = 1, 2, 3, 4,$ and 5. If we were to redraw the figure by hand, we would erase the

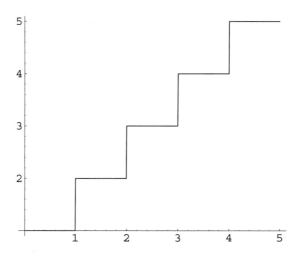

Figure 2-2 $s(t) = 1 + s(t - 1), 0 \leq t \leq 5$

vertical line segments, and then for emphasis place open dots at $(1, 1)$, $(2, 2)$, $(3, 3)$, $(4, 4)$, and $(5, 5)$ and then closed dots at $(1, 2)$, $(2, 3)$, $(3, 4)$, $(4, 5)$, and $(5, 6)$.

■

Entering `Options[Plot]` lists all `Plot` options and their default values. The most frequently used options include `PlotStyle`, `DisplayFunction`, `AspectRatio`, `PlotRange`, `PlotLabel`, and `AxesLabel`.

1. `PlotStyle` controls the color and thickness of a plot. `PlotStyle-> GrayLevel[w]`, where $0 \leq w \leq 1$ instructs Mathematica to generate the plot in `GrayLevel[w]`. `GrayLevel[0]` corresponds to black and `GrayLevel[1]` corresponds to white. Color plots can be generated using `RGBColor`. `RGBColor[1,0,0]` corresponds to red, `RGBColor[0,1,0]` corresponds to green, and `RGBColor[0,0,1]` corresponds to blue. `PlotStyle->Dashing[{a1,a2,...,an}]` indicates that successive segments be dashed with repeating lengths of $a_1, a_2, \ldots, a_n$. The thickness of the plot is controlled with `PlotStyle->Thickness[w]`, where w is the fraction of the total width of the graphic. For a single plot, the `PlotStyle` options are combined with `PlotStyle->{{option1, option2, ... , optionn}}`.
2. A plot is not displayed when the option `DisplayFunction-> Identity` is included. Including the option `DisplayFunction->$ DisplayFunction` in `Show` or `Plot` commands instructs Mathematica to display graphics.

3. The ratio of height to width of a plot is controlled by `AspectRatio`. The default is $1/$`GoldenRatio`. Generally, a plot is drawn to scale when the option `AspectRatio->Automatic` is included in the `Plot` or `Show` command.

4. `PlotRange` controls the horizontal and vertical axes. `PlotRange->{c,d}` specifies that the vertical axis displayed corresponds to the interval $c \le y \le d$ while `PlotRange->{{a,b},{c,d}}` specifes that the horizontal axis displayed corresponds to the interval $a \le x \le b$ and that the vertical axis displayed corresponds to the interval $c \le y \le d$.

5. `PlotLabel->"titleofplot"` labels the plot `titleofplot`.

6. `AxesLabel->{"xaxislabel","yaxislabel"}` labels the x-axis with `xaxislabel` and the y-axis with `yaxislabel`.

EXAMPLE 2.3.3: Graph $y = \sin x$, $y = \cos x$, and $y = \tan x$ together with their inverse functions.

Be sure you have completed the previous example immediately before entering the following commands.

SOLUTION: In p2 and p3, we use `Plot` to graph $y = \sin^{-1} x$ and $y = x$, respectively. Neither plot is displayed because we include the option `Display Function->Identity`. p1, p2, and p3 are displayed together with `Show` in Figure 2-3. The plot is shown to scale; the graph of $y = \sin x$ is in black, $y = \sin^{-1} x$ is in gray, and $y = x$ is dashed.

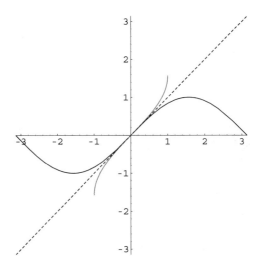

Figure 2-3 $y = \sin x$, $y = \sin^{-1} x$, and $y = x$

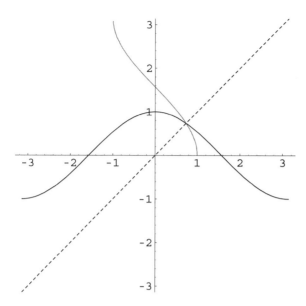

Figure 2-4 $y = \cos x$, $y = \cos^{-1} x$, and $y = x$

```
In[110]:= p2 = Plot[ArcSin[x], {x, -1, 1},
              PlotStyle → GrayLevel[0.3],
              DisplayFunction → Identity];

          p3 = Plot[x, {x, -π, 2π},
              PlotStyle → Dashing[{0.01}],
              DisplayFunction → Identity];

          p4 = Show[p1, p2, p3, PlotRange → {{-π, π}, {-π, π}},
              AspectRatio → Automatic]
```

The command `Plot[{f1[x], f2[x], ..., fn[x]}, {x, a, b}]` plots $f_1(x)$, $f_2(x)$, ..., $f_n(x)$ together for $a \le x \le b$. Simple `PlotStyle` options are incorporated with `PlotStyle->{option1, option2, ..., optionn}` where `optioni` corresponds to the plot of $f_i(x)$. Multiple options are incorporated using `PlotStyle->{{options1}, {options2}, ..., {optionsn}}` where `optionsi` are the options corresponding to the plot of $f_i(x)$.

In the following, we use `Plot` to graph $y = \cos x$, $y = \cos^{-1} x$, and $y = x$ together. Mathematica generates several error messages because the interval $[-\pi, \pi]$ contains numbers not in the domain of $y = \cos^{-1} x$. Nevertheless, Mathematica displays the plot correctly in Figure 4-36.

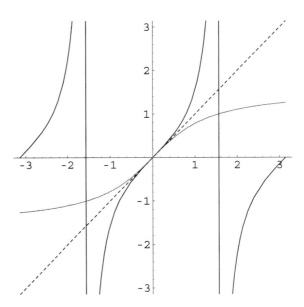

Figure 2-5 $y = \tan x$, $y = \tan^{-1} x$, and $y = x$

The plot is shown to scale; the graph of $y = \cos x$ is in black, $y = \cos^{-1} x$ is in gray, and $y = x$ is dashed.

```
In[111]:= r4 = Plot[{Cos[x], ArcCos[x], x}, {x, -π, π},
              PlotStyle → {GrayLevel[0], GrayLevel[0.3],
              Dashing[{0.01}]},
              PlotRange → {-π, π}, AspectRatio → Automatic]
```

Plot :: plnr : arccos[x] is not a machine - size real number at x =
-3.14159.

Plot :: plnr : arccos[x] is not a machine - size real number at x =
-2.8867.

Plot :: plnr : arccos[x] is not a machine - size real number at x =
-2.60872.

General :: stop : Further output of Plot :: plnr will be suppressed
during this calculation.

We use the same idea to graph $y = \tan x$, $y = \tan^{-1} x$, and $y = x$ in Figure 2-5.

```
In[112]:= q4 = Plot[{Tan[x], ArcTan[x], x}, {x, -π, π},
              PlotStyle → {GrayLevel[0], GrayLevel[0.3],
              Dashing[{0.01}]},
              PlotRange → {-π, π}, AspectRatio → Automatic]
```

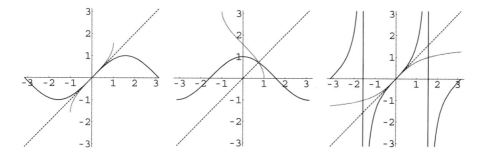

Figure 2-6   The elementary trigonometric functions and their inverses

Use Show together with GraphicsArray to display graphics in rectangular arrays. Entering

    In[113]:= **Show[GraphicsArray[{p4,r4,q4}]]**

shows the three plots p4, r4, and q4 in a row as shown in Figure 2-6.

■

The previous example illustrates the graphical relationship between a function
and its inverse.

EXAMPLE 2.3.4 (Inverse functions): $f(x)$ and $g(x)$ are **inverse functions** if

$$f(g(x)) = g(f(x)) = x.$$

If $f(x)$ and $g(x)$ are inverse functions, their graphs are symmetric about
the line $y = x$. The command

        Composition[f1,f2,f3,...,fn,x]

computes the composition

$$(f_1 \circ f_2 \circ \cdots \circ f_n)(x) = f_1(f_2(\cdots(f_n(x)))).$$

For two functions $f(x)$ and $g(x)$, it is usually easiest to compute the composition $f(g(x))$ with f[g[x]] or f[x]//g.
    Show that

$$f(x) = \frac{-1 - 2x}{-4 + x} \qquad \text{and} \qquad g(x) = \frac{4x - 1}{x + 2}$$

are inverse functions.

$f(x)$ and $g(x)$ are not returned because a semi-colon is included at the end of each command.

SOLUTION: After defining $f(x)$ and $g(x)$,

$$In[114] := \text{f}[\text{x\_}] = \frac{-1 - 2\,\text{x}}{-4 + \text{x}};$$

$$\text{g}[\text{x\_}] = \frac{4\,\text{x} - 1}{\text{x} + 2};$$

we compute and simplify the compositions $f(g(x))$ and $g(f(x))$. Because both results are x, $f(x)$ and $g(x)$ are inverse functions.

$$In[115] := \text{f}[\text{g}[\text{x}]]$$

$$Out[115] = \frac{-1 - \frac{2\,(-1+4\,\text{x})}{2+\text{x}}}{-4 + \frac{-1+4\,\text{x}}{2+\text{x}}}$$

$$In[116] := \text{Simplify}[\text{f}[\text{g}[\text{x}]]]$$

$$Out[116] = \text{x}$$

$$In[117] := \text{Simplify}[\text{g}[\text{f}[\text{x}]]]$$

$$Out[117] = \text{x}$$

To see that the graphs of $f(x)$ and $g(x)$ are symmetric about the line $y = x$, we use `Plot` to graph $f(x)$, $g(x)$, and $y = x$ together in Figure 2-7.

$$In[118] := \text{Plot}[\{\text{f}[\text{x}], \text{g}[\text{x}], \text{f}[\text{g}[\text{x}]]\}, \{\text{x}, -10, 10\},$$
$$\text{PlotStyle} \rightarrow \{\text{GrayLevel}[0], \text{GrayLevel}[0.3],$$
$$\text{Dashing}[\{0.01\}]\}, \text{PlotRange} \rightarrow \{-10, 10\},$$
$$\text{AspectRatio} \rightarrow \text{Automatic}]$$

In the plot, observe that the graphs of $f(x)$ and $g(x)$ are symmetric about the line $y = x$. The plot also illustrates that the domain and range of a function and its inverse are interchanged: $f(x)$ has domain $(-\infty, 4) \cup (4, \infty)$ and range $(-\infty, -2) \cup (-2, \infty)$; $g(x)$ has domain $(-\infty, -2) \cup (-2, \infty)$ and range $(-\infty, 4) \cup (4, \infty)$.

■

For repeated compositions of a function with itself, `Nest[f,x,n]` computes the composition

$$\underbrace{(f \circ f \circ f \circ \cdots \circ f)(x)}_{n \text{ times}} = \underbrace{(f\,(f\,(f\,\cdots)))(x)}_{n \text{ times}} = f^n(x).$$

EXAMPLE 2.3.5: Graph $f(x)$, $f^{10}(x)$, $f^{20}(x)$, $f^{30}(x)$, $f^{40}(x)$, and $f^{50}(x)$ if $f(x) = \sin x$ for $0 \le x \le 2\pi$.

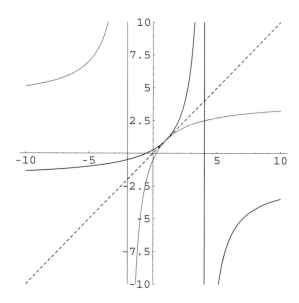

Figure 2-7 $f(x)$ in black, $g(x)$ in gray, and $y = x$ dashed

SOLUTION: After defining $f(x) = \sin x$,

```
In[119] := f[x_] = Sin[x]
Out[119] = Sin[x]
```

we graph $f(x)$ in p1 with Plot

```
In[120] := p1 = Plot[f[x], {x, 0, 2π},
              DisplayFunction → Identity];
```

and then illustrate the use of Nest by computing $f^5(x)$.

```
In[121] := Nest[f, x, 5]
Out[121] = Sin[Sin[Sin[Sin[Sin[x]]]]]
```

Next, we use Table together with Nest to create the list of functions

$$\left\{ f^{10}(x),\ f^{20}(x),\ f^{30}(x),\ f^{40}(x),\ f^{50}(x) \right\}.$$

Because the resulting output is rather long, we include a semi-colon at the end of the Table command to suppress the resulting output.

```
In[122] := toplot = Table[Nest[f, x, n], {n, 10, 50, 10}];
```

In grays, we compute a list of GrayLevel[i] for five equally spaced values of i between 0.2 and 0.8. We then graph the functions in toplot

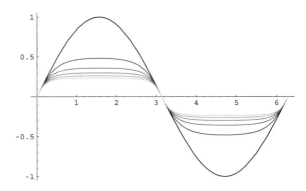

Figure 2-8 $f(x)$ in black; the graphs of $f^{10}(x)$, $f^{20}(x)$, $f^{30}(x)$, $f^{40}(x)$, and $f^{50}(x)$ are successively lighter – the graph of $f^{50}(x)$ is the lightest

on the interval $[0, 2\pi]$ with Plot. The graphs are shaded according to grays and named p2. Evaluate[toplot] causes toplot to be evaluated before the Plot command. It is important: if you do not evaluate toplot first with Evaluate, Mathematica attempts to plot toplot. Since toplot is not a function of a single variable, Mathematica generates error messages and an empty plot. When Mathematica evaluates toplot first, Mathematica understands that toplot is a list of functions and graphs each as expected.

Finally, we use Show together with the option

```
DisplayFunction->$DisplayFunction
```

to display p1 and p2 together in Figure 2-8.

```
In[123]:= grays = Table[GrayLevel[i], {i, 0.2, 0.8, 0.6/4}];

          p2 = Plot[Evaluate[toplot], {x, 0, 2π},
                 PlotStyle → grays,
                 DisplayFunction → Identity];

          Show[p1, p2, DisplayFunction → $DisplayFunction]
```

In the plot, we see that repeatedly composing sine with itself has a flattening effect on $y = \sin x$.

■

The command

```
ListPlot[{{x1,y1}, {x2,y2}, ..., {xn,yn}}]
```

plots the list of points $\{(x_1, y_1), (x_2, y_2), \ldots, (x_n, y_n)\}$. The size of the points in the resulting plot is controlled with the option `PlotStyle->PointSize[w]`, where w is the fraction of the total width of the graphic. For two-dimensional graphics, the default value is 0.008.

Remark. The command

$$\texttt{ListPlot[\{y1,y2,..,yn\}]}$$

plots the list of points $\{(1, y_1), (2, y_2), \ldots, (n, y_n)\}$.

EXAMPLE 2.3.6: Graph $y = \dfrac{\sqrt{9 - x^2}}{x^2 - 4}$.

SOLUTION: We use `Plot` to generate the basic graph of y shown in Figure 4-38(a). Observe that Mathematica generates several error messages, which is because we have instructed Mathematica to plot the function on an interval that contains numbers not in the domain of the function.

Mathematica's error messages do not always mean that you have made a mistake entering a command.

$$In[124] := \texttt{p1 = Plot[Sqrt[9 - x^2]/(x^2 - 4), \{x, -5, 5\}]}$$

Plot :: plnr : $\frac{\sqrt{9-x^2}}{-4+x^2}$ is not a machine − size real number at $x = -5$ . .
Plot :: plnr : $\frac{\sqrt{9-x^2}}{-4+x^2}$ is not a machine − size real number at $x = -4.59433$.
Plot :: plnr : $\frac{\sqrt{9-x^2}}{-4+x^2}$ is not a machine − size real number at $x = -4.15191$.
General :: stop : Further output of Plot :: plnr will be suppressed during this calculation .

Observe that the domain of y is $[-3, -2) \cup (-2, 2) \cup (2, 3]$. A better graph of y is obtained by plotting y for $-3 \leq x \leq 3$ and shown in Figure 4-38(b). We then use the `PlotRange` option to specify that the displayed horizontal axis corresponds to $-7 \leq x \leq 7$ and that the displayed vertical axis corresponds to $-7 \leq y \leq 7$. The graph is drawn to scale because we include the option `AspectRatio->Automatic`. In this case, Mathematica does not generate any error messages. Mathematica uses a point-plotting scheme to generate graphs. Coincidentally, Mathematica happens to not sample $x = \pm 2$ so does not generate any error messages.

$$In[125] := \texttt{p2 = Plot[Sqrt[9 - x^2]/(x^2 - 4), \{x, -3, 3\},}$$
$$\texttt{PlotRange} \rightarrow \texttt{\{\{-7, 7\}, \{-7, 7\}\},}$$
$$\texttt{AspectRatio} \rightarrow \texttt{Automatic]}$$

To see the endpoints in the plot, we use `ListPlot` to plot the points $(-3, 0)$ and $(3, 0)$. The points are slightly enlarged in Figure 4-38(c) because we increase their size using `PointSize`.

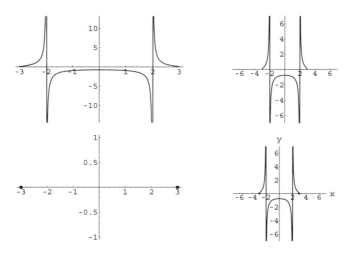

Figure 2-9   (from left to right) (a)–(d) The four plots p1, p2, p3, and p4 combined into a single graphic

In[126]:= **p3 = ListPlot[{{-3, 0}, {3, 0}},**
                PlotStyle → PointSize[0.02]]

Finally, we use Show to display p2 and p3 together in Figure 4-38(d), where we have labeled the axes using the AxesLabel option.

In[127]:= **p4 = Show[p2, p3, AxesLabel → {"x", "y"}]**

The sequence of plots shown in Figure 4-38, which combines p1, p2, p3, and p4 into a single graphic, is generated using Show together with GraphicsArray.

In[128]:= **Show[GraphicsArray[{{p1, p2}, {p3, p4}}]]**

■

When graphing functions involving odd roots, Mathematica's results may be surprising to the beginner. The key is to load the **RealOnly** package located in the **Miscellaneous** folder (or directory) first.

EXAMPLE 2.3.7: Graph $y = x^{1/3}(x - 2)^{2/3}(x + 1)^{4/3}$.

SOLUTION: Entering

In[129] := **p1 = Plot[x^(1/3) (x-2)^(2/3) (x+1)^(4/3), {x, -2, 3}]**

Plot :: plnr : $(-2 + x)^{2/3} x^{1/3} (1 + x)^{4/3}$ is not a machine-size real number at $x = -2..$

Plot :: plnr : $(-2 + x)^{2/3} x^{1/3} (1 + x)^{4/3}$ is not a machine-size real number at $x = -1.79717$.

Plot :: plnr : $(-2 + x)^{2/3} x^{1/3} (1 + x)^{4/3}$ is not a machine-size real number at $x = -1.57596$.

General :: stop : Further output of Plot :: plnr will be suppressed during this calculation.

not only produces many error messages but does not produce the graph we expect (see Figure 2-10(a)) because many of us consider $y = x^{1/3}(x - 2)^{2/3}(x + 1)^{4/3}$ to be a real-valued function with domain $(-\infty, \infty)$. Generally, Mathematica does return a real number when computing the odd root of a negative number. For example, $x^3 = -1$ has three solutions

In[130] := **s1 = Solve[x^3 + 1 == 0]**

Out[130] = $\{\{x \to -1\}, \{x \to (-1)^{1/3}\}, \{x \to -(-1)^{2/3}\}\}$

In[131] := **N[s1]**

Out[131] = $\{\{x \to -1.\}, \{x \to 0.5 + 0.866025 \, i\}, \{x \to 0.5 - 0.866025 \, i\}\}$

When computing an odd root of a negative number, Mathematica has many choices (as illustrated above) and chooses a root with positive imaginary part – the result is not a real number.

In[132] := **N[(-1)^(1/3)]**

Out[132] = $0.5 + 0.866025 \, i$

Solve is discussed in more detail in the next section.

N[number] returns an approximation of number.

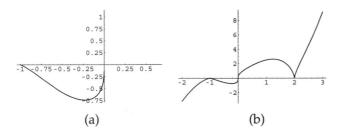

(a)   (b)

Figure 2-10   (a) and (b) Two plots of $y = x^{1/3}(x - 2)^{2/3}(x + 1)^{4/3}$

To obtain real values when computing odd roots of negative numbers, load the **RealOnly** package that is located in the **Miscellaneous** folder or directory.

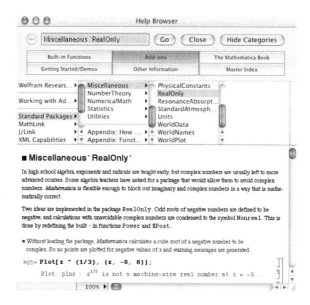

After the **RealOnly** package has been loaded, reentering the `Plot` command produces the expected graph. See Figure 2-10(b).

```
In[133]:= << Miscellaneous`RealOnly`
```

```
In[134]:= p2 = Plot[x^(1/3) (x - 2)^(2/3) (x + 1)^(4/3),
               {x, -2, 3}]
```

```
In[135]:= Show[GraphicsArray[{p1, p2}]]
```

∎

2.3.2 Parametric and Polar Plots in Two Dimensions

To graph the parametric equations $x = x(t)$, $y = y(t)$, $a \le t \le b$, use

`ParametricPlot` has the same options as `Plot`.

```
ParametricPlot[{x[t],y[t]},{t,a,b}]
```

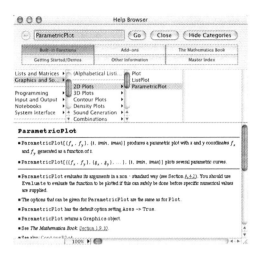

and to graph the polar function $r = r(\theta)$, $\alpha \le \theta \le \beta$, use

```
PolarPlot[r[theta],{theta,alpha,beta}].
```

The `PolarPlot` function is contained in the `Graphics` package which is located in the **Graphics** directory, so load the **Graphics** package by entering `<<Graphics'Graphics'` before using the `PolarPlot` function.

EXAMPLE 2.3.8 (The Unit Circle): The **unit circle** is the set of points (x, y) exactly 1 unit from the origin, $(0, 0)$, and, in rectangular coordinates, has equation $x^2 + y^2 = 1$. The unit circle is the classic example of a relation that is neither a function of x nor a function of y. The top half of the unit circle is given by $y = \sqrt{1 - x^2}$ and the bottom half is given by $y = -\sqrt{1 - x^2}$.

```
In[136]:= p1 = Plot[{Sqrt[1 - x^2], -Sqrt[1 - x^2]},
              {x, -1, 1},
              PlotRange → {{-3/2, 3/2}, {-3/2, 3/2}},
              AspectRatio → Automatic,
              DisplayFunction → Identity];
```

Each point (x, y) on the unit circle is a function of the angle, t, that subtends the x-axis, which leads to a parametric representation of the unit circle, $\begin{cases} x = \cos t, \\ y = \sin t, \end{cases}$ $0 \le t \le 2\pi$, which we graph with `ParametricPlot`.

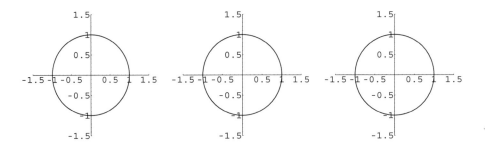

Figure 2-11   The unit circle generated with `Plot`, `ParametricPlot`, and `PolarPlot`

```
In[137]:= p2 = ParametricPlot[{Cos[t], Sin[t]}, {t, 0, 2π},
              PlotRange → {{-3/2, 3/2}, {-3/2, 3/2}},
              AspectRatio → Automatic,
              DisplayFunction → Identity];
```

Using the change of variables $x = r \cos t$ and $y = r \sin t$ to convert from rectangular to polar coordinates, a polar equation for the unit circle is $r = 1$. After loading the **Graphics** package, we use `PolarPlot` to graph $r = 1$.

```
In[138]:= << Graphics`Graphics`
```

```
In[139]:= p3 = PolarPlot[1, {t, 0, 2π},
              PlotRange → {{-3/2, 3/2}, {-3/2, 3/2}},
              AspectRatio → Automatic,
              DisplayFunction → Identity];
```

We display `p1`, `p2`, and `p3` side-by-side using `Show` together with `GraphicsArray` in Figure 2-11. Of course, they all look the same.

```
In[140]:= Show[GraphicsArray[{p1, p2, p3}]]
```

EXAMPLE 2.3.9: Graph the parametric equations

$$\begin{cases} x = t + \sin 2t, \\ y = t + \sin 3t, \end{cases} \quad -2\pi \le t \le 2\pi.$$

SOLUTION: After defining x and y, we use `ParametricPlot` to graph the parametric equations in Figure 2-12.

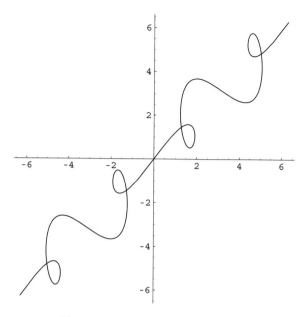

Figure 2-12 $(x(t), y(t)), -2\pi \le t \le 2\pi$

```
In[141]:= x[t_] = t + Sin[2t];
          y[t_] = t + Sin[3t];
          ParametricPlot[
            {x[t],y[t]}, {t,-2π, 2π},
            AspectRatio- > Automatic]
```

■

In the following example, the equations involve integrals.

Remark. Topics from calculus are discussed in Chapter 3. For now, we state that Integrate[f[x],{x,a,b}] attempts to evaluate $\int_a^b f(x)\,dx$.

EXAMPLE 2.3.10 (Cornu Spiral): The **Cornu spiral** (or **clothoid**) (see [11] and [20]) has parametric equations

$$x = \int_0^t \sin\left(\frac{1}{2}u^2\right) du \quad \text{and} \quad y = \int_0^t \cos\left(\frac{1}{2}u^2\right) du.$$

Graph the Cornu spiral.

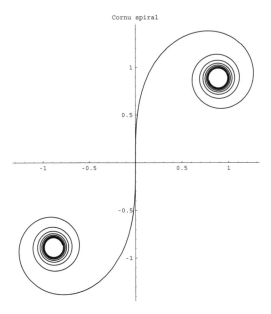

Figure 2-13   The Cornu spiral

SOLUTION: We begin by defining x and y. Notice that Mathematica can evaluate these integrals, even though the results are in terms of the FresnelS and FresnelC functions, which are defined in terms of integrals:

$$\text{FresnelS[t]} = \int_0^t \sin\left(\frac{\pi}{2}u^2\right) du \quad \text{and} \quad \text{FresnelC[t]} = \int_0^t \cos\left(\frac{\pi}{2}u^2\right) du.$$

```
In[142] := x[t_] = Integrate[Sin[u^2/2], {u, 0, t}]
```
$$\text{Out[142]} = \sqrt{\pi}\,\text{FresnelS}\left[\frac{t}{\sqrt{\pi}}\right]$$

```
In[143] := y[t_] = Integrate[Cos[u^2/2], {u, 0, t}]
```
$$\text{Out[143]} = \sqrt{\pi}\,\text{FresnelC}\left[\frac{t}{\sqrt{\pi}}\right]$$

We use ParametricPlot to graph the Cornu spiral in Figure 2-13. The option AspectRatio->Automatic instructs Mathematica to generate the plot to scale; PlotLabel->"Cornu spiral" labels the plot.

```
In[144] := ParametricPlot[{x[t],y[t]}, {t, -10, 10},
               AspectRatio → Automatic,
               PlotLabel- > "Cornu spiral"]
```

■

Observe that the graph of the polar equation $r = f(\theta)$, $\alpha \le \theta \le \beta$ is the same as the graph of the parametric equations

$$x = f(\theta) \cos \theta \quad \text{and} \quad y = f(\theta) \sin \theta, \quad \alpha \le \theta \le \beta$$

so both `ParametricPlot` and `PolarPlot` can be used to graph polar equations.

EXAMPLE 2.3.11: Graph (a) $r = \sin(8\theta/7)$, $0 \le \theta \le 14\pi$; (b) $r = \theta \cos \theta$, $-19\pi/2 \le \theta \le 19\pi/2$; (c) ("The Butterfly") $r = e^{\cos\theta} - 2\cos 4\theta + \sin^5(\theta/12)$, $0 \le \theta \le 24\pi$; and (d) ("The Lituus") $r^2 = 1/\theta$, $0.1 \le \theta \le 10\pi$.

SOLUTION: For (a) and (b) we use `ParametricPlot`. First define r and then use `ParametricPlot` to generate the graph of the polar curve. No graphics are displayed because we include the option `DisplayFunction->Identity` in each `ParametricPlot` command.

```
In[145]:= Clear[r]
          r[θ_] = Sin[8θ/7];
          pp1 = ParametricPlot[{r[θ] Cos[θ], r[θ] Sin[θ]},
                  {θ, 0, 14π}, AspectRatio → Automatic,
                  DisplayFunction → Identity];
```

For (b), we use the option `PlotRange->{{-30,30},{-30,30}}` to indicate that the range displayed on both the vertical and horizontal axes corresponds to the interval $[-30, 30]$. To help assure that the resulting graphic appears "smooth", we increase the number of points that Mathematica samples when generating the graph by including the option `PlotPoints->200`.

```
In[146]:= Clear[r]
          r[θ_] = θ Cos[θ];
          pp2 = ParametricPlot[{r[θ] Cos[θ], r[θ] Sin[θ]},
                  {θ, -19π/2, 19π/2},
                  PlotRange → {{-30, 30}, {-30, 30}},
                  AspectRatio → Automatic, PlotPoints → 200,
                  DisplayFunction → Identity];
```

For (c) and (d), we use `PolarPlot`. Using standard mathematical notation, we know that $\sin^5(\theta/12) = (\sin(\theta/12))^5$. However, when defining r with Mathematica, be sure you use the form `Sin(θ/12)^5`, not `Sin^5[θ/12]`, which Mathematica will not interpret in the way intended.

You do not need to reload the **Graphics** package if you have already loaded it during your current Mathematica session.

```
In[147]:= << Graphics`Graphics`
```

```
In[148]:= Clear[r]
          r[θ_] = Exp[Cos[θ]] - 2 Cos[4θ] + Sin[θ/12]^5;
          pp3 = PolarPlot[r[θ], {θ, 0, 24π},
                PlotPoints → 200,
                PlotRange → {{-4, 5}, {-4.5, 4.5}},
                AspectRatio → Automatic,
                DisplayFunction → Identity];
```

For (d), we graph $r^2 = 1/\theta$ by graphing $r = 1/\sqrt{\theta}$ and $r = -1/\sqrt{\theta}$ together with `PolarPlot`.

```
In[149]:= Clear[r]

          pp4 = PolarPlot[{Sqrt[1/θ], -Sqrt[1/θ]},
                {θ, 0.1, 10π},
                AspectRatio → Automatic,
                PlotRange → All,
                DisplayFunction → Identity];
```

Finally, we use `Show` together with `GraphicsArray` to display all four graphs as a graphics array in Figure 2-14. pp1 and pp2 are shown in the first row; pp3 and pp4 in the second.

```
In[150]:= Show[GraphicsArray[{{pp1, pp2}, {pp3, pp4}}]]
```

∎

2.3.3 Three-Dimensional and Contour Plots; Graphing Equations

An elementary function of two variables, $z = f(x, y) = $ *expression in x and y*, is typically defined using the form

```
f[x_,y_]=expression in x and y.
```

Once a function has been defined, a basic graph is generated with `Plot3D`:

```
Plot3D[f[x,y],{x,a,b},{y,c,d}]
```

graphs $f(x, y)$ for $a \leq x \leq b$ and $c \leq y \leq d$.

For details regarding `Plot3D` and its options enter `?Plot3D` or `??Plot3D` or access the **Help Browser** to obtain information about the `Plot3D` command, as we do here.

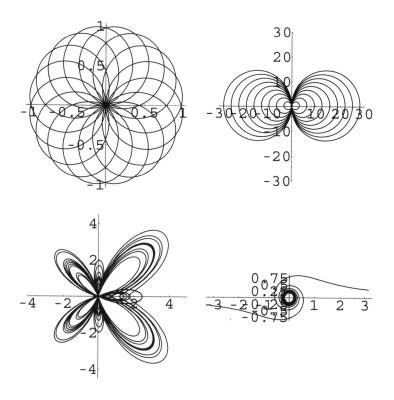

Figure 2-14   Graphs of four polar equations

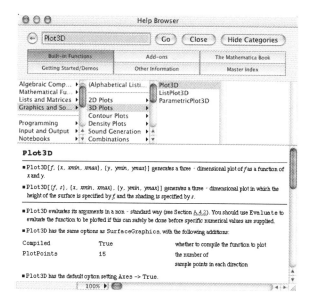

Graphs of several level curves of $z = f(x, y)$ are generated with

$$\texttt{ContourPlot[f[x,y],\{x,a,b\},\{y,c,d\}]}.$$

For details regarding `ContourPlot` and its options enter `?ContourPlot` or `??ContourPlot` or access the **Help Browser**.

EXAMPLE 2.3.12: Let

$$f(x, y) = \frac{x^2 y}{x^4 + 4y^2}.$$

(a) Calculate $f(1, -1)$. (b) Graph $f(x, y)$ and several contour plots of $f(x, y)$ on a region containing $(0, 0)$.

SOLUTION: After defining $f(x, y)$, we evaluate $f(1, -1) = -1/5$.

```
In[151] := f[x_, y_] = x^2y/(x^4 + 4y^2)
```
$$Out[151] = \frac{x^2\ y}{x^4 + 4\ y^2}$$
```
In[152] := f[1, -1]
```
$$Out[152] = -\frac{1}{5}$$

Next, we use `Plot3D` to graph $f(x, y)$ for $-1/2 \le x \le 1/2$ and $-1/2 \le y \le 1/2$ in Figure 2-15. We illustrate the use of the `Axes`, `Boxed`, and `PlotPoints` options.

```
In[153] := Plot3D[f[x,y],{x,-1/2,1/2},{y,-1/2,1/2},
              Axes- >Automatic, Boxed- >False,
              PlotPoints- >{50,50}]
```

Two contour plots are generated with `ContourPlot`. The second illustrates the use of the `PlotPoints`, `Frame`, `ContourShading`, `Axes`, and `AxesOrigin` options. (See Figure 2-16.)

```
In[154] := ContourPlot[f[x,y],{x,-1/2,1/2},
              {y,-1/2,1/2},PlotPoints- >{50,50}]

In[155] := ContourPlot[f[x,y],{x,-1/2,1/2},
              {y,-1/2,1/2},PlotPoints- >{60,60},
              Frame- >False, ContourShading- >
              False, Axes- >Automatic,
              AxesOrigin- >{0,0}]
```

∎

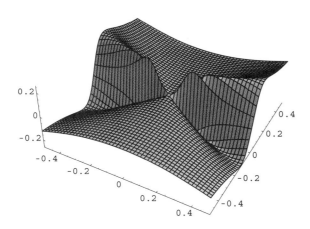

Figure 2-15   Three-dimensional plot of $f(x, y)$

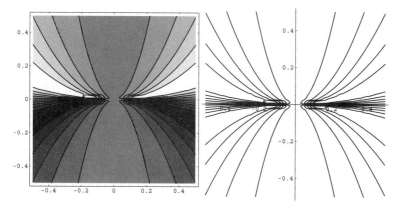

Figure 2-16   Two contour plots of $f(x, y)$

The ViewPoint option can be changed by going to the Mathematica menu, selecting **Input** and then **3D ViewPoint Selector...** at which point the following window appears.

Various perspectives can be adjusted by clicking and dragging the bounding box. When a satisfactory ViewPoint is found, select **Paste** and the ViewPoint will be pasted into the Mathematica notebook at the location of the cursor.

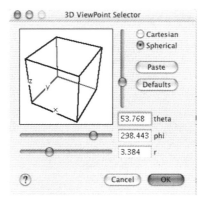

Figure 2-17 shows four different views of the graph of $g(x, y) = x \sin y + y \sin x$ for $0 \le x \le 5\pi$ and $0 \le y \le 5\pi$. The options AxesLabel, BoxRatios, ViewPoint, PlotPoints, Shading, and Mesh are also illustrated.

In[156] := **Clear[g]**
       g[x_, y_] = x Sin[y] + y Sin[x];

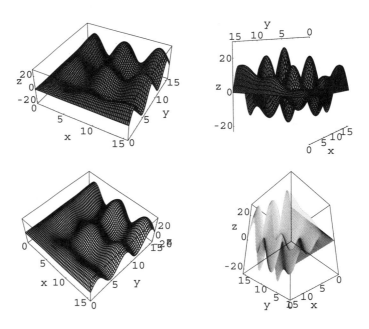

Figure 2-17   Four different plots of $g(x, y) = x \sin y + y \sin x$ for $0 \le x \le 5\pi$

```
In[157]:= p1 = Plot3D[g[x, y], {x, 0, 5π}, {y, 0, 5π},
              PlotPoints → 60, AxesLabel → {"x", "y", "z"},
              DisplayFunction → Identity];

In[158]:= p2 = Plot3D[g[x, y], {x, 0, 5π}, {y, 0, 5π},
              PlotPoints → 60,
              ViewPoint- > {-2.846, -1.813, 0.245},
              Boxed → False, BoxRatios → {1, 1, 1},
              AxesLabel → {"x", "y", "z"},
              DisplayFunction → Identity];

In[159]:= p3 = Plot3D[g[x, y], {x, 0, 5π}, {y, 0, 5π},
              PlotPoints → 60,
              ViewPoint- > {1.488, -1.515, 2.634},
              AxesLabel → {"x", "y", "z"}, Shading → False,
              DisplayFunction → Identity];

In[160]:= p4 = Plot3D[g[x, y], {x, 0, 5π}, {y, 0, 5π},
              PlotPoints → 60, AxesLabel → {"x", "y", "z"},
              Mesh → False, BoxRatios → {2, 2, 3},
              ViewPoint- > {-1.736, 1.773, -2.301},
              DisplayFunction → Identity];

In[161]:= Show[GraphicsArray[{{p1, p2}, {p3, p4}}]]
```

ContourPlot is especially useful when graphing equations. The graph of the equation $f(x, y) = C$, where C is a constant, is the same as the contour plot of $z = f(x, y)$ corresponding to C. That is, the graph of $f(x, y) = C$ is the same as the level curve of $z = f(x, y)$ corresponding to $z = C$.

EXAMPLE 2.3.13: Graph the unit circle, $x^2 + y^2 = 1$.

SOLUTION: We first graph $z = x^2 + y^2$ for $-4 \leq x \leq 4$ and $-4 \leq y \leq 4$ with Plot3D in Figure 2-18.

```
In[162]:= Plot3D[x^2 + y^2, {x, -4, 4}, {y, -4, 4}]
```

The graph of $x^2 + y^2 = 1$ is the graph of $z = x^2 + y^2$ corresponding to $z = 1$. We use ContourPlot together with the Contours option to graph this equation in Figure 2-19.

```
In[163]:= ContourPlot[x^2 + y^2, {x, -3/2, 3/2},
              {y, -3/2, 3/2}, Contours- > {1},
              ContourShading- > False]
```

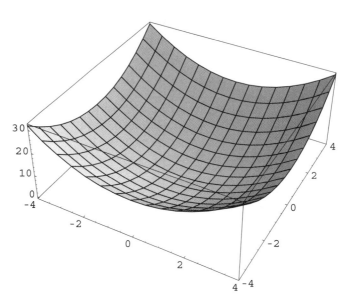

Figure 2-18   Three-dimensional plot of $z = x^2 + y^2$

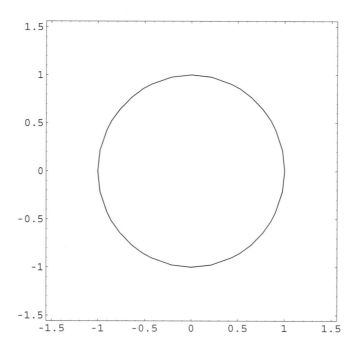

Figure 2-19   The unit circle, $x^2 + y^2 = 1$

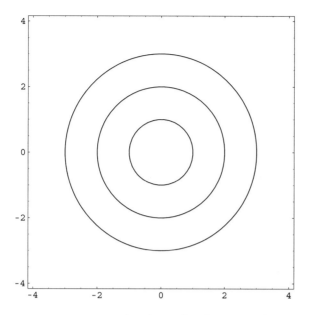

Figure 2-20 Graphs of $x^2 + y^2 = 1$, $x^2 + y^2 = 4$, and $x^2 + y^2 = 9$

Multiple graphs can be generated as well. As an illustration, we graph $x^2 + y^2 = C$ for $C = 1, 4$, and 9 in Figure 2-20.

```
In[164]:= ContourPlot[x^2+y^2, {x,-4,4}, {y,-4,4},
          Contours- > {1, 4, 9}, ContourShading- > False,
          PlotPoints- > {50, 50}]
```

■

As an alternative to using `ContourPlot` to graph equations, you can also use the `ImplicitPlot` function which is contained in the **ImplicitPlot** package located in the **Graphics** folder (or directory).

After loading the **ImplicitPlot** package by entering `<<Graphics`ImplicitPlot`, the command

$$\texttt{ImplicitPlot}[\texttt{equation}, \{\texttt{x,x0,x1}\}]$$

graphs the equation, `equation`, from $x = x_0$ to $x = x_1$. (Recall that a double equals sign (==) must be used to separate the left and right-hand sides of an equation.) The set of y-values displayed may be specified by entering the command using the form

$$\texttt{ImplicitPlot}[\texttt{equation}, \{\texttt{x,x0,x1}\}, \{\texttt{y,y0,y1}\}].$$

When graphing relatively simple equations, like those solvable using Solve, it is not necessary to specify the y-values in the ImplicitPlot command. When Solve cannot solve an equation, it is usually necessary to specify both the x and y-values. In these cases, ImplicitPlot uses the same method to produce the graph as ContourPlot. However, ContourPlot may produce better results.

EXAMPLE 2.3.14: Graph the equation $y^2 - 2x^4 + 2x^6 - x^8 = 0$ for $-1.5 \le x \le 1.5$.

SOLUTION: After loading the **ImplicitPlot** package, we define eq to be the equation $y^2 - 2x^4 + 2x^6 - x^8 = 0$ and then use ImplicitPlot to graph eq for $-1.5 \le x \le 1.5$ in Figure 2-21.

```
In[165]:= << Graphics`ImplicitPlot`

In[166]:= eq = y^2 - x^4 + 2 x^6 - x^8 == 0;

In[167]:= ImplicitPlot[eq, {x, -1.5, 1.5},
              Ticks → {{-1, 1}, {-1, 1}}]
```

∎

Equations can be plotted together, as with the command Plot, with

$$\text{ImplicitPlot}[\{eq1, eq2, \ldots, eqn\}, \{x, x0, x1\}]$$

or

$$\text{ImplicitPlot}[\{eq1, eq2, \ldots, eqn\}, \{x, x0, x1\}, \{y, 01, y1\}].$$

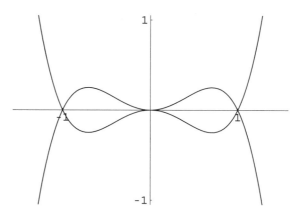

Figure 2-21   Plot of $y^2 - 2x^4 + 2x^6 - x^8 = 0$

EXAMPLE 2.3.15: Graph the equations $x^2 + y^2 = 1$ and $4x^2 - y^2 = 1$ for $-1.5 \leq x \leq 1.5$.

SOLUTION: We use `ImplicitPlot` to graph the equations together on the same axes in Figure 2-22. The graph of $x^2 + y^2 = 1$ is the unit circle while the graph of $4x^2 - y^2 = 1$ is a hyperbola.

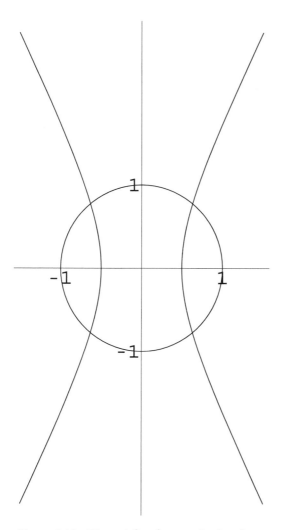

Figure 2-22   Plots of $x^2 + y^2 = 1$ and $4x^2 - y^2 = 1$

$$In[168] := \texttt{ImplicitPlot}\left[\left\{x^2 + y^2 == 1, 4\,x^2 - y^2 == 1\right\},\right.$$
$$\left.\{x, -1.5, 1.5\}, \texttt{Ticks} \rightarrow \{\{-1, 1\}, \{-1, 1\}\}\right]$$

■

Also see Example 2.3.19.

EXAMPLE 2.3.16 (Conic Sections): A **conic section** is a graph of the equation

$$Ax^2 + Bxy + Cy^2 + Dx + Ey + F = 0.$$

Except when the conic is degenerate, the conic $Ax^2 + Bxy + Cy^2 + Dx + Ey + F = 0$ is a (an)

1. **Ellipse** or **circle** if $B^2 - 4AC < 0$;
2. **Parabola** if $B^2 - 4AC = 0$; or
3. **Hyperbola** if $B^2 - 4AC > 0$.

Graph the conic section $ax^2 + bxy + cy^2 = 1$ for $-4 \le x \le 4$ and for a, b, and c equal to all possible combinations of -1, 1, and 2.

SOLUTION: We begin by defining conic to be the equation $ax^2 + bxy + cy^2 = 1$ and then use Permutations to produce all possible orderings of the list of numbers $\{-1, 1, 2\}$, naming the resulting output vals.

Permutations[list]
returns a list of all possible
orderings of the list list.

```
In[169] := Clear[a, b, c]
           conic = a x² + b x y + c y² == 1;

In[170] := vals = Permutations[{-1, 1, 2}]
Out[170] = {{-1, 1, 2}, {-1, 2, 1}, {1, -1, 2},
            {1, 2, -1}, {2, -1, 1}, {2, 1, -1}}
```

Next we define the function p. Given a1, b1, and c1, p defines toplot to be the equation obtained by replacing a, b, and c in conic by a1, b1, and c1, respectively. Then, toplot is graphed for $-4 \le x \le 4$. p returns a graphics object which is not displayed because the option DisplayFunction->Identity is included in the ImplicitPlot command.

```
In[171] := p[{a1_, b1_, c1_}] := Module[{toplot},
              toplot = conic/.{a → a1, b → b1, c → c1};
              ImplicitPlot[toplot, {x, -4, 4},
              Ticks → None, DisplayFunction → Identity]]
```

We then use Map to compute p for each ordered triple in vals. The resulting output, named graphs, is a set of six graphics objects.

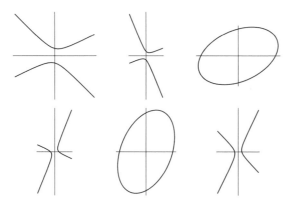

Figure 2-23   Plots of six conic sections

```
In[172]:= graphs = Map[p, vals]
Out[172]= {-Graphics-, -Graphics-, -Graphics-,
              -Graphics-, -Graphics-, -Graphics-}
```

Partition is then used to partition graphs into three element subsets. The resulting array of graphics objects named toshow is displayed with Show and GraphicsArray in Figure 2-23.

```
In[173]:= toshow = Partition[graphs, 3];
           Show[GraphicsArray[toshow]]
```

∎

2.3.4 Parametric Curves and Surfaces in Space

The command

$$\text{ParametricPlot3D}[\{x[t],y[t],z[t]\},\{t,a,b\}]$$

generates the three-dimensional curve $\begin{cases} x = x(t), \\ y = y(t), \\ z = z(t), \end{cases}$ $a \le t \le b$ and the command

$$\text{ParametricPlot3D}[\{x[u,v],y[u,v],z[u,v]\},\{u,a,b\},\{v,c,d\}]$$

plots the surface $\begin{cases} x = x(u, v), \\ y = y(u, v), \\ z = z(u, v), \end{cases}$ $a \le u \le b, c \le v \le d.$

Entering `Information[ParametricPlot3D]` or `??ParametricPlot3D`
returns a description of the `ParametricPlot3D` command along with a list of
options and their current settings.

EXAMPLE 2.3.17 (Umbilic Torus NC): A parametrization of **umbilic
torus NC** is given by $\mathbf{r}(s, t) = x(s, t)\mathbf{i} + y(s, t)\mathbf{j} + z(s, t)\mathbf{k}$, $-\pi \le s \le \pi$, $-\pi \le
t \le \pi$, where

$$x = \left[7 + \cos\left(\frac{1}{3}s - 2t\right) + 2\cos\left(\frac{1}{3}s + t\right)\right]\sin s$$

$$y = \left[7 + \cos\left(\frac{1}{3}s - 2t\right) + 2\cos\left(\frac{1}{3}s + t\right)\right]\cos s$$

and

$$z = \sin\left(\frac{1}{3}s - 2t\right) + 2\sin\left(\frac{1}{3}s + t\right).$$

Graph the torus.

SOLUTION: We define x, y, and z.

```
In[174]:= x[s_, t_] = (7 + Cos[1/3s - 2t] + 2 Cos[1/3s + t]) Sin[s];
          y[s_, t_] = (7 + Cos[1/3s - 2t] + 2 Cos[1/3s + t]) Cos[s];
          z[s_, t_] = Sin[1/3s - 2t] + 2 Sin[1/3s + t];
```

The torus is then graphed with `ParametricPlot3D` in Figure 2-24.
We illustrate the use of the `PlotPoints` option.

```
In[175]:= ParametricPlot3D[{x[s, t], y[s, t], z[s, t]},
             {s, -π, π}, {t, -π, π}, PlotPoints- > {40, 40}]
```

∎

This example is explored in
detail in Sections 8.2 and 11.4
of Gray's *Modern Differential
Geometry of Curves and
Surfaces*, [11], an
indispensable reference for
those who use Mathematica's
graphics extensively.

EXAMPLE 2.3.18 (Gray's Torus Example): A parametrization of
an **elliptical torus** is given by

$$x = (a + b\cos v)\cos u, \quad y = (a + b\cos v)\sin u, \quad z = c\sin v$$

For positive integers p and q, the curve with parametrization

$$x = (a + b\cos qt)\cos pt, \quad y = (a + b\cos qt)\sin pt, \quad z = c\sin qt$$

winds around the elliptical torus and is called a **torus knot**.
  Plot the torus if $a = 8$, $b = 3$, and $c = 5$ and then graph the torus knots
for $p = 2$ and $q = 5$, $p = 1$ and $q = 10$, and $p = 2$ and $q = 3$.

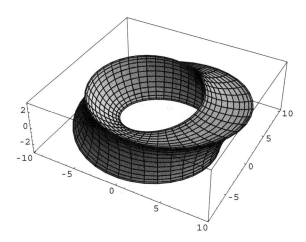

Figure 2-24   Umbilic torus

SOLUTION: We begin by defining `torus` and `torusknot`.

```
In[176]:= torus[a_, b_, c_][p_, q_][u_, v_] :=
            {(a + b Cos[u]) Cos[v],
            (a + b Cos[u]) Sin[v], c Sin[u]}
```

```
In[177]:= torusknot[a_, b_, c_][p_, q_][t_] :=
            {(a + b Cos[q t]) Cos[p t],
            (a + b Cos[q t]) Sin[p t], c Sin[q t]}
```

Next, we use `ParametricPlot3D` to generate all four graphs

```
In[178]:= pp1 = ParametricPlot3D[Evaluate[torus[8, 3, 5]
                [2, 5][u, v]], {u, 0, 2π}, {v, 0, 2π},
                PlotPoints → 60, DisplayFunction → Identity];
```

```
In[179]:= pp2 = ParametricPlot3D[Evaluate[torusknot
                [8, 3, 5][2, 5][t]], {t, 0, 3π},
                PlotPoints → 200,
                DisplayFunction → Identity];
```

```
In[180]:= pp3 = ParametricPlot3D[Evaluate[torusknot
                [8, 3, 5][1, 10][t]], {t, 0, 3π},
                PlotPoints → 200,
                DisplayFunction → Identity];
```

```
In[181]:= pp4 = ParametricPlot3D[Evaluate[torusknot
                [8, 3, 5][2, 3][t]], {t, 0, 3π},
                PlotPoints → 200,
                DisplayFunction → Identity];
```

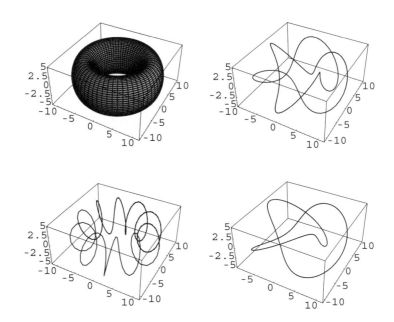

Figure 2-25   (a) An elliptical torus. (b) This knot is also known as the trefoil knot. (c) The curve generated by `torusknot [8,3,5] [2,3] [1,10]` is not a knot. (d) The torus knot with $p = 2$ and $q = 3$

and show the result as a `GraphicsArray` with `Show` and `GraphicsArray` in Figure 2-25. As when plotting lists of functions, we are careful to evaluate the list of functions to be evaluated in each `ParametricPlot3D` first with `Evaluate`.

> $In[182]:=$ **Show[GraphicsArray[{{pp1, pp2}, {pp3, pp4}}]]**

∎

Also see Example 2.3.16.

EXAMPLE 2.3.19 (Quadric Surfaces): The **quadric surfaces** are the three-dimensional objects corresponding to the conic sections in two dimensions. A **quadric surface** is a graph of

$$Ax^2 + By^2 + Cz^2 + Dxy + Exz + Fyz + Gx + Hy + Iz + J = 0,$$

where $A, B, C, D, E, F, G, H, I$, and J are constants.

The intersection of a plane and a quadric surface is a conic section.

Several of the basic quadric surfaces, in standard form, and a parametrization of the surface are listed in the following table.

Name	Parametric Equations
Ellipsoid $$\frac{x^2}{a^2} + \frac{y^2}{b^2} + \frac{z^2}{c^2} = 1$$	$\begin{cases} x = a\cos t \cos r, \\ y = b\cos t \sin r, \qquad -\pi/2 \le t \le \pi/2, -\pi \le r \le \pi \\ z = c\sin t, \end{cases}$
Hyperboloid of One Sheet $$\frac{x^2}{a^2} + \frac{y^2}{b^2} - \frac{z^2}{c^2} = 1$$	$\begin{cases} x = a\sec t \cos r, \\ y = b\sec t \sin r, \qquad -\pi/2 < t < \pi/2, -\pi \le r \le \pi \\ z = c\tan t, \end{cases}$
Hyperboloid of Two Sheets $$\frac{x^2}{a^2} - \frac{y^2}{b^2} - \frac{z^2}{c^2} = 1$$	$\begin{cases} x = a\sec t, \\ y = b\tan t \cos r, \qquad -\pi/2 < t < \pi/2 \text{ or } \pi/2 < t < \\ z = c\tan t \sin r, \\ 3\pi/2, -\pi \le r \le \pi \end{cases}$

Graph the ellipsoid with equation $\frac{1}{16}x^2 + \frac{1}{4}y^2 + z^2 = 1$, the hyperboloid
of one sheet with equation $\frac{1}{16}x^2 + \frac{1}{4}y^2 - z^2 = 1$, and the hyperboloid of
two sheets with equation $\frac{1}{16}x^2 - \frac{1}{4}y^2 - z^2 = 1$.

SOLUTION: A parametrization of the ellipsoid with equation $\frac{1}{16}x^2 + \frac{1}{4}y^2 + z^2 = 1$ is given by

$$x = 4\cos t \cos r, \quad y = 2\cos t \sin r, \quad z = \sin t, \quad -\pi/2 \le t \le \pi/2, -\pi \le r \le \pi,$$

which is graphed with `ParametricPlot3D`.

```
In[183]:= Clear[x, y, z]
          x[t_, r_] = 4 Cos[t] Cos[r];
          y[t_, r_] = 2 Cos[t] Sin[r];
          z[t_, r_] = Sin[t];
          pp1 = ParametricPlot3D[{x[t, r], y[t, r], z[t, r]},
              {t, -π/2, π/2}, {r, -π, π}, PlotPoints → 30,
              DisplayFunction → Identity];
```

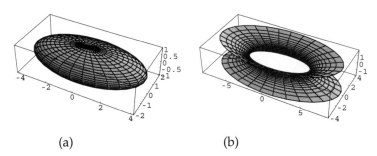

(a)   (b)

Figure 2-26   (a) Plot of $\frac{1}{16}x^2 + \frac{1}{4}y^2 + z^2 = 1$. (b) Plot of $\frac{1}{16}x^2 + \frac{1}{4}y^2 - z^2 = 1$

A parametrization of the hyperboloid of one sheet with equation $\frac{1}{16}x^2 + \frac{1}{4}y^2 - z^2 = 1$ is given by

$$x = 4 \sec t \cos r, \quad y = 2 \sec t \sin r, \quad z = \tan t, \quad -\pi/2 < t < \pi/2, \ -\pi \le r \le \pi.$$

Because $\sec t$ and $\tan t$ are undefined if $t = \pm\pi/2$, we use
ParametricPlot3D to graph these parametric equations on a subinterval of $[-\pi/2, \pi/2]$, $[-\pi/3, \pi/3]$.

```
In[184]:= Clear[x,y,z]
          x[t_,r_] = 4 Sec[t] Cos[r];
          y[t_,r_] = 2 Sec[t] Sin[r];
          z[t_,r_] = Tan[t];
          pp2 = ParametricPlot3D[{x[t,r],y[t,r],z[t,r]},
               {t,-π/3,π/3},{r,-π,π},PlotPoints → 30,
               DisplayFunction → Identity];
```

pp1 and pp2 are shown together in Figure 2-26 using Show and GraphicsArray.

```
In[185]:= Show[GraphicsArray[{pp1,pp2}]]
```

For (c), we take advantage of the ContourPlot3D command, which is located in the **ContourPlot3D** package contained in the **Graphics** folder (or directory). After the **ContourPlot3D** package has been loaded by entering <<Graphics`ContourPlot3D`, the command

```
ContourPlot3D[f[x,y,z],{x,a,b},{y,c,d},{z,u,v}]
```

attempts to graph the level surface of $w = f(x, y, z)$ corresponding to $w = 0$.

After loading the **ContourPlot3D** package, we use ContourPlot3D to graph the equation $\frac{1}{16}x^2 - \frac{1}{4}y^2 - z^2 - 1 = 0$ in Figure 2-27, illustrating the use of the PlotPoints, Axes, AxesLabel, and BoxRatios options.

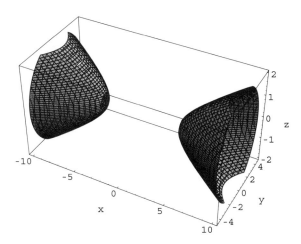

Figure 2-27   Plot of $\frac{1}{16}x^2 - \frac{1}{4}y^2 - z^2 = 1$ generated with ContourPlot3D

```
In[186]:= << Graphics`ContourPlot3D`

In[187]:= ContourPlot3D[x^2/16 - y^2/4 - z^2 - 1,
            {x, -10, 10}, {y, -8, 8}, {z, -2, 2},
            PlotPoints → {8, 8},
            Axes → Automatic, AxesLabel → {"x", "y", "z"},
            BoxRatios → {2, 1, 1}]
```

■

2.4 Solving Equations

2.4.1 Exact Solutions of Equations

Mathematica can find exact solutions to many equations and systems of equations, including exact solutions to polynomial equations of degree four or less. Because a single equals sign "=" is used to name objects and assign values in Mathematica, equations in Mathematica are of the form

```
left-hand side==right-hand side.
```

The "double-equals" sign "==" between the left-hand side and right-hand side specifies that the object is an equation. For example, to represent the equation

$3x + 7 = 4$ in Mathematica, type `3x+7==4`. The command `Solve[lhs==rhs,x]` solves the equation `lhs = rhs` for x. If the only unknown in the equation `lhs = rhs` is x and Mathematica does not need to use inverse functions to solve for x, the command `Solve[lhs==rhs]` solves the equation `lhs = rhs` for x. Hence, to solve the equation $3x + 7 = 4$, both the commands `Solve[3x+7==4]`+ and `Solve[3x+7==4,x]`+ return the same result.

EXAMPLE 2.4.1: Solve the equations $3x + 7 = 4$, $\dfrac{x^2 - 1}{x - 1} = 0$, and $x^3 + x^2 + x + 1 = 0$.

SOLUTION: In each case, we use `Solve` to solve the indicated equation. Be sure to include the double equals sign "`==`" between the left and right-hand sides of each equation. Thus, the result of entering

> *In[188]:=* **Solve[3x + 7 == 4]**
>
> *Out[188]=* $\{\{x \to -1\}\}$

means that the solution of $3x + 7 = 4$ is $x = -1$ and the result of entering

> *In[189]:=* **Solve$\left[\dfrac{x^2 - 1}{x - 1} == 0\right]$**
>
> *Out[189]=* $\{\{x \to -1\}\}$

means that the solution of $\dfrac{x^2 - 1}{x - 1} = 0$ is $x = -1$. On the other hand, the
equation $x^3 + x^2 + x + 1 = 0$ has two imaginary roots. We see that entering

> $In[190] := $ **Solve$\left[x^3 + x^2 + x + 1 == 0\right]$**
>
> $Out[190] = \{\{x \rightarrow -1\}, \{x \rightarrow -i\}, \{x \rightarrow i\}\}$

yields all three solutions. Thus, the solutions of $x^3 + x^2 + x + 1 = 0$ are
$x = -1$ and $x = \pm i$. Remember that the Mathematica symbol I represents
the complex number $i = \sqrt{-1}$. In general, Mathematica can find the
exact solutions of any polynomial equation of degree four or less.

■

Observe that the results of a Solve command are a **list**.

Mathematica can also solve equations involving more than one variable for one
variable in terms of other unknowns.

Lists and tables are discussed in more detail in Chapter 4.

EXAMPLE 2.4.2: (a) Solve the equation $v = \pi r^2 / h$ for h. (b) Solve the
equation $a^2 + b^2 = c^2$ for a.

SOLUTION: These equations involve more than one unknown so we
must specify the variable for which we are solving in the Solve commands. Thus, entering

> $In[191] := $ **Solve$\left[v == \dfrac{\pi r^2}{h}, h\right]$**
>
> $Out[191] = \left\{\left\{h \rightarrow \dfrac{\pi r^2}{v}\right\}\right\}$

solves the equation $v = \pi r^2 / h$ for h. (Be sure to include a space or *
between π and r.) Similarly, entering

> $In[192] := $ **Solve$\left[a^2 + b^2 == c^2, a\right]$**
>
> $Out[192] = \left\{\left\{a \rightarrow -\sqrt{-b^2 + c^2}\right\}, \left\{a \rightarrow \sqrt{-b^2 + c^2}\right\}\right\}$

solves the equation $a^2 + b^2 = c^2$ for a.

■

If Mathematica needs to use inverse functions to solve an equation, you must be
sure to specify the variable(s) for which you want Mathematica to solve.

EXAMPLE 2.4.3: Find a solution of $\sin^2 x - 2 \sin x - 3 = 0$.

SOLUTION: When the command `Solve[Sin[x]^2-2Sin[x]-3==0]` is entered, Mathematica solves the equation for `Sin[x]`. However, when the command

$$\texttt{Solve[Sin[x]^2-2Sin[x]-3==0,x]}$$

is entered, Mathematica attempts to solve the equation for x. In this case, Mathematica succeeds in finding one solution. In fact, this equation has infinitely many solutions of the form $x = \frac{1}{2}(4k - 1)\pi$, $k = 0, \pm 1, \pm 2, ...$; $\sin x = 3$ has no solutions.

$In[193]:=$ **Solve$\left[\text{Sin}[x]^2 - 2\,\text{Sin}[x] - 3 == 0\right]$**

$Out[193]=$ $\{\{\text{Sin}[x] \to -1\}, \{\text{Sin}[x] \to 3\}\}$

$In[194]:=$ **Solve$\left[\text{Sin}[x]^2 - 2\,\text{Sin}[x] - 3 == 0, x\right]$**

Solve : : ifun :

  Inverse functions are being used by
    Solve, so some solutions may not be found.

$Out[194]=$ $\left\{\left\{x \to -\dfrac{\pi}{2}\right\}, \{x \to \text{ArcSin}[3]\}\right\}$

∎

The example indicates that it is especially important to be careful when dealing with equations involving trigonometric functions.

EXAMPLE 2.4.4: Let $f(\theta) = \sin 2\theta + 2 \cos \theta$, $0 \le \theta \le 2\pi$. (a) Solve $f'(\theta) = 0$. (b) Graph $f(\theta)$ and $f'(\theta)$.

SOLUTION: After defining $f(\theta)$, we use D to compute $f'(\theta)$ and then use Solve to solve $f'(\theta) = 0$.

`D[f[x],x]` computes $f'(x)$; `D[f[x],{x,n}]` computes $f^{(n)}(x)$. Topics from calculus are discussed in more detail in Chapter 3.

$In[195]:=$ **f[θ_] = Sin[2θ] + 2 Cos[θ]**

$Out[195]=$ $2 \text{ Cos}[\theta] + \text{Sin}[2\,\theta]$

$In[196]:=$ **df = D[f[θ], θ]**

$Out[196]=$ $2 \text{ Cos}[2\,\theta] - 2 \text{ Sin}[\theta]$

$In[197]:=$ **Solve[df == 0, θ]**

```
Solve :: "ifun" :  "Inversefunctionsarebeingused
    bySolve, sosomesolutionsmaynotbefound."
```
$$Out\,[197] = \left\{\left\{\theta \to -\frac{\pi}{2}\right\}, \left\{\theta \to \frac{\pi}{6}\right\}, \left\{\theta \to \frac{5\,\pi}{6}\right\}\right\}$$

Notice that $-\pi/2$ is not between 0 and 2π. Moreover, $\pi/6$ and $5\pi/6$ are *not* the only solutions of $f'(\theta) = 0$ between 0 and 2π. Proceeding by hand, we use the identity $\cos 2\theta = 1 - 2\sin^2\theta$ and factor:

$$2\cos 2\theta - 2\sin\theta = 0$$
$$1 - 2\sin^2\theta - \sin\theta = 0$$
$$2\sin^2\theta + \sin\theta - 1 = 0$$
$$(2\sin\theta - 1)(\sin\theta + 1) = 0$$

so $\sin\theta = 1/2$ or $\sin\theta = -1$. Because we are assuming that $0 \le \theta \le 2\pi$, we obtain the solutions $\theta = \pi/6$, $5\pi/6$, or $3\pi/2$. We perform the same steps with Mathematica.

expression /. x->y
replaces all occurrences of *x*
in *expression* by *y*.

```
In[198] := s1 = TrigExpand[df]
```
$$Out\,[198] = 2\ \text{Cos}[\theta]^2 - 2\ \text{Sin}[\theta] - 2\ \text{Sin}[\theta]^2$$

```
In[199] := s2 = s1/. Cos[θ]^2- >1 - Sin[θ]^2
```
$$Out\,[199] = -2\ \text{Sin}[\theta] - 2\ \text{Sin}[\theta]^2 + 2\ \left(1 - \text{Sin}[\theta]^2\right)$$

```
In[200] := Factor[s2]
```
$$Out\,[200] = -2\ (1 + \text{Sin}[\theta])\ (-1 + 2\ \text{Sin}[\theta])$$

Finally, we graph $f(\theta)$ and $f'(\theta)$ with Plot in Figure 2-28. Note that the plot is drawn to scale because we include the option `AspectRatio->Automatic`.

```
In[201] := Plot[{f[θ], df}, {θ, 0, 2π},
            PlotStyle- > {GrayLevel[0], GrayLevel[0.3]},
            AspectRatio- > Automatic]
```

■

We can also use `Solve` to find the solutions, if any, of various types of systems of equations. Entering

```
Solve [{lhs1==rhs1, lhs2==rhs2}, {x, y}]
```

solves a system of two equations for x and y while entering

```
Solve [{lhs1==rhs1, lhs2==rhs2}]
```

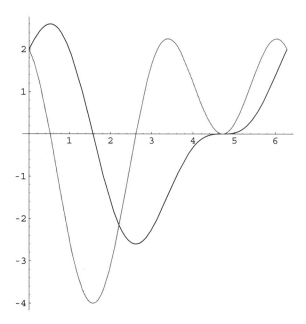

Figure 2-28   Graphs of $f(\theta)$ and $f'(\theta)$

attempts to solve the system of equations for all unknowns. In general, Solve can find the solutions to a system of linear equations. In fact, if the systems to be solved are inconsistent or dependent, Mathematica's output indicates so.

EXAMPLE 2.4.5: Solve each system:

(a) $\begin{cases} 3x - y = 4 \\ x + y = 2 \end{cases}$; (b) $\begin{cases} 2x - 3y + 4z = 2 \\ 3x - 2y + z = 0 \\ x + y - z = 1 \end{cases}$; (c) $\begin{cases} 2x - 2y - 2z = -2 \\ -x + y + 3z = 0 \\ -3x + 3y - 2z = 1 \end{cases}$; and

(d) $\begin{cases} -2x + 2y - 2z = -2 \\ 3x - 2y + 2z = 2 \\ x + 3y - 3z = -3 \end{cases}$.

SOLUTION: In each case we use Solve to solve the given system. For (a), the result of entering

$In[202] := $ **Solve[{3x - y == 4, x + y == 2}, {x, y}]**

$Out[202] = \left\{\left\{x \to \dfrac{3}{2}, y \to \dfrac{1}{2}\right\}\right\}$

means that the solution of $\begin{cases} 3x - y = 4 \\ x + y = 2 \end{cases}$ is $(x, y) = (3/2, 1/2)$. (b) We can verify that the results returned by Mathematica are correct. First, we name the system of equations sys and then use Solve to solve the system of equations naming the result sols.

```
In[203]:= sys = {2x - 3y + 4z == 2, 3x - 2y + z == 0,
            x + y - z == 1};
```

```
In[204]:= sols = Solve[sys, {x, y, z}]
```
$$Out[204]= \left\{\left\{x \to \frac{7}{10}, y \to \frac{9}{5}, z \to \frac{3}{2}\right\}\right\}$$

We verify the result by substituting the values obtained with Solve back into sys with ReplaceAll (/ .).

```
In[205]:= sys /. sols
```
```
Out[205]= {{True, True, True}}
```

means that the solution of $\begin{cases} 2x - 3y + 4z = 2 \\ 3x - 2y + z = 0 \\ x + y - z = 1 \end{cases}$ is $(x, y, z) = (7/10, 9/5, 3/2)$.

(c) When we use Solve to solve this system, Mathematica returns { }, which indicates that the system has no solution; the system is inconsistent.

```
In[206]:= Solve[{2x - 2y - 2z == -2,
            -x + y + 3z == 0, -3x + 3y - 2z == 1}]
```
```
Out[206]= {}
```

(d) On the other hand, when we use Solve to solve this system, Mathematica's result indicates that the system has infinitely many solutions. That is, all ordered triples of the form $\{(0, z - 1, z)|z\,\text{real}\}$ are solutions of the system.

```
In[207]:= Solve[{-2x + 2y - 2z == -2,
            3x - 2y + 2z == 2, x + 3y - 3z == -3}]
```
```
Solve :: svars :
   Equations may not give solutions
      for all "solve" variables.
   Out[207]= {{x → 0, y → -1 + z}}
```

■

We can often use Solve to find solutions of a nonlinear system of equations as well.

EXAMPLE 2.4.6: Solve the systems

(a) $\begin{cases} 4x^2 + y^2 = 4 \\ x^2 + 4y^2 = 4 \end{cases}$ and (b) $\begin{cases} \frac{1}{a^2}x^2 + \frac{1}{b^2}y^2 = 1 \\ y = mx \end{cases}$ (a, b greater than zero) for

x and y.

SOLUTION: The graphs of the equations are both ellipses. We use ContourPlot to graph each equation, naming the results cp1 and cp2, respectively, and then use Show to display both graphs together in Figure 2-29. The solutions of the system correspond to the intersection points of the two graphs.

$$In[208] := \textbf{cp1 = ContourPlot}\left[\textbf{4x}^2 + \textbf{y}^2 - \textbf{4,}\right.$$
$$\textbf{\{x, -3, 3\}, \{y, -3, 3\},}$$
$$\textbf{Contours} \rightarrow \textbf{\{0\},}$$
$$\textbf{ContourShading} \rightarrow \textbf{False,}$$
$$\textbf{PlotPoints} \rightarrow \textbf{60,}$$
$$\left.\textbf{DisplayFunction} \rightarrow \textbf{Identity}\right];$$

$$\textbf{cp2 = ContourPlot}\left[\textbf{x}^2 + \textbf{4y}^2 - \textbf{4,}\right.$$
$$\textbf{\{x, -3, 3\}, \{y, -3, 3\},}$$
$$\textbf{Contours} \rightarrow \textbf{\{0\},}$$
$$\textbf{ContourShading} \rightarrow \textbf{False,}$$
$$\textbf{PlotPoints} \rightarrow \textbf{60,}$$
$$\left.\textbf{DisplayFunction} \rightarrow \textbf{Identity}\right];$$

$$\textbf{Show[cp1, cp2, Frame} \rightarrow \textbf{False,}$$
$$\textbf{Axes} \rightarrow \textbf{Automatic, AxesOrigin} \rightarrow \textbf{\{0, 0\},}$$
$$\textbf{DisplayFunction} \rightarrow \textbf{\$DisplayFunction]}$$

Finally, we use Solve to find the solutions of the system.

$$In[209] := \textbf{Solve}\left[\left\{\textbf{4x}^2 + \textbf{y}^2 == \textbf{4, x}^2 + \textbf{4y}^2 == \textbf{4}\right\}\right]$$

$$Out[209] = \left\{\left\{x \rightarrow -\frac{2}{\sqrt{5}}, y \rightarrow -\frac{2}{\sqrt{5}}\right\},\right.$$
$$\left\{x \rightarrow -\frac{2}{\sqrt{5}}, y \rightarrow \frac{2}{\sqrt{5}}\right\},$$
$$\left.\left\{x \rightarrow \frac{2}{\sqrt{5}}, y \rightarrow -\frac{2}{\sqrt{5}}\right\}, \left\{x \rightarrow \frac{2}{\sqrt{5}}, y \rightarrow \frac{2}{\sqrt{5}}\right\}\right\}$$

For (b), we also use Solve to find the solutions of the system. However, because the unknowns in the equations are a, b, m, x, and y, we must

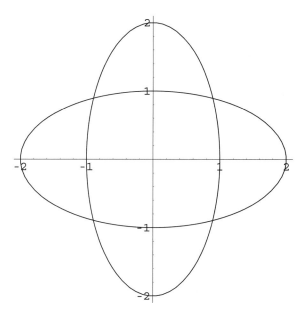

Figure 2-29   Graphs of $4x^2 + y^2 = 4$ and $x^2 + 4y^2 = 4$

specify that we want to solve for x and y in the Solve command.

$$In[210] := \text{Solve}\left[\left\{\frac{x^2}{a^2} + \frac{y^2}{b^2} == 1, y == mx\right\}, \{x, y\}\right]$$

$$Out[210] = \left\{\left\{y \to -\frac{a\,b\,m}{\sqrt{b^2 + a^2\,m^2}}, x \to -\frac{a\,b}{\sqrt{b^2 + a^2\,m^2}}\right\},\right.$$

$$\left.\left\{y \to \frac{a\,b\,m}{\sqrt{b^2 + a^2\,m^2}}, x \to \frac{a\,b}{\sqrt{b^2 + a^2\,m^2}}\right\}\right\}$$

■

Although Mathematica can find the exact solution to every polynomial equation of degree four or less, exact solutions to some equations may not be meaningful. In those cases, Mathematica can provide approximations of the exact solutions using either the N[expression] or the expression // N commands.

EXAMPLE 2.4.7: Approximate the solutions to the equations (a) $x^4 - 2x^2 = 1 - x$; and (b) $1 - x^2 = x^3$.

SOLUTION: Each of these is a polynomial equation with degree less than five so Solve will find the exact solutions of each equation.

However, the solutions are quite complicated so we use N to obtain approximate solutions of each equation. For (a), entering

$In[211] :=$ **N$\left[$Solve$\left[x^4 - 2x^2 == 1 - x\right]\right]$**
$Out[211] =$ $\{\{x \to 0.182777 - 0.633397\,i\},$
        $\{x \to 0.182777 + 0.633397\,i\},$
        $\{x \to -1.71064\}, \{x \to 1.34509\}\}$

first finds the exact solutions of the equation $x^4 - 2x^2 = 1 - x$ and then computes approximations of those solutions. The resulting output is the list of approximate solutions. For (b), entering

$In[212] :=$ **Solve$\left[1 - x^2 == x^3, x\right]$//N**
$Out[212] =$ $\{\{x \to 0.754878\},$
        $\{x \to -0.877439 + 0.744862\,i\},$
        $\{x \to -0.877439 - 0.744862\,i\}\}$

first finds the exact solutions of the equation $1 - x^2 = x^3$ and then computes approximations of those solutions. The resulting output is the list of approximate solutions.

■

2.4.2 Approximate Solutions of Equations

When solving an equation is either impractical or impossible, Mathematica provides several functions including FindRoot, NRoots, and NSolve to approximate solutions of equations. NRoots and NSolve numerically approximate the roots of any polynomial equation. The command

NRoots [poly1==poly2,x]

approximates the solutions of the polynomial equation poly1==poly2, where both poly1 and poly2 are polynomials in x. The syntax for NSolve is the same as the syntax of NRoots.

FindRoot attempts to approximate a root to an equation provided that a "reasonable" guess of the root is given. FindRoot works on functions other than polynomials. The command

FindRoot [lhs==rhs,{x,firstguess}]

searches for a numerical solution to the equation lhs==rhs, starting with $x =$ firstguess. To locate more than one root, FindRoot must be used several times. One way of obtaining firstguess (for real-valued solutions) is to graph both lhs and rhs with Plot, find the point(s) of intersection, and estimate the x-coordinates of the point(s) of intersection. Generally, NRoots is easier to use than FindRoot when trying to approximate the roots of a polynomial.

EXAMPLE 2.4.8: Approximate the solutions of
$x^5 + x^4 - 4x^3 + 2x^2 - 3x - 7 = 0$.

SOLUTION: Because $x^5 + x^4 - 4x^3 + 2x^2 - 3x - 7 = 0$ is a polynomial equation, we may use NRoots to approximate the solutions of the equation. Thus, entering

$In[213]:=$ **NRoots$\left[\mathbf{x^5 + x^4 - 4x^3 + 2x^2 - 3x - 7 == 0,}\right.$**

   $\left.\mathbf{x}\right]$

$Out[213]=$ x == -2.74463 || x == -0.880858 ||

   x == 0.41452 - 1.19996 i ||

   x == 0.41452 + 1.19996 i || x == 1.79645

approximates the solutions of $x^5 + x^4 - 4x^3 + 2x^2 - 3x - 7 = 0$. The symbol || appearing in the result represents "or".

   We obtain equivalent results with NSolve.

$In[214]:=$ **NSolve$\left[\mathbf{x^5 + x^4 - 4x^3 + 2x^2 - 3x - 7 == 0,}\right.$**

   $\left.\mathbf{x}\right]$

$Out[214]=$ {{x → -2.74463}, {x → -0.880858},

   {x → 0.41452 - 1.19996 i},

   {x → 0.41452 + 1.19996 i}, {x → 1.79645}}

FindRoot may also be used to approximate each root of the equation. However, to use FindRoot, we must supply an initial approximation of the solution that we wish to approximate. The real solutions of $x^5 + x^4 - 4x^3 + 2x^2 - 3x - 7 = 0$ correspond to the values of x where the graph of $f(x) = x^5 + x^4 - 4x^3 + 2x^2 - 3x - 7$ intersects the x-axis. We use Plot to graph $f(x)$ in Figure 2-30.

$In[215]:=$ **Plot$\left[\mathbf{x^5 + x^4 - 4x^3 + 2x^2 - 3x - 7,}\right.$**

   $\left.\mathbf{\{x, -3, 2\}}\right]$

We see that the graph intersects the x-axis near $x \approx -2.5$, -1, and 1.5. We use these values as initial approximations of each solution. Thus, entering

$In[216]:=$ **FindRoot$\left[\mathbf{x^5 + x^4 - 4x^3 + 2x^2 - 3x - 7 == 0,}\right.$**

   $\left.\mathbf{\{x, -2.5\}}\right]$

$Out[216]=$ {x → -2.74463}

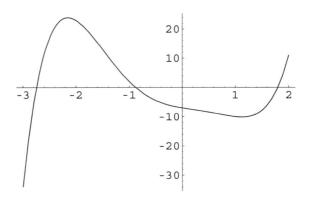

Figure 2-30   Graph of $f(x) = x^5 + x^4 - 4x^3 + 2x^2 - 3x - 7$

approximates the solution near −2.5, entering

$$In[217] := \textbf{FindRoot}\left[\textbf{x}^5 + \textbf{x}^4 - \textbf{4x}^3 + \textbf{2x}^2 - \textbf{3x} - \textbf{7} == \textbf{0},\right.$$
$$\left.\{\textbf{x, -1}\}\right]$$
$$Out[217] = \{x \rightarrow -0.880858\}$$

approximates the solution near −1, and entering

$$In[218] := \textbf{FindRoot}\left[\textbf{x}^5 + \textbf{x}^4 - \textbf{4x}^3 + \textbf{2x}^2 - \textbf{3x} - \textbf{7} == \textbf{0},\right.$$
$$\left.\{\textbf{x, 2}\}\right]$$
$$Out[218] = \{x \rightarrow 1.79645\}$$

approximates the solution near 1.5. Note that FindRoot may be used to approximate complex solutions as well. To obtain initial guesses, observe that the solutions of $f(z) = 0, z = x+iy, x, y$ real, are the level curves of $w = |f(z)|$ that are points. In Figure 2-31, we use ContourPlot to graph various level curves of $w = |f(x+iy)|, -2 \le x \le 2, -2 \le y \le 2$. In the plot, observe that the two complex solutions occur at $x \pm iy \approx 0.5 \pm 1.2i$.

```
In[219] := f[z_] = z^5 + z^4 - 4z^3 + 2z^2 - 3z - 7;
```

```
In[220] := ContourPlot[Abs[f[x + I y]], {x, -2, 2},
              {y, -2, 2}, ContourShading → False,
              Contours → 60,
              PlotPoints → 200,
              Frame → False, Axes → Automatic,
              AxesOrigin → {0, 0}]
```

Thus, entering

```
In[221] := FindRoot[f[z] == 0, {z, 0.5 + 1.2I}]
Out[221] = {z → 0.41452 + 1.19996 i}
```

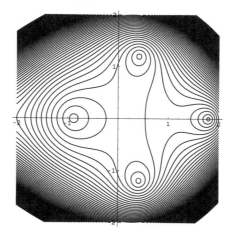

Figure 2-31   Level curves of $w = |f(x + iy)|$, $-2 \le x \le 2$, $-2 \le y \le 2$

approximates the solution near $x + iy \approx 0.5 + 1.2i$. For polynomials with real coefficients, complex solutions occur in conjugate pairs so the other complex solution is approximately $0.41452 - 1.19996i$.

■

To approximate points in a two-dimensional graphic, first move the cursor within the graphics cell and click once. Notice that a box appears around the graph as shown in the following screen shot.

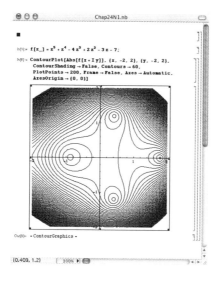

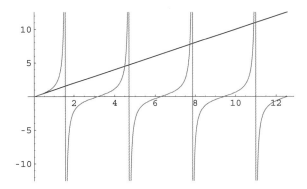

Figure 2-32 $y = x$ and $y = \tan x$

Next, press and hold down the command-key. As you move the cursor within the graphics cell, notice that the thermometer at the bottom of the screen has changed to ordered pairs approximating the location of the cursor within the graphics cell.

EXAMPLE 2.4.9: Find the first three nonnegative solutions of $x = \tan x$.

SOLUTION: We attempt to solve $x = \tan x$ with `Solve`.

```
In[222]:= Solve[x == Tan[x], x]
Solve :: "tdep" :  "Theequationsappeartoinvolve
   transcendentalfunctionsofthevariablesin
   anessentiallynon-algebraicway."
Out[222]= Solve[x == Tan[x], x]
```

We next graph $y = x$ and $y = \tan x$ together in Figure 2-32.

```
In[223]:= Plot[{x, Tan[x]}, {x, 0, 4π},
            PlotRange- > {-4π, 4π},
            PlotStyle- > {GrayLevel[0], GrayLevel[0.3]}]
```

Remember that vertical lines are never the graphs of functions. In this case, they represent the vertical asymptotes at odd multiples of $\pi/2$.

In the graph, we see that $x = 0$ is a solution. This is confirmed with `FindRoot`.

```
In[224]:= FindRoot[x == Tan[x], {x, 0}]
Out[224]= {x → 0.}
```

The second solution is near 4 while the third solution is near 7. Using `FindRoot` together with these initial approximations locates the second two solutions.

```
In[225]:= FindRoot[x == Tan[x], {x, 4}]
Out[225]= {x → 4.49341}

In[226]:= FindRoot[x == Tan[x], {x, 7}]
Out[226]= {x → 7.72525}
```

∎

FindRoot can also be used to approximate solutions to systems of equations. (Although NRoots can solve a polynomial equation, NRoots cannot be used to solve a system of polynomial equations.) When approximations of solutions of systems of equations are desired, use either Solve and N together, when possible, or FindRoot.

EXAMPLE 2.4.10: Approximate the solutions to the system of equations $\begin{cases} x^2 + 4xy + y^2 = 4 \\ 5x^2 - 4xy + 2y^2 = 8 \end{cases}$.

SOLUTION: We begin by using ContourPlot to graph each equation in Figure 2-33. From the resulting graph, we see that $x^2 + 4xy + y^2 = 4$ is a hyperbola, $5x^2 - 4xy + 2y^2 = 8$ is an ellipse, and there are four solutions to the system of equations.

```
In[227]:= cp1 = ContourPlot[x² + 4xy + y² - 4,
              {x, -4, 4}, {y, -4, 4},
              Contours → {0}, PlotPoints → 60,
              ContourShading → False,
              DisplayFunction → Identity];

          cp2 = ContourPlot[5x² - 4xy + 2y² - 8,
              {x, -4, 4}, {y, -4, 4},
              Contours → {0}, PlotPoints → 60,
              ContourStyle- >Dashing[{0.01}],
              ContourShading → False,
              DisplayFunction → Identity];

          Show[cp1, cp2, Frame → False,
              Axes → Automatic, AxesOrigin → {0, 0},
              DisplayFunction → $DisplayFunction]
```

From the graph we see that possible solutions are $(0, 2)$ and $(0, -2)$. In fact, substituting $x = 0$ and $y = -2$ and $x = 0$ and $y = 2$ into each equation

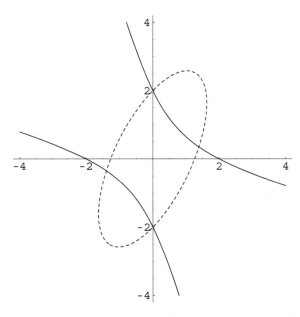

Figure 2-33   Graphs of $x^2 + 4xy + y^2 = 4$ and $5x^2 - 4xy + 2y^2 = 8$

verifies that these points are both exact solutions of the equation. The remaining two solutions are approximated with FindRoot.

```
In[228]:= FindRoot[{x² + 4xy + y² == 4,
              5x² - 4xy + 2y² == 8}, {x, 1}, {y, 0.25}]
Out[228]= {x → 1.39262, y → 0.348155}

In[229]:= FindRoot[{x² + 4xy + y² == 4,
              5x² - 4xy + 2y² == 8}, {x, -1}, {y, -0.25}]
Out[229]= {x → -1.39262, y → -0.348155}
```

■

Calculus

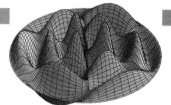

Chapter 3 introduces Mathematica's built-in calculus commands. The examples used to illustrate the various commands are similar to examples routinely done in first-year calculus courses.

3.1 Limits

One of the first topics discussed in calculus is that of limits. Mathematica can be used to investigate limits graphically and numerically. In addition, the Mathematica command

$$\texttt{Limit[f[x],x->a]}$$

attempts to compute the limit of $y = f(x)$ as x approaches a, $\lim_{x \to a} f(x)$, where a can be a finite number, ∞ (Infinity), or $-\infty$ (-Infinity). The arrow "->" is obtained by typing a minus sign "-" followed by a greater than sign ">".

Remark. To define a function of a single variable, $f(x) = expression\,in\,x$, enter `f[x_]=expression in x`. To generate a basic plot of $y = f(x)$ for $a \le x \le b$, enter `Plot[f[x],{x,a,b}]`.

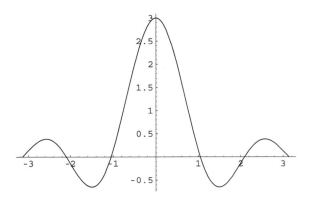

Figure 3-1   Graph of $f(x) = \frac{\sin 3x}{x}$ on the interval $[-\pi, \pi]$

3.1.1 Using Graphs and Tables to Predict Limits

EXAMPLE 3.1.1: Use a graph and table of values to investigate
$\lim_{x\to 0} \dfrac{\sin 3x}{x}$.

Clear [f] clears all prior
definitions of f, if any.
Clearing function definitions
before defining new ones
helps eliminate any possible
confusion and/or ambiguities.

SOLUTION: We clear all prior definitions of f, define $f(x) = \dfrac{\sin 3x}{x}$, and then graph $y = f(x)$ on the interval $[-\pi, \pi]$ with Plot.

```
In[230]:= Clear[f]

        f[x_] = Sin[3x]/x;
        Plot[f[x], {x, -π, π}]
```

From the graph shown in Figure 3-1, we might, correctly, conclude that $\lim_{x\to 0} \frac{\sin 3x}{x} = 3$. Further evidence that $\lim_{x\to 0} \frac{\sin 3x}{x} = 3$ can be obtained by computing the values of $f(x)$ for values of x "near" 0. In the following, we use Random to define xvals to be a table of 6 "random" real numbers. The first number in xvals is between -1 and 1, the second between $-1/10$ and $1/10$, and so on.

Random [Real, {a,b}]
returns a "random" real
number between a and b.
Because we are generating
"random" numbers, your
results will differ from those
obtained here.

```
In[231]:= xvals = Table[Random[Real,
             {-1/10^n, 1/10^n}], {n, 0, 5}]

Out[231]= {0.244954, 0.0267254,
           0.00433248, -0.000864136,
           -0.0000995605, 1.8335410^{-6}}
```

We then use `Map` to compute the value of $f(x)$ for each x in `xvals`.

`Map[f,{x1,x2,...,xn}]` returns the list $\{f(x_1), f(x_2), ..., f(x_n)\}$.

$In[232]:=$ **`Map[f,xvals]`**

$Out[232]=$ `{2.73719,2.99679,`
           `2.99992,3.,`
           `3.,3.}`

From these values, we might again correctly deduce that $\lim_{x\to 0} \frac{\sin 3x}{x} = 3$. Of course, these results do not prove that $\lim_{x\to 0} \frac{\sin 3x}{x} = 3$ but they are helpful in convincing us that $\lim_{x\to 0} \frac{\sin 3x}{x} = 3$.

■

3.1.2 Computing Limits

Some limits involving rational functions can be computed by factoring the numerator and denominator.

EXAMPLE 3.1.2: Compute $\lim_{x\to -9/2} \dfrac{2x^2 + 25x + 72}{72 - 47x - 14x^2}$.

SOLUTION: We define `frac1` to be the rational expression $\dfrac{2x^2 + 25x + 72}{72 - 47x - 14x^2}$. We then attempt to compute the value of `frac1` if $x = -9/2$ by using `ReplaceAll (/.)` to evaluate `frac1` if $x = -9/2$ but see that it is undefined.

$In[233]:=$ **`frac1 = (2x^2 + 25x + 72)/(72 - 47x - 14x^2);`**
           `frac1/.x- >-9/2`
`Power :: "infy" :`
   `"Infiniteexpression10 encountered."`
`∞ :: "indet" : "Indeterminateexpression0`

`InterpretationBox[ComplexInfinity,`
       `DirectedInfinity[]]encountered."`
$Out[233]=$ `Indeterminate`

Factoring the numerator and denominator with `Factor`, `Numerator`, and `Denominator`, we see that

$$\lim_{x\to -9/2} \frac{2x^2 + 25x + 72}{72 - 47x - 14x^2} = \lim_{x\to -9/2} \frac{(x+8)(2x+9)}{(8-7x)(2x+9)} = \lim_{x\to -9/2} \frac{x+8}{8-7x}.$$

The fraction $(x + 8)/(8 - 7x)$ is named `frac2` and the limit is evaluated by computing the value of `frac2` if $x = -9/2$.

> In[234]:= **Factor[Numerator[frac1]]**
>
> Out[234]= (8 + x) (9 + 2 x)

> In[235]:= **Factor[Denominator[frac1]]**
>
> Out[235]= - (9 + 2 x) (-8 + 7 x)

> In[236]:= **frac2 = Simplify[frac1]**
>
> Out[236]= $\dfrac{8 + x}{8 - 7\ x}$

`Simplify[expression]`
attempts to simplify
expression.

> In[237]:= **frac2/.x- > -9/2**
>
> Out[237]= $\dfrac{7}{79}$

We conclude that

$$\lim_{x\to -9/2} \frac{2x^2 + 25x + 72}{72 - 47x - 14x^2} = \frac{7}{79}.$$

∎

We can also use the `Limit` command to evaluate frequently encountered limits.

$$\texttt{Limit[f[x],x->a]}$$

attempts to compute $\lim_{x\to a} f(x)$.

Thus, entering

> In[238]:= **Limit[(2x^2 + 25x + 72)/(72 - 47x - 14x^2),x- > -9/2]**
>
> Out[238]= $\dfrac{7}{79}$

computes $\lim_{x\to -9/2} \frac{2x^2+25x+72}{72-47x-14x^2} = 7/79$.

EXAMPLE 3.1.3: Calculate each limit: (a) $\lim_{x\to -5/3} \dfrac{3x^2 - 7x - 20}{21x^2 + 14x - 35}$;

(b) $\lim_{x\to 0} \dfrac{\sin x}{x}$; (c) $\lim_{x\to\infty}\left(1 + \dfrac{1}{x}\right)^x$; (d) $\lim_{x\to 0}\dfrac{e^{3x} - 1}{x}$; (e) $\lim_{x\to\infty} e^{-2x}\sqrt{x}$;

and (f) $\lim_{x\to 1^+}\left(\dfrac{1}{\ln x} - \dfrac{1}{x - 1}\right)$.

SOLUTION: In each case, we use `Limit` to evaluate the indicated limit. Entering

> In[239]:= **Limit[(3x^2-7x-20)/(21x^2+14x-35),x- > -5/3]**
>
> Out[239]= $\dfrac{17}{56}$

computes

$$\lim_{x \to -5/3} \frac{3x^2 - 7x - 20}{21x^2 + 14x - 35} = \frac{17}{56};$$

and entering

```
In[240]:= Limit[Sin[x]/x, x- > 0]
Out[240]= 1
```

computes

$$\lim_{x \to 0} \frac{\sin x}{x} = 1.$$

Mathematica represents ∞ by `Infinity`. Thus, entering

```
In[241]:= Limit[(1 + 1/x)^x, x- > ∞]
Out[241]= e
```

computes

$$\lim_{x \to \infty} \left(1 + \frac{1}{x}\right)^x = e.$$

Entering

```
In[242]:= Limit[(Exp[3x] - 1)/x, x- > 0]
Out[242]= 3
```

computes

$$\lim_{x \to 0} \frac{e^{3x} - 1}{x} = 3.$$

Entering

```
In[243]:= Limit[Exp[-2x] Sqrt[x], x- > ∞]
Out[243]= 0
```

computes $\lim_{x \to \infty} e^{-2x} \sqrt{x} = 0$, and entering

```
In[244]:= Limit[1/Log[x] - 1/(x - 1), x- > 1]
Out[244]= 1/2
```

computes

$$\lim_{x \to 1^+} \left(\frac{1}{\ln x} - \frac{1}{x - 1}\right) = \frac{1}{2}.$$

Because $\ln x$ is undefined for $x \le 0$, a right-hand limit is mathematically necessary, even though Mathematica's `Limit` function computes the limit correctly without the distinction.

■

We can often use the `Limit` command to compute symbolic limits.

EXAMPLE 3.1.4: If $P is compounded n times per year at an annual interest rate of r, the value of the account, A, after t years is given by

$$A = \left(1 + \frac{r}{n}\right)^{nt}.$$

The formula for continuously compounded interest is obtained by taking the limit of this expression as $t \to \infty$.

SOLUTION: The formula for continuously compounded interest, $A = Pe^{rt}$, is obtained using Limit.

```
In[245] := Limit[p(1 + r/n)^(n t), n- > ∞]
Out[245] = e^r t p
```

■

3.1.3 One-Sided Limits

In some cases, Mathematica can compute certain one-sided limits. The command

```
                Limit[f[x],x->a,Direction->1]
```

attempts to compute $\lim_{x \to a^-} f(x)$ while

```
                Limit[f[x],x->a,Direction->-1]
```

attempts to compute $\lim_{x \to a^+} f(x)$.

EXAMPLE 3.1.5: Compute (a) $\lim_{x \to 0^+} |x|/x$; (b) $\lim_{x \to 0^-} |x|/x$; (c) $\lim_{x \to 0^+} e^{-1/x}$; and (d) $\lim_{x \to 0^-} e^{-1/x}$.

SOLUTION: Even though $\lim_{x \to 0} |x|/x$ does not exist, $\lim_{x \to 0^+} |x|/x = 1$ and $\lim_{x \to 0^-} |x|/x = -1$, as we see using Limit together with the Direction->1 and Direction->-1 options, respectively.

```
In[246] := Limit[Abs[x]/x, x- > 0]
Out[246] = 1

In[247] := Limit[Abs[x]/x, x- > 0, Direction- > -1]
Out[247] = 1
```

```
In[248]:= Limit[Abs[x]/x, x- > 0, Direction- > 1]
Out[248]= -1
```

The `Direction->-1` and `Direction->1` options are used to calculate the correct values for (c) and (d), respectively. For (c), we have:

```
In[249]:= Limit[1/x, x- > 0]
Out[249]= ∞
```

```
In[250]:= Limit[1/x, x- > 0, Direction- > -1]
Out[250]= ∞
```

```
In[251]:= Limit[1/x, x- > 0, Direction- > 1]
Out[251]= -∞
```

Technically, $\lim_{x\to 0} e^{-1/x}$ does not exist (see Figure 3-2) so the following is incorrect.

```
In[252]:= Limit[Exp[-1/x], x- > 0]
Out[252]= 0
```

However, using `Limit` together with the `Direction` option gives the correct left and right limits.

```
In[253]:= Limit[Exp[-1/x], x- > 0, Direction- > 1]
Out[253]= ∞
```

```
In[254]:= Limit[Exp[-1/x], x- > 0, Direction- > -1]
Out[254]= 0
```

We confirm these results by graphing $y = e^{-1/x}$ with `Plot` in Figure 3-2.

```
In[255]:= Plot[Exp[-1/x], {x, -3/2, 3/2},
              PlotRange- > {{-1, 1}, {0, 10}}]
```

■

The `Limit` command together with the `Direction->1` and `Direction->-1` options is a "fragile" command and should be used with caution because its results are unpredictable, especially for the beginner. It is wise to check or confirm results using a different technique for nearly all problems faced by the beginner.

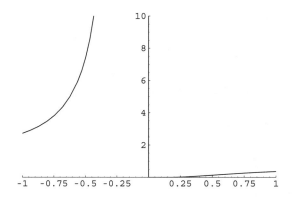

Figure 3-2   Graph of $y = e^{-1/x}$ on the interval $[-3/2, 3/2]$

3.2 Differential Calculus

3.2.1 Definition of the Derivative

Definition 1. *The **derivative** of $y = f(x)$ is*

$$y' = f'(x) = \frac{dy}{dx} = \lim_{h \to 0} \frac{f(x + h) - f(x)}{h}, \tag{3.1}$$

provided the limit exists.

The `Limit` command can be used along with `Simplify` to compute the derivative of a function using the definition of the derivative.

EXAMPLE 3.2.1: Use the definition of the derivative to compute the derivative of (a) $f(x) = x + 1/x$, (b) $g(x) = 1/\sqrt{x}$, and (c) $h(x) = \sin 2x$.

SOLUTION: For (a) and (b), we first define f and g, compute the difference quotient, $(f(x + h) - f(x))/h$, simplify the difference quotient with `Simplify`, and use `Limit` to calculate the derivative.

```
In[256]:= f[x_] = x + 1/x;
          s1 = (f[x + h] - f[x])/h
```
$$Out[256] = \frac{h - \frac{1}{x} + \frac{1}{h+x}}{h}$$

```
In[257]:= s2 = Simplify[s1]
```
$$Out[257] = \frac{-1 + h\ x + x^2}{x\ (h + x)}$$

In[258]:= **Limit[s2,h- > 0]**

Out[258]= $\dfrac{-1 + x^2}{x^2}$

In[259]:= **g[x_] = 1/Sqrt[x]**
s1 = (g[x + h] - g[x])/h

Out[259]= $\dfrac{1}{\sqrt{x}}$

Out[259]= $\dfrac{-\frac{1}{\sqrt{x}} + \frac{1}{\sqrt{h+x}}}{h}$

In[260]:= **s2 = Together[s1]**

Out[260]= $\dfrac{\sqrt{x} - \sqrt{h + x}}{h \ \sqrt{x} \ \sqrt{h + x}}$

In[261]:= **Limit[s2,h- > 0]**

Out[261]= $-\dfrac{1}{2 \ x^{3/2}}$

For (c), we define h and then use TrigExpand to simplify the difference quotient. We use Limit to compute the derivative. The result indicates that $\frac{d}{dx} (\sin 2x) = 2 \cos 2x$.

In[262]:= **h[x_] = Sin[2x];**
s2 = (h[x + h] - h[x])/h

Out[262]= $\dfrac{- \text{Sin}[2 \ x] + \text{Sin}[2 \ (h + x)]}{h}$

In[263]:= **s2 = TrigExpand[s2]**

Out[263]= $\dfrac{1}{h} \big(2 \ \text{Cos}[h] \ \text{Cos}[x]^2 \ \text{Sin}[h] - 2 \ \text{Cos}[x] \ \text{Sin}[x] +$

$2 \ \text{Cos}[h]^2 \ \text{Cos}[x] \ \text{Sin}[x] -$

$2 \ \text{Cos}[x] \ \text{Sin}[h]^2 \ \text{Sin}[x] -$

$2 \ \text{Cos}[h] \ \text{Sin}[h] \ \text{Sin}[x]^2\big)$

In[264]:= **s3 = Limit[s2,h- > 0]**

Out[264]= $2 \ \text{Cos}[2 \ x]$

■

If the derivative of $y = f(x)$ exists at $x = a$, a geometric interpretation of $f'(a)$ is that $f'(a)$ is the slope of the line tangent to the graph of $y = f(x)$ at the point $(a, f(a))$.

To motivate the definition of the derivative, many calculus texts choose a value of x, $x = a$, and then draw the graph of the secant line passing through the points $(a, f(a))$ and $(a + h, f(a + h))$ for "small" values of h to show that as h approaches 0, the secant line approaches the tangent line. An equation of the secant line passing through the points $(a, f(a))$ and $(a + h, f(a + h))$ is given by

$$y - f(a) = \frac{f(a + h) - f(a)}{(a + h) - a} (x - a) \quad \text{or} \quad y = \frac{f(a + h) - f(a)}{h} (x - a) + f(a).$$

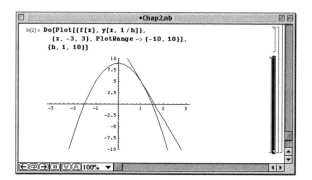

Figure 3-3   An animation

EXAMPLE 3.2.2: If $f(x) = 9 - 4x^2$, graph $f(x)$ together with the secant line containing $(1, f(1))$ and $(1 + h, f(1 + h))$ for various values of h.

SOLUTION: We define $f(x) = 9 - 4x^2$ and $y(x, h)$ to be a function returning the line containing $(1, f(1))$ and $(1 + h, f(1 + h))$.

```
In[265] := f[x_] = 9 - 4x^2;
           y[x_, h_] = (f[1 + h] - f[1])/h(x - 1) + f[1];
```

In the following, we use Do to show the graphs of $f(x)$ and $y(x, h)$ for $h = 1, 2, \ldots, 9$. The resulting animation can be played and controlled from the Mathematica menu. (See Figure 3-3.)

```
In[266] := Do[Plot[{f[x], y[x, 1/h]},
              {x, -3, 3}, PlotRange- > {-10, 10}],
              {h, 1, 10}]
```

If instead the command is entered as

```
In[267] := listofgraphics = Table[Plot[{f[x], y[x, 1/h]},
              {x, -3, 3}, PlotRange- > {-10, 10},
              DisplayFunction- > Identity], {h, 1, 10}]
```

```
In[268] := toshow = Partition[listofgraphics, 3]
```

```
In[269] := Show[GraphicsArray[toshow]]
```

the result is displayed as a graphics array. (See Figure 3-4.)

■

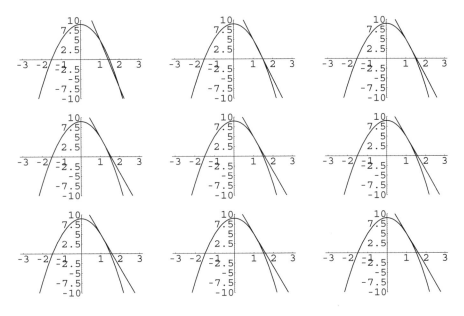

Figure 3-4   A graphics array

3.2.2 Calculating Derivatives

The functions D and ' are used to differentiate functions. Assuming that $y = f(x)$ is differentiable,

1. D[f[x],x] computes and returns $f'(x) = df/dx$,
2. f'[x] computes and returns $f'(x) = df/dx$,
3. f''[x] computes and returns $f^{(2)}(x) = d^2 f/dx^2$, and
4. D[f[x],{x,n}] computes and returns $f^{(n)}(x) = d^n f/dx^n$.

Mathematica knows the numerous differentiation rules, including the product, quotient, and chain rules. Thus, entering

    In[270]:= **Clear[f,g]**
                D[f[x]g[x],x]

    Out[270]= g[x] f'[x] + f[x] g'[x]

shows us that $\frac{d}{dx}(f(x) \cdot g(x)) = f'(x)g(x) + f(x)g'(x)$; entering

    In[271]:= **Together[D[f[x]/g[x],x]]**

    Out[271]= $\dfrac{g[x] \ f'[x] - f[x] \ g'[x]}{g[x]^2}$

shows us that $\frac{d}{dx}(f(x)/g(x)) = (f'(x)g(x) - f(x)g'(x))/(g(x))^2$; and entering

```
In[272]:= D[f[g[x]],x]
```

```
Out[272]= f'[g[x]] g'[x]
```

shows us that $\frac{d}{dx}(f(g(x))) = f'(g(x))g'(x)$.

EXAMPLE 3.2.3: Compute the first and second derivatives of (a) $y = x^4 + \frac{4}{3}x^3 - 3x^2$, (b) $f(x) = 4x^5 - \frac{5}{2}x^4 - 10x^3$, (c) $y = \sqrt{e^{2x} + e^{-2x}}$, and (d) $y = (1 + 1/x)^x$.

SOLUTION: For (a), we use D.

```
In[273]:= D[x^4 + 4/3x^3 - 3x^2, x]
Out[273]= -6 x + 4 x² + 4 x³
```

```
In[274]:= D[x^4 + 4/3x^3 - 3x^2, {x, 2}]
Out[274]= -6 + 8 x + 12 x²
```

For (b), we first define f and then use ′ together with Factor to calculate and factor $f'(x)$ and $f''(x)$.

```
In[275]:= f[x_] = 4x^5 - 5/2x^4 - 10x^3;
          Factor[f'[x]]

          Factor[f''[x]]
Out[275]= 10 x² (1 + x) (-3 + 2 x)
Out[275]= 10 x (-6 - 3 x + 8 x²)
```

For (c), we use Simplify together with D to calculate and simplify y' and y''.

```
In[276]:= D[Sqrt[Exp[2x] + Exp[-2x]], x]
```

$$Out[276] = \frac{-2\, e^{-2\,x} + 2\, e^{2\,x}}{2\, \sqrt{e^{-2\,x} + e^{2\,x}}}$$

```
In[277]:= D[Sqrt[Exp[2x] + Exp[-2x]], {x, 2}]//Simplify
```

$$Out[277] = \frac{\sqrt{e^{-2\,x} + e^{2\,x}}\, (1 + 6\, e^{4\,x} + e^{8\,x})}{(1 + e^{4\,x})^2}$$

By hand, (d) would require logarithmic differentiation. The second derivative would be particularly difficult to compute by hand. Mathematica quickly computes and simplifies each derivative.

$In[278] :=$ **Simplify[D[(1 + 1/x)^x, x]]**

$Out[278] = \dfrac{\left(1 + \frac{1}{x}\right)^x \left(-1 + (1 + x)\ \text{Log}\left[1 + \frac{1}{x}\right]\right)}{1 + x}$

$In[279] :=$ **Simplify[D[(1 + 1/x)^x, {x, 2}]]**

$Out[279] = \dfrac{\left(1 + \frac{1}{x}\right)^x \left(-1 + x - 2\ x\ (1 + x)\ \text{Log}\left[1 + \frac{1}{x}\right] + x\ (1 + x)^2\ \text{Log}\left[1 + \frac{1}{x}\right]^2\right)}{x\ (1 + x)^2}$

■

The command Map[f, list] applies the function f to each element of the list list. Thus, if you are computing the derivatives of a large number of functions, you can use Map together with D.

Map and operations on lists are discussed in more detail in Chapter 4.

Remark. A built-in Mathematica function is **threadable** if f[list] returns the same result as Map[f, list]. Many familiar functions like D and Integrate are threadable.

EXAMPLE 3.2.4: Compute the first and second derivatives of $\sin x$, $\cos x$, $\tan x$, $\sin^{-1} x$, $\cos^{-1} x$, and $\tan^{-1} x$.

SOLUTION: Notice that lists are contained in braces. Thus, entering

$In[280] :=$ **Map[D[#, x] &,**
         {Sin[x], Cos[x], Tan[x], ArcSin[x], ArcCos[x],
         ArcTan[x]}]

$Out[280] = \Big\{ \text{Cos}[x], -\text{Sin}[x], \text{Sec}[x]^2,$
$\qquad \dfrac{1}{\sqrt{1 - x^2}}, -\dfrac{1}{\sqrt{1 - x^2}}, \dfrac{1}{1 + x^2} \Big\}$

computes the first derivative of the three trigonometric functions and their inverses. In this case, we have applied a *pure function* to the list of trigonometric functions and their inverses. Given an argument #, D[#, x] & computes the derivative of # with respect to x. The & symbol is used to mark the end of a pure function. Similarly, entering

$In[281] :=$ **Map[D[#, {x, 2}] &,**
         {Sin[x], Cos[x], Tan[x], ArcSin[x], ArcCos[x],
         ArcTan[x]}]

$Out[281] = \Big\{ -\text{Sin}[x], -\text{Cos}[x], 2\ \text{Sec}[x]^2\ \text{Tan}[x], \dfrac{x}{(1 - x^2)^{3/2}},$
$\qquad -\dfrac{x}{(1 - x^2)^{3/2}}, -\dfrac{2\ x}{(1 + x^2)^2} \Big\}$

computes the second derivative of the three trigonometric functions and their inverses. Because D is threadable, the same results are obtained with

In[282]:= **D[**
       {Sin[x],Cos[x],Tan[x],ArcSin[x],
       ArcCos[x],ArcTan[x]},
       x]

Out[282]= $\left\{ \text{Cos}[x], -\text{Sin}[x], \text{Sec}[x]^2, \dfrac{1}{\sqrt{1-x^2}}, -\dfrac{1}{\sqrt{1-x^2}}, \dfrac{1}{1+x^2} \right\}$

In[283]:= **D[**
       {Sin[x],Cos[x],Tan[x],ArcSin[x],
       ArcCos[x],ArcTan[x]},
       {x,2}]

Out[283]= $\left\{ -\text{Sin}[x], -\text{Cos}[x], 2\ \text{Sec}[x]^2\ \text{Tan}[x], \dfrac{x}{(1-x^2)^{3/2}}, \right.$

$\left. -\dfrac{x}{(1-x^2)^{3/2}}, -\dfrac{2\ x}{(1+x^2)^2} \right\}$

∎

3.2.3 Implicit Differentiation

If an equation contains two variables, x and y, implicit differentiation can be carried out by explicitly declaring y to be a function of x, $y = y(x)$, and using D or by using the Dt command.

EXAMPLE 3.2.5: Find $y' = dy/dx$ if (a) $\cos(e^{xy}) = x$ and (b) $\ln(x/y) + 5xy = 3y$.

SOLUTION: For (a) we illustrate the use of D. Notice that we are careful to specifically indicate that $y = y(x)$. First we differentiate with respect to x

In[284]:= **s1 = D[Cos[Exp[x y[x]]] - x, x]**

Out[284]= BoxData($-1-e^{x\ y[x]}\ \text{Sin}[e^{x\ y[x]}]\ (y[x]+x\ y'[x])$)

and then we solve the resulting equation for $y' = dy/dx$ with Solve.

In[285]:= **Solve[s1 == 0, y'[x]]**

Out[285]= BoxData($\{\{y'[x] \rightarrow -$

$\dfrac{e^{-x\ y[x]}\ \text{Csc}[e^{x\ y[x]}]\ (1+e^{x\ y[x]}\ \text{Sin}[e^{x\ y[x]}]\ y[x])}{x}\}\}$)

For (b), we use Dt. When using Dt, we interpret Dt [x] = 1 and Dt [y] = $y' = dy/dx$. Thus, entering

In[286]:= **s2 = Dt[Log[x/y] + 5x y - 3y]**

Out[286]= $5 \ y \ Dt[x] - 3 \ Dt[y] + 5 \ x \ Dt[y] + \dfrac{y \left(\frac{Dt[x]}{y} - \frac{x \ Dt[y]}{y^2} \right)}{x}$

In[287]:= **s3 = s2/.{Dt[x]- >1, Dt[y]- > dydx}**

Out[287]= $-3 \ dydx + 5 \ dydx \ x + 5 \ y + \dfrac{\left(- \frac{dydx \ x}{y^2} + \frac{1}{y} \right) \ y}{x}$

In[288]:= **Solve[s3 == 0, dydx]**

Out[288]= $\left\{ \left\{ dydx \to - \dfrac{y \ (1 + 5 \ x \ y)}{x \ (-1 - 3 \ y + 5 \ x \ y)} \right\} \right\}$

shows us that if $\ln(x/y) + 5xy = 3y$,

$$y' = \frac{dy}{dx} = -\frac{(1 + 5xy)y}{(5xy - 3y - 1)x}.$$

■

3.2.4 Tangent Lines

If $f'(a)$ exists, we interpret $f'(a)$ to be the slope of the line tangent to the graph of $y = f(x)$ at the point $(a, f(a))$. An equation of the tangent is given by

$$y - f(a) = f'(a)(x - a) \quad \text{or} \quad y = f'(a)(x - a) + f(a).$$

EXAMPLE 3.2.6: Find an equation of the line tangent to the graph of

$$f(x) = \sin x^{1/3} + \cos^{1/3} x$$

at the point with x-coordinate $x = 5\pi/3$.

SOLUTION: Because we will be graphing a function involving odd roots of negative numbers, we begin by loading the **RealOnly** package contained in the **Miscellaneous** folder (or directory). We then define $f(x)$ and compute $f'(x)$.

In[289]:= **<< Miscellaneous`RealOnly`**

In[290]:= **f[x_] = Sin[x^(1/3)] + Cos[x]^(1/3);**

In[291]:= **f'[x]**

Out[291]= $\dfrac{Cos[x^{1/3}]}{3 \ x^{2/3}} - \dfrac{Sin[x]}{3 \ Cos[x]^{2/3}}$

Then, the slope of the line tangent to the graph of $f(x)$ at the point with x-coordinate $x = 5\pi/3$ is

$$Out[291] = \frac{1}{2^{1/3}\sqrt{3}} + \frac{\text{Cos}\left[\left(\frac{5\pi}{3}\right)^{1/3}\right]}{3^{1/3}\,(5\pi)^{2/3}}$$

$$In[292] := \textbf{f'[5}\pi\textbf{/3]//N}$$

$$Out[292] = 0.440013$$

while the y-coordinate of the point is

$$In[293] := \textbf{f[5}\,\pi\textbf{/3]}$$

$$Out[293] = \frac{1}{2^{1/3}} + \text{Sin}\left[\left(\frac{5\pi}{3}\right)^{1/3}\right]$$

$$In[294] := \textbf{f[5}\,\pi\textbf{/3]//N}$$

$$Out[294] = 1.78001$$

Thus, an equation of the line tangent to the graph of $f(x)$ at the point with x-coordinate $x = 5\pi/3$ is

$$y - \left(\frac{1}{\sqrt[3]{2}} + \sin\sqrt[3]{5\pi/3}\right) = \left(\frac{\cos\sqrt[3]{5\pi/3}}{\sqrt[3]{3}\sqrt[3]{25\pi^2}} + \frac{1}{\sqrt[3]{2}\sqrt{3}}\right)\left(x - \frac{5\pi}{3}\right),$$

as shown in Figure 3-5.

```
In[295] := p1 = Plot[f[x], {x, 0, 4π},
                DisplayFunction → Identity];

           p2 = ListPlot[{{5π/3, f[5π/3]}//N},
                PlotStyle → PointSize[0.03],
                DisplayFunction → Identity];

           p3 = Plot[f'[5π/3] (x - 5π/3) + f[5π/3], {x, 0, 4π},
                PlotStyle- > GrayLevel[0.6],
                DisplayFunction → Identity];

In[296] := Show[p1, p2, p3, AspectRatio- > Automatic,
           DisplayFunction → $DisplayFunction]
```

■

EXAMPLE 3.2.7: Find an equation of the line tangent to the graph of $f(x) = 9 - 4x^2$ at the point $(1, f(1))$.

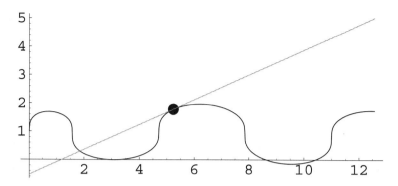

Figure 3-5 $f(x) = \sin x^{1/3} + \cos^{1/3} x$ together with its tangent at the point $(5\pi/3, f(5\pi/3))$

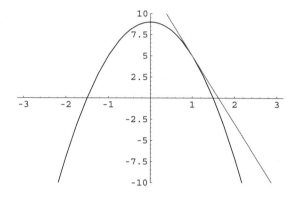

Figure 3-6 $f(x)$ together with its tangent at $(1, f(1))$

SOLUTION: After defining f, we see that $f(1) = 5$ and $f'(1) = -8$

```
In[297]:= f[x_] = 9 - 4x^2;
          f[1]

          f'[1]
Out[297]= 5

Out[297]= -8
```

so an equation of the line tangent to $y = f(x)$ at the point $(1, 5)$ is $y - 5 = -8(x - 1)$ or $y = -8x + 13$. We can visualize the tangent at $(1, f(1))$ with Plot. (See Figure 3-6.)

```
In[298]:= Plot[{f[x], f'[1](x - 1) + f[1]}, {x, -3, 3},
            PlotStyle- > {GrayLevel[0], GrayLevel[0.3]},
            PlotRange- > {-10, 10}]
```

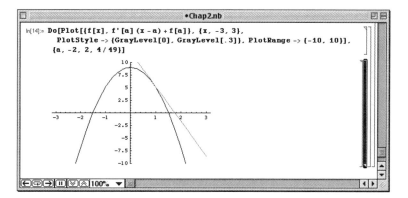

Figure 3-7   An animation

In addition, we can view a sequence of lines tangent to the graph of a function for a sequence of x values using Do. In the following, we use Do to generate graphs of $y = f(x)$ and $y = f'(a)(x-a) + f(a)$ for 50 equally spaced values of a between -3 and 3. (See Figure 3-7.)

```
In[299] := Do[Plot[{f[x], f'[a] (x - a) + f[a]}, {x, -3, 3}, }
              PlotStyle- > {GrayLevel[0], GrayLevel[0.3]},
              PlotRange- > {-10, 10}], {a, -2, 2, 4/49}]
```

On the other hand,

```
In[300] := listofgraphics = Table[
              Plot[{f[x], f'[a] (x - a) + f[a]}, {x, -3, 3},
              PlotStyle- > {GrayLevel[0], GrayLevel[0.3]},
              PlotRange- > {-10, 10},
              DisplayFunction- > Identity], {a, -2, 2, 4/8}];
              toshow = Partition[listofgraphics, 3];
              Show[GraphicsArray[toshow]]
```

graphs $y = f(x)$ and $y = f'(a)(x - a) + f(a)$ for nine equally spaced values of a between -3 and 3 and displays the result as a graphics array. (See Figure 3-8.)
In the graphs, notice that where the tangent lines have positive slope ($f'(x) > 0$), $f(x)$ is increasing while where the tangent lines have negative slope ($f'(x) < 0$), $f(x)$ is decreasing.

■

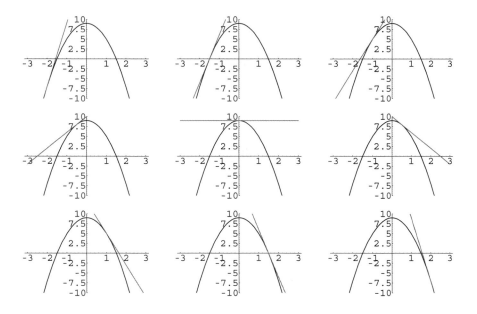

Figure 3-8 $f(x)$ together with various tangents

Tangent Lines of Implicit Functions

EXAMPLE 3.2.8: Find equations of the tangent line and normal line to the graph of $x^2y - y^3 = 8$ at the point $(-3, 1)$. Find and simplify $y'' = d^2y/dx^2$.

SOLUTION: We will evaluate $y' = dy/dx$ if $x = -3$ and $y = 1$ to determine the slope of the tangent line at the point $(-3, 1)$. Note that we cannot (easily) solve $x^2y - y^3 = 8$ for y so we use implicit differentiation to find $y' = dy/dx$:

By the product and chain rules,
$\frac{d}{dx}(x^2y) = \frac{d}{dx}(x^2)y + x^2\frac{d}{dx}(y) = 2x \cdot y + x^2 \cdot \frac{dy}{dx} = 2xy + x^2y'$.

$$\frac{d}{dx}\left(x^2y - y^3\right) = \frac{d}{dx}(8)$$
$$2xy + x^2y' - 3y^2y' = 0$$
$$y' = \frac{-2xy}{x^2 - 3y^2}.$$

```
In[301]:= eq = x^2y - y^3 == 8
```
$Out[301]= x^2\,y - y^3 == 8$

```
In[302]:= s1 = Dt[eq]
Out[302]= 2 x y Dt[x] + x² Dt[y] - 3 y² Dt[y] == 0

In[303]:= s2 = s1/.Dt[x] → 1
Out[303]= 2 x y + x² Dt[y] - 3 y² Dt[y] == 0

In[304]:= s3 = Solve[s2, Dt[y]]
Out[304]= {{Dt[y] → - (2 x y)/(x² - 3 y²)}}
```

Lists are discussed in more detail in Chapter 4.

Notice that s3 is a **list**. The formula for $y' = dy/dx$ is the second part of the first part of the first part of s3 and extracted from s3 with

```
In[305]:= s3[[1, 1, 2]]
Out[305]= - (2 x y)/(x² - 3 y²)
```

We then use ReplaceAll (/.) to find that the slope of the tangent at $(-3, 1)$ is

```
In[306]:= s3[[1, 1, 2]]/.{x → -3, y → 1}
Out[306]= 1
```

The slope of the normal is $-1/1 = -1$. Equations of the tangent and normal are given by

$$y - 1 = 1(x + 3) \qquad \text{and} \qquad y - 1 = -1(x + 3),$$

respectively. See Figure 3-9.

```
In[307]:= cp1 = ContourPlot[x^2y - y^3 - 8, {x, -5, 5}, {y, -5, 5},
                Contours → {0}, ContourShading → False,
                PlotPoints → 200, DisplayFunction → Identity];

          p1 = ListPlot[{{-3, 1}},
                PlotStyle → PointSize[0.03],
                DisplayFunction → Identity];

          p2 = Plot[{(x + 3) + 1, -(x + 3) + 1}, {x, -5, 5},
                PlotStyle → GrayLevel[0.3],
                DisplayFunction → Identity];


          Show[cp1, p1, p2, Frame → False, Axes → Automatic,
              AxesOrigin → {0, 0}, AspectRatio → Automatic,
              DisplayFunction → $DisplayFunction]
```

To find $y'' = d^2y/dx^2$, we proceed as follows.

```
In[308]:= s4 = Dt[s3[[1, 1, 2]]]//Simplify
```

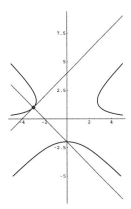

Figure 3-9   Graphs of $x^2y - y^3 = 8$ (in black) and the tangent and normal at $(-3, 1)$ (in gray)

$$Out[308] = -\frac{2\,(x^2 + 3\,y^2)\,(-y\,Dt[x] + x\,Dt[y])}{(x^2 - 3\,y^2)^2}$$

$In[309]:= $ **s5 = s4/.Dt[x] → 1/.s3[[1]]//Simplify**

$$Out[309] = \frac{6\,y\,(x^2 - y^2)\,(x^2 + 3\,y^2)}{(x^2 - 3\,y^2)^3}$$

The result means that

$$y'' = \frac{d^2y}{dx^2} = \frac{6\left(x^2y - y^3\right)\left(x^2 + 3y^2\right)}{\left(x^2 - 3y^2\right)^3}.$$

Because $x^2y - y^3 = 8$, the second derivative is further simplified to

$$y'' = \frac{d^2y}{dx^2} = \frac{48\left(x^2 + 3y^2\right)}{\left(x^2 - 3y^2\right)^3}.$$

∎

Parametric Equations and Polar Coordinates

For the parametric equations $\{x = f(t), y = g(t)\}$, $t \in I$,

$$y' = \frac{dy}{dx} = \frac{dy/dt}{dx/dt} = \frac{g'(t)}{f'(t)}$$

and

$$y'' = \frac{d^2y}{dx^2} = \frac{d}{dx}\frac{dy}{dx} = \frac{d/dt(dy/dx)}{dx/dt}.$$

If $\{x = f(t), y = g(t)\}$ has a tangent line at the point $(f(a), g(a))$, parametric equations of the tangent are given by

$$x = f(a) + tf'(a) \qquad \text{and} \qquad y = g(a) + tg'(a). \qquad (3.2)$$

If $f'(a), g'(a) \neq 0$, we can eliminate the parameter from (3.2)

$$\frac{x - f(a)}{f'(a)} = \frac{y - g(a)}{g'(a)}$$

$$y - g(a) = \frac{g'(a)}{f'(a)}(x - f(a))$$

and obtain an equation of the tangent line in point-slope form.

```
In[310]:= l = Solve[x[a] + t x'[a] == cx, t]

          r = Solve[y[a] + t y'[a] == cy, t]
```

$$Out[310]= \text{BoxData}\left(\left\{\left\{t \to -\frac{-cx + x[a]}{x'[a]}\right\}\right\}\right)$$

$$Out[310]= \text{BoxData}\left(\left\{\left\{t \to -\frac{-cy + y[a]}{y'[a]}\right\}\right\}\right)$$

EXAMPLE 3.2.9 (The Cycloid): The **cycloid** has parametric equations

$$x = t - \sin t \qquad \text{and} \qquad y = 1 - \cos t.$$

Graph the cycloid together with the line tangent to the graph of the cycloid at the point $(x(a), y(a))$ for various values of a between -2π and 4π.

SOLUTION: After defining x and y we use $'$ to compute dy/dt and dx/dt. We then compute $dy/dx = (dy/dt)/(dx/dt)$ and d^2y/dx^2.

```
In[311]:= x[t_] = t - Sin[t];
          y[t_] = 1 - Cos[t];
          dx = x'[t]

          dy = y'[t]

          dydx = dy/dx
```
$$Out[311]= 1 - \text{Cos}[t]$$
$$Out[311]= \text{Sin}[t]$$
$$Out[311]= \frac{\text{Sin}[t]}{1 - \text{Cos}[t]}$$

```
In[312]:= dypdt = Simplify[D[dydx, t]]
```
$$Out[312]= \frac{1}{-1 + \text{Cos}[t]}$$

```
In[313]:= secondderiv = Simplify[dypdt/dx]
```
$$Out[313] = -\frac{1}{(-1 + \text{Cos}[t])^2}$$

We then use `ParametricPlot` to graph the cycloid for $-2\pi \le t \le 4\pi$, naming the resulting graph `p1`.

```
In[314]:= p1 = ParametricPlot[{x[t],y[t]},{t,-2π,4π},
              PlotStyle- > {{GrayLevel[0],Thickness[0.01]}},
              DisplayFunction- > Identity];
```

Next, we use `Table` to define `toplot` to be 40 tangent lines (3.2) using equally spaced values of a between -2π and 4π. We then graph each line `toplot` and name the resulting graph `p2`. Finally, we show `p1` and `p2` together with the `Show` function. The resulting plot is shown to scale because the lengths of the x and y-axes are equal and we include the option `AspectRatio->1`. In the graphs, notice that on intervals for which dy/dx is defined, dy/dx is a decreasing function and, consequently, $d^2y/dx^2 < 0$. (See Figure 3-10.)

```
In[315]:= toplot = Table[
              {x[a] + t x'[a],y[a] + t y'[a]}, {a, -2π, 4π, 6π/39}];
           p2 = ParametricPlot[Evaluate[toplot],
              {t, -2, 2}, PlotStyle- > GrayLevel[0.5],
              DisplayFunction- > Identity];
           Show[p1, p2, AspectRatio- > 1, PlotRange- > {-3π, 3π},
              DisplayFunction- > $DisplayFunction]
```

■

EXAMPLE 3.2.10 (Orthogonal Curves): Two lines L_1 and L_2 with slopes m_1 and m_2, respectively, are **orthogonal** if their slopes are negative reciprocals: $m_1 = -1/m_2$.

Extended to curves, we say that the curves C_1 and C_2 are **orthogonal** at a point of intersection if their respective tangent lines to the curves at that point are orthogonal.

Show that the family of curves with equation $x^2 + 2xy - y^2 = C$ is orthogonal to the family of curves with equation $y^2 + 2xy - x^2 = C$.

SOLUTION: We begin by defining `eq1` and `eq2` to be the left-hand sides of the equations $x^2 + 2xy - y^2 = C$ and $y^2 + 2xy - x^2 = C$, respectively.

```
In[316]:= eq1 = x^2 + 2x y - y^2;
           eq2 = y^2 + 2x y - x^2;
```

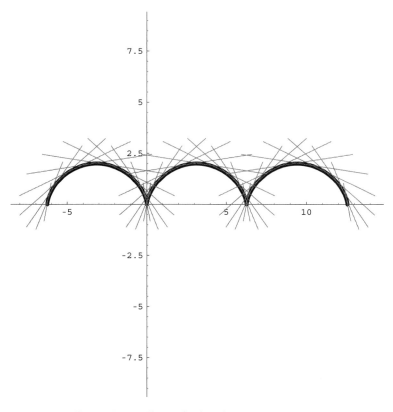

Figure 3-10   The cycloid with various tangents

We then use Dt to differentiate and Solve to find $y' = dy/dx$. Because the derivatives are negative reciprocals, we conclude that the curves are orthogonal. We confirm this graphically by graphing several members of each family with ContourPlot and showing the results together. (See Figure 3-11.)

```
In[317]:= BoxData({s1 = Dt[eq1]/.{Dt[x]->1, Dt[y]->dydx},
              Solve[s1 == 0, dydx]})
```

$$Out[317] = 2\ x + 2\ dydx\ x + 2\ y - 2\ dydx\ y$$

$$Out[317] = \left\{\left\{dydx \to -\frac{x+y}{x-y}\right\}\right\}$$

```
In[318]:= BoxData({s2 = Dt[eq2]/.{Dt[x]->1, Dt[y]->dydx},
              Solve[s2 == 0, dydx]})
```

$$Out[318] = -2\ x + 2\ dydx\ x + 2\ y + 2\ dydx\ y$$

$$Out[318] = \left\{\left\{dydx \to -\frac{-x+y}{x+y}\right\}\right\}$$

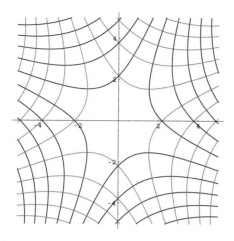

Figure 3-11 $x^2 + 2xy - y^2 = C$ and $y^2 + 2xy - x^2 = C$ for various values of C

```
In[319]:= cp1 = ContourPlot[eq1, {x, -5, 5}, {y, -5, 5},
              ContourShading- > False,
              ContourStyle- > GrayLevel[0],
              Frame- > False, Axes- > Automatic,
              AxesOrigin- > {0, 0},
              DisplayFunction- > Identity, PlotPoints- > 60];
          cp2 = ContourPlot[eq2, {x, -5, 5}, {y, -5, 5},
              ContourShading- > False,
              ContourStyle- > GrayLevel[0.4],
              Frame- > False, Axes- > Automatic,
              AxesOrigin- > {0, 0},
              DisplayFunction- > Identity, PlotPoints- > 60];
          Show[cp1, cp2, DisplayFunction- > $DisplayFunction]
```

■

EXAMPLE 3.2.11 (Theorem 1. The Mean-Value Theorem for Derivatives): If $y = f(x)$ is continuous on $[a, b]$ and differentiable on (a, b) then there is at least one value of c between a and b for which

$$f'(c) = \frac{f(b) - f(a)}{b - a} \qquad \text{or, equivalently,} \qquad f(b) - f(a) = f'(c)(b - a). \qquad (3.3)$$

Find all number(s) c that satisfy the conclusion of the Mean-Value Theorem for $f(x) = x^2 - 3x$ on the interval $[0, 7/2]$.

SOLUTION: By the power rule, $f'(x) = 2x - 3$. The slope of the secant containing $(0, f(0))$ and $(7/2, f(7/2))$ is

$$\frac{f(7/2) - f(0)}{7/2 - 0} = \frac{1}{2}.$$

Solving $2x - 3 = 1/2$ for x gives us $x = 7/4$.

```
In[320] := f[x_] = x^2 - 3x
```
$$Out[320] = -3x + x^2$$

```
In[321] := Solve[f'[x] == 0, x]
```
$$Out[321] = \left\{\left\{x \to \frac{3}{2}\right\}\right\}$$

```
In[322] := Solve[f'[x] == (f[7/2] - f[0])/(7/2 - 0)]
```
$$Out[322] = \left\{\left\{x \to \frac{7}{4}\right\}\right\}$$

$x = 7/4$ satisfies the conclusion of the Mean-Value Theorem for $f(x) = x^2 - 3x$ on the interval $[0, 7/2]$, as shown in Figure 3-12.

```
In[323] := p1 = Plot[f[x], {x, -2, 4},
              DisplayFunction → Identity];

          p2 = Plot[f[x], {x, 0, 7/2},
              PlotStyle → Thickness[0.02],
              DisplayFunction → Identity];

          p3 = ListPlot[{{0, f[0]}, {7/4, f[7/4]},
              {7/2, f[7/2]}},
              PlotStyle → PointSize[0.05],
              DisplayFunction → Identity];

          p4 = Plot[{f'[7/4] (x - 7/4) + f[7/4],
              (f[7/2] - f[0])/(7/2 - 0)x},
              {x, -2, 4}, PlotStyle → {Dashing[{0.01}],
              Dashing[{0.02}]},
              DisplayFunction → Identity];

          Show[p1, p2, p3, p4, DisplayFunction →
              $DisplayFunction, AspectRatio → Automatic]
```

■

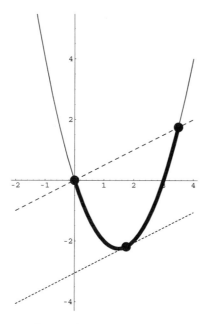

Figure 3-12   Graphs of $f(x) = x^2 - 3x$, the secant containing $(0, f(0))$ and $(7/2, f(7/2))$, and the tangent at $(7/4, f(7/4))$

3.2.5 The First Derivative Test and Second Derivative Test

Examples 3.2.7 and 3.2.9 illustrate the following properties of the first and second derivative.

Theorem 2. *Let* $y = f(x)$ *be continuous on* $[a, b]$ *and differentiable on* (a, b).

1. *If* $f'(x) = 0$ *for all* x *in* (a, b), *then* $f(x)$ *is constant on* $[a, b]$.
2. *If* $f'(x) > 0$ *for all* x *in* (a, b), *then* $f(x)$ *is increasing on* $[a, b]$.
3. *If* $f'(x) < 0$ *for all* x *in* (a, b), *then* $f(x)$ *is decreasing on* $[a, b]$.

For the second derivative, we have the following theorem.

Theorem 3. *Let* $y = f(x)$ *have a second derivative on* (a, b).

1. *If* $f''(x) > 0$ *for all* x *in* (a, b), *then the graph of* $f(x)$ *is concave up on* (a, b).
2. *If* $f''(x) < 0$ *for all* x *in* (a, b), *then the graph of* $f(x)$ *is concave down on* (a, b).

The **critical points** correspond to those points on the graph of $y = f(x)$ where the tangent line is horizontal or vertical; the number $x = a$ is a **critical number** if $f'(a) = 0$ or $f'(x)$ does not exist if $x = a$. The **inflection points** correspond to those points on the graph of $y = f(x)$ where the graph of $y = f(x)$ is neither concave up nor concave down. Theorems 2 and 3 help establish the first derivative test and second derivative test.

Theorem 4 (First Derivative Test). *Let $x = a$ be a critical number of a function $y = f(x)$ continuous on an open interval I containing $x = a$. If $f(x)$ is differentiable on I, except possibly at $x = a$, $f(a)$ can be classified as follows.*

1. *If $f'(x)$ changes from positive to negative at $x = a$, then $f(a)$ is a **relative maximum**.*
2. *If $f'(x)$ changes from negative to positive at $x = a$, then $f(a)$ is a **relative minimum**.*

Theorem 5 (Second Derivative Test). *Let $x = a$ be a critical number of a function $y = f(x)$ and suppose that $f''(x)$ exists on an open interval containing $x = a$.*

1. *If $f''(a) < 0$, then $f(a)$ is a relative maximum.*
2. *If $f''(a) > 0$, then $f(a)$ is a relative minimum.*

EXAMPLE 3.2.12: Graph $f(x) = 3x^5 - 5x^3$.

SOLUTION: We begin by defining $f(x)$ and then computing and factoring $f'(x)$ and $f''(x)$.

```
In[324]:= f[x_] = 3x^5 - 5x^3;
          d1 = Factor[f'[x]]
          d2 = Factor[f''[x]]
Out[324]= 15 (-1 + x) x² (1 + x)
Out[324]= 30 x (-1 + 2 x²)
```

By inspection, we see that the critical numbers are $x = 0, 1$, and -1 while $f''(x) = 0$ if $x = 0, 1/\sqrt{2}$, or $-1/\sqrt{2}$. Of course, these values can also be found with `Solve` as done next in `cns` and `ins`, respectively.

```
In[325]:= cns = Solve[d1 == 0]
          ins = Solve[d2 == 0]
Out[325]= {{x → -1}, {x → 0}, {x → 0}, {x → 1}}
Out[325]= {{x → 0}, {x → -1/√2}, {x → 1/√2}}
```

We find the critical and inflection points by using /. (`Replace All`) to compute $f(x)$ for each value of x in `cns` and `ins`, respectively. The result means that the critical points are $(0, 0)$, $(1, -2)$ and $(-1, 2)$; the inflection points are $(0, 0)$, $(1/\sqrt{2}, -7\sqrt{2}/8)$, and $(-1/\sqrt{2}, 7\sqrt{2}/8)$. We also see that $f''(0) = 0$ so Theorem 5 cannot be used to classify $f(0)$.

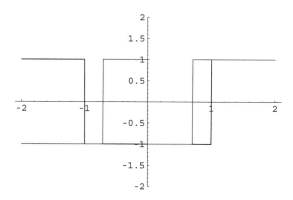

Figure 3-13   Graphs of $|f'(x)|/f'(x)$ and $|f''(x)|/f''(x)$

On the other hand, $f''(1) = 30 > 0$ and $f''(-1) = -30 < 0$ so by Theorem 5, $f(1) = -2$ is a relative minimum and $f(-1) = 2$ is a relative maximum.

```
In[326]:= cps = {x, f[x]}/.cns
          f''[x]/.cns
          ips = {x, f[x]}/.ins
Out[326]= {{-1, 2}, {0, 0}, {0, 0}, {1, -2}}
Out[326]= {-30, 0, 0, 30}
```
$$Out[326]= \left\{\{0, 0\}, \left\{-\frac{1}{\sqrt{2}}, \frac{7}{4\sqrt{2}}\right\}, \left\{\frac{1}{\sqrt{2}}, -\frac{7}{4\sqrt{2}}\right\}\right\}$$

We can graphically determine the intervals of increase and decrease by noting that if $f'(x) > 0$ ($f'(x) < 0$), $|f'(x)|/f'(x) = 1$ ($|f'(x)|/f'(x) = -1$). Similarly, the intervals for which the graph is concave up and concave down can be determined by noting that if $f''(x) > 0$ ($f''(x) < 0$), $|f''(x)|/f''(x) = 1$ ($|f''(x)|/f''(x) = -1$). We use Plot to graph $|f'(x)|/f'(x)$ and $|f''(x)|/f''(x)$ in Figure 3-13.

```
In[327]:= Plot[{Abs[d1]/d1, Abs[d2]/d2}, {x, -2, 2},
            PlotStyle- > {GrayLevel[0], GrayLevel[0.3]},
            PlotRange- > {-2, 2}]
```

From the graph, we see that $f'(x) > 0$ for x in $(-\infty, -1) \cup (1, \infty)$, $f'(x) < 0$ for x in $(-1, 1)$, $f''(x) > 0$ for x in $(-1/\sqrt{2}, 0) \cup (1/\sqrt{2}, \infty)$, and $f''(x) < 0$ for x in $(-\infty, -1/\sqrt{2}) \cup (0, 1/\sqrt{2})$. Thus, the graph of $f(x)$ is

- increasing and concave down for x in $(-\infty, -1)$,
- decreasing and concave down for x in $(-1, -1/\sqrt{2})$,
- decreasing and concave up for x in $(-1/\sqrt{2}, 0)$,
- decreasing and concave down for x in $(0, 1\sqrt{2})$,

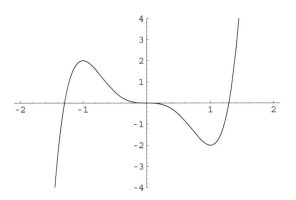

Figure 3-14 $f(x)$ for $-2 \le x \le 2$ and $-4 \le y \le 4$

- decreasing and concave up for x in $(1/\sqrt{2}, 1)$, and
- increasing and concave up for x in $(1, \infty)$.

We also see that $f(0) = 0$ is neither a relative minimum nor maximum. To see all points of interest, our domain must contain -1 and 1 while our range must contain -2 and 2. We choose to graph $f(x)$ for $-2 \le x \le 2$; we choose the range displayed to be $-4 \le y \le 4$. (See Figure 3-14.)

```
In[328]:= Plot[f[x],{x,-2,2},PlotRange->{-4,4}]
```

■

Remember to be especially careful when working with functions that involve odd roots.

EXAMPLE 3.2.13: Graph $f(x) = (x - 2)^{2/3}(x + 1)^{1/3}$.

SOLUTION: We begin by defining $f(x)$ and then computing and simplifying $f'(x)$ and $f''(x)$ with $'$ and Simplify.

```
In[329]:= f[x_] = (x - 2)^(2/3)(x + 1)^(1/3);
          d1 = Simplify[f'[x]]

          d2 = Simplify[f''[x]]
```

$$Out[329] = \frac{x}{(-2 + x)^{1/3}\,(1 + x)^{2/3}}$$

$$Out[329] = -\frac{2}{(-2 + x)^{4/3}\,(1 + x)^{5/3}}$$

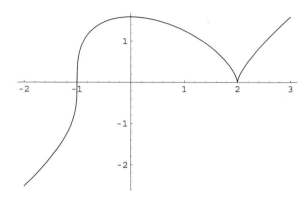

Figure 3-15 $f(x)$ for $-2 \leq x \leq 3$

By inspection, we see that the critical numbers are $x = 0$, 2, and -1. We cannot use Theorem 5 to classify $f(2)$ and $f(-1)$ because $f''(x)$ is undefined if $x = 2$ or -1. On the other hand, $f''(0) < 0$ so $f(0) = 2^{2/3}$ is a relative maximum. By hand, we make a sign chart to see that the graph of $f(x)$ is

- increasing and concave up on $(-\infty, -1)$,
- increasing and concave down on $(-1, 0)$,
- decreasing and concave down on $(0, 2)$, and
- increasing and concave down on $(2, \infty)$.

Hence, $f(-1) = 0$ is neither a relative minimum nor maximum while $f(2) = 0$ is a relative minimum by Theorem 4. To graph $f(x)$, we load the RealOnly package and then use Plot to graph $f(x)$ for $-2 \leq x \leq 3$ in Figure 3-15.

```
In[330]:= << Miscellaneous`RealOnly`

           f[0]

           Plot[f[x], {x, -2, 3}]
Out[330]= 2^2/3
```

■

The previous examples illustrate that if $x = a$ is a critical number of $f(x)$ and $f'(x)$ makes a *simple change in sign* from positive to negative at $x = a$, then $(a, f(a))$ is a relative maximum. If $f'(x)$ makes a simple change in sign from negative to positive at $x = a$, then $(a, f(a))$ is a relative minimum. Mathematica is especially useful in investigating interesting functions for which this may not be the case.

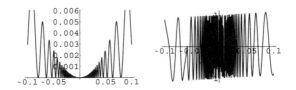

Figure 3-16 $f(x) = \left[x\sin\left(\frac{1}{x}\right)\right]^2$ and $f'(x)$ for $-0.1 \le x \le 0.1$

EXAMPLE 3.2.14: Consider

$$f(x) = \begin{cases} x^2 \sin^2\left(\frac{1}{x}\right), x \ne 0 \\ 0, x = 0. \end{cases}$$

$x = 0$ is a critical number because $f'(x)$ does not exist if $x = 0$. $(0,0)$ is both a relative and absolute minimum, even though $f'(x)$ does not make a simple change in sign at $x = 0$, as illustrated in Figure 3-16.

```
In[331]:= f[x_] = x^2 Sin[1/x]^2;

           f'[x]//Factor
```

$$Out[331]= -2\,\text{Sin}\left[\frac{1}{x}\right]\left(\text{Cos}\left[\frac{1}{x}\right] - x\,\text{Sin}\left[\frac{1}{x}\right]\right)$$

```
In[332]:= p1 = Plot[f[x], {x, -0.1, 0.1},
               DisplayFunction → Identity];

           p2 = Plot[f'[x], {x, -0.1, 0.1},
               DisplayFunction → Identity];

           Show[GraphicsArray[{p1, p2}]]
```

Notice that the derivative "oscillates" infinitely many times near $x = 0$, so the first derivative test cannot be used to classify $(0,0)$.

3.2.6 Applied Max/Min Problems

Mathematica can be used to assist in solving maximization/minimization problems encountered in a differential calculus course.

EXAMPLE 3.2.15: A woman is located on one side of a body of water 4 miles wide. Her position is directly across from a point on the other side of the body of water 16 miles from her house, as shown in the following figure.

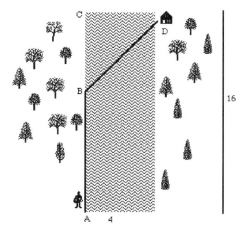

If she can move across land at a rate of 10 miles per hour and move over water at a rate of 6 miles per hour, find the least amount of time for her to reach her house.

SOLUTION: From the figure, we see that the woman will travel from *A* to *B* by land and then from *B* to *D* by water. We wish to find the least time for her to complete the trip.

Let x denote the distance BC, where $0 \le x \le 16$. Then, the distance AB is given by $16 - x$ and, by the Pythagorean theorem, the distance BD is given by $\sqrt{x^2 + 4^2}$. Because rate × time = distance, time = distance/rate. Thus, the time to travel from *A* to *B* is $\frac{1}{10}(16 - x)$, the time to travel from *B* to *D* is $\frac{1}{6}\sqrt{x^2 + 16}$, and the total time to complete the trip, as a function of x, is

$$time(x) = \frac{1}{10}(16 - x) + \frac{1}{6}\sqrt{x^2 + 16}, \quad 0 \le x \le 16.$$

We must minimize the function *time*. First, we define `time` and then verify that `time` has a minimum by graphing `time` on the interval [0, 16] in Figure 3-17.

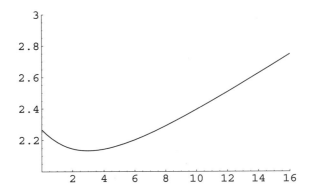

Figure 3-17   Plot of $time(x) = \frac{1}{10}(16 - x) + \frac{1}{6}\sqrt{x^2 + 16}, \quad 0 \leq x \leq 16$

```
In[333]:= Clear[time]
          time[x_] = 16 - x   +  1  √(x² + 16);
                     ------      ---
                       10         6
          Plot[time[x], {x, 0, 16},
          PlotRange → {{0, 16}, {2, 3}}]
```

Next, we compute the derivative of time and find the values of x for which the derivative is 0 with Solve. The resulting output is named critnums.

```
In[334]:= Together[time'[x]]
```

$$Out[334] = \frac{5\,x - 3\,\sqrt{16 + x^2}}{30\,\sqrt{16 + x^2}}$$

```
In[335]:= critnums = Solve[time'[x] == 0]
Out[335]= {{x → 3}}
```

At this point, we can calculate the minimum time by calculating time[3].

```
In[336]:= time[3]
Out[336]= 32
          --
          15
```

Alternatively, we demonstrate how to find the value of time[x] for the value(s) listed in critnums.

```
In[337]:= time[x]/.x → 3
Out[337]= 32
          --
          15
```

Regardless, we see that the minimum time to complete the trip is 32/15 hours.

■

One of the more interesting applied max/min problems is the *beam problem*. We present two solutions.

EXAMPLE 3.2.16 (The Beam Problem): Find the exact length of the longest beam that can be carried around a corner from a hallway 2 feet wide to a hallway that is 3 feet wide. (See Figure 3-18.)

SOLUTION: We assume that the beam has negligible thickness. Our first approach is algebraic. Using Figure 3-18, which is generated with

Graphics primitives like Point, Line, and Text are discussed in more detail in Chapter 7.

```
In[338]:= p1 = Plot[x + 2, {x, 0, 4},
              PlotStyle- > Thickness[0.01],
              PlotRange- > {0, 6}]
```

```
In[339]:= p2 = Graphics[Line[{{1, 0}, {1, f[1]},
              {4, f[1]}, {4, f[4]}, {4, f[4]},
              {0, f[4]}, {0, 0}, {1, 0}}]]
```

```
In[340]:= p3 = Graphics[{Text["3", {0.5, 0.2}],
              {Text["2", {3.8, 4.5}]}}]
```

```
In[341]:= p4 = Graphics[{Dashing[{0.01, 0.01}],
              Line[{{0, f[0]}, {1, f[0]}}]}]
```

```
In[342]:= p5 = Graphics[{Text["θ", {0.5, 2.25}],
              Text["θ", {1.5, 3.25}]}]
```

```
In[343]:= p6 = Graphics[{Text["y", {0.9, 2.35}],
              Text["x", {2.5, 3.25}]}]
```

```
In[344]:= Show[p1, p2, p3, p4, p5, p6, Axes- > None]
```

and the Pythagorean theorem, the total length of the beam is

$$L = \sqrt{2^2 + x^2} + \sqrt{y^2 + 3^2}.$$

By similar triangles,

$$\frac{y}{3} = \frac{2}{x} \quad \text{so} \quad y = \frac{6}{x}$$

and the length of the beam, L, becomes

$$L(x) = \sqrt{4 + x^2} + \sqrt{9 + \frac{36}{x^2}}, \quad 0 < x < \infty.$$

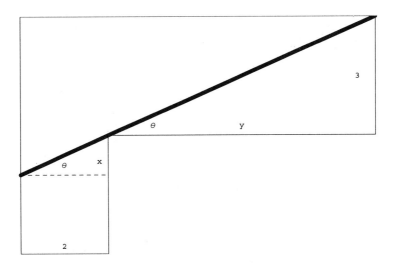

Figure 3-18   The length of the beam is found using similar triangles

$In[345] := $ **Clear[l]**

$$l[x\_] = \text{Sqrt}[2^\wedge2 + x^\wedge2] + \text{Sqrt}[y^\wedge2 + 3^\wedge2] /. y- > 6/x$$

$Out[345] = \sqrt{9 + \dfrac{36}{x^2}} + \sqrt{4 + x^2}$

Observe that the length of the longest beam is obtained by *minimizing*
L. (Why?)

We ignore negative values because length must be nonnegative.

Differentiating gives us

$In[346] := $ **l'[x]**

$Out[346] = -\dfrac{36}{\sqrt{9 + \frac{36}{x^2}}\, x^3} + \dfrac{x}{\sqrt{4 + x^2}}$

and solving $L'(x) = 0$ for x results in

$In[347] := $ **Solve$\left[-12\sqrt{4 + x^2} + x^4 \sqrt{\dfrac{4 + x^2}{x^2}} == 0, x\right]$**

$Out[347] = \{\{x \to -2I\}, \{x \to 2I\}, \{x \to -2^{2/3}3^{1/3}\}, \{x \to 2^{2/3}3^{1/3}\}\}$

so $x = 2^{2/3}3^{1/3} \approx 2.29$.

$In[348] := $ **N$\left[2^{2/3}\,3^{1/3}\right]$**

$Out[348] = 2.28943$

$In[349] := $ **l$\left[2^{2/3}\,3^{1/3}\right]$**

$Out[349] = \sqrt{9 + 3\,2^{2/3}\,3^{1/3}} + \sqrt{4 + 2\,2^{1/3}\,3^{2/3}}$

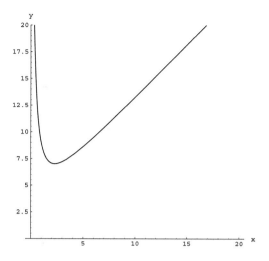

Figure 3-19   Graph of $L(x)$

```
In[350]:= l[2^(2/3) 3^(1/3)] //FullSimplify
```

$$Out[350] = \sqrt{13 + 9\, 2^{2/3}\, 3^{1/3} + 6\, 2^{1/3}\, 3^{2/3}}$$

```
In[351]:= N[%]
```

```
Out[351]= 7.02348
```

It follows that the length of the beam is $L(2^{2/3}3^{1/3}) = \sqrt{9 + 3 \cdot 2^{2/3} \cdot 3^{1/3}} + \sqrt{4 + 2 \cdot 2^{1/3} \cdot 3^{2/3}} = \sqrt{13 + 9 \cdot 2^{2/3} \cdot 3^{1/3} + 6 \cdot 2^{1/3} \cdot 3^{2/3}} \approx 7.02$. See Figure 3-19.

```
In[352]:= Plot[l[x], {x, 0, 20}, PlotRange- > {0, 20},
             AspectRatio- > Automatic,
             AxesLabel- > {"x", "y"}]
```

Our second approach uses right triangle trigonometry. In terms of θ, the length of the beam is given by

$$L(\theta) = 2\csc\theta + 3\sec\theta, \quad 0 < \theta < \pi/2.$$

Differentiating gives us

$$L'(\theta) = -2\csc\theta\cot\theta + 3\sec\theta\tan\theta.$$

To avoid typing the θ symbol, we define L as a function of t.

```
In[353]:= l[t_] = 2 Csc[t] + 3 Sec[t]
```

```
Out[353]= 2 Csc[t] + 3 Sec[t]
```

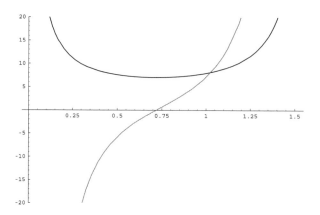

Figure 3-20   Graph of $L(\theta)$ and $L'(\theta)$

We now solve $L'(\theta) = 0$. First multiply through by $\sin\theta$ and then by $\tan\theta$.

$$3\sec\theta\tan\theta = 2\csc\theta\cot\theta$$

$$\tan^2\theta = \frac{2}{3}\cot\theta$$

$$\tan^3\theta = \frac{2}{3}$$

$$\tan\theta = \sqrt[3]{\frac{2}{3}}.$$

In this case, observe that we cannot compute θ exactly. However, we do not need to do so. Let $0 < \theta < \pi/2$ be the unique solution of $\tan\theta = \sqrt[3]{2/3}$. See Figure 3-20. Using the identity $\tan^2\theta + 1 = \sec^2\theta$, we find that $\sec\theta = \sqrt{1 + \sqrt[3]{4/9}}$. Similarly, because $\cot\theta = \sqrt[3]{3/2}$ and $\cot^2\theta + 1 = \csc^2\theta$, $\csc\theta = \sqrt[3]{3/2}\sqrt{1 + \sqrt[3]{4/9}}$. Hence, the length of the beam is

$$L(\theta) = 2\sqrt[3]{\frac{3}{2}}\sqrt{1 + \sqrt[3]{\frac{4}{9}}} + 3\sqrt{1 + \sqrt[3]{\frac{4}{9}}} \approx 7.02.$$

```
In[354] := Plot[{l[t], l'[t]}, {t, 0, π/2},
            PlotRange- > {-20, 20},
            PlotStyle- > {GrayLevel[0], GrayLevel[0.4]}]
```

■

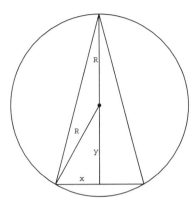

Figure 3-21   Cross-section of a right circular cone inscribed in a sphere

In the next two examples, the constants do not have specific numerical values.

EXAMPLE 3.2.17: Find the volume of the right circular cylinder of maximum volume that can be inscribed in a sphere of radius R.

SOLUTION: Try to avoid three-dimensional figures unless they are absolutely necessary. For this problem, a cross-section of the situation is sufficient. See Figure 3-21, which is created with

```
In[355]:= p1 = ParametricPlot[{Cos[t], Sin[t]}, {t, 0, 2π},
              DisplayFunction → Identity];

          p2 = Graphics[
              {Line[{{0, 1}, {Cos[4π/3], Sin[4π/3]},
                  {Cos[5π/3], Sin[5π/3]}, {0, 1}}],
                  PointSize[0.02], Point[{0, 0}],
                  Line[{{Cos[4π/3], Sin[4π/3]}, {0, 0}, {0, 1}}],
                  Line[{{0, 0}, {0, Sin[4π/3]}}]}];

          p3 = Graphics[{Text["R", {-0.256, -0.28}],
                  Text["R", {-0.04, 0.5}],
                  Text["y", {-0.04, -0.5}],
                  Text["x", {-0.2, -0.8}]}];

          Show[p1, p2, p3, AspectRatio- > Automatic, Ticks- > None,
            Axes- > None, DisplayFunction → $DisplayFunction]
```

The volume, V, of a right circular cone with radius r and height h is $V = \frac{1}{3}\pi r^2 h$. Using the notation in Figure 3-21, the volume is given by

$$V = \frac{1}{3}\pi x^2 (R + y). \tag{3.4}$$

However, by the Pythagorean theorem, $x^2 + y^2 = R^2$ so $x^2 = R^2 - y^2$ and equation (3.4) becomes

$$V = \frac{1}{3}\pi \left(R^2 - y^2 \right)(R + y) = \frac{1}{3}\pi \left(R^3 + R^2 y - Ry^2 - y^3 \right), \tag{3.5}$$

```
In[356] := s1 = Expand[(r^2 - y^2)(r + y)]
Out[356] = r³ + r² y - r y² - y³
```

where $0 \le y \le R$. $V(y)$ is continuous on $[0, R]$ so it will have a minimum and maximum value on this interval. Moreover, the minimum and maximum values either occur at the endpoints of the interval or at the critical numbers on the interior of the interval. Differentiating equation (3.5) with respect to y gives us

Remember that R is a constant.

$$\frac{dV}{dy} = \frac{1}{3}\pi \left(R^2 - 2Ry - 3y^2 \right) = \frac{1}{3}\pi (R - 3y)(R + y)$$

```
In[357] := s2 = D[s1, y]
Out[357] = r² - 2 r y - 3 y²
```

and we see that $dV/dy = 0$ if $y = \frac{1}{3}R$ or $y = -R$.

```
In[358] := Factor[s2]
Out[358] = (r - 3 y) (r + y)
```

```
In[359] := Solve[s2 == 0, y]
Out[359] = {{y → -r}, {y → r/3}}
```

We ignore $y = -R$ because $-R$ is not in the interval $[0, R]$. Note that $V(0) = V(R) = 0$. The maximum volume of the cone is

$$V\left(\frac{1}{3}R\right) = \frac{1}{3}\pi \cdot \frac{32}{27}R^3 = \frac{32}{81}\pi R^2 \approx 1.24R^3.$$

```
In[360] := s3 = s1/.y->r/3//Together
         32 r³
Out[360] = ─────
           27
```

```
In[361] := s3 * 1/3 π
         32 π r³
Out[361] = ───────
            81
```

```
In[362] := N[%]
Out[362] = 1.24112 r³
```

■

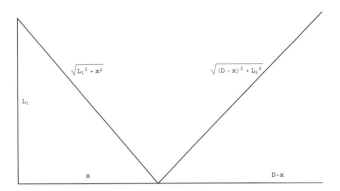

Figure 3-22   When the wire is stayed to minimize the length, the result is two similar triangles

---

EXAMPLE 3.2.18 (The Stayed-Wire Problem): Two poles D feet apart with heights L_1 feet and L_2 feet are to be stayed by a wire as shown in Figure 3-22. Find the minimum amount of wire required to stay the poles, as illustrated in Figure 3-22, which is generated with

```
In[363]:= p1 = Graphics[Line[{{0,0}, {0,4}, {3.5,0},
            {9,5.5}, {9,0}, {0,0}}]];

        p2 = Graphics[{Text[L₁, {0.2,2}],
            Text[L₂, {8.8,2.75}], Text["x", {1.75,0.2}],
            Text["x", {1.75,0.2}],
            Text[√(L₁² + x²), {1.75,2.75}],
            Text[√((D - x)² + L₂²), {5.5,2.75}],
            Text["D - x", {6.5,0.2}]}]

        Show[p1, p2]
```

SOLUTION: Using the notation in Figure 3-22, the length of the wire, L, is

$$L(x) = \sqrt{L_1{}^2 + x^2} + \sqrt{L_2{}^2 + (D - x)^2}, \qquad 0 \le x \le D. \qquad (3.6)$$

In the special case that $L_1 = L_2$, the length of the wire to stay the beams is minimized when the wire is placed halfway between the two beams, at a distance $D/2$ from each beam. Thus, we assume that the lengths of the beams are different; we assume that $L_1 < L_2$, as illustrated in Figure 3-22. We compute $L'(x)$ and then solve $L'(x) = 0$.

```
In[364]:= Clear[1]
          l[x_] = Sqrt[x^2 + 11^2] + Sqrt[(d - x)^2 + 12^2]
```
$$Out[364]= \sqrt{12^2 + (d - x)^2} + \sqrt{11^2 + x^2}$$

```
In[365]:= l'[x]//Together
```
$$Out[365]= \frac{\sqrt{12^2 + (d - x)^2}\, x - d\, \sqrt{11^2 + x^2} + x\, \sqrt{11^2 + x^2}}{\sqrt{12^2 + (d - x)^2}\, \sqrt{11^2 + x^2}}$$

```
In[366]:= 1[0]//PowerExpand
```
$$Out[366]= 11 + \sqrt{d^2 + 12^2}$$

```
In[367]:= 1[d]//PowerExpand
```
$$Out[367]= \sqrt{d^2 + 11^2} + 12$$

```
In[368]:= l'[x]//Together
```
$$Out[368]= \frac{\sqrt{12^2 + (d - x)^2}\, x - d\, \sqrt{11^2 + x^2} + x\, \sqrt{11^2 + x^2}}{\sqrt{12^2 + (d - x)^2}\, \sqrt{11^2 + x^2}}$$

```
In[369]:= Solve[l'[x] == 0, x]
```
$$Out[369]= \left\{\left\{x \to \frac{d\, 11}{11 - 12}\right\}, \left\{x \to \frac{d\, 11}{11 + 12}\right\}\right\}$$

The result indicates that $x = L_1 D/(L_1 + L_2)$ minimizes $L(x)$. Moreover, the triangles formed by minimizing L are similar triangles.

```
In[370]:= 11/(d 11/(11 + 12))//Simplify
```
$$Out[370]= \frac{11 + 12}{d}$$

```
In[371]:= 12/(d - d 11/(11 + 12))//Simplify
```
$$Out[371]= \frac{11 + 12}{d}$$

■

3.2.7 Antidifferentiation

3.2.7.1 Antiderivatives

$F(x)$ is an **antiderivative** of $f(x)$ if $F'(x) = f(x)$. The symbol

$$\int f(x)\, dx$$

means "find all antiderivatives of $f(x)$." Because all antiderivatives of a given function differ by a constant, we usually find an antiderivative, $F(x)$, of $f(x)$ and then write

$$\int f(x)\,dx = F(x) + C,$$

where C represents an arbitrary constant. The command

```
Integrate[f[x],x]
```

attempts to find an antiderivative, $F(x)$, of $f(x)$. Mathematica does not include the "+C" that we include when writing $\int f(x)\,dx = F(x) + C$. In the same way as D can differentiate many functions, Integrate can antidifferentiate many functions. However, antidifferentiation is a fundamentally difficult procedure so it is not difficult to find functions $f(x)$ for which the command Integrate[f[x],x] returns unevaluated.

EXAMPLE 3.2.19: Evaluate each of the following antiderivatives: (a) $\int \frac{1}{x^2} e^{1/x}\,dx$, (b) $\int x^2 \cos x\,dx$, (c) $\int x^2 \sqrt{1+x^2}\,dx$, (d) $\int \frac{x^2 - x + 2}{x^3 - x^2 + x - 1}\,dx$, and (e) $\int \frac{\sin x}{x}\,dx$.

SOLUTION: Entering

```
In[372]:= Integrate[1/x^2 Exp[1/x],x]
```
$$Out[372]= -e^{\frac{1}{x}}$$

shows us that $\int \frac{1}{x^2} e^{1/x}\,dx = -e^{1/x} + C$. Notice that Mathematica does not automatically include the arbitrary constant, C. When computing several antiderivatives, you can use Map to apply Integrate to a list of antiderivatives. However, because Integrate is threadable,

```
Map[Integrate[#,x]&,list]
```

returns the same result as Integrate[list,x], which we illustrate to compute (b), (c), and (d).

```
In[373]:= Integrate[{x^2 Cos[x], x^2 Sqrt[1+x^2],
             (x^2-x+2)/(x^3-x^2+x-1)}, x]
```
$$Out[373]= \Big\{2\,x\,\text{Cos}[x] - 2\,\text{Sin}[x] + x^2\,\text{Sin}[x],$$
$$\sqrt{1+x^2}\,\Big(\frac{x}{8} + \frac{x^3}{4}\Big) - \frac{\text{ArcSinh}[x]}{8},$$
$$-\,\text{ArcTan}[x] + \text{Log}[-1+x]\Big\}$$

For (e), we see that there is not a "closed form" antiderivative of $\int \frac{\sin x}{x} dx$ and the result is given in terms of a definite integral, the **sine integral function**:

$$Si(x) = \int_0^x \frac{\sin t}{t} dt.$$

```
In[374]:= Integrate[Sin[x]/x, x]
Out[374]= SinIntegral[x]
```

■

u-Substitutions

Usually, the first antidifferentiation technique discussed is the method of *u*-**substitution**. Suppose that $F(x)$ is an antiderivative of $f(x)$. Given

$$\int f(g(x)) g'(x) dx,$$

we let $u = g(x)$ so that $du = g'(x) dx$. Then,

$$\int f(g(x)) g'(x) dx = \int f(u) du = F(u) + C = F(g(x)) + C,$$

where $F(x)$ is an antiderivative of $f(x)$. After mastering *u*-substitutions, the **integration by parts formula**,

$$\int u \, dv = uv - \int v \, du, \tag{3.7}$$

is introduced.

EXAMPLE 3.2.20: Evaluate $\int 2^x \sqrt{4^x - 1} \, dx$.

SOLUTION: We use Integrate to evaluate the antiderivative. Notice that the result is *very* complicated.

```
In[375]:= Integrate[2^x Sqrt[4^x - 1], x]
```
$$Out[375]= \frac{2^{1+x} \sqrt{-1 + 4^x}}{2 \text{ Log}[2] + \text{Log}[4]} - \left(2^x \sqrt{1 - 4^x} \text{ Hypergeometric2F1}\left[\frac{1}{2}, \frac{\text{Log}[2]}{\text{Log}[4]}, 1 + \frac{\text{Log}[2]}{\text{Log}[4]}, 4^x\right] \text{ Log}[4]\right) / \left(\sqrt{-1 + 4^x} \text{ Log}[2] \ (2 \text{ Log}[2] + \text{Log}[4])\right)$$

Proceeding by hand, we let $u = 2^x$. Then, $du = 2^x \ln 2 \, dx$ or, equivalently, $\frac{1}{\ln 2} du = 2^x \, dx$

```
In[376]:= D[2^x, x]
```

$$Out[376] = 2^x \ Log[2]$$

so $\int 2^x \sqrt{4^x - 1} \, dx = \frac{1}{\ln 2} \int \sqrt{u^2 - 1} \, du$. We now use $\texttt{Integrate}$ to evaluate $\int \sqrt{u^2 - 1} \, du$

```
In[377]:= s1 = Integrate[Sqrt[u^2 - 1], u]
```

$$Out[377] = \frac{1}{2} \ u \ \sqrt{-1 + u^2} - \frac{1}{2} \ Log \left[u + \sqrt{-1 + u^2} \right]$$

and then $\texttt{/.}$ $(\texttt{ReplaceAll})$/ to replace u with 2^x.

```
In[378]:= s1 /.u->2^x
```

$$Out[378] = 2^{-1+x} \ \sqrt{-1 + 2^{2 \ x}} - \frac{1}{2} \ Log \left[2^x + \sqrt{-1 + 2^{2 \ x}} \right]$$

Clearly, proceeding by hand results in a significantly simpler antiderivative than using $\texttt{Integrate}$ directly.

■

3.3 Integral Calculus

3.3.1 Area

In integral calculus courses, the definite integral is frequently motivated by investigating the area under the graph of a positive continuous function on a closed interval. Let $y = f(x)$ be a nonnegative continuous function on an interval $[a, b]$ and let n be a positive integer. If we divide $[a, b]$ into n subintervals of equal length and let $[x_{k-1}, x_k]$ denote the kth subinterval, the length of each subinterval is $(b - a)/n$ and $x_k = a + k\frac{b-a}{n}$. The area bounded by the graphs of $y = f(x)$, $x = a$, $x = b$, and the y-axis can be approximated with the sum

$$\sum_{k=1}^{n} f(x_k^*) \frac{b - a}{n}, \tag{3.8}$$

where $x_k^* \in [x_{k-1}, x_k]$. Typically, we take $x_k^* = x_{k-1} = a + (k-1)\frac{b-a}{n}$ (the left endpoint of the kth subinterval), $x_k^* = x_{k-1} = a + k\frac{b-a}{n}$ (the right endpoint of the kth subinterval), or $x_k^* = \frac{1}{2}(x_{k-1} + x_k) = a + \frac{1}{2}(2k - 1)\frac{b-a}{n}$ (the midpoint of the kth subinterval).

For these choices of $x_k{}^*$, (3.8) becomes

$$\frac{b-a}{n}\sum_{k=1}^{n}f\left(a+(k-1)\frac{b-a}{n}\right) \tag{3.9}$$

$$\frac{b-a}{n}\sum_{k=1}^{n}f\left(a+k\frac{b-a}{n}\right), \text{ and} \tag{3.10}$$

$$\frac{b-a}{n}\sum_{k=1}^{n}f\left(a+\frac{1}{2}(2k-1)\frac{b-a}{n}\right), \tag{3.11}$$

respectively. If $y = f(x)$ is increasing on $[a, b]$, (3.9) is an under approximation and (3.10) is an upper approximation: (3.9) corresponds to an approximation of the area using n inscribed rectangles; (3.10) corresponds to an approximation of the area using n circumscribed rectangles. If $y = f(x)$ is decreasing on $[a, b]$, (3.10) is an under approximation and (3.9) is an upper approximation: (3.10) corresponds to an approximation of the area using n inscribed rectangles; (3.9) corresponds to an approximation of the area using n circumscribed rectangles.

In the following example, we define the functions `leftsum[f[x],a,b,n]`, `middlesum[f[x],a,b,n]`, and `rightsum[f[x],a,b,n]` to compute (3.9), (3.11), and (3.10), respectively, and `leftbox[f[x],a,b,n]`, `middlebox[f[x], a,b,n]`, and `rightbox[f[x],a,b,n]` to generate the corresponding graphs. After you have defined these functions, you can use them with functions $y = f(x)$ that you define.

Remark. To define a function of a single variable, $f(x) = expression\ in\ x$, enter `f[x_]=expression in x`. To generate a basic plot of $y = f(x)$ for $a \le x \le b$, enter `Plot[f[x],{x,a,b}]`.

EXAMPLE 3.3.1: Let $f(x) = 9 - 4x^2$. Approximate the area bounded by the graph of $y = f(x)$, $x = 0$, $x = 3/2$, and the y-axis using (a) 100 inscribed and (b) 100 circumscribed rectangles. (c) What is the exact value of the area?

SOLUTION: We begin by defining and graphing $y = f(x)$ in Figure 3-23.

```
In[379]:= f[x_] = 9 - 4x^2;
          Plot[f[x], {x, 0, 3/2}]
```

The first derivative, $f'(x) = -8x$ is negative on the interval so $f(x)$ is decreasing on $[0, 3/2]$. Thus, an approximation of the area using 100

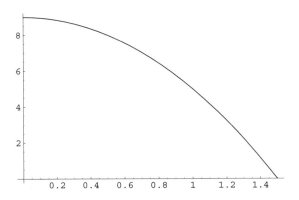

Figure 3-23 $f(x)$ for $0 \le x \le 3/2$

inscribed rectangles is given by (3.10) while an approximation of the area using 100 circumscribed rectangles is given by (3.9). After defining leftsum, rightsum, and middlesum, these values are computed using leftsum and rightsum. The use of middlesum is illustrated as well. Approximations of the sums are obtained with N.

N[number] returns a numerical approximation of number.

```
In[380]:= leftsum[f_, a_, b_, n_] := Module[{},
            (b - a)/n Sum[f/.x- > a + (k - 1) (b - a)/n,
            {k, 1, n}]];

            rightsum[f_, a_, b_, n_] := Module[{},
            (b - a)/n Sum[f/.x- > a + k (b - a)/n, {k, 1, n}]];

            middlesum[f_, a_, b_, n_] := Module[{},
            (b - a)/n Sum[f/.x- > a + 1/2 (2k - 1) (b - a)/n,
            {k, 1, n}]];

In[381]:= l100 = leftsum[f[x], 0, 3/2, 100]

            N[l100]

            r100 = rightsum[f[x], 0, 3/2, 100]

            N[r100]

            m100 = middlesum[f[x], 0, 3/2, 100]

            N[m100]
```

$$Out[381] = \frac{362691}{40000}$$
$$Out[381] = 9.06728$$
$$Out[381] = \frac{357291}{40000}$$

Out [381]= 8.93228

$$Out [381]= \frac{720009}{80000}$$

Out [381]= 9.00011

Observe that these three values appear to be close to 9. In fact, 9 is the exact value of the area of the region bounded by $y = f(x)$, $x = 0$, $x = 3/2$, and the y-axis. To help us see why this is true, we define leftbox, middlebox, and rightbox, and then use these functions to visualize the situation using $n = 4$, 16, and 32 rectangles in Figure 3-24.

It is not important that you understand the syntax of these three functions at this time. Once you have entered the code, you can use them to visualize the process for your own functions, $y = f(x)$.

```
In[382]:= leftbox[f_, a_, b_, n_, opts___] :=
        Module[{z, p1, recs, ls},
        z[k_] = a + (b - a)k/n;
        p1 = Plot[f, {x, a, b},
            PlotStyle- > {{Thickness[0.01],
                GrayLevel[0.3]}},
            DisplayFunction- > Identity];
        recs = Table[Rectangle[
            {z[k - 1], 0}, {z[k], f/.x- > z[k - 1]}], {k, 1, n}];
        ls = Table[Line[{{z[k - 1], 0}, {z[k - 1],
            f/.x- > z[k - 1]}, {z[k], f/.x- > z[k - 1]},
            {z[k], 0}}], {k, 1, n}];
        Show[Graphics[{GrayLevel[0.8], recs}],
            Graphics[ls], p1, opts, Axes- > Automatic,
            DisplayFunction- > $DisplayFunction]]
```

```
In[383]:= rightbox[f_, a_, b_, n_, opts___] :=
        Module[{z, p1, recs, ls},
        z[k_] = a + (b - a)k/n;
        p1 = Plot[f, {x, a, b},
            PlotStyle- > {{Thickness[0.01],
                GrayLevel[0.3]}},
            DisplayFunction- > Identity];
        recs = Table[Rectangle[
            {z[k - 1], 0}, {z[k], f/.x- > z[k]}], {k, 1, n}];
        ls = Table[Line[{{z[k - 1], 0}, {z[k - 1],
            f/.x- > z[k]}, {z[k], f/.x- > z[k]},
            {z[k], 0}}], {k, 1, n}];
        Show[Graphics[{GrayLevel[0.8], recs}],
            Graphics[ls], p1, opts, Axes- > Automatic,
            DisplayFunction- > $DisplayFunction]]
```

```
In[384]:= middlebox[f_, a_, b_, n_, opts___] :=
            Module[{z, p1, recs, ls},
          z[k_] = a + (b - a) k/n;
          p1 = Plot[f, {x, a, b},
               PlotStyle- > {{Thickness[0.01],
                GrayLevel[0.3]}},
               DisplayFunction- > Identity];
          recs = Table[Rectangle[{z[k - 1], 0},
               {z[k], f/.x- > 1/2 (z[k - 1] + z[k])}], {k, 1, n}];
          ls = Table[Line[
               {{z[k - 1], 0}, {z[k - 1], f/.x- > 1/2 (z[k - 1] +
               z[k])}, {z[k], f/.x- > 1/2 (z[k - 1] + z[k])},
               {z[k], 0}}], {k, 1, n}];
          Show[Graphics[{GrayLevel[0.8], recs}],
               Graphics[ls], p1, opts, Axes- > Automatic,
               DisplayFunction- > $DisplayFunction]]

In[385]:= somegraphs = {{leftbox[f[x], 0, 3/2, 4,
               DisplayFunction- > Identity],
               middlebox[f[x], 0, 3/2, 4,
               DisplayFunction- > Identity],
               rightbox[f[x], 0, 3/2, 4,
               DisplayFunction- > Identity]},
              {leftbox[f[x], 0, 3/2, 16,
               DisplayFunction- > Identity],
               middlebox[f[x], 0, 3/2, 16,
               DisplayFunction- > Identity],
               rightbox[f[x], 0, 3/2, 16,
               DisplayFunction- > Identity]},
              {leftbox[f[x], 0, 3/2, 32,
               DisplayFunction- > Identity],
               middlebox[f[x], 0, 3/2, 32,
               DisplayFunction- > Identity],
               rightbox[f[x], 0, 3/2, 32,
               DisplayFunction- > Identity]}};
            Show[GraphicsArray[somegraphs]]
```

Notice that as n increases, the under approximations increase while the upper approximations decrease.

These graphs help convince us that the limit of the sum as $n \rightarrow \infty$ of the areas of the inscribed and circumscribed rectangles is the same. We compute the exact value of (3.9) with leftsum, evaluate and simplify the sum with Simplify, and compute the limit as $n \rightarrow \infty$ with Limit. We see that the limit is 9.

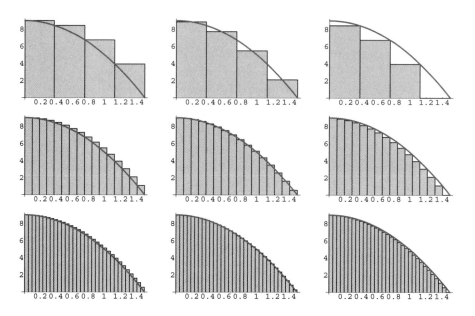

Figure 3-24 $f(x)$ with 4, 16, and 32 rectangles

$In[386]:=$ **ls = leftsum[f[x], 0, 3/2, n]**

ls2 = Simplify[ls]

Limit[ls2, n- > ∞]

$Out[386]=$ $-\dfrac{27\left(n-n^3-n\ (1+n)+\frac{1}{6}\ n\ (1+n)\ (1+2\ n)\right)}{2\ n^3}$

$Out[386]=$ $\dfrac{9\ (-1+3\ n+4\ n^2)}{4\ n^2}$

$Out[386]=$ 9

Similar calculations are carried out for (3.10) and again we see that the limit is 9. We conclude that the exact value of the area is 9.

$In[387]:=$ **rs = rightsum[f[x], 0, 3/2, n]**

rs2 = Simplify[rs]

Limit[rs2, n- > ∞]

$Out[387]=$ $-\dfrac{27\left(-n^3+\frac{1}{6}\ n\ (1+n)\ (1+2\ n)\right)}{2\ n^3}$

$Out[387]=$ $\dfrac{9\ (-1-3\ n+4\ n^2)}{4\ n^2}$

$Out[387]=$ 9

For illustrative purposes, we confirm this result with `middlesum`.

```
In[388]:= ms = middlesum[f[x],0,3/2,n]

        ms2 = Simplify[ms]

        Limit[ms2,n->∞]
```

$$Out[388] = -\frac{27 \left(n - 4\ n^3 - 2\ n\ (1+n) + \frac{2}{3}\ n\ (1+n)\ (1+2\ n)\right)}{8\ n^3}$$

$$Out[388] = 9 + \frac{9}{8\ n^2}$$

$$Out[388] = 9$$

∎

3.3.2 The Definite Integral

In integral calculus courses, we formally learn that the **definite integral** of the function $y = f(x)$ from $x = a$ to $x = b$ is

$$\int_a^b f(x)\,dx = \lim_{|P|\to 0} \sum_{k=1}^n f(x_k^*)\,\Delta x_k, \tag{3.12}$$

provided that the limit exists. In equation (3.12), $P = \{a = x_0 < x_1 < x_2 < \cdots < x_n = b\}$ is a partition of $[a, b]$, $|P|$ is the **norm** of P,

$$|P| = \max\{x_k - x_{k-1}|k = 1, 2, \ldots, n\},$$

$\Delta x_k = x_k - x_{k-1}$, and $x_k^* \in [x_{k-1}, x_k]$.

The *Fundamental Theorem of Calculus* provides the fundamental relationship between differentiation and integration.

Theorem 6 (The Fundamental Theorem of Calculus). *Suppose that $y = f(x)$ is continuous on $[a, b]$.*

1. *If $F(x) = \int_a^x f(t)\,dt$, then F is an antiderivative of f: $F'(x) = f(x)$.*
2. *If G is any antiderivative of f, then $\int_a^b f(x)\,dx = G(b) - G(a)$.*

Mathematica's `Integrate` command can compute many definite integrals. The command

$$\text{Integrate}[\text{f}[\text{x}],\{\text{x},\text{a},\text{b}\}]$$

attempts to compute $\int_a^b f(x)\,dx$. Because integration is a fundamentally difficult procedure, it is easy to create integrals for which the exact value cannot be found explicitly. In those cases, use N to obtain an approximation of its value or obtain a numerical approximation of the integral directly with

$$\text{NIntegrate}[f[x],\{x,a,b\}].$$

EXAMPLE 3.3.2: Evaluate (a) $\int_1^4 (x^2+1)/\sqrt{x}\,dx$; (b) $\int_0^{\sqrt{\pi/2}} x\cos x^2\,dx$; (c) $\int_0^\pi e^{2x}\sin^2 2x\,dx$; (d) $\int_0^1 \frac{2}{\sqrt{\pi}}e^{-x^2}\,dx$; and (e) $\int_{-1}^0 \sqrt[3]{u}\,du$.

SOLUTION: We evaluate (a)-(c) directly with Integrate.

```
In[389] := Integrate[(x^2 + 1)/Sqrt[x], {x, 1, 4}]
```

$$Out[389] = \frac{72}{5}$$

```
In[390] := Integrate[x Cos[x^2], {x, 0, Sqrt[π/2]}]
```

$$Out[390] = \frac{1}{2}$$

```
In[391] := Integrate[Exp[2x] Sin[2x]^2, {x, 0, π}]
```

$$Out[391] = -\frac{1}{5} + \frac{e^{2\pi}}{5}$$

For (d), the result returned is in terms of the **error function**, Erf[x], which is defined by the integral

$$\text{Erf}[x] = \frac{2}{\sqrt{\pi}}\int_0^x e^{-t^2}\,dt.$$

```
In[392] := Integrate[2/Sqrt[π] Exp[-x^2], {x, 0, 1}]
Out[392] = Erf[1]
```

We use N to obtain an approximation of the value of the definite integral.

```
In[393] := Integrate[2/Sqrt[π] Exp[-x^2], {x, 0, 1}]//N
Out[393] = 0.842701
```

(e) Recall that Mathematica does not return a real number when we compute odd roots of negative numbers so the following result would be surprising to many students in an introductory calculus course because it is complex.

See Chapter 2, Example 2.1.3.

```
In[394] := Integrate[u^(1/3), {u, -1, 0}]
```

$$Out[394] = \frac{3}{4}\,(-1)^{1/3}$$

Therefore, we load the **RealOnly** package contained in the **Miscellaneous** directory so that Mathematica returns the real-valued third root of *u*.

$In[395]:= <<$ **Miscellaneous`RealOnly`**

$In[396]:=$ **Integrate[u^(1/3),{u,-1,0}]**

$Out[396]= -\dfrac{3}{4}$

∎

Improper integrals are computed using `Integrate` in the same way as other definite integrals.

EXAMPLE   3.3.3:   Evaluate   (a) $\int_0^1 \dfrac{\ln x}{\sqrt{x}}\,dx$;   (b) $\int_{-\infty}^{\infty} \dfrac{2}{\sqrt{\pi}} e^{-x^2}\,dx$;

(c) $\int_1^{\infty} \dfrac{1}{x\sqrt{x^2-1}}\,dx$;   (d) $\int_0^{\infty} \dfrac{1}{x^2+x^4}\,dx$;   (e) $\int_2^4 \dfrac{1}{\sqrt[3]{(x-3)^2}}\,dx$;   and

(f) $\int_{-\infty}^{\infty} \dfrac{1}{x^2+x-6}\,dx$.

SOLUTION: (a) This is an improper integral because the integrand is discontinuous on the interval [0, 1] but we see that the improper integral converges to −4.

$In[397]:=$ **Integrate[Log[x]/Sqrt[x],{x,0,1}]**

$Out[397]= -4$

(b) This is an improper integral because the interval of integration is infinite but we see that the improper integral converges to 2.

$In[398]:=$ **Integrate[2/Sqrt[π] Exp[-x^2],{x,-∞,∞}]**

$Out[398]= 2$

(c) This is an improper integral because the integrand is discontinuous on the interval of integration and because the interval of integration is infinite but we see that the improper integral converges to $\pi/2$.

$In[399]:=$ **Integrate[1/(x Sqrt[x^2 - 1]),{x,1,∞}]**

$Out[399]= \dfrac{\pi}{2}$

(d) As with (c), this is an improper integral because the integrand is discontinuous on the interval of integration and because the interval of integration is infinite but we see that the improper integral diverges to ∞.

```
Integrate[1/(x^2+x^4), {x, 0, Infinity}]

Integrate::idiv : Integral of  1/(x²+x⁴)  does not converge on {0, ∞}.

∫₀^∞ 1/(x²+x⁴) dx
```

(e) Recall that Mathematica does not return a real number when we compute odd roots of negative numbers so the following result would be surprising to many students in an introductory calculus course because it contains imaginary numbers.

$$In[400] := \textbf{Integrate[1/(x - 3)\^(2/3), \{x, 2, 4\}]}$$

$$Out[400] = 3 - 3 \ (-1)^{1/3}$$

You do not need to reload the **RealOnly** package if you have already loaded it during your *current* Mathematica session.

Therefore, we load the **RealOnly** package contained in the **Miscellaneous** directory so that Mathematica returns the real-valued third root of $x - 3$.

$$In[401] := \textbf{<< Miscellaneous`RealOnly`}$$

$$In[402] := \textbf{Integrate[1/(x - 3)\^(2/3), \{x, 2, 4\}]}$$

$$Out[402] = 6$$

(f) In this case, Mathematica warns us that the improper integral diverges.

```
s1 = Integrate[1/(x^2+x-6), {x, -Infinity, Infinity}]

Integrate::idiv : Integral of  1/(-6+x+x²)  does not converge on {-∞, ∞}.

∫₋∞^∞ 1/(-6+x+x²) dx
```

To help us understand why the improper integral diverges, we note that $\frac{1}{x^2+x-6} = \frac{1}{5}\left(\frac{1}{x-2} - \frac{1}{x+3}\right)$ and

$$\int \frac{1}{x^2 + x - 6}\,dx = \int \frac{1}{5}\left(\frac{1}{x - 2} - \frac{1}{x + 3}\right)dx = \frac{1}{5}\ln\left(\frac{x - 2}{x + 3}\right) + C.$$

$$In[403] := \textbf{Integrate[1/(x\^2 + x - 6), x]}$$

$$Out[403] = \frac{1}{5} \, \text{Log}[-2 + x] - \frac{1}{5} \, \text{Log}[3 + x]$$

Hence the integral is improper because the interval of integration is infinite and because the integrand is discontinuous on the interval of integration so

$$\int_{-\infty}^{\infty} \frac{1}{x^2 + x - 6} \, dx = \int_{-\infty}^{-4} \frac{1}{x^2 + x - 6} \, dx + \int_{-4}^{-3} \frac{1}{x^2 + x - 6} \, dx$$
$$+ \int_{-3}^{0} \frac{1}{x^2 + x - 6} \, dx + \int_{0}^{2} \frac{1}{x^2 + x - 6} \, dx \qquad (3.13)$$
$$+ \int_{2}^{3} \frac{1}{x^2 + x - 6} \, dx + \int_{3}^{\infty} \frac{1}{x^2 + x - 6} \, dx$$

Evaluating each of these integrals,

```
Integrate[1/(x^2 + x - 6), x]

1/5 Log[-2 + x] - 1/5 Log[3 + x]

Integrate[1/(x^2 + x - 6), {x, -Infinity, -4}]

Integrate::idiv : Integral of ──────── does not converge on {-∞, -4}.
                               -6 + x + x²

∫  ──────── dx
-4 -6 + x + x²
-∞

Integrate[1/(x^2 + x - 6), {x, -4, -3}]

Integrate::idiv : Integral of ──────── does not converge on {-4, -3}.
                               -6 + x + x²

∫  ──────── dx
-3 -6 + x + x²
-4

Integrate[1/(x^2 + x - 6), {x, -3, 0}]

Integrate::idiv : Integral of ──────── does not converge on {-3, 0}.
                               -6 + x + x²

∫  ──────── dx
0 -6 + x + x²
-3

Integrate[1/(x^2 + x - 6), {x, 0, 2}]

Integrate::idiv : Integral of ──────── does not converge on {0, 2}.
                               -6 + x + x²

∫  ──────── dx
2 -6 + x + x²
0
```

```
Integrate[1/(x^2 + x - 6), {x, 2, 3}]

Integrate::idiv : Integral of ──────── does not converge on {2, 3}.
                               -6 + x + x²

∫  ──────── dx
3 -6 + x + x²
2

Integrate[1/(x^2 + x - 6), {x, 3, Infinity}]

Integrate::idiv : Integral of ──────── does not converge on {3, ∞}.
                               -6 + x + x²

∫  ──────── dx
∞ -6 + x + x²
3
```

we conclude that the improper integral diverges because at least one of the improper integrals in (3.13) diverges.

∎

In many cases, Mathematica can help illustrate the steps carried out when computing integrals using standard methods of integration like u-substitutions and integration by parts.

EXAMPLE 3.3.4: Evaluate (a) $\int_e^{e^3} \dfrac{1}{x\sqrt{\ln x}}\,dx$ and (b) $\int_0^{\pi/4} x\sin 2x\,dx$.

SOLUTION: (a) We let $u = \ln x$. Then, $du = \frac{1}{x}dx$ so $\int_e^{e^3} \frac{1}{x\sqrt{\ln x}}\,dx = \int_1^3 \frac{1}{\sqrt{u}}du = \int_1^3 u^{-1/2}du$, which we evaluate with `Integrate`.

The new lower limit of integration is 1 because if $x = e$, $u = \ln e = 1$. The new upper limit of integration is 3 because if $x = e^3$, $u = \ln e^3 = 3$.

```
In[404]:= Integrate[1/Sqrt[u], {u, 1, 3}]
Out[404]= -2 + 2 √3
```

To evaluate (b), we let $u = x \Rightarrow du = dx$ and $dv = \sin 2x\,dx \Rightarrow v = -\frac{1}{2}\cos 2x$.

```
In[405]:= u = x;
          dv = Sin[2x];

In[406]:= du = D[x, x]

          v = Integrate[Sin[2x], x]
Out[406]= 1
             1
Out[406]= - ─  Cos[2 x]
             2

In[407]:= v du
             1
Out[407]= - ─  Cos[2 x]
             2
```

The results mean that

$$\int_0^{\pi/4} x\sin 2x\,dx = -\frac{1}{2}x\cos 2x\Big|_0^{\pi/4} + \frac{1}{2}\int_0^{\pi/4}\cos 2x\,dx$$

$$= 0 + \frac{1}{2}\int_0^{\pi/4}\cos 2x\,dx.$$

The resulting indefinite integral is evaluated with `Integrate`

```
In[408]:= u v - Integrate[v du, x]
             1                1
Out[408]= - ─  x  Cos[2 x] + ─  Sin[2 x]
             2                4

In[409]:= Integrate[x Sin[2x], x]
             1
Out[409]= ─  (-2 x  Cos[2 x] + Sin[2 x])
             4
```

and the definite integral is evaluated with `Integrate`.

```
Integrate[x Sin[2 x], {x, 0, Pi / 4}]
 1
 ─
 4
```

■

3.3.3 Approximating Definite Integrals

Because integration is a fundamentally difficult procedure, Mathematica is unable to compute a "closed form" of the value of many definite integrals. In these cases, numerical integration can be used to obtain an approximation of the definite integral using N together with Integrate or NIntegrate:

$$\text{NIntegrate}[f[x],\{x,a,b\}]$$

attempts to approximate $\int_a^b f(x)\,dx$.

EXAMPLE 3.3.5: Evaluate $\int_0^{\sqrt[3]{\pi}} e^{-x^2} \cos x^3\,dx$.

SOLUTION: In this case, Mathematica is unable to evaluate the integral with Integrate.

```
In[410] := i1 = Integrate[Exp[-x^2] Cos[x^3],
             {x, 0, π^(1/3)}]
```
$$Out[410] = \int_0^{\pi^{1/3}} e^{-x^2} \; Cos[x^3]\,dx$$

An approximation is obtained with N.

```
In[411] := N[i1]
Out[411] = 0.701566
```

Instead of using Integrate followed by N, you can use NIntegrate to numerically evaluate many integrals.

$$\text{NIntegrate}[f[x],\{x,a,b\}]$$

attempts to approximate $\int_a^b f(x)\,dx$. Thus, entering

```
In[412] := NIntegrate[Exp[-x^2] Cos[x^3], {x, 0, π^(1/3)}]
Out[412] = 0.701566
```

returns the same result as that obtained using Integrate followed by N.

■

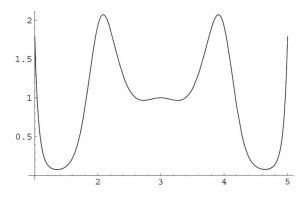

Figure 3-25 $f(x)$ for $1 \le x \le 5$

In some cases, you may wish to investigate particular numerical methods that can be used to approximate integrals. If needed you can redefine the functions leftsum, middlesum, and rightsum that were discussed previously. In addition we define the functions simpson, which implements Simpson's rule, and trapezoid, which implements the trapezoidal rule, in the following example that can be used to investigate approximations of definite integrals using those numerical methods.

EXAMPLE 3.3.6: Let $f(x) = e^{-(x-3)^2 \cos(4(x-3))}$. (a) Graph $y = f(x)$ on the interval $[1, 5]$. Use (b) Simpson's rule with $n = 4$, (c) the trapezoidal rule with $n = 4$, and (d) the midpoint rule with $n = 4$ to approximate $\int_1^5 f(x)\,dx$.

SOLUTION: We define f, and then graph $y = f(x)$ on the interval $[1, 5]$ with Plot in Figure 3-25.

```
In[413]:= f[x_] = Exp[-(x - 3)^2 Cos[4 (x - 3)]];
          Plot[f[x], {x, 1, 5}]
```

After defining simpson and trapezoid,

```
In[414]:= simpson[f_, a_, b_, n_] := Module[{z, h},
          h = (b - a)/n; z[k_] = a + h k; f0 = f/.x- > z[0];
          fn = f/.x- > z[n]; h/3 (f0 + fn) +
            h/3 Sum[(3 + (-1)^(k + 1))f /. x- > z[k],
            {k, 1, n - 1}]
          ]
```

```
In[415]:= trapezoid[f_, a_, b_, n_] := Module[{z, h},
          h = (b - a)/n; z[k_] = a + h k;
          f0 = f/.x- > z[0]; fn = f/.x- > z[n];
          h/2 (f0 + fn) + h Sum[f /. x- > z[k],
            {k, 1, n - 1}]
          ]
```

we use these functions and `middlesum`, which was defined earlier, to approximate $\int_1^5 f(x)\,dx$ using $n = 4$ rectangles. In each case, N is used to evaluate the sum.

> Be sure to redefine `middlesum` if you have not already used it during your current *Mathematica* session before executing the following commands.

```
In[416]:= s1 = simpson[f[x], 1, 5, 4]

          N[s1]

          t1 = trapezoid[f[x], 1, 5, 4]

          N[t1]

          m1 = middlesum[f[x], 1, 5, 4]

          N[m1]
```
$$Out[416]= \frac{2}{3}\ e^{-4\ Cos[8]} + \frac{1}{3}\ \left(2 + 8\ e^{-Cos[4]}\right)$$
$$Out[416]= 6.9865$$
$$Out[416]= 1 + 2\ e^{-Cos[4]} + e^{-4\ Cos[8]}$$
$$Out[416]= 6.63468$$
$$Out[416]= 2\ e^{-\frac{Cos[2]}{4}} + 2\ e^{-\frac{9\ Cos[6]}{4}}$$
$$Out[416]= 2.44984$$

We obtain an accurate approximation of the value of the integral using `NIntegrate`.

```
In[417]:= NIntegrate[f[x], {x, 1, 5}]
Out[417]= 3.761
```

Notice that with $n = 4$ rectangles, the midpoint rule gives the best approximation. However, as n increases, Simpson's rule gives a better approximation as we see using $n = 50$ rectangles.

```
In[418]:= simpson[f[x], 1, 5, 50]//N
          trapezoid[f[x], 1, 5, 50]//N
          middlesum[f[x], 1, 5, 50]//N
Out[418]= 3.76445
Out[418]= 3.7913
Out[418]= 3.74623
```

∎

3.3.4 Area

Suppose that $y = f(x)$ and $y = g(x)$ are continuous on $[a, b]$ and that $f(x) \geq g(x)$ for $a \leq x \leq b$. The **area** of the region bounded by the graphs of $y = f(x)$, $y = g(x)$, $x = a$, and $x = b$ is

$$A = \int_a^b [f(x) - g(x)] \, dx. \tag{3.14}$$

EXAMPLE 3.3.7: Find the area between the graphs of $y = \sin x$ and $y = \cos x$ on the interval $[0, 2\pi]$.

SOLUTION: We graph $y = \sin x$ and $y = \cos x$ on the interval $[0, 2\pi]$ in Figure 3-26. The graph of $y = \cos x$ is gray.

```
In[419]:= Plot[{Sin[x],Cos[x]},{x,0,2π},
            PlotStyle->{GrayLevel[0],GrayLevel[0.3]},
            AspectRatio->Automatic]
```

To find the upper and lower limits of integration, we must solve the equation $\sin x = \cos x$ for x.

```
In[420]:= Solve[Sin[x] == Cos[x],x]

Solve :: "ifun" :  "Inversefunctionsarebeingused
      bySolve, sosomesolutionsmaynotbefound."

Set :: "write" :  "TagPowerinDownValues[
      Power]isProtected."
```

$$Out[420] = \left\{ \left\{ x \to -\frac{3\,\pi}{4} \right\}, \left\{ x \to \frac{\pi}{4} \right\} \right\}$$

Thus, for $0 \leq x \leq 2\pi$, $\sin x = \cos x$ if $x = \pi/4$ or $x = 5\pi/4$. Hence, the area of the region between the graphs is given by

$$A = \int_0^{\pi/4} [\cos x - \sin x] \, dx + \int_{\pi/4}^{5\pi/4} [\sin x - \cos x] \, dx + \int_{5\pi/4}^{2\pi} [\cos x - \sin x] \, dx. \tag{3.15}$$

Notice that if we take advantage of symmetry we can simplify (3.15) to

$$A = 2 \int_{\pi/4}^{5\pi/4} [\sin x - \cos x] \, dx. \tag{3.16}$$

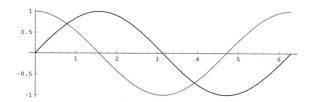

Figure 3-26 $y = \sin x$ and $y = \cos x$ on the interval $[0, 2\pi]$

We evaluate (3.16) with `Integrate` to see that the area is $4\sqrt{2}$.

```
In[421]:= 2 Integrate[Sin[x] - Cos[x], {x, π/4, 5π/4}]
Out[421]= 4 √2
```

■

In cases when we cannot calculate the points of intersection of two graphs exactly, we can frequently use `FindRoot` to approximate the points of intersection.

EXAMPLE 3.3.8: Let

$$p(x) = \frac{3}{10}x^5 - 3x^4 + 11x^3 - 18x^2 + 12x + 1$$

and

$$q(x) = -4x^3 + 28x^2 - 56x + 32.$$

Approximate the area of the region bounded by the graphs of $y = p(x)$ and $y = q(x)$.

SOLUTION: After defining p and q, we graph them on the interval $[-1, 5]$ in Figure 3-27 to obtain an initial guess of the intersection points of the two graphs.

```
In[422]:= p[x_] = 3/10x^5 - 3x^4 + 11x^3 - 18x^2 + 12x + 1;
          q[x_] = -4x^3 + 28x^2 - 56x + 32;
          Plot[{p[x], q[x]}, {x, -1, 5}, PlotRange- > {-15, 20},
            PlotStyle- > {GrayLevel[0], GrayLevel[0.3]}]
```

The x-coordinates of the three intersection points are the solutions of the equation $p(x) = q(x)$. Although Mathematica can solve this equation exactly, approximate solutions are more useful for the problem and obtained with `FindRoot`.

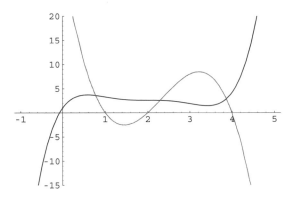

Figure 3-27 p and q on the interval $[-1, 5]$

```
In[423]:= FindRoot[p[x] == q[x], {x, 1}]

          FindRoot[p[x] == q[x], {x, 2}]

          FindRoot[p[x] == q[x], {x, 4}]
Out[423]= {x → 0.772058}

Out[423]= {x → 2.29182}

Out[423]= {x → 3.86513}
```

All three FindRoot commands can be combined together if we use Map as illustrated next.

```
In[424]:= intpts = Map[FindRoot[p[x] == q[x],
              {x, #}]&, {1, 2, 4}]
Out[424]= {{x → 0.772058}, {x → 2.29182},
              {x → 3.86513}}

In[425]:= intpts[[1, 1, 2]]

Out[425]= 0.772058
```

Using the roots to the equation $p(x) = q(x)$ and the graph we see that $p(x) \geq q(x)$ for $0.772 \leq x \leq 2.292$ and $q(x) \geq p(x)$ for $2.292 \leq x \leq 3.865$. Hence, an approximation of the area bounded by p and q is given by the sum

$$\int_{0.772}^{2.292} [p(x) - q(x)] \, dx + \int_{2.292}^{3.865} [q(x) - p(x)] \, dx.$$

These two integrals are computed with NIntegrate

```
In[426]:= intone = NIntegrate[p[x] - q[x],
              {x, intpts[[1, 1, 2]], intpts[[2, 1, 2]]}]

          inttwo = NIntegrate[q[x] - p[x],
              x, intpts[[2, 1, 2]], intpts[[3, 1, 2]]}]
Out[426]= 5.26912
Out[426]= 6.92599
```

and added to see that the area is approximately 12.195.

```
In[427]:= intone + inttwo
Out[427]= 12.1951
```

■

Parametric Equations

If the curve, C, defined parametrically by $x = x(t), y = y(t), a \le t \le b$ is a nonnegative continuous function of x and $x(a) < x(b)$ the area under the graph of C and above the x-axis is

Graphically, y is a function of x, $y = y(x)$, if the graph of $y = y(x)$ passes the vertical line test.

$$\int_{x(a)}^{x(b)} y\,dx = \int_a^b y(t)x'(t)\,dt.$$

EXAMPLE 3.3.9 (The Astroid): Find the area enclosed by the **astroid** $x = \sin^3 t, y = \cos^3 t, 0 \le t \le 2\pi$.

SOLUTION: We begin by defining x and y and then graphing the astroid with ParametricPlot in Figure 3-28.

```
In[428]:= x[t_] = Sin[t]^3;
          y[t_] = Cos[t]^3;
          ParametricPlot[
              {x[t], y[t]}, {t, 0, 2π}, AspectRatio-> Automatic]
```

Observe that $x(0) = 0$ and $x(\pi/2) = 1$ and the graph of the astroid in the first quadrant is given by $x = \sin^3 t, y = \cos^3 t, 0 \le t \le \pi/2$. Hence, the area of the astroid in the first quadrant is given by

$$\int_0^{\pi/2} y(t)x'(t)\,dt = 3\int_0^{\pi/2} \sin^2 t \cos^4 t\,dt$$

and the total area is given by

$$A = 4\int_0^{\pi/2} y(t)x'(t)\,dt = 12\int_0^{\pi/2} \sin^2 t \cos^4 t\,dt = \frac{3}{8}\pi \approx 1.178,$$

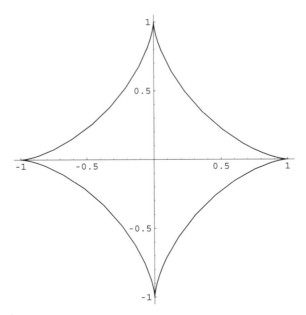

Figure 3-28   The astroid $x = \sin^3 t, y = \cos^3 t, 0 \le t \le 2\pi$

which is computed with Integrate and then approximated with N.

$In\,[429] :=$ **area = 4 Integrate[y[t] x´[t], {t, 0, π/2}]**

$Out\,[429] = \dfrac{3\;\pi}{8}$

$In\,[430] :=$ **N[area]**

$Out\,[430] =$ 1.1781

■

Polar Coordinates

For problems involving "circular symmetry" it is often easier to work in polar coordinates. The relationship between (x, y) in rectangular coordinates and (r, θ) in polar coordinates is given by

$$x = r \cos \theta \qquad y = r \sin \theta$$

and

$$r^2 = x^2 + y^2 \qquad \tan \theta = \frac{y}{x}.$$

If $r = f(\theta)$ is continuous and nonnegative for $\alpha \le \theta \le \beta$, then the **area** A of the region enclosed by the graphs of $r = f(\theta), \theta = \alpha$, and $\theta = \beta$ is

$$A = \frac{1}{2} \int_{\alpha}^{\beta} [f(\theta)]^2 \; d\theta = \frac{1}{2} \int_{\alpha}^{\beta} r^2 \; d\theta.$$

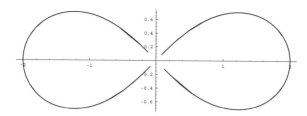

Figure 3-29   The lemniscate

EXAMPLE 3.3.10 (Lemniscate of Bernoulli): The **lemniscate of Bernoulli** is given by

$$\left(x^2 + y^2\right)^2 = a^2\left(x^2 - y^2\right),$$

where a is a constant. (a) Graph the lemniscate of Bernoulli if $a = 2$. (b) Find the area of the region bounded by the lemniscate of Bernoulli.

SOLUTION: This problem is much easier solved in polar coordinates so we first convert the equation from rectangular to polar coordinates with `ReplaceAll (/.)` and then solve for r with `Solve`.

```
In[431]:= lofb = (x^2 + y^2)^2 == a^2(x^2 - y^2);
          topolar = lofb/.{x- > r Cos[t], y- > r Sin[t]}
Out[431]= (r^2 Cos[t]^2 + r^2 Sin[t]^2)^2 ==
          a^2 (r^2 Cos[t]^2 - r^2 Sin[t]^2)

In[432]:= Solve[topolar, r]//Simplify
Out[432]= {{r → 0}, {r → 0}, {r → -a √Cos[2 t]},
          {r → a √Cos[2 t]}}
```

These results indicate that an equation of the lemniscate in polar coordinates is $r^2 = a^2 \cos 2\theta$. The graph of the lemniscate is then generated in Figure 3-29 using `PolarPlot`, which is contained in the **Graphics** package located in the **Graphics** directory.

```
In[433]:= << Graphics`Graphics`

          PolarPlot[{-2 Sqrt[Cos[2t]],
            2Sqrt[Cos[2t]]}, {t, 0, 2π}]
```

The portion of the lemniscate in quadrant one is obtained by graphing $r = 2\cos 2\theta$, $0 \le \theta \le \pi/4$.

```
In[434]:= PolarPlot[2Sqrt[Cos[2t]], {t, 0, π/4}]
```

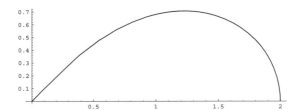

Figure 3-30   The portion of the lemniscate in quadrant 1

Then, taking advantage of symmetry, the area of the lemniscate is given by

$$A = 2 \cdot \frac{1}{2} \int_{-\pi/4}^{\pi/4} r^2 \, d\theta = 2 \int_{0}^{\pi/4} r^2 \, d\theta = 2 \int_{0}^{\pi/4} a^2 \cos 2\theta \, d\theta = a^2,$$

which we calculate with `Integrate`.

```
In[435]:= Integrate[2 a^2 Cos[2 t], {t, 0, π/4}]
Out[435]= a²
```

■

3.3.5 Arc Length

Let $y = f(x)$ be a function for which $f'(x)$ is continuous on an interval $[a, b]$. Then the **arc length** of the graph of $y = f(x)$ from $x = a$ to $x = b$ is given by

$$L = \int_{a}^{b} \sqrt{\left(\frac{dy}{dx}\right)^2 + 1} \, dx. \tag{3.17}$$

The resulting definite integrals used for determining arc length are usually difficult to compute because they involve a radical. In these situations, Mathematica is helpful with approximating solutions to these types of problems.

EXAMPLE 3.3.11: Find the length of the graph of

$$y = \frac{x^4}{8} + \frac{1}{4x^2}$$

from (a) $x = 1$ to $x = 2$ and from (b) $x = -2$ to $x = -1$.

SOLUTION: With no restrictions on the value of x, $\sqrt{x^2 2} = |x|$. Notice that Mathematica does not automatically algebraically simplify $\sqrt{\left(\frac{dy}{dx}\right)^2 + 1}$ because Mathematica does not know if x is positive or negative.

```
In[436] := y[x_] = x^4/8 + 1/(4x^2);
           i1 = Factor[y'[x]^2 + 1]
```
$$Out[436] = \frac{(1 + x^2)^2 \ (1 - x^2 + x^4)^2}{4 \ x^6}$$

```
In[437] := i2 = PowerExpand[Sqrt[i1]]
```
$$Out[437] = \frac{(1 + x^2) \ (1 - x^2 + x^4)}{2 \ x^3}$$

In fact, for (b), x is negative so

$$\frac{1}{2}\sqrt{\frac{\left(x^6 + 1\right)^2}{x^6}} = -\frac{1}{2}\frac{x^6 + 1}{x^3}.$$

Mathematica simplifies

$$\frac{1}{2}\sqrt{\frac{\left(x^6 + 1\right)^2}{x^6}} = \frac{1}{2}\frac{x^6 + 1}{x^3}$$

PowerExpand[expr] simplifies radicals in the expression expr assuming that all variables are positive.

and correctly evaluates the arc length integral (3.17) for (a).

```
In[438] := Integrate[Sqrt[y'[x]^2 + 1], {x, 1, 2}]
```
$$Out[438] = \frac{33}{16}$$

For (b), we compute the arc length integral (3.17).

```
In[439] := Integrate[Sqrt[y'[x]^2 + 1], {x, -2, -1}]
```
$$Out[439] = \frac{33}{16}$$

As we expect, both values are the same.

∎

Parametric Equations

If the smooth curve, C, defined parametrically by $x = x(t)$, $y = y(t)$, $t \in [a, b]$ is traversed exactly once as t increases from $t = a$ to $t = b$, the arc length of C is given by

C is **smooth** if both $x'(t)$ and $y'(t)$ are continuous on (a, b) and not simultaneously zero for $t \in (a, b)$.

$$L = \int_a^b \sqrt{\left(\frac{dx}{dt}\right)^2 + \left(\frac{dy}{dt}\right)^2} \, dt. \tag{3.18}$$

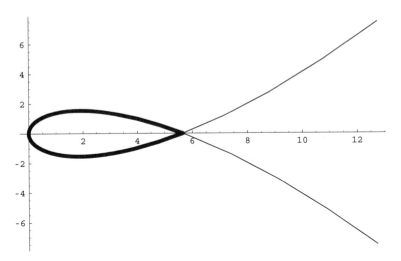

Figure 3-31 $x = \sqrt{2}t^2, y = 2t - \frac{1}{2}t^3$

EXAMPLE 3.3.12: Find the length of the graph of $x = \sqrt{2}t^2, y = 2t - \frac{1}{2}t^3$, $-2 \le t \le 2$.

SOLUTION: For illustrative purposes, we graph $x = \sqrt{2}t^2, y = 2t - \frac{1}{2}t^3$ for $-3 \le t \le 3$ (in black) and $-2 \le t \le 2$ (in thick black) in Figure 3-31.

```
In[440]:= x[t_] = t^2 Sqrt[2]; y[t_] = 2t - 1/2t^3;
          p1 = ParametricPlot[{x[t],y[t]},
              {t,-3,3}, DisplayFunction- > Identity];
          p2 = ParametricPlot[{x[t],y[t]}, {t,-2,2},
              PlotStyle- > Thickness[0.01],
              DisplayFunction- > Identity];
          Show[p1, p2, DisplayFunction- >
              $DisplayFunction, PlotRange- > All]
```

Mathematica is able to compute the exact value of the arc length (3.18) although the result is quite complicated.

```
In[441]:= Factor[x'[t]^2 + y'[t]^2]
```

$$Out[441]= \frac{1}{4}\left(4 - 4\ t + 3\ t^2\right)\left(4 + 4\ t + 3\ t^2\right)$$

```
In[442]:= i1 = Integrate[2 Sqrt[x'[t]^2+y'[t]^2], {t, 0, 2}]
```

$$Out[442] = \frac{1}{18} \left(96 \sqrt{3} - \left(32 \ i \right. \right.$$

$$\sqrt{6 \left(1 + \frac{1}{2} \left(2 - 4 \ i \ \sqrt{2} \right) \right)} \ \sqrt{1 + \frac{1}{2} \left(2 + 4 \ i \ \sqrt{2} \right)}$$

$$\text{EllipticE}\left[i \ \text{ArcSinh}\left[\sqrt{\frac{1}{2} \left(2 + 4 \ i \ \sqrt{2} \right)} \right],$$

$$\left. -1 + \frac{1}{9} \left(2 - 4 \ i \ \sqrt{2} \right) \right] \right) / \left(\left(2 + 4 \ i \ \sqrt{2} \right)^{3/2} \right.$$

$$\left(-1 + \frac{1}{9} \left(2 - 4 \ i \ \sqrt{2} \right) \right) - \frac{1}{\sqrt{2 + 4 \ i \ \sqrt{2}}} \left(16 \ i \right.$$

$$\sqrt{6 \left(1 + \frac{1}{2} \left(2 - 4 \ i \ \sqrt{2} \right) \right)} \ \sqrt{1 + \frac{1}{2} \left(2 + 4 \ i \ \sqrt{2} \right)}$$

$$\text{EllipticF}\left[i \ \text{ArcSinh}\left[\sqrt{\frac{1}{2} \left(2 + 4 \ i \ \sqrt{2} \right)} \right],$$

$$\left. -1 + \frac{1}{9} \left(2 - 4 \ i \ \sqrt{2} \right) \right] \right) + \left(32 \ i \right.$$

$$\sqrt{6 \left(1 + \frac{1}{2} \left(2 - 4 \ i \ \sqrt{2} \right) \right)} \ \sqrt{1 + \frac{1}{2} \left(2 + 4 \ i \ \sqrt{2} \right)}$$

$$\text{EllipticF}\left[i \ \text{ArcSinh}\left[\sqrt{\frac{1}{2} \left(2 + 4 \ i \ \sqrt{2} \right)} \right],$$

$$\left. -1 + \frac{1}{9} \left(2 - 4 \ i \ \sqrt{2} \right) \right] \right) / \left(\left(2 + 4 \ i \ \sqrt{2} \right)^{3/2} \right.$$

$$\left. \left. \left. \left(-1 + \frac{1}{9} \left(2 - 4 \ i \ \sqrt{2} \right) \right) \right) \right) \right)$$

A more meaningful approximation is obtained with N or using NIntegrate.

```
In[443]:= N[i1]
```

$$Out[443] = 13.7099 - 1.18424 \ 10^{-15} \ i$$

```
In[444]:= NIntegrate[2 Sqrt[x'[t]^2+y'[t]^2], {t, 0, 2}]
```

$$Out[444] = 13.7099$$

We conclude that the arc length is approximately 13.71.

∎

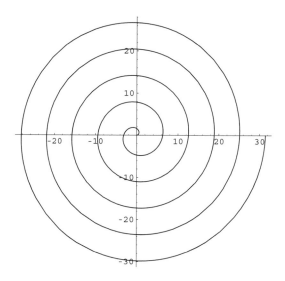

Figure 3-32 $r = \theta$ for $0 \le \theta \le 10\pi$

Polar Coordinates

If the smooth polar curve C given by $r = f(\theta)$, $\alpha \le \theta \le \beta$ is traversed exactly once as θ increases from α to β, the arc length of C is given by

$$L = \int_\alpha^\beta \sqrt{\left(\frac{dr}{d\theta}\right)^2 + r^2}\, d\theta \tag{3.19}$$

EXAMPLE 3.3.13: Find the length of the graph of $r = \theta$, $0 \le \theta \le 10\pi$.

SOLUTION: We begin by defining r and then graphing r with `PolarPlot` in Figure 3-32.

```
In[445]:= << Graphics`Graphics`

          r[t_] = t;
          PolarPlot[r[t], {t, 0, 10π},
            AspectRatio- > Automatic]
```

Using (3.19), the length of the graph of r is given by $\int_0^{10\pi} \sqrt{1 + \theta^2}\, d\theta$. The exact value is computed with `Integrate`

```
In[446]:= ev = Integrate[Sqrt[r'[t]^2+r[t]^2], {t, 0, 10π}]
```
$$Out[446] = 5\,\pi\,\sqrt{1 + 100\,\pi^2} + \frac{1}{2}\,\mathrm{ArcSinh}[10\,\pi]$$

and then approximated with N.

> *In[447]:=* **N[ev]**
>
> *Out[447]=* 495.801

We conclude that the length of the graph is approximately 495.8.

■

3.3.6 Solids of Revolution

Volume

Let $y = f(x)$ be a nonnegative continuous function on $[a, b]$. The **volume** of the solid of revolution obtained by revolving the region bounded by the graphs of $y = f(x)$, $x = a$, $x = b$, and the x-axis about the x-axis is given by

$$V = \pi \int_a^b [f(x)]^2 \, dx. \tag{3.20}$$

If $0 \le a < b$, the **volume** of the solid of revolution obtained by revolving the region bounded by the graphs of $y = f(x)$, $x = a$, $x = b$, and the x-axis about the y-axis is given by

$$V = 2\pi \int_a^b x f(x) \, dx. \tag{3.21}$$

EXAMPLE 3.3.14: Let $g(x) = x \sin^2 x$. Find the volume of the solid obtained by revolving the region bounded by the graphs of $y = g(x)$, $x = 0$, $x = \pi$, and the x-axis about (a) the x-axis; and (b) the y-axis.

SOLUTION: After defining g, we graph g on the interval $[0, \pi]$ in Figure 3-33.

> *In[448]:=* **g[x_] = x Sin[x]^2;**
>   **Plot[g[x], {x, 0, π}, AspectRatio- > Automatic]**

The volume of the solid obtained by revolving the region about the x-axis is given by equation (3.20) while the volume of the solid obtained by revolving the region about the y-axis is given by equation (3.21). These integrals are computed with Integrate and named xvol and yvol, respectively. N is used to approximate each volume.

> *In[449]:=* **xvol = Integrate[π g[x]^2, {x, 0, π}]**
>
>   **N[xvol]**

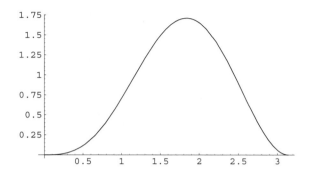

Figure 3-33 $g(x)$ for $0 \le x \le \pi$

$Out[449]= \dfrac{1}{256} \ \pi \ \left(-60 \ \pi + 32 \ \pi^3\right)$

$Out[449]=$ 9.86295

$In[450]:=$ **yvol = Integrate[2 π x g[x], {x, 0, π}]**

        N[yvol]

$Out[450]= \dfrac{1}{12} \ \pi \ \left(-6 \ \pi + 4 \ \pi^3\right)$

$Out[450]=$ 27.5349

We can use `ParametricPlot3D` to visualize the resulting solids by parametrically graphing the equations given by

$$\begin{cases} x = r \cos t \\ y = r \sin t \\ z = g(r) \end{cases}$$

for r between 0 and π and t between $-\pi$ and π to visualize the graph of the solid obtained by revolving the region about the y-axis and by parametrically graphing the equations given by

$$\begin{cases} x = r \\ y = g(r) \cos t \\ z = g(r) \sin t \end{cases}$$

for r between 0 and π and t between $-\pi$ and π to visualize the graph of the solid obtained by revolving the region about the x-axis. (See Figures 3-34 and 3-35.) In this case, we identify the z-axis as the y-axis. Notice that we are simply using polar coordinates for the x and y-coordinates, and the height above the x,y-plane is given by $z = g(r)$ because r is replacing x in the new coordinate system.

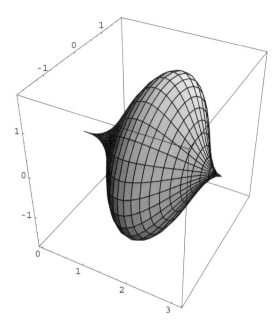

Figure 3-34   *g*(*x*) revolved about the *x*-axis

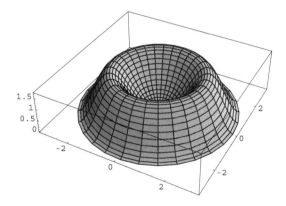

Figure 3-35   *g*(*x*) revolved about the *y*-axis

In[451]:= **ParametricPlot3D[{r, g[r] Cos[t], g[r] Sin[t]},**
　　　　　{r, 0, π}, {t, 0, 2π}, PlotPoints- > {30, 30}]

In[452]:= **ParametricPlot3D[{r Cos[t], r Sin[t], g[r]},**
　　　　　{r, 0, π}, {t, 0, 2π}, PlotPoints- > {30, 30}]

■

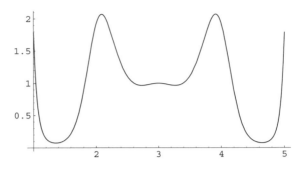

Figure 3-36 $f(x)$ for $1 \leq x \leq 5$

We now demonstrate a volume problem that requires the method of disks.

EXAMPLE 3.3.15: Let $f(x) = e^{-(x-3)\cos[4(x-3)]}$. Approximate the volume of the solid obtained by revolving the region bounded by the graphs of $y = f(x)$, $x = 1$, $x = 5$, and the x-axis about the x-axis.

SOLUTION: Proceeding as in the previous example, we first define and graph f on the interval $[1, 5]$ in Figure 3-36.

```
In[453]:= f[x_] = Exp[-(x - 3)^2 Cos[4(x - 3)]];
          Plot[f[x], {x, 1, 5}, AspectRatio- > Automatic]
```

In this case, an approximation is desired so we use NIntegrate to approximate the integral $V = \int_1^5 \pi [f(x)]^2 \, dx$.

```
In[454]:= NIntegrate[π f[x]^2, {x, 1, 5}]

Out[454]= 16.0762
```

In the same manner as before, ParametricPlot3D can be used to visualize the resulting solid by graphing the set of equations given parametrically by

$$\begin{cases} x = r \\ y = f(r) \cos t \\ z = f(r) \sin t \end{cases}$$

for r between 1 and 5 and t between 0 and 2π. In this case, polar coordinates are used in the y,z-plane with the distance from the x-axis given by $f(x)$. Because r replaces x in the new coordinate system, $f(x)$ becomes $f(r)$ in these equations. See Figure 3-37.

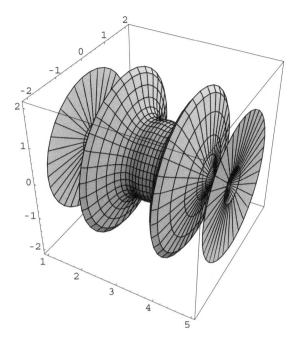

Figure 3-37 $f(x)$ revolved about the x-axis

```
In[455]:= ParametricPlot3D[{r, f[r] Cos[t], f[r] Sin[t]},
            {r, 1, 5}, {t, 0, 2π}, PlotPoints- > {45, 35}]
```

■

Surface Area

Let $y = f(x)$ be a nonnegative function for which $f'(x)$ is continuous on an interval $[a, b]$. Then the **surface area** of the solid of revolution obtained by revolving the region bounded by the graphs of $y = f(x)$, $x = a$, $x = b$, and the x-axis about the x-axis is given by

$$SA = 2\pi \int_a^b f(x)\sqrt{1 + [f'(x)]^2}\, dx. \tag{3.22}$$

EXAMPLE 3.3.16 (Gabriel's Horn): Gabriel's horn is the solid of revolution obtained by revolving the area of the region bounded by $y = 1/x$ and the x-axis for $x \geq 1$ about the x-axis. Show that the surface area of Gabriel's horn is infinite but that its volume is finite.

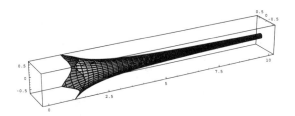

Figure 3-38   A portion of Gabriel's horn

SOLUTION: After defining $f(x) = 1/x$, we use `ParametricPlot3D` to visualize a portion of Gabriel's horn in Figure 3-38.

```
In[456]:= f[x_] = 1/x;
          ParametricPlot3D[{r, f[r] Cos[t], f[r] Sin[t]},
          {r, 1, 10}, {t, 0, 2π}, PlotPoints- > {40, 40},
          ViewPoint- > {-1.509,  -2.739,  1.294}]
```

Using equation (3.22), the surface area of Gabriel's horn is given by the improper integral

$$SA = 2\pi \int_1^\infty \frac{1}{x} \sqrt{1 + \frac{1}{x^4}}\, dx = 2\pi \lim_{L \to \infty} \int_1^L \frac{1}{x} \sqrt{1 + \frac{1}{x^4}}\, dx.$$

```
In[457]:= step1 = Integrate[2 π f[x] Sqrt[1 + f'[x]^2],
          {x, 1, cap1}]
Integrate :: "gener" :  "Unabletocheckconvergence"
Out[457]= -π (- √2 + ArcSinh[1]) +
```

$$\sqrt{1 + \frac{1}{cap1^4}}\; \pi \left(-1 + \frac{cap1^2\ ArcSinh[cap1^2]}{\sqrt{1 + cap1^4}}\right)$$

```
In[458]:= Limit[step1, cap1- > ∞]
Out[458]= ∞
```

On the other hand, using equation (3.20) the volume of Gabriel's horn is given by the improper integral

$$SA = 2\pi \int_1^\infty \frac{1}{x^2}\, dx = \pi \lim_{L \to \infty} \int_1^L \frac{1}{x^2}\, dx,$$

which converges to π.

```
In[459]:= step1 = Integrate[π f[x]^2, {x, 1, cap1}]
Out[459]= π - π/cap1
```

```
In[460]:= Limit[step1, cap1- > ∞]
Out[460]= π

In[461]:= Integrate[π f[x]^2, {x, 1, ∞}]
Out[461]= π
```

■

3.4 Series

3.4.1 Introduction to Sequences and Series

Sequences and series are usually discussed in the third quarter or second semester of introductory calculus courses. Most students find that it is one of the most difficult topics covered in calculus. A **sequence** is a function with domain consisting of the positive integers. The **terms** of the sequence $\{a_n\}$ are $a_1, a_2, a_3, \ldots$. The nth term is a_n; the $(n+1)$st term is a_{n+1}. If $\lim_{n\to\infty} a_n = L$, we say that $\{a_n\}$ **converges** to L. If $\{a_n\}$ does not converge, $\{a_n\}$ **diverges**. We can sometimes prove that a sequence converges by applying the following theorem.

Theorem 7. *Every bounded monotonic sequence converges.*

In particular, Theorem 7 gives us the following special cases.

> A sequence $\{a_n\}$ is monotonic if $\{a_n\}$ is increasing ($a_{n+1} \geq a_n$ for all n) or decreasing ($a_{n+1} \leq a_n$ for all n).

   1. If $\{a_n\}$ has positive terms and is eventually decreasing, $\{a_n\}$ converges.
   2. If $\{a_n\}$ has negative terms and is eventually increasing $\{a_n\}$ converges.

After you have defined a sequence, use `Table` to compute the first few terms of the sequence.

   1. `Table[a[n],{n,1,m}]` returns the list $\{a_1, a_2, a_3, \ldots, a_m\}$.
   2. `Table[a[n],{n,k,m}]` returns $\{a_k, a_{k+1}, a_{k+2}, \ldots, a_m\}$.

EXAMPLE 3.4.1: If $a_n = \dfrac{50^n}{n!}$, show that $\lim_{n\to\infty} a_n = 0$.

SOLUTION: We remark that the symbol $n!$ in the denominator of a_n represents the **factorial sequence**:

$$n! = n \cdot (n-1) \cdot (n-2) \cdots 2 \cdot 1.$$

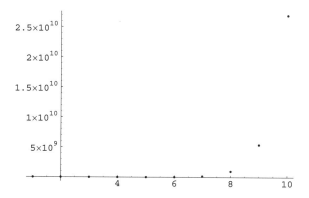

Figure 3-39   The first few terms of a_n

We begin by defining a_n and then computing the first few terms of the sequence with Table.

```
In[462]:= a[n_] = 50^n/n!;
          afewterms = Table[a[n], {n, 1, 10}]

          N[afewterms]
```

$$Out[462]= \left\{50, 1250, \frac{62500}{3}, \frac{781250}{3}, \frac{7812500}{3}, \frac{195312500}{9},\right.$$
$$\frac{9765625000}{63}, \frac{61035156250}{63}, \frac{3051757812500}{567},$$
$$\left.\frac{15258789062500}{567}\right\}$$

$$Out[462]= \{50., 1250., 20833.3,$$
$$260417., 2.60417\,10^6,$$
$$2.17014\,10^7, 1.5501\,10^8,$$
$$9.68812\,10^8, 5.38229\,10^9,$$
$$2.69114\,10^{10}\}$$

The first few terms increase in magnitude. In fact, this is further confirmed by graphing the first few terms of the sequence with ListPlot in Figure 3-39. Based on the graph and the values of the first few terms we might incorrectly conclude that the sequence diverges.

```
In[463]:= ListPlot[afewterms]
```

However, notice that

$$a_{n+1} = \frac{50}{n+1}a_n \Rightarrow \frac{a_{n+1}}{a_n} = \frac{50}{n+1}.$$

Because $50/(n+1) < 1$ for $n > 49$, we conclude that the sequence is decreasing for $n > 49$. Because it has positive terms, it is bounded below

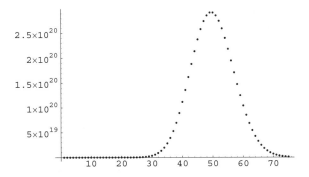

Figure 3-40 The first 75 terms of a_n

by 0 so the sequence converges by Theorem 7. Let $L = \lim_{n\to\infty} a_n$. Then,

$$\lim_{n\to\infty} a_{n+1} = \lim_{n\to\infty} \frac{50}{n+1} a_n$$

$$L = \lim_{n\to\infty} \frac{50}{n+1} \cdot L$$

$$L = 0.$$

When we graph a larger number of terms, it is clear that the limit is 0. (See Figure 3-40.) It is a good exercise to show that for any real value of x, $\lim_{n\to\infty} \frac{x^n}{n!} = 0$.

```
In[464]:= ListPlot[Evaluate[Table[a[k], {k, 1, 75}]]]
```

■

An **infinite series** is a series of the form

$$\sum_{k=1}^{\infty} a_k \tag{3.23}$$

where $\{a_n\}$ is a sequence. The nth **partial sum** of (3.23) is

$$s_n = \sum_{k=1}^{n} a_k = a_1 + a_2 + \cdots + a_n. \tag{3.24}$$

Notice that the partial sums of the series (3.23) form a sequence $\{s_n\}$. Hence, we say that the infinite series (3.23) **converges** to L if the sequence of partial sums $\{s_n\}$ converges to L and write

$$\sum_{k=1}^{\infty} a_k = L.$$

The infinite series (3.23) **diverges** if the sequence of partial sums diverges. Given the infinite series (3.23),

$$\texttt{Sum[a[k],\{k,1,n\}]}$$

calculates the nth partial sum (3.24). In *some* cases, if the infinite series (3.23) converges,

$$\texttt{Sum[a[k],\{k,1,Infinity\}]}$$

can compute the value of the infinite sum. You should think of the Sum function as a "fragile" command and be certain to carefully examine its results.

EXAMPLE 3.4.2: Determine whether each series converges or diverges. If the series converges, find its sum. (a) $\sum_{k=1}^{\infty}(-1)^{k+1}$ (b) $\sum_{k=2}^{\infty}\dfrac{2}{k^2-1}$ (c) $\sum_{k=0}^{\infty}ar^k$.

SOLUTION: For (a), we compute the nth partial sum (3.24) in sn with Sum.

$$In[465]:= \textbf{sn = Sum[(-1)\^{}(k+1),\{k,1,n\}]}$$
$$Out[465]= \frac{1}{2}\ (1-(-1)^n)$$

Notice that the odd partial sums are 1:

$$s_{2n+1} = \frac{1}{2}\left((-1)^{2n+1+1}+1\right) = \frac{1}{2}(1+1) = 1$$

while the even partial sums are 0:

$$s_{2n} = \frac{1}{2}\left((-1)^{2n+1}+1\right) = \frac{1}{2}(-1+1) = 0.$$

We confirm that the limit of the partial sums does not exist with Limit. Mathematica's result indicates that it cannot determine the limit.

$$In[466]:= \textbf{Limit[sn,n->∞]}$$
$$Out[466]= \text{Limit}\left[\frac{1}{2}\ (1-(-1)^n),n\to\infty\right]$$

However, when we attempt to compute the infinite sum with Sum, Mathematica is able to determine that the sum diverges.

$$In[467]:= \textbf{Sum[(-1)\^{}(k+1),\{k,1,∞\}]}$$
$$\texttt{Sum :: "div" : "Sumdoesnotconverge."}$$
$$Out[467]= \sum_{k=1}^{\infty}(-1)^{k+1}$$

Thus, the series diverges.

For (b), we have a *telescoping series*. Using partial fractions,

$$\sum_{k=2}^{\infty} \frac{2}{k^2 - 1} = \sum_{k=2}^{\infty} \left(\frac{1}{k-1} - \frac{1}{k+1} \right)$$

$$= \left(1 - \frac{1}{3} \right) + \left(\frac{1}{2} - \frac{1}{4} \right) + \left(\frac{1}{3} - \frac{1}{5} \right) + \cdots + \left(\frac{1}{n-2} - \frac{1}{n} \right)$$

$$+ \left(\frac{1}{n-1} - \frac{1}{n+1} \right) + \cdots$$

we see that the nth partial sum is given by

$$s_n = \frac{3}{2} - \frac{1}{n} - \frac{1}{n+1}$$

and $s_n \to 3/2$ as $n \to \infty$ so the series converges to 3/2:

$$\sum_{k=2}^{\infty} \frac{2}{k^2 - 1} = \frac{3}{2}.$$

We perform the same steps with Mathematica using Sum, Apart, and Limit.

> *Apart* computes the partial fraction decomposition of a rational expression.

```
In[468]:= sn = Sum[1/(k - 1) - 1/(k + 1), {k, 2, n}]
```
$$Out[468] = \frac{(-1 + n) \ (2 + 3 \ n)}{2 \ n \ (1 + n)}$$

```
In[469]:= Apart[sn]
```
$$Out[469] = \frac{3}{2} - \frac{1}{n} - \frac{1}{1 + n}$$

```
In[470]:= Limit[sn, n->∞]
```
$$Out[470] = \frac{3}{2}$$

(c) A series of the form $\sum_{k=0}^{\infty} ar^k$ is called a **geometric series**. We compute the nth partial sum of the geometric series with Sum.

```
In[471]:= sn = Sum[a r^k, {k, 0, n}]
```
$$Out[471] = \frac{a \ (-1 + r^{1+n})}{-1 + r}$$

When using Limit to determine the limit of s_n as $n \to \infty$, we see that Mathematica returns the limit unevaluated because Mathematica does not know the value of r.

```
In[472]:= Limit[sn, n->∞]
```
$$Out[472] = Limit\left[\frac{a \ (-1 + r^{1+n})}{-1 + r}, n \to \infty \right]$$

In fact, the geometric series diverges if $|r| \geq 1$ and converges if $|r| < 1$. Observe that if we simply compute the sum with Sum, Mathematica returns $a/(1 - r)$ which is correct if $|r| < 1$ but incorrect if $|r| \geq 1$.

> *In[473]* := **Sum[a r^k, {k, 0, ∞}]**
>
> *Out[473]* = $\dfrac{a}{1 - r}$

However, the result of entering

> *In[474]* := **Sum[(-5/3)^k, {k, 0, ∞}]**
>
> Sum :: "div" : "Sumdoesnotconverge."
>
> *Out[474]* = $\displaystyle\sum_{k=0}^{\infty} \left(-\frac{5}{3} \right)^{k}$

is correct because the series $\sum_{k=0}^{\infty} \left(-\frac{5}{3}\right)^k$ is geometric with $|r| = 5/3 \geq 1$ and, consequently, diverges. Similarly,

> *In[475]* := **Sum[9 (1/10)^k, {k, 1, ∞}]**
>
> *Out[475]* = 1

is correct because $\sum_{k=1}^{\infty} 9\left(\frac{1}{10}\right)^k$ is geometric with $a = 9/10$ and $r = 1/10$ so the series converges to

$$\frac{a}{1 - r} = \frac{9/10}{1 - 1/10} = 1.$$

■

3.4.2 Convergence Tests

Frequently used convergence tests are stated in the following theorems.

Theorem 8 (The Divergence Test). *Let $\sum_{k=1}^{\infty} a_k$ be an infinite series. If $\lim_{k\to\infty} a_k \neq 0$, then $\sum_{k=1}^{\infty} a_k$ diverges.*

Theorem 9 (The Integral Test). *Let $\sum_{k=1}^{\infty} a_k$ be an infinite series with positive terms. If $f(x)$ is a decreasing continuous function for which $f(k) = a_k$ for all k, then $\sum_{k=1}^{\infty} a_k$ and $\int_{1}^{\infty} f(x)\,dx$ either both converge or both diverge.*

Theorem 10 (The Ratio Test). *Let $\sum_{k=1}^{\infty} a_k$ be an infinite series with positive terms and let $\rho = \lim_{k\to\infty} \frac{a_{k+1}}{a_k}$.*

1. *If $\rho < 1$, $\sum_{k=1}^{\infty} a_k$ converges.*
2. *If $\rho > 1$, $\sum_{k=1}^{\infty} a_k$ diverges.*
3. *If $\rho = 1$, the Ratio Test is inconclusive.*

Theorem 11 (The Root Test). *Let $\sum_{k=1}^{\infty} a_k$ be an infinite series with positive terms and let $\rho = \lim_{k \to \infty} \sqrt[k]{a_k}$.*

1. *If $\rho < 1$, $\sum_{k=1}^{\infty} a_k$ converges.*
2. *If $\rho > 1$, $\sum_{k=1}^{\infty} a_k$ diverges.*
3. *If $\rho = 1$, the Root Test is inconclusive.*

Theorem 12 (The Limit Comparison Test). *Let $\sum_{k=1}^{\infty} a_k$ and $\sum_{k=1}^{\infty} b_k$ be infinite series with positive terms and let $L = \lim_{k \to \infty} \frac{a_k}{b_k}$. If $0 < L < \infty$, then either both series converge or both series diverge.*

EXAMPLE 3.4.3: Determine whether each series converges or diverges.

(a) $\sum_{k=1}^{\infty} \left(1 + \frac{1}{k}\right)^k$; (b) $\sum_{k=1}^{\infty} \frac{1}{k^p}$; (c) $\sum_{k=1}^{\infty} \frac{k}{3^k}$; (d) $\sum_{k=1}^{\infty} \frac{(k!)^2}{(2k)!}$; (e) $\sum_{k=1}^{\infty} \left(\frac{k}{4k+1}\right)^k$;

(f) $\sum_{k=1}^{\infty} \frac{2\sqrt{k}+1}{(\sqrt{k}+1)(2k+1)}$.

SOLUTION: (a) Using `Limit`, we see that the limit of the terms is $e \neq 0$ so the series diverges by the the Divergence Test, Theorem 8.

```
In[476]:= Limit[(1 + 1/k)^k, k- > ∞]

Out[476]= e
```

It is a very good exercise to show that the limit of the terms of the series is e by hand. Let $L = \lim_{k \to \infty} \left(1 + \frac{1}{k}\right)^k$. Take the logarithm of each side of this equation and apply L'Hôpital's rule:

$$\ln L = \lim_{k \to \infty} \ln\left(1 + \frac{1}{k}\right)^k$$

$$\ln L = \lim_{k \to \infty} k \ln\left(1 + \frac{1}{k}\right)$$

$$\ln L = \lim_{k \to \infty} \frac{\ln\left(1 + \frac{1}{k}\right)}{\frac{1}{k}}$$

$$\ln L = \lim_{k \to \infty} \frac{\frac{1}{1 + \frac{1}{k}} \cdot -\frac{1}{k^2}}{-\frac{1}{k^2}}$$

$$\ln L = 1.$$

Exponentiating yields $L = e^{\ln L} = e^1 = e$. (b) A series of the form $\sum_{k=1}^{\infty} \frac{1}{k^p}$ is called a *p*-**series**. Let $f(x) = x^{-p}$. Then, $f(x)$ is continuous and decreasing for $x \geq 1$, $f(k) = k^{-p}$ and

$$\int_1^{\infty} x^{-p} dx = \begin{cases} \infty, \text{ if } p \leq 1 \\ 1/(p-1), \text{ if } p > 1 \end{cases}$$

so the *p*-series converges if $p > 1$ and diverges if $p \leq 1$. If $p = 1$, the series $\sum_{k=1}^{\infty} \frac{1}{k}$ is called the **harmonic series**.

```
In[477]:= s1 = Integrate[x^(-p), {x, 1, ∞}]
```

$$Out[477]= \text{If}\left[\text{Re}[p] > 1, \frac{1}{-1+p}, \int_1^{\infty} x^{-p} dx\right]$$

(c) Let $f(x) = x \cdot 3^{-x}$. Then, $f(k) = k \cdot 3^{-k}$ and $f(x)$ is decreasing for $x > 1/\ln 3$.

```
In[478]:= f[x_] = x 3^(-x);
          Factor[f'[x]]
```

$$Out[478]= -3^{-x} \ (-1 + x \ \text{Log}[3])$$

```
In[479]:= Solve[-1 + x Log[3] == 0]
```

$$Out[479]= \left\{\left\{x \rightarrow \frac{1}{\text{Log}[3]}\right\}\right\}$$

Using `Integrate`, we see that the improper integral $\int_1^{\infty} f(x) \, dx$ converges.

```
In[480]:= ival = Integrate[f[x], {x, 1, ∞}]

          N[ival]
```

$$Out[480]= \frac{1 + \text{Log}[3]}{3 \ \text{Log}[3]^2}$$

$$Out[480]= 0.579592$$

Thus, by the Integral Test, Theorem 9, we conclude that the series converges. Note that when applying the Integral Test, if the improper integral converges its value is *not* the value of the sum of the series. In this case, we see that Mathematica is able to evaluate the sum with `Sum` and the series converges to 3/4.

```
In[481]:= Sum[k 3^(-k), {k, 1, ∞}]
```

$$Out[481]= \frac{3}{4}$$

(d) If a_k contains factorials, the Ratio Test is a good first test to try. After defining a_k we compute

$$\lim_{k \to \infty} \frac{a_{k+1}}{a_k} = \lim_{k \to \infty} \frac{\dfrac{[(k+1)!]^2}{[2(k+1)]}}{\dfrac{(k!)^2}{(2k)!}}$$

$$= \lim_{k \to \infty} \frac{(k+1)! \cdot (k+1)!}{k! \cdot k!} \frac{(2k)!}{(2k+2)!}$$

$$= \lim_{k \to \infty} \frac{(k+1)^2}{(2k+2)(2k+1)} = \lim_{k \to \infty} \frac{(k+1)}{2(2k+1)} = \frac{1}{4}.$$

Because $1/4 < 1$, the series converges by the Ratio Test. We confirm these results with Mathematica.

Remark. Use FullSimplify instead of Simplify to simplify expressions involving factorials.

```
In[482]:= a[k_] = (k!)^2/(2k)!;
          s1 = FullSimplify[a[k + 1]/a[k]]
```
$$Out[482] = \frac{1 + k}{2 + 4\ k}$$

```
In[483]:= Limit[s1, k- > ∞]
```
$$Out[483] = \frac{1}{4}$$

We illustrate that we can approximate the sum using N and Sum as follows.

```
In[484]:= ev = Sum[a[k], {k, 1, ∞}]
```
$$Out[484] = \frac{1}{27}\left(9 + 2\ \sqrt{3}\ \pi\right)$$

```
In[485]:= N[ev]
```
$$Out[485] = 0.7364$$

(e) Because

$$\lim_{k \to \infty} \sqrt[k]{\left(\frac{k}{4k+1}\right)^k} = \lim_{k \to \infty} \frac{k}{4k+1} = \frac{1}{4} < 1,$$

the series converges by the Root Test.

```
In[486]:= a[k_] = (k/(4k + 1))^k;
          Limit[a[k]^(1/k), k- > ∞]
```
$$Out[486] = \frac{1}{4}$$

As with (d), we can approximate the sum with N and Sum.

```
In[487]:= ev = Sum[a[k], {k, 1, ∞}]
          ∞
Out[487]= ∑ a[k]
          k=1
```

```
In[488]:= N[ev]
Out[488]= 0.265757
```

(f) We use the Limit Comparison Test and compare the series to $\sum_{k=1}^{\infty} \frac{\sqrt{k}}{k\sqrt{k}} = \sum_{k=1}^{\infty} \frac{1}{k}$, which diverges because it is a p-series with $p = 1$. Because

$$0 < \lim_{k \to \infty} \frac{\frac{2\sqrt{k}+1}{(\sqrt{k}+1)(2k+1)}}{\frac{1}{k}} = 1 < \infty$$

and the harmonic series diverges, the series diverges by the Limit Comparison Test.

```
In[489]:= a[k_] = (2Sqrt[k] + 1)/((Sqrt[k] + 1)(2k + 1));
          b[k_] = 1/k;
          Limit[a[k]/b[k], k- > ∞]
Out[489]= 1
```

∎

3.4.3 Alternating Series

An **alternating series** is a series of the form

$$\sum_{k=1}^{\infty} (-1)^k a_k \qquad \text{or} \qquad \sum_{k=1}^{\infty} (-1)^{k+1} a_k \tag{3.25}$$

where $\{a_k\}$ is a sequence with positive terms.

Theorem 13 (Alternating Series Test). *If $\{a_k\}$ is decreasing and $\lim_{k \to \infty} a_k = 0$, the alternating series (3.25) converges.*

The alternating series (3.25) **converges absolutely** if $\sum_{k=1}^{\infty} a_k$ converges.

Theorem 14. *If the alternating series (3.25) converges absolutely, it converges.*

If the alternating series (3.25) converges but does not converge absolutely, we say that it **conditionally converges**.

EXAMPLE 3.4.4: Determine whether each series converges or diverges. If the series converges, determine whether the convergence is conditional or absolute. (a) $\sum_{k=1}^{\infty} \frac{(-1)^{k+1}}{k}$; (b) $\sum_{k=1}^{\infty}(-1)^{k+1}\frac{(k+1)!}{4^k(k!)^2}$; (c) $\sum_{k=1}^{\infty}(-1)^{k+1}\left(1+\frac{1}{k}\right)^k$.

SOLUTION: (a) Because $\{1/k\}$ is decreasing and $1/k \to 0$ as $k \to \infty$, the series converges. The series does not converge absolutely because the harmonic series diverges. Hence, $\sum_{k=1}^{\infty} \frac{(-1)^{k+1}}{k}$, which is called the **alternating harmonic series**, converges conditionally. We see that this series converges to $\ln 2$ with Sum.

```
In[490]:= a[k_] = (-1)^(k + 1)/k;
          Sum[a[k], {k, 1, ∞}]
Out[490]= Log[2]
```

(b) We test for absolute convergence first using the Ratio Test. Because

$$\lim_{k\to\infty} \frac{\dfrac{((k+1)+1)!}{4^{k+1}[(k+1)!]^2}}{\dfrac{(k+2)!}{4^k(k!)^2}} = \lim_{k\to\infty} \frac{k+2}{4(k+1)^2} = 0 < 1,$$

```
In[491]:= a[k_] = (k + 1)!/(4^k (k!)^2);
          s1 = FullSimplify[a[k + 1]/a[k]]

          Limit[s1, k- > ∞]
                 2 + k
Out[491]= ─────────
          4 (1 + k)²
Out[491]= 0
```

the series converges absolutely by the Ratio Test. Absolute convergence implies convergence so the series converges. (c) Because $\lim_{k\to\infty}\left(1+\frac{1}{k}\right)^k = e$, $\lim_{k\to\infty}(-1)^{k+1}\left(1+\frac{1}{k}\right)^k$ does not exist, so the series diverges by the Divergence Test. We confirm that the limit of the terms is not zero with Limit.

```
In[492]:= a[k_] = (-1)^(k + 1) (1 + 1/k)^k;
          Sum[a[k], {k, 1, ∞}]
Sum :: "div" :  "Sumdoesnotconverge."
                 ∞
Out[492]= ─────  a[k]
                k=1
```

$$In[493] := \texttt{Limit[a[k], k-> \infty]}$$

$$Out[493] = \texttt{Limit}\left[(-1)^{1+k} \left(1 + \frac{1}{k} \right)^k, k \to \infty \right]$$

■

3.4.4 Power Series

Let x_0 be a number. A **power series** in $x - x_0$ is a series of the form

$$\sum_{k=0}^{\infty} a_k (x - x_0)^k. \tag{3.26}$$

A fundamental problem is determining the values of x, if any, for which the power series converges.

Theorem 15. *For the power series (3.26), exactly one of the following is true.*

1. *The power series converges absolutely for all values of x. The interval of convergence is $(-\infty, \infty)$.*
2. *There is a positive number r so that the series converges absolutely if $x_0 - r < x < x_0 + r$. The series may or may not converge at $x = x_0 - r$ and $x = x_0 + r$. The interval of convergence will be one of $(x_0 - r, x_0 + r)$, $[x_0 - r, x_0 + r)$, $(x_0 - r, x_0 + r]$, or $[x_0 - r, x_0 + r]$.*
3. *The series converges only if $x = x_0$. The interval of convergence is $\{x_0\}$.*

EXAMPLE 3.4.5: Determine the interval of convergence for each of the following power series. (a) $\sum_{k=0}^{\infty} \frac{(-1)^k}{(2k+1)!} x^{2k+1}$ (b) $\sum_{k=0}^{\infty} \frac{k!}{1000^k} (x-1)^k$;

(c) $\sum_{k=1}^{\infty} \frac{2^k}{\sqrt{k}} (x-4)^k$.

SOLUTION: (a) We test for absolute convergence first using the Ratio Test. Because

$$\lim_{k \to \infty} \left| \frac{\dfrac{(-1)^{k+1}}{(2(k+1)+1)!} x^{2(k+1)+1}}{\dfrac{(-1)^k}{(2k+1)!} x^{2k+1}} \right| = \lim_{k \to \infty} \frac{1}{2(k+1)(2k+3)} x^2 = 0 < 1$$

```
In[494] := a[x_, k_] = (-1)^k /(2k + 1) !x^(2k + 1);
          s1 = FullSimplify[a[x, k + 1]/a[x, k]]

          Limit[s1, k- > ∞]
```

$$Out[494] = -\frac{x^2}{6 + 10\ k + 4\ k^2}$$

$$Out[494] = 0$$

for all values of x, we conclude that the series converges absolutely for all values of x; the interval of convergence is $(-\infty, \infty)$. In fact, we will see later that this series converges to $\sin x$:

$$\sin x = \sum_{k=0}^{\infty} \frac{(-1)^{k+1}}{(2k+1)!}x^{2k+1} = x - \frac{1}{3!}x^3 + \frac{1}{5!}x^5 - \frac{1}{7!}x^7 + \dots,$$

which means that the partial sums of the series converge to $\sin x$. Graphically, we can visualize this by graphing partial sums of the series together with the graph of $y = \sin x$. Note that the partial sums of a series are a recursively defined function: $s_n = s_{n-1} + a_n$, $s_0 = a_0$. We use this observation to define p to be the nth partial sum of the series. We use the form p[x_, n_] :=p[x,n] =... so that Mathematica "remembers" the partial sums computed. That is, once p[x,3] is computed, Mathematica need not recompute p[x,3] when computing p[x,4].

```
In[495] := Clear[p]

          p[x_, 0] = a[0];
          p[x_, n_] := p[x, n] = p[x, n - 1] + a[x, n]

In[496] := p[x, 2]
```

$$Out[496] = x - \frac{x^3}{6} + \frac{x^5}{120}$$

In Figure 3-41 we graph $p_n(x) = \sum_{k=0}^{n} \frac{(-1)^k}{(2k+1)!}x^{2k+1}$ together with $y = \sin x$ for $n = 1, 5$, and 10. In the graphs, notice that as n increases, the graphs of $p_n(x)$ more closely resemble the graph of $y = \sin x$.

```
In[497] := Plot[{Sin[x], p[x, 1], p[x, 5], p[x, 10]},
          {x, -2π, 2π}, PlotRange- > {-π, π},
          AspectRatio- > Automatic,
          PlotStyle- > {GrayLevel[0], GrayLevel[0.3],
            Dashing[{0.01}], {GrayLevel[0.3],
            Dashing[{0.01}]}}]
```

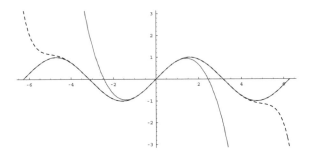

Figure 3-41 $y = \sin x$ together with the graphs of $p_1(x)$, $p_5(x)$, and $p_{10}(x)$

(b) As in (a), we test for absolute convergence first using the Ratio Test:

$$\lim_{k \to \infty} \left| \frac{\dfrac{(k+1)k!}{1000^{k+1}}(x-1)^{k+1}}{\dfrac{k!}{1000^k}(x-1)^k} \right| = \frac{1}{1000}(k+1)|x-1| = \begin{cases} 0, \text{ if } x = 1 \\ \infty, \text{ if } x \neq 1. \end{cases}$$

```
In[498]:= a[x_, k_] = k!/1000^k (x - 1)^k;
          s1 = FullSimplify[a[x, k + 1]/a[x, k]]

          Limit[s1, k- > ∞]
```
$$Out[498] = \frac{(1 + k)\ (-1 + x)}{1000}$$
```
Out[498]= Indeterminate
```

Be careful of your interpretation of the result of the Limit command because Mathematica does not consider the case $x = 1$ separately: if $x = 1$ the limit is 0. Because $0 < 1$ the series converges by the Ratio Test.

The series converges only if $x = 1$; the interval of convergence is $\{1\}$. You should observe that if you graph several partial sums for "small" values of n, you might incorrectly conclude that the series converges.
(c) Use the Ratio Test to check absolute convergence first:

$$\lim_{k \to \infty} \left| \frac{\dfrac{2^{k+1}}{\sqrt{k+1}}(x-4)^{k+1}}{\dfrac{2^k}{\sqrt{k}}(x-4)^k} \right| = \lim_{k \to \infty} 2\sqrt{\frac{k}{k+1}}|x-4| = 2|x-4|.$$

By the Ratio Test, the series converges absolutely if $2|x-4| < 1$. We solve this inequality for x with InequalitySolve to see that $2|x-4| < 1$ if $7/2 < x < 9/2$.

```
In[499]:= a[x_, k_] = 2^k /Sqrt[k] (x - 4)^k;
          s1 = FullSimplify[Abs[a[x, k + 1]/a[x, k]]]

          Limit[s1, k- > ∞]
```

$$Out[499] = 2 \; Abs\left[\sqrt{\frac{k}{1+k}} \; (-4+x)\right]$$

$$Out[499] = 2 \; Abs[-4+x]$$

```
In[500]:= << Algebra`InequalitySolve`

          InequalitySolve[2 Abs[x - 4] < 1, x]
```

$$Out[500] = \frac{7}{2} < x < \frac{9}{2}$$

We check $x = 7/2$ and $x = 9/2$ separately. If $x = 7/2$, the series becomes $\sum_{k=1}^{\infty}(-1)^k \frac{1}{\sqrt{k}}$, which converges conditionally.

```
In[501]:= Simplify[a[x, k]/.x- > 7/2]
```

$$Out[501] = \frac{(-1)^k}{\sqrt{k}}$$

On the other hand, if $x = 9/2$, the series is $\sum_{k=1}^{\infty} \frac{1}{\sqrt{k}}$, which diverges. We conclude that the interval of convergence is $[7/2, 9/2)$.

```
In[502]:= Simplify[a[x, k]/.x- > 9/2]
```

$$Out[502] = \frac{1}{\sqrt{k}}$$

3.4.5 Taylor and Maclaurin Series

Let $y = f(x)$ be a function with derivatives of all orders at $x = x_0$. The **Taylor series** for $f(x)$ about $x = x_0$ is

$$\sum_{k=0}^{\infty} \frac{f^{(k)}(x_0)}{k!} (x - x_0)^k. \tag{3.27}$$

The **Maclaurin series** for $f(x)$ is the Taylor series for $f(x)$ about $x = 0$. If $y = f(x)$ has derivatives up to at least order n at $x = x_0$, the nth degree **Taylor polynomial** for $f(x)$ about $x = x_0$ is

$$p_n(x) = \sum_{k=0}^{n} \frac{f^{(k)}(x_0)}{k!} (x - x_0)^k. \tag{3.28}$$

The nth degree **Maclaurin polynomial** for $f(x)$ is the nth degree Taylor polynomial for $f(x)$ about $x = 0$. Generally, finding Taylor and Maclaurin series using the definition is a tedious task at best.

EXAMPLE 3.4.6: Find the first few terms of (a) the Maclaurin series and (b) the Taylor series about $x = \pi/4$ for $f(x) = \tan x$.

SOLUTION: (a) After defining $f(x) = \tan x$, we use Table together with /. and D to compute $f^{(k)}(0)/k!$ for $k = 0, 1, \ldots, 8$.

```
In[503]:= f[x_] = Tan[x];
          Table[
            {k, D[f[x], {x, k}], D[f[x], {x, k}]/.x- > 0},
            {k, 0, 8}]
```

```
Out[503]= {{0, Tan[x], 0}, {1, Sec[x]², 1},
            {2, 2 Sec[x]² Tan[x], 0},
            {3, 2 Sec[x]⁴ + 4 Sec[x]² Tan[x]², 2},
            {4, 16 Sec[x]⁴ Tan[x] + 8 Sec[x]² Tan[x]³, 0},
            {5, 16 Sec[x]⁶ + 88 Sec[x]⁴ Tan[x]²+
                16 Sec[x]² Tan[x]⁴, 16},
            {6, 272 Sec[x]⁶ Tan[x]+
                416 Sec[x]⁴ Tan[x]³ + 32 Sec[x]² Tan[x]⁵, 0},
            {7, 272 Sec[x]⁸ + 2880 Sec[x]⁶ Tan[x]²+
                1824 Sec[x]⁴ Tan[x]⁴+
                64 Sec[x]² Tan[x]⁶, 272},
            {8, 7936 Sec[x]⁸ Tan[x]+
                24576 Sec[x]⁶ Tan[x]³+
                7680 Sec[x]⁴ Tan[x]⁵+
                128 Sec[x]² Tan[x]⁷, 0}}
```

Using the values in the table, we apply the definition to see that the Maclaurin series is

$$\sum_{k=0}^{\infty} \frac{f^{(k)}(0)}{k!} x^k = x + \frac{1}{3}x^3 + \frac{2}{15}x^5 + \frac{17}{315}x^7 + \ldots$$

For (b), we repeat (a) using $x = \pi/4$ instead of $x = 0$

```
In[504]:= f[x_] = Tan[x];
          Table[
            {k, D[f[x], {x, k}], D[f[x], {x, k}]/.x- > π/4},
            {k, 0, 8}]
```

$$Out[504]= \{\{0, \text{Tan}[\text{x}], 1\}, \{1, \text{Sec}[\text{x}]^2, 2\},$$
$$\{2, 2\ \text{Sec}[\text{x}]^2\ \text{Tan}[\text{x}], 4\},$$
$$\{3, 2\ \text{Sec}[\text{x}]^4 + 4\ \text{Sec}[\text{x}]^2\ \text{Tan}[\text{x}]^2, 16\},$$
$$\{4, 16\ \text{Sec}[\text{x}]^4\ \text{Tan}[\text{x}] + 8\ \text{Sec}[\text{x}]^2\ \text{Tan}[\text{x}]^3, 80\},$$
$$\{5, 16\ \text{Sec}[\text{x}]^6 + 88\ \text{Sec}[\text{x}]^4\ \text{Tan}[\text{x}]^2 +$$
$$16\ \text{Sec}[\text{x}]^2\ \text{Tan}[\text{x}]^4, 512\},$$
$$\{6, 272\ \text{Sec}[\text{x}]^6\ \text{Tan}[\text{x}] + 416\ \text{Sec}[\text{x}]^4\ \text{Tan}[\text{x}]^3 +$$
$$32\ \text{Sec}[\text{x}]^2\ \text{Tan}[\text{x}]^5, 3904\},$$
$$\{7, 272\ \text{Sec}[\text{x}]^8 + 2880\ \text{Sec}[\text{x}]^6\ \text{Tan}[\text{x}]^2 +$$
$$1824\ \text{Sec}[\text{x}]^4\ \text{Tan}[\text{x}]^4 +$$
$$64\ \text{Sec}[\text{x}]^2\ \text{Tan}[\text{x}]^6, 34816\},$$
$$\{8, 7936\ \text{Sec}[\text{x}]^8\ \text{Tan}[\text{x}] +$$
$$24576\ \text{Sec}[\text{x}]^6\ \text{Tan}[\text{x}]^3 +$$
$$7680\ \text{Sec}[\text{x}]^4\ \text{Tan}[\text{x}]^5 +$$
$$128\ \text{Sec}[\text{x}]^2\ \text{Tan}[\text{x}]^7, 354560\}\}$$

and then apply the definition to see that the Taylor series about $x = \pi/4$ is

$$\sum_{k=0}^{\infty} \frac{f^{(k)}(x_0)}{k!}(x - x_0)^k = 1 + 2\left(x - \frac{\pi}{4}\right) + 2\left(x - \frac{\pi}{4}\right)^2 + \frac{8}{3}\left(x - \frac{\pi}{4}\right)^3 +$$

$$\frac{10}{3}\left(x - \frac{\pi}{4}\right)^4 + \frac{64}{15}\left(x - \frac{\pi}{4}\right)^5 + \frac{244}{45}\left(x - \frac{\pi}{4}\right)^6 + \dots$$

From the series, we can see various Taylor and Maclaurin polynomials. For example, the third Maclaurin polynomial is

$$p_3(x) = x + \frac{1}{3}x^3$$

and the 4th degree Taylor polynomial about $x = \pi/4$ is

$$p_4(x) = 1 + 2\left(x - \frac{\pi}{4}\right) + 2\left(x - \frac{\pi}{4}\right)^2 + \frac{8}{3}\left(x - \frac{\pi}{4}\right)^3 + \frac{10}{3}\left(x - \frac{\pi}{4}\right)^4.$$

■

The command

$$\text{Series}[f[x], \{x, x0, n\}]$$

computes (3.27) to (at least) order $n - 1$. Because of the O-term in the result that represents the terms that are omitted from the power series for $f(x)$ expanded about the point $x = x_0$, the result of entering a Series command is not a function that can be evaluated if x is a particular number. We remove the remainder (O-) term of the power series `Series[f[x],{x,x0,n}]` with the command `Normal` and can then evaluate the resulting polynomial for particular values of x.

EXAMPLE 3.4.7: Find the first few terms of the Taylor series for $f(x)$ about $x = x_0$. (a) $f(x) = \cos x, x = 0$; (b) $f(x) = 1/x^2, x = 1$.

SOLUTION: Entering

```
In[505]:= Series[Cos[x],{x,0,4}]
```
$$Out[505]= 1 - \frac{x^2}{2} + \frac{x^4}{24} + O[x]^5$$

computes the Maclaurin series to order 4. Entering

```
In[506]:= Series[Cos[x],{x,0,14}]
```
$$Out[506]= 1 - \frac{x^2}{2} + \frac{x^4}{24} - \frac{x^6}{720} + \frac{x^8}{40320} - \frac{x^{10}}{3628800} + \frac{x^{12}}{479001600} -$$
$$\frac{x^{14}}{87178291200} + O[x]^{15}$$

computes the Maclaurin series to order 14. In this case, the Maclaurin series for $\cos x$ converges to $\cos x$ for all real x. To graphically see this, we define the function p. Given n, p[n] returns the Maclaurin polynomial of degree n for $\cos x$.

```
In[507]:= p[n_]:= Series[Cos[x],{x,0,n}]//Normal

In[508]:= p[8]
```
$$Out[508]= 1 - \frac{x^2}{2} + \frac{x^4}{24} - \frac{x^6}{720} + \frac{x^8}{40320}$$

We then graph $\cos x$ together with the Maclaurin polynomial of degree $n = 2, 4, 8$, and 16 on the interval $[-3\pi/2, 3\pi/2]$ in Figure 3-42. Notice that as n increases, the graph of the Maclaurin polynomial more closely resembles the graph of $\cos x$. We would see the same pattern if we increased the length of the interval and the value of n.

```
In[509]:= somegraphs = Table[Plot[Evaluate[{Cos[x],
              p[2^n]}],{x,-3π/2,3π/2},
              PlotRange->{-3π/2,3π/2},
              AspectRatio->Automatic,
              PlotStyle->{GrayLevel[0],
              GrayLevel[0.3]},
              DisplayFunction->Identity],
           {n,1,4}]
```

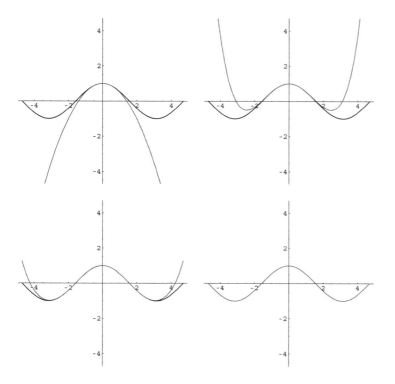

Figure 3-42   Graphs of $y = \cos x$ together with its second, fourth, eighth, and sixteenth Maclaurin polynomials

```
Out[509]= BoxData({-Graphics-, -Graphics-,
                    -Graphics-, -Graphics-})

In[510]:= toshow = Partition[somegraphs, 2]
Out[510]= BoxData({{-Graphics-, -Graphics-},
                    {-Graphics-, -Graphics-}})

In[511]:= Show[GraphicsArray[toshow]]
```

(b) After defining $f(x) = 1/x^2$, we compute the first 10 terms of the Taylor series for $f(x)$ about $x = 1$ with Series.

```
In[512]:= f[x_] = 1/x^2;
          p10 = Series[f[x], {x, 1, 10}]
Out[512]= 1 - 2 (x - 1) + 3 (x - 1)² - 4 (x - 1)³ +
            5 (x - 1)⁴ - 6 (x - 1)⁵ + 7 (x - 1)⁶ - 8 (x - 1)⁷ +
            9 (x - 1)⁸ - 10 (x - 1)⁹ + 11 (x - 1)¹⁰ +
            O[x - 1]¹¹
```

In this case, the pattern for the series is relatively easy to see: the Taylor series for $f(x)$ about $x = 1$ is

$$\sum_{k=0}^{\infty} (-1)^k (k+1)(x-1)^k.$$

This series converges absolutely if

$$\lim_{k \to \infty} \left| \frac{(-1)^{k+1}(k+2)(x-1)^{k+1}}{(-1)^k(k+1)(x-1)^k} \right| = |x-1| < 1$$

or $0 < x < 2$. The series diverges if $x = 0$ and $x = 2$. In this case, the series converges to $f(x)$ on the interval $(0, 2)$.

```
In[513]:= a[x_, k_] = (-1)^k (k + 1) (x - 1)^k;
          s1 = FullSimplify[Abs[a[x, k + 1]/a[x, k]]]
```
$$Out[513]= \text{Abs}\left[\frac{(2+k) \ (-1+x)}{1+k} \right]$$

```
In[514]:= s2 = Limit[s1, k- > ∞]
```
$$Out[514]= \text{Abs}[-1 + x]$$

```
In[515]:= << Algebra`InequalitySolve`

          InequalitySolve[s2 < 1, x]
```
$$Out[515]= 0 < x < 2$$

To see this, we graph $f(x)$ together with the Taylor polynomial for $f(x)$ about $x = 1$ of degree n for large n. Regardless of the size of n, the graphs of $f(x)$ and the Taylor polynomial closely resemble each other on the interval $(0, 2)$–but not at the endpoints or outside the interval. (See Figure 3-43.)

```
In[516]:= p[n_] := Series[f[x], {x, 1, n + 1}]//Normal

In[517]:= Plot[Evaluate[{f[x], p[16]}],
             {x, 0, 2}, PlotRange- > {-5, 45},
             PlotStyle- > {GrayLevel[0], GrayLevel[0.3]}]
```

■

3.4.6 Taylor's Theorem

Taylor's theorem states the relationship between $f(x)$ and the Taylor series for $f(x)$ about $x = x_0$.

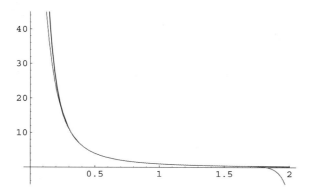

Figure 3-43   Graph of $f(x)$ together with the sixteenth degree Taylor polynomial about $x = 1$

Theorem 16 (Taylor's Theorem). *Let $y = f(x)$ have (at least) $n + 1$ derivatives on an interval I containing $x = x_0$. Then, for every number $x \in I$, there is a number z between x and x_0 so that*

$$f(x) = p_n(x) + R_n(x),$$

where $p_n(x)$ is given by equation (3.28) and

$$R_n(x) = \frac{f^{(n+1)}(z)}{(n + 1)!} (x - x_0)^{n+1}. \tag{3.29}$$

EXAMPLE 3.4.8: Use Taylor's theorem to show that

$$\sin x = \sum_{k=0}^{\infty} \frac{(-1)^k}{(2k + 1)!} x^{2k+1}$$

SOLUTION: Let $f(x) = \sin x$. Then, for each value of x, there is a number z between 0 and x so that $\sin x = p_n(x) + R_n(x)$ where $p_n(x) = \sum_{k=0}^{n} \frac{f^{(k)}(0)}{k!} x^k$ and $R_n(x) = \frac{f^{(n+1)}(z)}{(n+1)!} x^{n+1}$. Regardless of the value of n, $f^{(n+1)}(z)$ is one of $\sin z, -\sin z, \cos z,$ or $-\cos z$, which are all bounded by 1. Then,

$$|\sin x - p_n(x)| = \left| \frac{f^{(n+1)}(z)}{(n + 1)!} x^{n+1} \right|$$

$$|\sin x - p_n(x)| \le \frac{1}{(n + 1)!} |x|^{n+1}$$

and $\frac{x^n}{n!} \to 0$ as $n \to \infty$ for all real values of x.

You should remember that the number z in $R_n(x)$ is guaranteed to exist by Taylor's theorem. However, from a practical point of view, you would rarely (if ever) need to compute the z value for a particular x value.

For illustrative purposes, we show the difficulties. Suppose we wish to approximate $\sin \pi/180$ using the Maclaurin polynomial of degree 4, $p_4(x) = x - \frac{1}{6}x^3$, for $\sin x$. The fourth remainder is

The Maclaurin polynomial of degree 4 for $\sin x$ is
$\sum_{k=0}^{4} \frac{f^{(k)}(0)}{k!} x^4 =$
$0 + x + 0 \cdot x^2 + \frac{-1}{3!}x^3 + 0 \cdot x^4.$

$$R_4(x) = \frac{1}{120} \cos z x^5.$$

```
In[518]:= f[x_] = Sin[x];
          r5 = D[f[z],{z,5}]/5! x^5
```
$$Out[518] = \frac{1}{120} \ x^5 \ \text{Cos}[z]$$

If $x = \pi/180$ there is a number z between 0 and $\pi/180$ so that

$$\left| R_4\left(\frac{\pi}{180}\right) \right| = \frac{1}{120} \cos z \left(\frac{\pi}{180}\right)^5$$

$$\leq \frac{1}{120}\left(\frac{\pi}{180}\right)^5 \approx 0.135 \times 10^{-10},$$

which shows us that the maximum the error can be is $\frac{1}{120}\left(\frac{\pi}{180}\right)^5 \approx 0.135 \times 10^{-10}$.

```
In[519]:= maxerror = N[1/120 * (π/180)^5]
```
$$Out[519] = 1.3496\,10^{-11}$$

Abstractly, the exact error can be computed. By Taylor's theorem, z satisfies

$$f\left(\frac{\pi}{180}\right) = p_4\left(\frac{\pi}{180}\right) + R_4\left(\frac{\pi}{180}\right)$$

$$\sin \frac{\pi}{180} = \frac{1}{180}\pi - \frac{1}{34992000}\pi^3 + \frac{1}{22674816000000}\pi^5 \cos z$$

$$0 = \frac{1}{180}\pi - \frac{1}{34992000}\pi^3 + \frac{1}{22674816000000}\pi^5 \cos z - \sin \frac{\pi}{180}.$$

We graph the right-hand side of this equation with Plot in Figure 3-44. The exact value of z is the z-coordinate of the point where the graph intersects the z-axis.

```
In[520]:= p4 = Series[f[x],{x,0,4}]//Normal
```
$$Out[520] = x - \frac{x^3}{6}$$

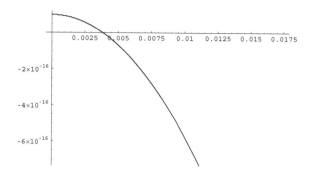

Figure 3-44   Finding z

$In[521]:=$ **exval = Sin[π/180]**

p4b = p4/.x- > π/180

r5b = r5/.x- > π/180

$Out[521]=$ $\mathrm{Sin}\left[\dfrac{\pi}{180}\right]$

$Out[521]=$ $\dfrac{\pi}{180} - \dfrac{\pi^3}{34992000}$

$Out[521]=$ $\dfrac{\pi^5\ \mathrm{Cos}[z]}{22674816000000}$

$In[522]:=$ **toplot = r5b + p4b - exval;**
Plot[toplot, {z, 0, π/180}]

We can use FindRoot to approximate z, if we increase the number of digits carried in floating point calculations with WorkingPrecision.

$In[523]:=$ **exz = FindRoot[toplot == 0, {z, 0.004},**
WorkingPrecision- > 32]
$Out[523]=$ $\{z \to 0.0038086149165541607949333316330124\}$

Alternatively, we can compute the exact value of z with Solve

$In[524]:=$ **cz = Solve[toplot == 0, z]**
Solve :: "ifun" :   "Inversefunctionsarebeingused
bySolve, sosomesolutionsmaynotbefound."
$Out[524]=$ $\left\{\left\{z \to -\mathrm{ArcCos}\right.\right.$
$\left[\dfrac{648000\ \left(-194400\ \pi + \pi^3 + 34992000\ \mathrm{Sin}\left[\frac{\pi}{180}\right]\right)}{\pi^5}\right]\Big\}$,
$\left\{z \to \mathrm{ArcCos}\right.$
$\left[\dfrac{648000\ \left(-194400\ \pi + \pi^3 + 34992000\ \mathrm{Sin}\left[\frac{\pi}{180}\right]\right)}{\pi^5}\right]\Big\}\Big\}$

and then approximate the result with N.

$$In[525]:= \textbf{N[cz]}$$

$$Out[525]= \{\{z \to -0.00384232\},$$
$$\{z \to 0.00384232\}\}$$

■

3.4.7 Other Series

In calculus, we learn that the power series $f(x) = \sum_{k=0}^{\infty} a_k (x - x_0)^k$ is differentiable and integrable on its interval of convergence. However, for series that are not power series this result is not generally true. For example, in more advanced courses, we learn that the function

$$f(x) = \sum_{k=0}^{\infty} \frac{1}{2^k} \sin\left(3^k x\right)$$

is continuous for all values of x but nowhere differentiable. We can use Mathematica to help us see why this function is not differentiable. Let

$$f_n(x) = \sum_{k=0}^{n} \frac{1}{2^k} \sin\left(3^k x\right).$$

Notice that $f_n(x)$ is defined recursively by $f_0(x) = \sin x$ and $f_n(x) = f_{n-1}(x) + \frac{1}{2^n} \sin\left(3^n x\right)$. We use Mathematica to recursively define $f_n(x)$.

$$In[526]:= \textbf{f[n\_] := f[n] = f[n - 1] + Sin[3^n x]/2^n;}$$
$$\textbf{f[0] = Sin[x];}$$

We define $f_n(x)$ using the form

$$\texttt{f[n\_]:=f[n]=...}$$

so that Mathematica "remembers" the values it computes. Thus, to compute f[5], Mathematica uses the previously computed values, namely f[4], to compute f[5]. Note that we can produce the same results by defining $f_n(x)$ with the command

$$\texttt{f[n\_]:=...}$$

However, the disadvantage of defining $f_n(x)$ in this manner is that Mathematica does not "remember" the previously computed values and thus takes longer to compute $f_n(x)$ for larger values of n.

Next, we use Table to generate $f_3(x)$, $f_6(x)$, $f_9(x)$, and $f_{12}(x)$.

```
In[527]:= posums = Table[f[n], {n, 3, 12, 3}]
```

$$Out[527]= \left\{ \text{Sin}[x] + \frac{1}{2} \text{Sin}[3 \ x] + \frac{1}{4} \text{Sin}[9 \ x] + \frac{1}{8} \text{Sin}[27 \ x], \right.$$

$$\text{Sin}[x] + \frac{1}{2} \text{Sin}[3 \ x] + \frac{1}{4} \text{Sin}[9 \ x] + \frac{1}{8} \text{Sin}[27 \ x] +$$

$$\frac{1}{16} \text{Sin}[81 \ x] + \frac{1}{32} \text{Sin}[243 \ x] +$$

$$\frac{1}{64} \text{Sin}[729 \ x], \text{Sin}[x] + \frac{1}{2} \text{Sin}[3 \ x] +$$

$$\frac{1}{4} \text{Sin}[9 \ x] + \frac{1}{8} \text{Sin}[27 \ x] + \frac{1}{16} \text{Sin}[81 \ x] +$$

$$\frac{1}{32} \text{Sin}[243 \ x] + \frac{1}{64} \text{Sin}[729 \ x] +$$

$$\frac{1}{128} \text{Sin}[2187 \ x] + \frac{1}{256} \text{Sin}[6561 \ x] +$$

$$\frac{1}{512} \text{Sin}[19683 \ x], \text{Sin}[x] + \frac{1}{2} \text{Sin}[3 \ x] +$$

$$\frac{1}{4} \text{Sin}[9 \ x] + \frac{1}{8} \text{Sin}[27 \ x] + \frac{1}{16} \text{Sin}[81 \ x] +$$

$$\frac{1}{32} \text{Sin}[243 \ x] + \frac{1}{64} \text{Sin}[729 \ x] +$$

$$\frac{1}{128} \text{Sin}[2187 \ x] + \frac{1}{256} \text{Sin}[6561 \ x] +$$

$$\frac{1}{512} \text{Sin}[19683 \ x] + \frac{\text{Sin}[59049 \ x]}{1024} +$$

$$\left. \frac{\text{Sin}[177147 \ x]}{2048} + \frac{\text{Sin}[531441 \ x]}{4096} \right\}$$

We now graph each of these functions and show the results as a graphics array with GraphicsArray in Figure 3-45.

```
In[528]:= somegraphs = Map[Plot[#, {x, 0, 3π},
                    DisplayFunction- > Identity]&, posums];
          toshow = Partition[somegraphs, 2];
          Show[GraphicsArray[toshow]]
```

From these graphs, we see that for large values of n, the graph of $f_n(x)$, although actually smooth, appears "jagged" and thus we might suspect that $f(x) = \lim_{n \to \infty} f_n(x) = \sum_{k=0}^{\infty} \frac{1}{2^k} \sin(3^k x)$ is indeed continuous everywhere but nowhere differentiable.

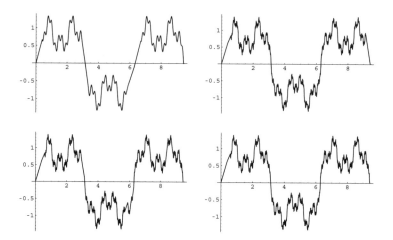

Figure 3-45   Approximating a function that is continuous everywhere but nowhere differentiable

3.5 Multi-Variable Calculus

Mathematica is useful in investigating functions involving more than one variable. In particular, the graphical analysis of functions that depend on two (or more) variables is enhanced with the help of Mathematica's graphics capabilities.

3.5.1 Limits of Functions of Two Variables

Mathematica's graphics and numerical capabilities are helpful in investigating limits of functions of two variables.

EXAMPLE 3.5.1: Show that the limit $\lim_{(x,y)\to(0,0)} \dfrac{x^2 - y^2}{x^2 + y^2}$ does not exist.

SOLUTION: We begin by defining $f(x, y) = \frac{x^2-y^2}{x^2+y^2}$. Next, we use Plot3D to graph $z = f(x, y)$ for $-1/2 \le x \le 1/2$ and $-1/2 \le y \le 1/2$. ContourPlot is used to graph several level curves on the same rectangle. (See Figure 3-46.) (To define a function of two variables, $f(x, y) = expression\ in\ x\ and\ y$, enter f[x_,y_]=expression in x and y. Plot3D[f[x,y],

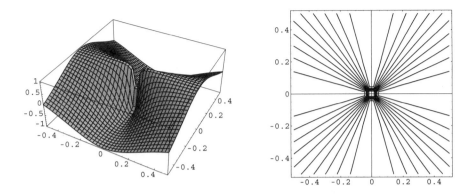

Figure 3-46 (a) Three-dimensional and (b) contour plots of $f(x, y)$

$\{a, x, b\}, \{y, c, d\}]$ generates a basic graph of $z = f(x, y)$ for $a \leq x \leq b$ and $c \leq y \leq d$)

```
In[529]:= f[x_, y_] = (x^2 - y^2)/(x^2 + y^2);
          p1 = Plot3D[f[x, y], {x, -0.5, 0.5},
             {y, -0.5, 0.5}, PlotPoints- > {40, 40},
             DisplayFunction- > Identity];
          p2 = ContourPlot[f[x, y],
             {x, -0.5, 0.5}, {y, -0.5, 0.5},
             PlotPoints- > 40,
             ContourShading- > False, Axes- > Automatic,
             AxesOrigin- > {0, 0},
             DisplayFunction- > Identity];
          Show[GraphicsArray[{p1, p2}]]
```

From the graph of the level curves, we suspect that the limit does not exist because we see that near $(0, 0)$, $z = f(x, y)$ attains many different values. We obtain further evidence that the limit does not exist by computing the value of $z = f(x, y)$ for various points chosen randomly near $(0, 0)$. We use `Table` and `Random` to generate 13 ordered triples $(x, y, f(x, y))$ for x and y "close to" 0. Because `Random` is included in the calculation, your results will almost certainly be different from those here. The first column corresponds to the x-coordinate, the second column the y-coordinate, and the third column the value of $z = f(x, y)$.

```
In[530]:= r[n_] := {Random[Real, {-10^(-n), 10^(-n)}],
             Random[Real, {-10^(-n), 10^(-n)}]}

In[531]:= r[1]
Out[531]= {5.25152, 9.37514}
```

In[532] := **toevaluate = Table[r[n], {n, 1, 15}]**

Out[532]= {{0.043922, 0.0768676},
        {-0.00775639, 0.0039307},
        {-0.0000561454,
          -0.0000790007},
        {0.0000536954, 0.0000373069},
        {3.24752 10^{-6}, 7.41243 10^{-6}},
        {1.70105 10^{-7}, -6.6412 10^{-7}},
        {-1.15231 10^{-9}, -8.69882 10^{-8}},
        {3.85914 10^{-9}, 4.18814 10^{-9}},
        {-4.07047 10^{-10},
          7.06248 10^{-10}},
        {8.19068 10^{-11}, 4.65551 10^{-11}},
        {1.67581 10^{-12},
          -8.23982 10^{-12}},
        {8.47593 10^{-13},
          -8.23785 10^{-13}},
        {7.28361 10^{-14},
          -5.92658 10^{-14}},
        {6.23232 10^{-15},
          -2.16855 10^{-15}},
        {-2.15493 10^{-16},
          4.86343 10^{-16}}}

In[533] := **Map[f[#[[1]], #[[2]]]&, toevaluate]**

Out[533]= {-0.507731,
        0.591324, -0.328828,
        0.348863, -0.677926,
        -0.876866, -0.999649,
        -0.0816327, -0.501298,
        0.511638, -0.920559,
        0.0284831, 0.203308,
        0.784009, -0.671783}

From the third column, we see that $z = f(x, y)$ does not appear to approach any particular value for points chosen randomly near $(0, 0)$. In fact, along the line $y = mx$ we see that

We choose lines of the form $y = mx$ because near $(0, 0)$ the level curves of $z = f(x, y)$ look like lines of the form $y = mx$.

$$f(x, y) = f(x, mx) = \frac{1 - m^2}{1 + m^2}.$$

Hence as $(x, y) \to (0, 0)$ along $y = mx$, $f(x, y) = f(x, mx) \to \frac{1-m^2}{1+m^2}$. Thus, $f(x, y)$ does not have a limit as $(x, y) \to (0, 0)$.

```
In[534]:= v1 = Simplify[f[x, m x]]

         v1 /.m- > 0

         v1/.m- > 1

         v1 /. m- > 1/2
```

$$Out[534] = \frac{1 - m^2}{1 + m^2}$$

$$Out[534] = 1$$

$$Out[534] = 0$$

$$Out[534] = \frac{3}{5}$$

∎

In some cases, you can establish that a limit does not exist by converting to polar coordinates. For example, in polar coordinates, $f(x, y) = \frac{x^2 - y^2}{x^2 + y^2}$ becomes $f(r \cos \theta, r \sin \theta) = 2 \cos^2 \theta - 1$

```
In[535]:= Simplify[f[r Cos[t], r Sin[t]]]
Out[535]= Cos[2 t]
```

and

$$\lim_{(x,y) \to (0,0)} f(x, y) = \lim_{r \to 0} f(r \cos \theta, r \sin \theta) = \lim_{r \to 0} 2 \cos^2 \theta - 1 = 2 \cos^2 \theta - 1 = \cos 2\theta$$

depends on θ.

3.5.2 Partial and Directional Derivatives

Partial derivatives of functions of two or more variables are computed with Mathematica using D. For $z = f(x, y)$,

1. `D[f[x,y],x]` computes $\frac{\partial f}{\partial x} = f_x(x, y)$,
2. `D[f[x,y],y]` computes $\frac{\partial f}{\partial y} = f_y(x, y)$,
3. `D[f[x,y],{x,n}]` computes $\frac{\partial^n f}{\partial x^n}$,
4. `D[f[x,y],y,x]` computes $\frac{\partial^2 f}{\partial y \partial x} = f_{xy}(x, y)$, and
5. `D[f[x,y],{x,n},{y,m}]` computes $\frac{\partial^{n+m} f}{\partial^n x \partial^m y}$.

The calculations are carried out similarly for functions of more than two variables.

EXAMPLE 3.5.2: Calculate $f_x(x, y)$, $f_y(x, y)$, $f_{xy}(x, y)$, $f_{yx}(x, y)$, $f_{xx}(x, y)$, and $f_{yy}(x, y)$ if $f(x, y) = \sin \sqrt{x^2 + y^2 + 1}$.

SOLUTION: After defining $f(x, y) = \sin \sqrt{x^2 + y^2 + 1}$,

$$In[536] := \texttt{f[x\_,y\_] = Sin[Sqrt[x\^2 + y\^2 + 1]];}$$

we illustrate the use of D to compute the partial derivatives. Entering

$$In[537] := \texttt{D[f[x,y],x]}$$
$$Out[537] = \frac{x \; \text{Cos}\left[\sqrt{1 + x^2 + y^2}\right]}{\sqrt{1 + x^2 + y^2}}$$

computes $f_x(x, y)$. Entering

$$In[538] := \texttt{D[f[x,y],y]}$$
$$Out[538] = \frac{y \; \text{Cos}\left[\sqrt{1 + x^2 + y^2}\right]}{\sqrt{1 + x^2 + y^2}}$$

computes $f_y(x, y)$. Entering

$$In[539] := \texttt{D[f[x,y],x,y]//Together}$$
$$Out[539] = \frac{-x \; y \; \text{Cos}\left[\sqrt{1 + x^2 + y^2}\right] - x \; y \; \sqrt{1 + x^2 + y^2} \; \text{Sin}\left[\sqrt{1 + x^2 + y^2}\right]}{(1 + x^2 + y^2)^{3/2}}$$

computes $f_{yx}(x, y)$. Entering

$$In[540] := \texttt{D[f[x,y],y,x]//Together}$$
$$Out[540] = \frac{-x \; y \; \text{Cos}\left[\sqrt{1 + x^2 + y^2}\right] - x \; y \; \sqrt{1 + x^2 + y^2} \; \text{Sin}\left[\sqrt{1 + x^2 + y^2}\right]}{(1 + x^2 + y^2)^{3/2}}$$

computes $f_{xy}(x, y)$. Remember that under appropriate assumptions, $f_{xy}(x, y) = f_{yx}(x, y)$. Entering

$$In[541] := \texttt{D[f[x,y],\{x,2\}]//Together}$$
$$Out[541] = \frac{1}{(1 + x^2 + y^2)^{3/2}} \left(\text{Cos}\left[\sqrt{1 + x^2 + y^2}\right] + y^2 \; \text{Cos}\left[\sqrt{1 + x^2 + y^2}\right] - \right.$$
$$\left. x^2 \; \sqrt{1 + x^2 + y^2} \; \text{Sin}\left[\sqrt{1 + x^2 + y^2}\right] \right)$$

computes $f_{xx}(x, y)$. Entering

$$In[542] := \texttt{D[f[x,y],\{y,2\}]//Together}$$
$$Out[542] = \frac{1}{(1 + x^2 + y^2)^{3/2}} \left(\text{Cos}\left[\sqrt{1 + x^2 + y^2}\right] + x^2 \; \text{Cos}\left[\sqrt{1 + x^2 + y^2}\right] - \right.$$
$$\left. y^2 \; \sqrt{1 + x^2 + y^2} \; \text{Sin}\left[\sqrt{1 + x^2 + y^2}\right] \right)$$

computes $f_{yy}(x, y)$.

∎

The **directional derivative** of $z = f(x, y)$ in the direction of the unit vector $\mathbf{u} = \cos\theta\,\mathbf{i} + \sin\theta\,\mathbf{j}$ is

$$D_{\mathbf{u}}f(x, y) = f_x(x, y)\cos\theta + f_y(x, y)\sin\theta,$$

The vectors $\mathbf{i}$ and $\mathbf{j}$ are defined by $\mathbf{i} = \langle 1, 0 \rangle$ and $\mathbf{j} = \langle 0, 1 \rangle$.

provided that $f_x(x, y)$ and $f_y(x, y)$ both exist.

If $f_x(x, y)$ and $f_y(x, y)$ both exist, the **gradient** of $f(x, y)$ is the vector-valued function

$$\nabla f(x, y) = f_x(x, y)\mathbf{i} + f_y(x, y)\mathbf{j} = \langle f_x(x, y), f_y(x, y) \rangle.$$

Calculus of vector-valued functions is discussed in more detail in Chapter 5.

Notice that if $\mathbf{u} = \langle \cos\theta, \sin\theta \rangle$,

$$D_{\mathbf{u}}f(x, y) = \nabla f(x, y) \cdot \langle \cos\theta, \sin\theta \rangle.$$

EXAMPLE 3.5.3: Let

$$f(x, y) = 6x^2y - 3x^4 - 2y^3.$$

(a) Find $D_{\mathbf{u}}f(x, y)$ in the direction of $\mathbf{v} = \langle 3, 4 \rangle$. (b) Compute

$$D_{\langle 3/5, 4/5 \rangle}f\left(\frac{1}{3}\sqrt{9 + 3\sqrt{3}}, 1\right).$$

(c) Find an equation of the line tangent to the graph of $6x^2y - 3x^4 - 2y^3 = 0$ at the point $\left(\frac{1}{3}\sqrt{9 + 3\sqrt{3}}, 1\right)$.

SOLUTION: After defining $f(x, y) = 6x^2y - 3x^4 - 2y^3$, we graph $z = f(x, y)$ with Plot3D in Figure 3-47, illustrating the PlotPoints, PlotRange, and ViewPoint options.

```
In[543]:= f[x_, y_] = 6x^2y - 3x^4 - 2y^3;
          Plot3D[f[x, y], {x, -2, 2},
             {y, -2, 3}, PlotPoints- > 50, PlotRange- >
                {{-2, 2}, {-2, 3}, {-2, 2}},
             BoxRatios- > {1, 1, 1},
             ViewPoint- > {1.887,  2.309,  1.6},
             ClipFill- > None]
```

(a) A unit vector, $\mathbf{u}$, in the same direction as $\mathbf{v}$ is

$$\mathbf{u} = \left\langle \frac{3}{\sqrt{3^2 + 4^2}}, \frac{4}{\sqrt{3^2 + 4^2}} \right\rangle = \left\langle \frac{3}{5}, \frac{4}{5} \right\rangle.$$

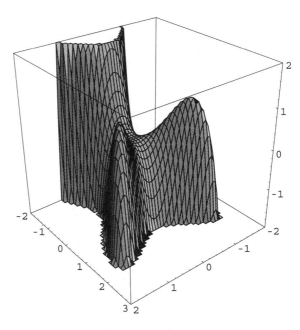

Figure 3-47 $f(x, y) = 6x^2y - 3x^4 - 2y^3$ for $-2 \le x \le 2$ and $-2 \le y \le 3$

In[544]:= **v = {3,4};**
       u = v/Sqrt[v.v]

Out[544]= $\left\{ \dfrac{3}{5}, \dfrac{4}{5} \right\}$

Then, $D_{\mathbf{u}}f(x, y) = \langle f_x(x, y), f_y(x, y) \rangle \cdot \mathbf{u}$, calculated in du.

In[545]:= **gradf = {D[f[x,y],x],D[f[x,y],y]}**

Out[545]= $\left\{ -12\ x^3 + 12\ x\ y,\ 6\ x^2 - 6\ y^2 \right\}$

In[546]:= **du = Simplify[grad.u]**

Out[546]= $-\dfrac{12}{5}\ \left(-2\ x^2 + 3\ x^3 - 3\ x\ y + 2\ y^2 \right)$

(b) $D_{\langle 3/5, 4/5 \rangle} f\left(\frac{1}{3}\sqrt{9 + 3\sqrt{3}}, 1 \right)$ is calculated by evaluating du if
$x = \frac{1}{3}\sqrt{9 + 3\sqrt{3}}$ and $y = 1$.

In[547]:= **du1 = du/.{x- >1/3Sqrt[9+3Sqrt[3]],y- >1}//Simplify**

Out[547]= $-\dfrac{4}{5}\ \sqrt{3}\ \left(-2 + \sqrt{3\ \left(3 + \sqrt{3} \right)} \right)$

(c) The gradient is evaluated if $x = \frac{1}{3}\sqrt{9 + 3\sqrt{3}}$ and $y = 1$.

```
In[548]:= nvec =
            gradf/.{x- > 1/3Sqrt[9 + 3Sqrt[3]], y- > 1}//Simplify
```

$Out[548] = \left\{ -4 \sqrt{3 + \sqrt{3}}, 2 \sqrt{3} \right\}$

Generally, $\nabla f(x, y)$ is perpendicular to the level curves of $z = f(x, y)$, so

$$\text{nvec} = \nabla f\left(\frac{1}{3}\sqrt{9 + 3\sqrt{3}}, 1\right) = \left\langle f_x\left(\frac{1}{3}\sqrt{9 + 3\sqrt{3}}, 1\right), f_y\left(\frac{1}{3}\sqrt{9 + 3\sqrt{3}}, 1\right) \right\rangle$$

is perpendicular to $f(x, y) = 0$ at the point $\left(\frac{1}{3}\sqrt{9 + 3\sqrt{3}}, 1\right)$. Thus, an equation of the line tangent to the graph of $f(x, y) = 0$ at the point $\left(\frac{1}{3}\sqrt{9 + 3\sqrt{3}}, 1\right)$ is

> An equation of the line L containing (x_0, y_0) and perpendicular to $\mathbf{n} = \langle a, b \rangle$ is $a(x - x_0) + b(y - y_0) = 0$.

$$f_x\left(\frac{1}{3}\sqrt{9 + 3\sqrt{3}}, 1\right)\left(x - \frac{1}{3}\sqrt{9 + 3\sqrt{3}}\right) + f_y\left(\frac{1}{3}\sqrt{9 + 3\sqrt{3}}, 1\right)(y - 1) = 0,$$

which we solve for y with Solve. We confirm this result by graphing $f(x, y) = 0$ using ContourPlot with the Contours->{0} option in conf and then graphing the tangent line in tanplot. tanplot and conf are shown together with Show in Figure 3-48.

```
In[549]:= conf = ContourPlot[
                f[x,y], {x,-2,2}, {y,-2,2}, Contours- > {0},
                PlotPoints- > 60, ContourShading- > False,
                Frame- > False, Axes- > Automatic,
                AxesOrigin- > {0,0},
                DisplayFunction- > Identity];
```

```
In[550]:= tanline = Solve[nvec[[1]]
                (x - 1/3Sqrt[9 + 3Sqrt[3]])+
                nvec[[2]](y - 1) == 0,
            y]
```

$Out[550] = \left\{\left\{y \to -\dfrac{-2 \sqrt{3} - 4 \sqrt{3 + \sqrt{3}} \left(-\frac{1}{3} \sqrt{9 + 3 \sqrt{3}} + x\right)}{2 \sqrt{3}}\right\}\right\}$

```
In[551]:= tanplot = Plot[Evaluate[y/.tanline],
                {x, -2, 2}, DisplayFunction- > Identity];
            Show[conf, tanplot, DisplayFunction- >
                $DisplayFunction, PlotRange- > {{-2, 2},
                {-2, 3}}, AspectRatio- > Automatic]
```

■

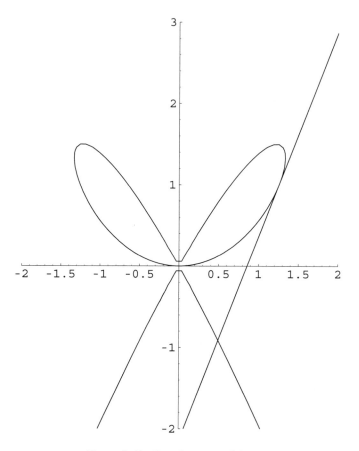

Figure 3-48   Level curves of $f(x, y)$

EXAMPLE 3.5.4: Let

$$f(x, y) = (y - 1)^2 e^{-(x+1)^2 - y^2} - \frac{10}{3}\left(-x^5 + \frac{1}{5}y - y^3\right)e^{-x^2 - y^2} - \frac{1}{9}e^{-x^2 - (y+1)^2}.$$

Calculate $\nabla f(x, y)$ and then graph $\nabla f(x, y)$ together with several level curves of $f(x, y)$.

SOLUTION: We begin by defining and graphing $z = f(x, y)$ with Plot3D in Figure 3-49.

```
In[552] := f[x_, y_] = (y - 1)^2 Exp[-(x + 1)^2 - y^2] -
              10/3 (-x^5 + 1/5y - y^3) Exp[-x^2 - y^2] -
              1/9 Exp[-x^2 - (y + 1)^2];
```

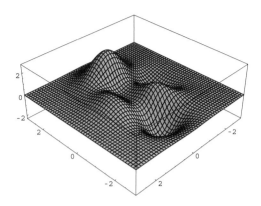

Figure 3-49 $f(x, y)$ for $-3 \le x \le 3$ and $-3 \le y \le 2$

```
In[553]:= Plot3D[f[x,y],{x,-3,3},{y,-3,3},PlotPoints->50,
            ViewPoint->{-1.99, 2.033, 1.833},
            PlotRange->All];
         conf = ContourPlot[f[x,y],{x,-3,3},
            {y,-3,3},PlotPoints->60,
            ContourShading->False,
            Frame->False, Axes->Automatic,
            AxesOrigin->{0,0},
            DisplayFunction->Identity];
```

In the three-dimensional plot, notice that z appears to have six relative extrema: three relative maxima and three relative minima. We also graph several level curves of $f(x, y)$ with ContourPlot and name the resulting graphic conf. The graphic is not displayed because we include the option DisplayFunction->Identity.

Next we calculate $f_x(x, y)$ and $f_y(x, y)$ using Simplify and D. The gradient is the vector-valued function $\langle f_x(x, y), f_y(x, y)\rangle$.

```
In[554]:= gradf = {D[f[x,y],x],D[f[x,y],y]}//Simplify
```

$$Out[554]= \{\frac{2}{9} \left(e^{-x^2-(1+y)^2} \ x + 75 \ e^{-x^2-y^2} \ x^4 - \right.$$
$$9 \ e^{-(1+x)^2-y^2} \ (1+x) \ (-1+y)^2 -$$
$$\left. 6 \ e^{-x^2-y^2} \ x \ \left(5 \ x^5 - y + 5 \ y^3\right)\right),$$
$$-\frac{2}{9} \ e^{1+x^2+y^2-2} \ \left(1+x+x^2+y+y^2\right) \ \left(-e^{2 \ x} + 9 \ e^{2 \ y} + \right.$$
$$3 \ e^{1+2 \ x+2 \ y} + e^{2 \ x} \ \left(-1+30 \ e^{1+2 \ y} \ x^5\right) \ y -$$
$$3 \ e^{2 \ y} \ \left(6+17 \ e^{1+2 \ x}\right) \ y^2 + 9 \ e^{2 \ y} \ y^3 +$$
$$\left. 30 \ e^{1+2 \ x+2 \ y} \ y^4\right)\}$$

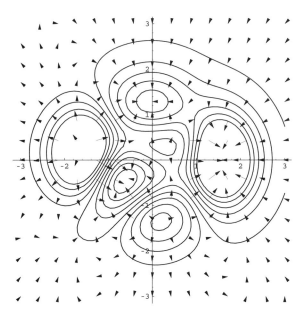

Figure 3-50   Contour plot of $f(x, y)$ along with several gradient vectors

To graph the gradient, we use `PlotGradientField`, which is contained in the `PlotField` package. We use `PlotGradientField` to graph the gradient naming the resulting graphic `gradfplot`. `gradfplot` and `conf` are displayed together using `Show`.

In[555]:= `<< Graphics'PlotField'`

```
gradfplot = PlotGradientField[f[x,y],
  x,-3,3},{y,-3,3},DisplayFunction- > Identity];
Show[conf,gradfplot,
  DisplayFunction- > $DisplayFunction]
```

In the result (see Figure 3-50), notice that the gradient is perpendicular to the level curves; the gradient is pointing in the direction of maximal increase of $z = f(x, y)$.

∎

Classifying Critical Points

Let $z = f(x, y)$ be a real-valued function of two variables with continuous second-order partial derivatives. A **critical point** of $z = f(x, y)$ is a point (x_0, y_0) in the interior of the domain of $z = f(x, y)$ for which

$$f_x(x_0, y_0) = 0 \qquad \text{and} \qquad f_y(x_0, y_0) = 0.$$

Critical points are classified by the *Second Derivatives* (or *Partials*) *test*.

Theorem 17 (Second Derivatives Test). *Let* (x_0, y_0) *be a critical point of a function* $z = f(x, y)$ *of two variables and let*

$$d = f_{xx}(x_0, y_0) \, f_{yy}(x_0, y_0) - \left[f_{xy}(x_0, y_0) \right]^2. \tag{3.30}$$

1. *If* $d > 0$ *and* $f_{xx}(x_0, y_0) > 0$, *then* $z = f(x, y)$ *has a **relative** (or **local**) **minimum** at* (x_0, y_0).
2. *If* $d > 0$ *and* $f_{xx}(x_0, y_0) < 0$, *then* $z = f(x, y)$ *has a **relative** (or **local**) **maximum** at* (x_0, y_0).
3. *If* $d < 0$, *then* $z = f(x, y)$ *has a **saddle point** at* (x_0, y_0).
4. *If* $d = 0$, *no conclusion can be drawn and* (x_0, y_0) *is called a **degenerate critical point**.*

EXAMPLE 3.5.5: Find the relative maximum, relative minimum, and saddle points of $f(x, y) = -2x^2 + x^4 + 3y - y^3$.

SOLUTION: After defining $f(x, y)$, the critical points are found with Solve and named critpts.

```
In[556]:= f[x_, y_] = -2x^2 + x^4 + 3y - y^3;
          critpts =
            Solve[{D[f[x, y], x] == 0, D[f[x, y], y] == 0}, {x, y}]
Out[556]= {{x → -1, y → -1}, {x → -1, y → 1}, {x → 0, y → -1},
           {x → 0, y → 1}, {x → 1, y → -1}, {x → 1, y → 1}}
```

We then define dfxx. Given (x_0, y_0), dfxx (x_0, y_0) returns the ordered quadruple x_0, y_0, equation (3.30) evaluated at (x_0, y_0), and $f_{xx}(x_0, y_0)$.

```
In[557]:= dfxx[x0_, y0_] =
            {x0, y0, D[f[x, y], {x, 2}]D[f[x, y], {y, 2}]-
              D[f[x, y], x, y]^2/.{x- >x0, y- >y0},
            D[f[x, y], {x, 2}]/.{x- >x0, y- >y0}}
Out[557]= {x0, y0, -6 (-4 + 12 x0^2) y0, -4 + 12 x0^2}
```

For example,

```
In[558]:= dfxx[0, 1]
Out[558]= {0, 1, 24, -4}
```

shows us that a relative maximum occurs at $(0, 1)$. We then use /. (ReplaceAll) to substitute the values in each element of critpts into dfxx.

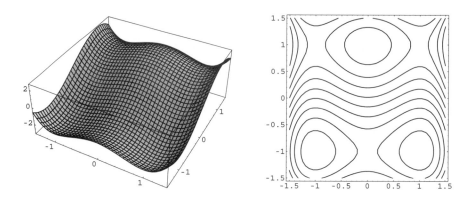

Figure 3-51   (a) Three-dimensional and (b) contour plots of $f(x, y)$

In[559] := **dfxx[x, y]/.critpts**

Out[559] = {{-1, -1, 48, 8}, {-1, 1, -48, 8}, {0, -1, -24, -4},
            {0, 1, 24, -4}, {1, -1, 48, 8}, {1, 1, -48, 8}}

From the result, we see that $(0, 1)$ results in a relative maximum, $(0, -1)$ results in a saddle, $(1, 1)$ results in a saddle, $(1, -1)$ results in a relative minimum, $(-1, 1)$ results in a saddle, and $(-1, -1)$ results in a relative minimum. We confirm these results graphically with a three-dimensional plot generated with `Plot3D` and a contour plot generated with `ContourPlot` in Figure 3-51.

In[560] := **p1 = Plot3D[f[x, y], {x, -3/2, 3/2}, {y, -3/2, 3/2},
            PlotPoints- > 40, DisplayFunction- > Identity];
          p2 = ContourPlot[f[x, y],
              {x, -3/2, 3/2}, {y, -3/2, 3/2}, PlotPoints- > 40,
              ContourShading- > False,
              DisplayFunction- > Identity];
          Show[GraphicsArray[{p1, p2}]]**

In the contour plot, notice that near relative extrema, the level curves look like circles while near saddles they look like hyperbolas.

∎

If the Second Derivatives Test fails, graphical analysis is especially useful.

EXAMPLE 3.5.6: Find the relative maximum, relative minimum, and saddle points of $f(x, y) = x^2 + x^2y^2 + y^4$.

SOLUTION: Initially we proceed in the exact same manner as in the previous example: we define $f(x, y)$ and compute the critical points. Several complex solutions are returned, which we ignore.

```
In[561]:= f[x_, y_] = x^2 + x^2y^2 + y^4;
          critpts =
            Solve[{D[f[x, y], x] == 0, D[f[x, y], y] == 0}, {x, y}]
```
$$Out[561] = \left\{ \{x \to 0, y \to 0\}, \left\{x \to -\sqrt{2}, y \to -i\right\}, \left\{x \to -\sqrt{2}, y \to i\right\}, \right.$$
$$\left\{x \to \sqrt{2}, y \to -i\right\}, \left\{x \to \sqrt{2}, y \to i\right\}, \{y \to 0, x \to 0\},$$
$$\left. \{y \to 0, x \to 0\} \right\}$$

We then compute the value of (3.30) at the real critical point, and the value of $f_{xx}(x, y)$ at this critical point.

```
In[562]:= dfxx[x0_, y0_] =
            {x0, y0, D[f[x, y], {x, 2}]D[f[x, y], {y, 2}]-
              D[f[x, y], x, y]^2/.{x- >x0, y- >y0},
            D[f[x, y], {x, 2}]/.{x- >x0, y- >y0}}
```
$$Out[562] = \left\{x0, y0, -16 \ x0^2 \ y0^2 + \left(2 + 2 \ y0^2\right) \ \left(2 \ x0^2 + 12 \ y0^2\right), 1 \right.$$
$$\left. 2 + 2 \ y0^2\right\}$$

```
In[563]:= dfxx[0, 0]
```
$$Out[563] = \{0, 0, 0, 2\}$$

The result shows us that the Second Derivatives Test fails at $(0, 0)$.

```
In[564]:= p1 = Plot3D[f[x, y], {x, -1, 1}, {y, -1, 1},
            PlotPoints- >40,
            DisplayFunction- >Identity,
            BoxRatios- >Automatic];
          p2 = ContourPlot[f[x, y], {x, -1, 1},
            {y, -1, 1}, PlotPoints- >40, Contours- >20,
            ContourShading- >False,
            DisplayFunction- >Identity];
          Show[GraphicsArray[{p1, p2}]]
```

However, the contour plot of $f(x, y)$ near $(0, 0)$ indicates that an extreme value occurs at $(0, 0)$. The three-dimensional plot shows that $(0, 0)$ is a relative minimum. (See Figure 3-52.)

■

Tangent Planes

Let $z = f(x, y)$ be a real-valued function of two variables. If both $f_x(x_0, y_0)$ and $f_y(x_0, y_0)$ exist, then an equation of the plane tangent to the graph of $z = f(x, y)$ at

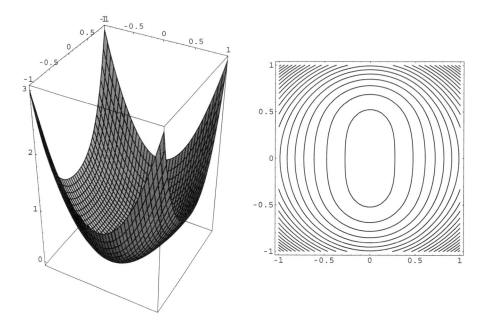

Figure 3-52   (a) Three-dimensional and (b) contour plots of $f(x, y)$

the point $(x_0, y_0, f(x_0, y_0))$ is given by

$$f_x(x_0, y_0)(x - x_0) + f_y(x_0, y_0)(y - y_0) - (z - z_0) = 0, \tag{3.31}$$

where $z_0 = f(x_0, y_0)$. Solving for z yields the function (of two variables)

$$z = f_x(x_0, y_0)(x - x_0) + f_y(x_0, y_0)(y - y_0) + z_0. \tag{3.32}$$

Symmetric equations of the line perpendicular to the surface $z = f(x, y)$ at the point (x_0, y_0, z_0) are given by

$$\frac{x - x_0}{f_x(x_0, y_0)} = \frac{y - y_0}{f_y(x_0, y_0)} = \frac{z - z_0}{-1} \tag{3.33}$$

and parametric equations are

$$\begin{cases} x = x_0 + f_x(x_0, y_0)\, t \\ y = y_0 + f_y(x_0, y_0)\, t \\ z = z_0 - t. \end{cases} \tag{3.34}$$

The plane tangent to the graph of $z = f(x, y)$ at the point $(x_0, y_0, f(x_0, y_0))$ is the "best" linear approximation of $z = f(x, y)$ near $(x, y) = (x_0, y_0)$ in the same way as the line tangent to the graph of $y = f(x)$ at the point $(x_0, f(x_0))$ is the "best" linear approximation of $y = f(x)$ near $x = x_0$.

EXAMPLE 3.5.7: Find an equation of the plane tangent and normal line to the graph of $f(x, y) = 4 - \frac{1}{4}\left(2x^2 + y^2\right)$ at the point $(1, 2, 5/2)$.

SOLUTION: We define $f(x, y)$ and compute $f_x(1, 2)$ and $f_y(1, 2)$.

```
In[565]:= f[x_, y_] = 4 - 1/4 (2x^2 + y^2);
          f[1, 2]

          dx = D[f[x, y], x] /. {x- > 1, y- > 2}

          dy = D[f[x, y], y] /. {x- > 1, y- > 2}
```

$$Out[565]= \frac{5}{2}$$

$$Out[565]= -1$$

$$Out[565]= -1$$

Using (3.32), an equation of the tangent plane is $z = -1(x - 1) - 1(y - 2) + f(1, 2)$. Using (3.34), parametric equations of the normal line are $x = 1 - t, y = 2 - t, z = f(1, 2) - t$. We confirm the result graphically by graphing $f(x, y)$ together with the tangent plane in p1 using `Plot3D`. We use `ParametricPlot3D` to graph the normal line in p2 and then display p1 and p2 together with `Show` in Figure 3-53.

```
In[566]:= p1 = Plot3D[f[x, y], {x, -1, 3}, {y, 0, 4},
              DisplayFunction- > Identity, PlotPoints- > 40];
          p2 = Plot3D[dx (x - 1) + dy (y - 2) + f[1, 2], {x, -1, 3},
              {y, 0, 4}, DisplayFunction- > Identity,
              PlotPoints- > 30];
          p3 = ParametricPlot3D[{1 + dx t, 2 + dy t, f[1, 2] - t},
              {t, -4, 4}, DisplayFunction- > Identity];
          Show[p1, p2, p3, PlotRange- > {{-1, 3}, {0, 4}, {0, 4}},
              BoxRatios- > Automatic,
              DisplayFunction- > $DisplayFunction]
```

Because $z = -1(x - 1) - 1(y - 2) + f(1, 2)$ is the "best" linear approximation of $f(x, y)$ near $(1, 2)$, the graphs are very similar near $(1, 2)$ as shown in the three-dimensional plot. We also expect the level curves of each near $(1, 2)$ to be similar, which is confirmed with `ContourPlot` in Figure 3-54.

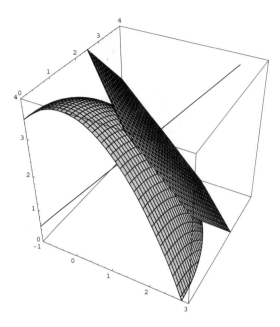

Figure 3-53   Graph of $f(x, y)$ with a tangent plane and normal line

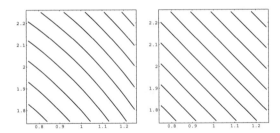

Figure 3-54   Zooming in near $(1, 2)$

```
In[567] := p4 = ContourPlot[f[x, y], {x, 0.75, 1.25},
              {y, 1.75, 2.25},
              ContourShading- > False,
              DisplayFunction- > Identity];
           p5 = ContourPlot[dx  (x - 1) + dy  (y - 2) + f[1, 2],
              {x, 0.75, 1.25}, {y, 1.75, 2.25},
              ContourShading- > False,
              DisplayFunction- > Identity];
           Show[GraphicsArray[{p4, p5}]]
```

■

Lagrange Multipliers

Certain types of optimization problems can be solved using the method of *Lagrange multipliers* that is based on the following theorem.

Theorem 18 (Lagrange's Theorem). *Let $z = f(x, y)$ and $z = g(x, y)$ be real-valued functions with continuous partial derivatives and let $z = f(x, y)$ have an extreme value at a point (x_0, y_0) on the smooth constraint curve $g(x, y) = 0$. If $\nabla g(x_0, y_0) \neq \mathbf{0}$, then there is a real number λ satisfying*

$$\nabla f(x_0, y_0) = \lambda \nabla g(x_0, y_0). \tag{3.35}$$

Graphically, the points (x_0, y_0) at which the extreme values occur correspond to the points where the level curves of $z = f(x, y)$ are tangent to the graph of $g(x, y) = 0$.

EXAMPLE 3.5.8: Find the maximum and minimum values of $f(x, y) = xy$ subject to the constraint $\frac{1}{4}x^2 + \frac{1}{9}y^2 = 1$.

SOLUTION: For this problem, $f(x, y) = xy$ and $g(x, y) = \frac{1}{4}x^2 + \frac{1}{9}y^2 - 1$. Observe that parametric equations for $\frac{1}{4}x^2 + \frac{1}{9}y^2 = 1$ are $x = 2\cos t$, $y = 3\sin t$, $0 \leq t \leq 2\pi$. In Figure 3-55, we use `ParametricPlot3D` to parametrically graph $g(x, y) = 0$ and $f(x, y)$ for x and y-values on the curve $g(x, y) = 0$ by graphing

$$\begin{cases} x = 2\cos t \\ y = 3\sin t \\ z = 0 \end{cases} \quad \text{and} \quad \begin{cases} x = 2\cos t \\ y = 3\sin t \\ z = x \cdot y = 6\cos t \sin t \end{cases}$$

for $0 \leq t \leq 2\pi$. Our goal is to find the minimum and maximum values in Figure 3-55 and the points at which they occur.

```
In[568]:= f[x_,y_] = x y;
          g[x_,y_] = x^2/4 + y^2/9 - 1;

In[569]:= s1 = ParametricPlot3D[{2 Cos[t],3 Sin[t],0},
             {t,0,2π},DisplayFunction- > Identity];
          s2 = ParametricPlot3D[{2 Cos[t],3 Sin[t],
             6 Cos[t] Sin[t]},
             {t,0,2π},DisplayFunction- > Identity];
          Show[s1,s2,BoxRatios- > Automatic,
             DisplayFunction- > $DisplayFunction]
```

To implement the method of Lagrange multipliers, we compute $f_x(x, y)$, $f_y(x, y)$, $g_x(x, y)$, and $g_y(x, y)$ with D.

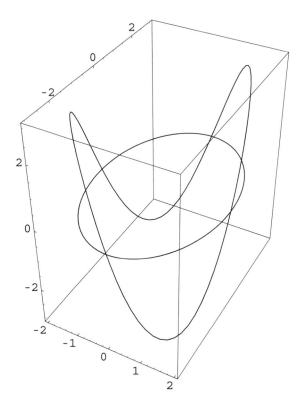

Figure 3-55 $f(x, y)$ on $g(x, y) = 0$

$In[570] := $ **fx = D[f[x,y],x]**

fy = D[f[x,y],y]

gx = D[g[x,y],x]

gy = D[g[x,y],y]

$Out[570] = $ y

$Out[570] = $ x

$Out[570] = $ $\dfrac{x}{2}$

$Out[570] = $ $\dfrac{2\,y}{9}$

Solve is used to solve the system of equations (3.35):

$$f_x(x, y) = \lambda g_x(x, y)$$
$$f_y(x, y) = \lambda g_y(x, y)$$
$$g(x, y) = 0$$

for x, y, and λ.

$In[571]:= \mathbf{vals = Solve[\{fx == \lambda\ gx,\ fy == \lambda\ gy,\ g[x,y] == 0\},}$
$\qquad \mathbf{\{x, y, \lambda\}]}$

$Out[571]= \left\{\left\{\lambda \to -3,\ x \to -\sqrt{2},\ y \to \dfrac{3}{\sqrt{2}}\right\},\right.$

$\qquad\qquad \left\{\lambda \to -3,\ x \to \sqrt{2},\ y \to -\dfrac{3}{\sqrt{2}}\right\},$

$\qquad\qquad \left\{\lambda \to 3,\ x \to -\sqrt{2},\ y \to -\dfrac{3}{\sqrt{2}}\right\},$

$\qquad\qquad \left.\left\{\lambda \to 3,\ x \to \sqrt{2},\ y \to \dfrac{3}{\sqrt{2}}\right\}\right\}$

The corresponding values of $f(x, y)$ are found using `ReplaceAll` (`/.`).

$In[572]:= \mathbf{n1 = \{x, y, f[x, y]\}/.vals}$

$Out[572]= \left\{\left\{-\sqrt{2},\ \dfrac{3}{\sqrt{2}},\ -3\right\},\ \left\{\sqrt{2},\ -\dfrac{3}{\sqrt{2}},\ -3\right\},\ \left\{-\sqrt{2},\ -\dfrac{3}{\sqrt{2}},\ 3\right\},\right.$

$\qquad\qquad \left.\left\{\sqrt{2},\ \dfrac{3}{\sqrt{2}},\ 3\right\}\right\}$

$In[573]:= \mathbf{N[n1]}$

$Out[573]= \{\{-1.41421, 2.12132, -3.\},$
$\qquad\qquad \{1.41421, -2.12132, -3.\},$
$\qquad\qquad \{-1.41421, -2.12132, 3.\},$
$\qquad\qquad \{1.41421, 2.12132, 3.\}\}$

We conclude that the maximum value $f(x, y)$ subject to the constraint
$g(x, y) = 0$ is 3 and occurs at $\left(\sqrt{2}, \frac{3}{2}\sqrt{2}\right)$ and $\left(-\sqrt{2}, -\frac{3}{2}\sqrt{2}\right)$. The minimum
value is -3 and occurs at $\left(-\sqrt{2}, \frac{3}{2}\sqrt{2}\right)$ and $\left(\sqrt{2}, -\frac{3}{2}\sqrt{2}\right)$. We graph several
level curves of $f(x, y)$ and the graph of $g(x, y) = 0$ with `ContourPlot`
and show the graphs together with `Show`. The minimum and maximum
values of $f(x, y)$ subject to the constraint $g(x, y) = 0$ occur at the points
where the level curves of $f(x, y)$ are tangent to the graph of $g(x, y) = 0$ as
illustrated in Figure 3-56.

```
In[574]:= cp1 = ContourPlot[f[x, y], {x, -3, 3},
            {y, -3, 3}, Contours- > 30, ContourShading- > False,
            PlotPoints- > 40, DisplayFunction- > Identity];
         cp2 = ContourPlot[
            g[x, y], {x, -3, 3}, {y, -3, 3}, Contours- > {0},
            ContourShading- > False,
            DisplayFunction- > Identity,
            ContourStyle- > Thickness[0.01]];
         Show[cp1, cp2, DisplayFunction- > $DisplayFunction]
```

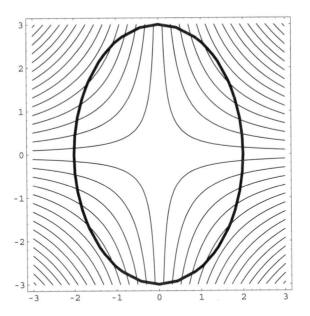

Figure 3-56   Level curves of $f(x, y)$ together with $g(x, y) = 0$

3.5.3 Iterated Integrals

The `Integrate` command, used to compute single integrals, is used to compute iterated integrals. The command

$$\texttt{Integrate[f[x,y],\{y,c,d\},\{x,a,b\}]}$$

attempts to compute the iterated integral

$$\int_c^d \int_a^b f(x, y) \, dx \, dy. \tag{3.36}$$

If Mathematica cannot compute the exact value of the integral, it is returned unevaluated, in which case numerical results may be more useful. The iterated integral (3.36) is numerically evaluated with the command `N` or

$$\texttt{NIntegrate[f[x,y],\{y,c,d\},\{x,a,b\}]}$$

EXAMPLE 3.5.9: Evaluate each integral: (a) $\int_2^4 \int_1^2 \left(2xy^2 + 3x^2y\right) dx \, dy$; (b) $\int_0^2 \int_{y^2}^{2y} \left(3x^2 + y^3\right) dx \, dy$; (c) $\int_0^\infty \int_0^\infty xye^{-x^2-y^2} \, dy \, dx$; (d) $\int_0^\pi \int_0^\pi e^{\sin xy} dx \, dy$.

SOLUTION: (a) First, we compute $\iint \left(2xy^2 + 3x^2y\right) dx\,dy$ with `Integrate`. Second, we compute $\int_2^4 \int_1^2 \left(2xy^2 + 3x^2y\right) dx\,dy$ with `Integrate`.

```
In[575]:= Integrate[2x y^2 + 3x^2 y, y, x]
```
$$Out[575] = \frac{x^3\,y^2}{2} + \frac{x^2\,y^3}{3}$$

```
In[576]:= Integrate[2x y^2 + 3x^2 y, {y, 2, 4}, {x, 1, 2}]
```
$$Out[576] = 98$$

(b) We illustrate the same commands as in (a), except we are integrating over a nonrectangular region.

```
In[577]:= Integrate[3x^2 + y^3, {x, y^2, 2y}]
```
$$Out[577] = 8\,y^3 + 2\,y^4 - y^5 - y^6$$

```
In[578]:= Integrate[3x^2 + y^3, y, {x, y^2, 2y}]
```
$$Out[578] = 2\,y^4 + \frac{2\,y^5}{5} - \frac{y^6}{6} - \frac{y^7}{7}$$

```
In[579]:= Integrate[3x^2 + y^3, {y, 0, 2}, {x, y^2, 2y}]
```
$$Out[579] = \frac{1664}{105}$$

(c) Improper integrals can be handled in the same way as proper integrals.

```
In[580]:= Integrate[x y Exp[-x^2 - y^2], x, y]
```
$$Out[580] = \frac{1}{4}\,e^{-x^2-y^2}$$

```
In[581]:= Integrate[x y Exp[-x^2 - y^2], {x, 0, ∞},
             {y, 0, ∞}]
```
$$Out[581] = \frac{1}{4}$$

(d) In this case, Mathematica cannot evaluate the integral exactly so we use `NIntegrate` to obtain an approximation.

```
In[582]:= Integrate[Exp[Sin[x y]], y, x]
```
$$Out[582] = \int\int e^{\sin[x\ y]}\,dx\,dy$$

```
In[583]:= NIntegrate[Exp[Sin[x y]], {y, 0, π}, {x, 0, π}]
```
$$Out[583] = 15.5092$$

■

Area, Volume, and Surface Area

Typical applications of iterated integrals include determining the area of a planar region, the volume of a region in three-dimensional space, or the surface area of a region in three-dimensional space. The area of the planar region R is given by

$$A = \iint_R dA. \tag{3.37}$$

If $z = f(x, y)$ has continuous partial derivatives on a closed region R, then the surface area of the portion of the surface that projects onto R is given by

$$SA = \iint_R \sqrt{\left(\frac{\partial f}{\partial x}\right)^2 + \left(\frac{\partial f}{\partial y}\right)^2 + 1} \, dA. \tag{3.38}$$

If $f(x, y) \geq g(x, y)$ on R, the volume of the region between the graphs of $f(x, y)$ and $g(x, y)$ is

$$V = \iint_R (f(x, y) - g(x, y)) \, dA. \tag{3.39}$$

EXAMPLE 3.5.10: Find the area of the region R bounded by the graphs of $y = 2x^2$ and $y = 1 + x^2$.

SOLUTION: We begin by graphing $y = 2x^2$ and $y = 1 + x^2$ with Plot in Figure 3-57. The x-coordinates of the intersection points are found with Solve.

```
In[584]:= Plot[{2x^2, 1 + x^2}, {x, -3/2, 3/2},
            PlotStyle- > {GrayLevel[0], GrayLevel[0.3]},
            AspectRatio- > Automatic]

In[585]:= Solve[2x^2 == 1 + x^2]
Out[585]= {{x → -1}, {x → 1}}
```

Using (3.37) and taking advantage of symmetry, the area of R is given by

$$A = \iint_R dA = 2 \int_0^1 \int_{2x^2}^{1+x^2} dy \, dx,$$

which we compute with Integrate.

```
In[586]:= 2 Integrate[1, {x, 0, 1}, {y, 2x^2, 1 + x^2}]
Out[586]= 4/3
```

We conclude that the area of R is 4/3.

∎

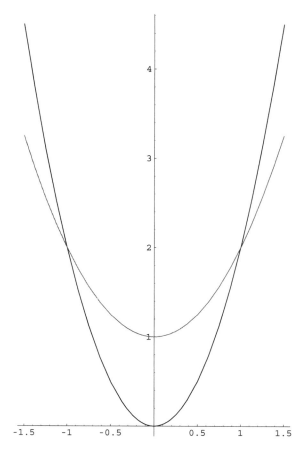

Figure 3-57 $y = 2x^2$ and $y = 1 + x^2$ for $-3/2 \le x \le 3/2$

If the problem exhibits "circular symmetry," changing to polar coordinates is often useful. If $R = \{(r, \theta) \,|\, a \le r \le b, \alpha \le \theta \le \beta \}$, then

$$\iint_R f(x, y)\, dA = \int_\alpha^\beta \int_a^b f(r \cos \theta, r \sin \theta)\, r\, dr\, d\theta.$$

EXAMPLE 3.5.11: Find the surface area of the portion of

$$f(x, y) = \sqrt{4 - x^2 - y^2}$$

that lies above the region $R = \{(x, y) \,|\, x^2 + y^2 \le 1 \}$.

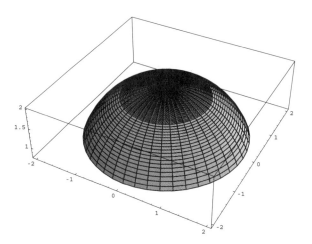

Figure 3-58   The portion of the graph of $f(x, y)$ above R

SOLUTION: First, observe that the domain of $f(x, y)$ is

$$\left\{(x, y)\Big| -\sqrt{4 - y^2} \le x \le \sqrt{4 - y^2}, -2 \le y \le 2\right\} = \{(r, \theta)|0 \le r \le 2, 0 \le \theta \le 2\pi\}.$$

Similarly,

$$R = \left\{(x, y)\Big| -\sqrt{1 - y^2} \le x \le \sqrt{1 - y^2}, -1 \le y \le 1\right\} = \{(r, \theta)|0 \le r \le 1, 0 \le \theta \le 2\pi\}.$$

With this observation, we use ParametricPlot3D to graph $f(x, y)$ in p1 and the portion of the graph of $f(x, y)$ above R in p2 and show the two graphs together with Show. We wish to find the area of the black region in Figure 3-58.

```
In[587]:= f[x_, y_] = Sqrt[4 - x^2 - y^2];
```

```
In[588]:= p1 = ParametricPlot3D[{r Cos[t], r Sin[t],
                f[r Cos[t], r Sin[t]]}, {r, 0, 2}, {t, 0, 2π},
                PlotPoints- > 45, DisplayFunction- > Identity];
          p2 = ParametricPlot3D[
                {r Cos[t], r Sin[t], f[r Cos[t], r Sin[t]],
                GrayLevel[0.3]}, {r, 0, 1}, {t, 0, 2π},
                PlotPoints- > 45, DisplayFunction- > Identity];
          Show[p1, p2, DisplayFunction- > $DisplayFunction,
            BoxRatios- > Automatic]
```

We compute $f_x(x, y)$, $f_y(x, y)$ and $\sqrt{[f_x(x, y)]^2 + [f_y(x, y)]^2 + 1}$ with D and Simplify.

```
In[589]:= fx = D[f[x,y],x]

      fy = D[f[x,y],y]
```

$$Out[589] = -\frac{x}{\sqrt{4 - x^2 - y^2}}$$

$$Out[589] = -\frac{y}{\sqrt{4 - x^2 - y^2}}$$

Then, using (3.38), the surface area is given by

$$
\begin{aligned}
SA &= \iint_R \sqrt{\left(\frac{\partial f}{\partial x}\right)^2 + \left(\frac{\partial f}{\partial y}\right)^2 + 1} \, dA \\
&= \iint_R \frac{2}{\sqrt{4 - x^2 - y^2}} \, dA \\
&= \int_{-1}^{1} \int_{-\sqrt{1-y^2}}^{\sqrt{1-y^2}} \frac{2}{\sqrt{4 - x^2 - y^2}} \, dx \, dy.
\end{aligned}
$$

(3.40)

However, notice that in polar coordinates,

$$R = \{(r, \theta) \,|\, 0 \le r \le 1, 0 \le \theta \le 2\pi\}$$

so in polar coordinates the surface area is given by

$$SA = \int_0^{2\pi} \int_0^1 \frac{2}{\sqrt{4 - r^2}} \, r \, dr \, d\theta,$$

```
In[590]:= s1 = Simplify[Sqrt[1 + fx^2 + fy^2]]
```

$$Out[590] = 2 \sqrt{-\frac{1}{-4 + x^2 + y^2}}$$

```
In[591]:= s2 = Simplify[s1 /.{x->r Cos[t],y->r Sin[t]}]
```

$$Out[591] = 2 \sqrt{\frac{1}{4 - r^2}}$$

which is much easier to evaluate than (3.40). We evaluate the iterated integral with Integrate

```
In[592]:= s3 = Integrate[r s2, {t, 0, 2π}, {r, 0, 1}]
```

$$Out[592] = 2 \left(4 - 2\sqrt{3}\right) \pi$$

```
In[593]:= N[s3]
Out[593]= 3.36715
```

and conclude that the surface area is $\left(8 - 4\sqrt{3}\right)\pi \approx 3.367$.

∎

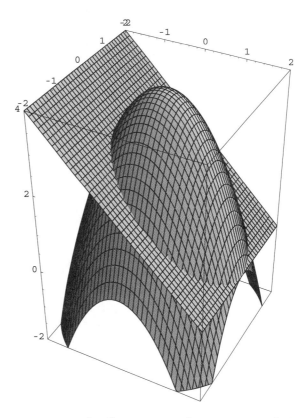

Figure 3-59 $z = 4 - x^2 - y^2$ and $z = 2 - x$ for $-2 \leq x \leq 2$ and $-2 \leq y \leq 2$

EXAMPLE 3.5.12: Find the volume of the region between the graphs of $z = 4 - x^2 - y^2$ and $z = 2 - x$.

SOLUTION: We begin by graphing $z = 4 - x^2 - y^2$ and $z = 2 - x$ together with Plot3D in Figure 3-59.

```
In[594]:= p1 = Plot3D[4 - x^2 - y^2, {x, -2, 2}, {y, -2, 2},
              PlotPoints- > 40, DisplayFunction- > Identity];
          p2 = Plot3D[2 - x, {x, -2, 2}, {y, -2, 2},
              PlotPoints- > 40, DisplayFunction- > Identity];
          Show[p1, p2, PlotRange- > {{-2, 2}, {-2, 2}, {-2, 4}},
            BoxRatios- >Automatic,
            DisplayFunction- > $DisplayFunction]
```

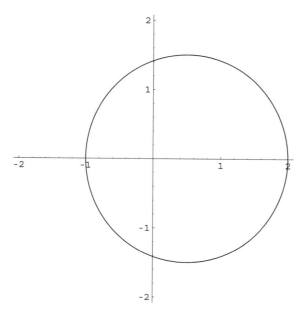

Figure 3-60   Graph of $4 - x^2 - y^2 = 2 - x$

The region of integration, R, is determined by graphing $4 - x^2 - y^2 = 2 - x$ in Figure 3-60.

```
In[595] := ContourPlot[4 - x^2 - y^2 - (2 - x), {x, -2, 2},
             {y, -2, 2}, Contours- > {0}, ContourShading- > False,
             PlotPoints- > 50,
             Frame- > False, Axes- > Automatic,
             AxesOrigin- > {0, 0}]
```

Completing the square shows us that

$$R = \left\{ (x, y) \left| \left(x - \frac{1}{2} \right)^2 + y^2 \le \frac{9}{4} \right. \right\}$$

$$= \left\{ (x, y) \left| \frac{1}{2} - \frac{1}{2}\sqrt{9 - 4y^2} \le x \le \frac{1}{2} + \frac{1}{2}\sqrt{9 - 4y^2}, -\frac{3}{2} \le y \le \frac{3}{2} \right. \right\}.$$

Thus, using (3.39), the volume of the solid is given by

$$V = \iint_R \left[\left(4 - x^2 - y^2 \right) - (2 - x) \right] dA$$

$$= \int_{-\frac{3}{2}}^{\frac{3}{2}} \int_{\frac{1}{2} - \frac{1}{2}\sqrt{9 - 4y^2}}^{\frac{1}{2} + \frac{1}{2}\sqrt{9 - 4y^2}} \left[\left(4 - x^2 - y^2 \right) - (2 - x) \right] dx\, dy,$$

which we evaluate with `Integrate`.

$In[596]:=$ **i1 = Integrate[(4 - x^2 - y^2) - (2 - x),**
{y, -3/2, 3/2}, {x, 1/2 - 1/2Sqrt[9 - 4y^2],
1/2 + 1/2 Sqrt[9 - 4y^2]}]

$Out[596]= \dfrac{81\ \pi}{32}$

$In[597]:=$ **N[i1]**

$Out[597]=$ 7.95216

We conclude that the volume is $\frac{81}{32}\pi \approx 7.952$.

■

Triple Iterated Integrals

Triple iterated integrals are calculated in the same manner as double iterated integrals.

EXAMPLE 3.5.13: Evaluate

$$\int_0^{\pi/4} \int_0^y \int_0^{y+z} (x + 2z)\sin y\, dx\, dz\, dy.$$

SOLUTION: Entering

$In[598]:=$ **i1 = Integrate[(x + 2z) Sin[y], {y, 0, π/4}, {z, 0, y},**
{x, 0, y + z}]

$Out[598]= -\dfrac{17}{\sqrt{2}} + \dfrac{17\ \pi}{4\ \sqrt{2}} + \dfrac{17\ \pi^2}{32\ \sqrt{2}} - \dfrac{17\ \pi^3}{384\ \sqrt{2}}$

calculates the triple integral exactly with Integrate.

An approximation of the exact value is found with N.

$In[599]:=$ **N[i1]**

$Out[599]=$ 0.157206

■

We illustrate how triple integrals can be used to find the volume of a solid when using spherical coordinates.

Figure 3-61   Mathematica's help for `SphericalPlot3D`

EXAMPLE 3.5.14: Find the volume of the torus with equation in spherical coordinates $\rho = \sin\phi$.

SOLUTION: We proceed by graphing the torus with `SphericalPlot3D` in Figure 3-62, which is contained in the **ParametricPlot3D** package that is located in the **Graphics** directory (see Figure 3-61).

```
In[600]:= << Graphics`ParametricPlot3D`

           SphericalPlot3D[
             Sin[φ], {φ, 0, π}, {θ, 0, 2π}, PlotPoints- > 40]
```

In general, the volume of the solid region D is given by

$$V = \iiint_D dV.$$

Thus, the volume of the torus is given by the triple iterated integral

$$V = \int_0^{2\pi} \int_0^{\pi} \int_0^{\sin\phi} \rho^2 \sin\phi \, d\rho \, d\phi \, d\theta,$$

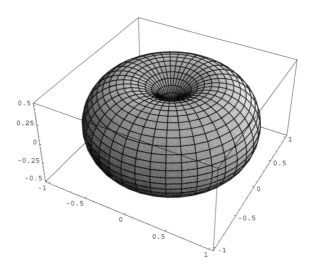

Figure 3-62   A graph of the torus

$In[601] :=$ **i1 = Integrate[ρ^2 Sin[ϕ], {θ, 0, 2π},**
                    {ϕ, 0, π}, {ρ, 0, Sin[ϕ]}]

$Out[601] = \dfrac{\pi^2}{4}$

$In[602] :=$ **N[i1]**

$Out[602] = 2.4674$

which we evaluate with Integrate. We conclude that the volume of
the torus is $\frac{1}{4}\pi^2 \approx 2.467$.

∎

Introduction to Lists and Tables 4

Chapter 4 introduces operations on lists and tables. The examples used to illustrate the various commands in this chapter are taken from calculus, business, dynamical systems, and engineering applications.

4.1 Lists and List Operations

4.1.1 Defining Lists

A **list** of n elements is a Mathematica object of the form

$$\texttt{list=\{a1,a2,a3,...,an\}.}$$

The *i*th element of the list is extracted from `list` with `list[[i]]`.

Elements of a list are separated by commas. Lists are always enclosed in braces `{...}` and each element of a list may be (almost any) Mathematica object – even other lists. Because lists are Mathematica objects, they can be named. For easy reference, we will usually name lists.

Lists can be defined in a variety of ways: they may be completely typed in, imported from other programs and text files, or they may be created with either the `Table` or `Array` commands. Given a function $f(x)$ and a number n, the command

1. `Table[f[i],{i,n}]` creates the list $\{f[1],\ldots,f[n]\}$;
2. `Table[f[i],{i,0,n}]` creates the list $\{f[0],\ldots,f[n]\}$;
3. `Table[f[i],{i,n,m}]` creates the list

$$\{f[n],f[n+1],\ldots,f[m-1],f[m]\};$$

4. `Table[f[i],{i,imin,imax,istep}]` creates the list

$$\{f[imin],f[imin+istep],f[imin+2*step],\ldots,f[imax]\};$$

and

5. `Array[f,n]` creates the list $\{f[1],\ldots,f[n]\}$.

In particular,

$$\texttt{Table[f[x],\{x,a,b,(b-a)/(n-1)\}]}$$

returns a list of $f(x)$ values for n equally spaced values of x between a and b;

$$\texttt{Table[\{x,f[x]\},\{x,a,b,(b-a)/(n-1)\}]}$$

returns a list of points $(x, f(x))$ for n equally spaced values of x between a and b.

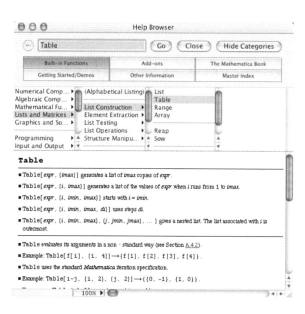

In addition to using `Table`, lists of numbers can be calculated using `Range`:

1. `Range[n]` generates the list $\{1,2, \ldots , n\}$;

2. Range [n1,n2] generates the list {n1, n1+1, ... , n2-1, n2}; and
3. Range [n1,n2,nstep] generates the list

$$\{n1, \ n1+nstep, n1+2*nstep, \ \ldots \ , \ n2-nstep, n2\}.$$

EXAMPLE 4.1.1: Use Mathematica to generate the list
{1,2,3,4,5,6,7,8,9,10}.

SOLUTION: Generally, a given list can be constructed in several ways. In fact, each of the following five commands generates the list {1,2,3,4,5,6,7,8,9,10}.

```
In[603]:= {1, 2, 3, 4, 5, 6, 7, 8, 9, 10}
Out[603]= {1, 2, 3, 4, 5, 6, 7, 8, 9, 10}
```

```
In[604]:= Table[i, {i, 10}]
Out[604]= {1, 2, 3, 4, 5, 6, 7, 8, 9, 10}
```

```
In[605]:= Table[i, {i, 1, 10}]
Out[605]= {1, 2, 3, 4, 5, 6, 7, 8, 9, 10}
```

```
In[606]:= Table[ i/2 , {i, 2, 20, 2}]
Out[606]= {1, 2, 3, 4, 5, 6, 7, 8, 9, 10}
```

```
In[607]:= Range[10]
Out[607]= {1, 2, 3, 4, 5, 6, 7, 8, 9, 10}
```

■

EXAMPLE 4.1.2: Use Mathematica to define listone to be the list of numbers {1, 3/2, 2, 5/2, 3, 7/2, 4}.

SOLUTION: In this case, we generate a list and name the result listone. As in Example 4.1.1, we illustrate that listone can be created in several ways.

```
In[608]:= listone = {1, 3/2, 2, 5/2, 3, 7/2, 4}
Out[608]= {1, 3/2, 2, 5/2, 3, 7/2, 4}
```

$$In[609] := \text{listone} = \text{Table}\left[i, \left\{i, 1, 4, \frac{1}{2}\right\}\right]$$

$$Out[609] = \left\{1, \frac{3}{2}, 2, \frac{5}{2}, 3, \frac{7}{2}, 4\right\}$$

Last, we define $i(n) = \frac{1}{2}n + \frac{1}{2}$ and use `Array` to create the table `listone`.

$$In[610] := \text{i[n\_]} = \frac{n}{2} + \frac{1}{2};$$

$$\text{listone} = \text{Array[i, 7]}$$

$$Out[610] = \left\{1, \frac{3}{2}, 2, \frac{5}{2}, 3, \frac{7}{2}, 4\right\}$$

■

EXAMPLE 4.1.3: Create a list of the first 25 prime numbers. What is the fifteenth prime number?

SOLUTION: The command `Prime[n]` yields the nth prime number. We use `Table` to generate a list of the ordered pairs `{n, Prime[n] }` for $n = 1, 2, 3, \ldots, 25$.

```
In[611] := list = Table[{n, Prime[n]}, {n, 1, 25}]
```

```
Out[611] = {{1, 2}, {2, 3}, {3, 5}, {4, 7}, {5, 11}, {6, 13},
           {7, 17}, {8, 19}, {9, 23}, {10, 29}, {11, 31},
           {12, 37}, {13, 41}, {14, 43}, {15, 47}, {16, 53},
           {17, 59}, {18, 61}, {19, 67}, {20, 71}, {21, 73},
           {22, 79}, {23, 83}, {24, 89}, {25, 97}}
```

The ith element of a list `list` is extracted from `list` with `list[[i]]`. From the resulting output, we see that the fifteenth prime number is 47.

```
In[612] := list[[15]]
```

```
Out[612] = {15, 47}
```

■

In addition, we can use `Table` to generate lists consisting of the same or similar objects.

EXAMPLE 4.1.4: (a) Generate a list consisting of five copies of the letter *a*. (b) Generate a list consisting of ten random integers between −10 and 10.

SOLUTION: Entering

```
In[613]:= Table[a, {5}]
Out[613]= {a, a, a, a, a}
```

generates a list consisting of five copies of the letter *a*. For (b), we use the command Random to generate the desired list. Because we are using Random, your results will certainly differ from those obtained here.

```
In[614]:= Table[Random[Integer, {-10, 10}], {10}]
Out[614]= {4, -2, -10, 2, 10, 0, 8, 7, -3, 0}
```

■

4.1.2 Plotting Lists of Points

Lists are plotted with ListPlot.

1. ListPlot[{{x1,y1},{x2,y2},...,{xn,yn}}] plots the list of points $\{(x_1, y_1), (x_2, y_2), \ldots, (x_n, y_n)\}$. The size of the points in the resulting plot is controlled with the option PlotStyle->PointSize[w], where *w* is the fraction of the total width of the graphic. For two-dimensional graphics, the default value is 0.008.
2. ListPlot[{y1,y2,..,yn}] plots the list of points $\{(1, y_1), (2, y_2), \ldots, (n, y_n)\}$.

EXAMPLE 4.1.5: Entering

```
In[615]:= t1 = Table[Sin[n], {n, 1, 1000}];

        ListPlot[t1]
```

creates a list consisting of sin *n* for *n* = 1, 2, ..., 1000 and then graphs the list of points (*n*, sin *n*) for *n* = 1, 2, ..., 1000. See Figure 4-1.

When a semi-colon is included at the end of a command, the resulting output is suppressed.

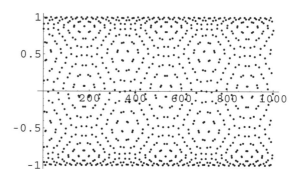

Figure 4-1   Plot of $(n, \sin n)$ for $n = 1, 2, \ldots, 1000$

EXAMPLE 4.1.6 (The Prime Difference Function and the Prime Number Theorem): In t1, we use Prime and Table to compute a list of the first 25, 000 prime numbers.

> *In[616]:=* **t1 = Table[Prime[n], {n, 1, 25000}];**

We use Length to verify that t1 has 25, 000 elements and Short to see an abbreviated portion of t1.

> *In[617]:=* **Length[t1]**
>
> *Out[617]=* 25000

> *In[618]:=* **Short[t1]**
>
> *Out[618]=* {2, 3, 5, 7, 11, 13, ≪24988≫, 287059, 287087,
>   287093, 287099, 287107, 287117}

First[list] returns the first element of list; Last[list] returns the last element of list.

You can also use Take to extract elements of lists.

1. Take[list,n] returns the first *n* elements of list;
2. Take[list,-n] returns the last *n* elements of list; and
3. Take[list,{n,m}] returns the *n*th through *m*th elements of list.

> *In[619]:=* **Take[t1, 5]**
>
> *Out[619]=* {2, 3, 5, 7, 11}

> *In[620]:=* **Take[t1, -5]**
>
> *Out[620]=* {287087, 287093, 287099, 287107, 287117}

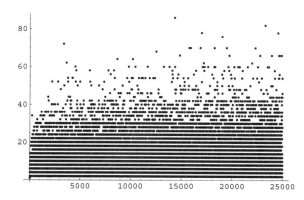

Figure 4-2   A plot of the difference, d_n, between successive prime numbers

```
In[621]:= Take[t1, {12501, 12505}]
Out[621]= {134059, 134077, 134081, 134087, 134089}
```

However, you can use Table together with Part ([[...]]) to obtain the same results as those obtained with Take.

```
In[622]:= Table[t1[[i]], {i, 1, 5}]

           Table[t1[[i]], {i, 24996, 25000}]

           Table[t1[[i]], {i, 12501, 12505}]
Out[622]= {2, 3, 5, 7, 11}
Out[622]= {287087, 287093, 287099, 287107, 287117}
Out[622]= {134059, 134077, 134081, 134087, 134089}
```

In t2, we compute the difference, d_n, between the successive prime numbers in t1. The result is plotted with ListPlot in Figure 4-2.

list[[i]] returns the *i*th element of list so list[[i + 1]] − list[[i]] computes the difference between the (*i* + 1)st and *i*th elements of list.

```
In[623]:= t2 = Table[t1[[i + 1]] - t1[[i]],
               {i, 1, Length[t1] - 1}];

In[624]:= Short[t2]
Out[624]= {1, 2, 2, 4, 2, 4, 2, 4, 6, 2, 6, 4, 2, 4, ≪24972≫,
           46, 8, 6, 12, 4, 44, 10, 2, 28, 6, 6, 8, 10}

In[625]:= ListPlot[t2, PlotRange → All]
```

Let $\pi(n)$ denote the number of primes less than n and $Li(x)$ denote the **logarithmic integral**:

$$\text{LogIntegral}[x] = Li(x) = \int_0^x \frac{1}{\ln t}\, dt.$$

We use Plot to graph *Li*(*x*) for $1 \leq x \leq 25,000$ in p1.

In[626]:= **p1 = Plot[LogIntegral[x], {x, 1, 25000}]**

The **Prime Number Theorem** states that

$$\pi(n) \sim Li(n).$$

(See [20].) In the following, we use Select and Length to define $\pi(n)$. Select[list,criteria] returns the elements of list for which criteria is true. Note that #<n is called a pure function: given an argument #, #<n is true if #<n and false otherwise. The & symbol marks the end of a pure function. Thus, given *n*, Select[t1,#<n&] returns a list of the elements of t1 less than *n*; Select[t1,#<n&]//Length returns the number of elements in the list.

In[627]:= **smallpi[n_] := Select[t1, # < n&]//Length**

For example,

In[628]:= **smallpi[100]**

Out[628]= 25

shows us that $\pi(100) = 25$. Note that because t1 contains the first 25,000 primes, smallpi[n] is valid for $1 \leq n \leq N$ where $\pi(N) = 25,000$. In t3, we compute $\pi(n)$ for $n = 1, 2, \ldots, 25,000$

In[629]:= **t3 = Table[smallpi[n], {n, 1, 25000}];**

In[630]:= **Short[t3]**

Out[630]= {0, 0, 1, 2, 2, 3, 3, 4, 4, 4, 4, ≪24978≫, 2762, 2762, 2762, 2762, 2762, 2762, 2762, 2762, 2762, 2762, 2762}

and plot the resulting list with ListPlot.

In[631]:= **p2 = ListPlot[t3, PlotStyle → GrayLevel[0.4]]**

p1 and p2 are displayed together with Show in Figure 4-3.

In[632]:= **Show[p1, p2]**

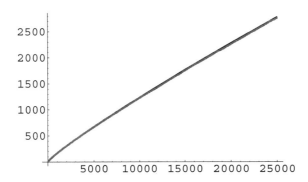

Figure 4-3   Graphs of *Li*(*x*) (in black) and $\pi(n)$ (in gray)

You can iterate recursively with `Table`. Both

```
In[633]:= t1 = Table[a[i, j], {j, 2, 10, 2}, {i, 1, 5}]
Out[633]= {{a[1, 2], a[2, 2], a[3, 2], a[4, 2], a[5, 2]},
           {a[1, 4], a[2, 4], a[3, 4], a[4, 4], a[5, 4]},
           {a[1, 6], a[2, 6], a[3, 6], a[4, 6], a[5, 6]},
           {a[1, 8], a[2, 8], a[3, 8], a[4, 8], a[5, 8]},
           {a[1, 10], a[2, 10], a[3, 10], a[4, 10], a[5, 10]}}

In[634]:= Length[t1]
Out[634]= 5
```

and

```
In[635]:= t2 = Table[Table[a[i, j], {i, 1, 5}], {j, 2, 10, 2}]
Out[635]= {{a[1, 2], a[2, 2], a[3, 2], a[4, 2], a[5, 2]},
           {a[1, 4], a[2, 4], a[3, 4], a[4, 4], a[5, 4]},
           {a[1, 6], a[2, 6], a[3, 6], a[4, 6], a[5, 6]},
           {a[1, 8], a[2, 8], a[3, 8], a[4, 8], a[5, 8]},
           {a[1, 10], a[2, 10], a[3, 10], a[4, 10], a[5, 10]}}
```

compute tables of a_{ij}. The outermost iterator is evaluated first: in this case, *i* is followed by *j* as in `t1` and the result is a list of lists. To eliminate the inner lists (that is, the braces), use `Flatten`. Generally, `Flatten[list,n]` flattens `list` (removes braces) to level *n*.

```
In[636]:= Flatten[t1]
Out[636]= {a[1, 2], a[2, 2], a[3, 2], a[4, 2], a[5, 2], a[1, 4],
           a[2, 4], a[3, 4], a[4, 4], a[5, 4], a[1, 6], a[2, 6],
           a[3, 6], a[4, 6], a[5, 6], a[1, 8], a[2, 8],
           a[3, 8], a[4, 8], a[5, 8], a[1, 10], a[2, 10],
           a[3, 10], a[4, 10], a[5, 10]}
```

The observation is especially important when graphing lists of points obtained by iterating Table. For example,

Length[list] returns the
number of elements in list.

```
In[637] := t1 = Table[{Sin[x + y], Cos[x - y]}, {x, 1, 5}, {y, 1, 5}]
Out[637]= {{{Sin[2], 1}, {Sin[3], Cos[1]}, {Sin[4], Cos[2]},
            {Sin[5], Cos[3]}, {Sin[6], Cos[4]}}, {{Sin[3], Cos[1]},
            {Sin[4], 1}, {Sin[5], Cos[1]}, {Sin[6], Cos[2]},
            {Sin[7], Cos[3]}}, {{Sin[4], Cos[2]}, {Sin[5], Cos[1]},
            {Sin[6], 1}, {Sin[7], Cos[1]}, {Sin[8], Cos[2]}},
            {{Sin[5], Cos[3]}, {Sin[6], Cos[2]}, {Sin[7], Cos[1]},
            {Sin[8], 1}, {Sin[9], Cos[1]}}, {{Sin[6], Cos[4]},
            {Sin[7], Cos[3]}, {Sin[8], Cos[2]}, {Sin[9], Cos[1]},
            {Sin[10], 1}}}
```

```
In[638] := Length[t1]
Out[638]= 5
```

is not a list of 25 points: t1 is a list of 5 lists each consisting of 5 points. t1 has two levels. For example, the 3rd element of the second level is

```
In[639] := t1[[3]]
Out[639]= {{Sin[4], Cos[2]}, {Sin[5], Cos[1]}, {Sin[6], 1},
           {Sin[7], Cos[1]}, {Sin[8], Cos[2]}}
```

and the 2nd element of the third level is

```
In[640] := t1[[3, 2]]
Out[640]= {Sin[5], Cos[1]}
```

To flatten t2 to level 1, we enter

```
In[641] := t2 = Flatten[t1, 1]
Out[641]= {{Sin[2], 1}, {Sin[3], Cos[1]}, {Sin[4], Cos[2]},
           {Sin[5], Cos[3]}, {Sin[6], Cos[4]}, {Sin[3], Cos[1]},
           {Sin[4], 1}, {Sin[5], Cos[1]}, {Sin[6], Cos[2]},
           {Sin[7], Cos[3]}, {Sin[4], Cos[2]}, {Sin[5], Cos[1]},
           {Sin[6], 1}, {Sin[7], Cos[1]}, {Sin[8], Cos[2]},
           {Sin[5], Cos[3]}, {Sin[6], Cos[2]}, {Sin[7], Cos[1]},
           {Sin[8], 1}, {Sin[9], Cos[1]}, {Sin[6], Cos[4]},
           {Sin[7], Cos[3]}, {Sin[8], Cos[2]},
           {Sin[9], Cos[1]}, {Sin[10], 1}}
```

and see the result is a list of points. These are plotted with ListPlot in Figure 4-4 (a). We also illustrate the use of the PlotStyle, PlotRange, and AspectRatio options in the ListPlot command.

> *In[642]:=* **lp1 = ListPlot[t2, PlotStyle → {PointSize[0.05],**
> **GrayLevel[0.5]}, PlotRange → {{-3/2, 3/2},**
> **{-3/2, 3/2}}, AspectRatio → Automatic]**

Increasing the number of points further illustrates the use of Flatten. Entering

> *In[643]:=* **t1 = Table[{Sin[x+y], Cos[x-y]}, {x, 1, 125}, {y, 1, 125}];**

> *In[644]:=* **Length[t1]**
> *Out[644]=* 125

results in a very long nested list. t1 has 125 elements each of which has 125 elements.

An abbreviated version is viewed with Short.

Short[list] yields an abbreviated version of list.

> *In[645]:=* **Short[t1]**
> *Out[645]=* {{{Sin[2], 1}, {Sin[3], Cos[1]}, ≪121≫,
> {Sin[125], Cos[123]}, {Sin[126], Cos[124]}},
> ≪123≫, {≪1≫}}

After using Flatten, we see with Length and Short that t2 contains 15,625 points,

> *In[646]:=* **t2 = Flatten[t1, 1];**

> *In[647]:=* **Length[t2]**
> *Out[647]=* 15625

> *In[648]:=* **Short[t2]**
> *Out[648]=* {{Sin[2], 1}, {Sin[3], Cos[1]}, ≪15621≫, 1
> {Sin[249], Cos[1]}, {Sin[250], 1}}

which are plotted with ListPlot in Figure 4-4 (b).

> *In[649]:=* **lp2 = ListPlot[t2, AspectRatio → Automatic]**

> *In[650]:=* **Show[GraphicsArray[{lp1, lp2}]]**

Remark. Mathematica is very flexible and most calculations can be carried out in more than one way. Depending on how you think, some sequences of calculations may make more sense to you than others, even if they are less efficient than the most efficient way to perform the desired calculations. Often, the difference in time required for Mathematica to perform equivalent – but different – calculations is quite small. For the beginner, we think it is wisest to work with familiar calculations first and then efficiency.

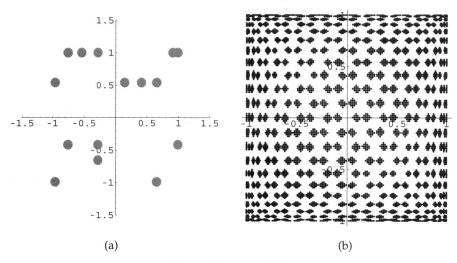

(a)   (b)

Figure 4-4   (a) and (b)

EXAMPLE 4.1.7 (Dynamical Systems): A sequence of the form $x_{n+1} = f(x_n)$ is called a **dynamical system**.

Sometimes, unusual behavior can be observed when working with dynamical systems. For example, consider the dynamical system with $f(x) = x + 2.5x(1 - x)$ and $x_0 = 1.2$. Note that we define x_n using the form $x[n_] := x[n] = \ldots$ so that Mathematica remembers the functional values it computes and thus avoids recomputing functional values previously computed. This is particularly advantageous when we compute the value of x_n for large values of n.

Observe that $x_{n+1} = f(x_n)$ can also be computed with $x_{n+1} = f^n(x_0)$.

```
In[651]:= Clear[f, x]

          f[x_] := x + 2.5 x (1 - x)

          x[n_] := x[n] = f[x[n - 1]]

          x[0] = 1.2;
```

In Figure 4-5, we see that the sequence oscillates between 0.6 and 1.2. We say that the dynamical system has a **2-cycle** because the values of the sequence oscillate between two numbers.

```
In[652]:= tb = Table[x[n], {n, 1, 200}];

In[653]:= Short[tb, 20]
```

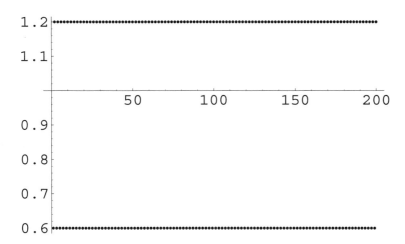

Figure 4-5   A 2-cycle

```
Out[653]= {0.6, 1.2, 0.6, 1.2, 0.6, 1.2, 0.6, 1.2, 0.6, 1.2,
           0.6, 1.2, 0.6, 1.2, 0.6, 1.2, 0.6, 1.2, 0.6, 1.2,
           0.6, 1.2, 0.6, 1.2, 0.6, 1.2, 0.6, 1.2, 0.6, 1.2,
           0.6, 1.2, 0.6, ≪134≫, 1.2, 0.6, 1.2, 0.6, 1.2,
           0.6, 1.2, 0.6, 1.2, 0.6, 1.2, 0.6, 1.2, 0.6, 1.2,
           0.6, 1.2, 0.6, 1.2, 0.599999, 1.2, 0.600001, 1.2,
           0.599999, 1.2, 0.600001, 1.2, 0.599999,
           1.2, 0.600002, 1.2, 0.599998, 1.2}

In[654]:= ListPlot[tb]
```

In Figure 4-6, we see that changing x_0 from 1.2 to 1.201 results in a 4-cycle.

```
In[655]:= Clear[f, x]

          f[x_] := x + 2.5 x (1 - x)

          x[n_] := x[n] = f[x[n - 1]]

          x[0] = 1.201;

In[656]:= tb = Table[x[n], {n, 1, 200}];

In[657]:= Short[tb, 20]
```

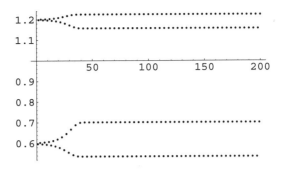

Figure 4-6   A 4-cycle

Out[657]= {0.597497, 1.19873, 0.603163, 1.20156, 0.596102,
             1.19801, 0.604957, 1.20242, 0.593943, 1.19688,
             0.607777, 1.20374, 0.590622, 1.19509, 0.612212,
             1.20573, 0.585585, 1.19227, 0.619168, 1.20867,
             0.578149, 1.18788, 0.629931, 1.21273, 0.567781,
             1.1813, 0.645888, ≪147≫, 0.701238, 1.225,
             0.535948, 1.15772, 0.701238, 1.225, 0.535948,
             1.15772, 0.701238, 1.225, 0.535948, 1.15772,
             0.701238, 1.225, 0.535948, 1.15772, 0.701238,
             1.225, 0.535948, 1.15772, 0.701238, 1.225,
             0.535948, 1.15772, 0.701238, 1.225}

In[658]:= **ListPlot[tb]**

The calculations indicate that the behavior of the system can change considerably for small changes in x_0. With the following, we adjust the definition of x so that x depends on $x_0 = c$: given c, $x_c(0) = c$.

In[659]:= **Clear[f, x]**

f[x_] := x + 2.5 x (1 - x)

x[c_][n_] := x[c][n] = f[x[c][n - 1]]//N

x[c_][0] := c//N;

In tb, we create a list of lists of the form $\{x_c(n) | n = 100, \ldots, 150\}$ for 150 equally spaced values of c between 0 and 1.5. Observe that Mathematica issues several error messages. When a Mathematica calculation is larger than the machine's precision, we obtain an Overflow[] warning. In numerical calculations, we interpret Overflow[] to correspond to ∞.

```
In[660]:= tb = Table[{c, x[c][n]}, {c, 0, 1.5, 0.01},
                {n, 100, 150}];
General :: ovfl : Overflow occurred in computation.
General :: ovfl : Overflow occurred in computation.
General :: ovfl : Overflow occurred in computation.
General :: stop : Further output of General :: ovfl will
    be suppressedduring this calculation.
```

We ignore the error messages and use Short to view an abbreviated form of tb.

```
In[661]:= Short[tb]
Out[661]= {{{0, 0.}, {0, 0.}, {0, 0.}, {0, 0.}, {0, 0.},
            ≪42≫, {0, 0.}, {0, 0.}, {0, 0.}, {0, 0.}}, ≪150≫}
```

We then use Flatten to convert tb to a list of points which are plotted with ListPlot in Figure 4-7 (a). Observe that even though Mathematica issues several warning messages, Mathematica is able to generate the plot.

```
In[662]:= tb2 = Flatten[tb, 1];

        f1 = ListPlot[tb2]
Graphics :: gptn : Coordinate Overflow[] in {1.4, Overflow[]}
    is not a floating – point number.
Graphics :: gptn : Coordinate Overflow[] in {1.4, Overflow[]}
    is not a floating – point number.
Graphics :: gptn : Coordinate Overflow[] in {1.4, Overflow[]}
    is not a floating – point number.
General :: stop : Further output of Graphics ::
    gptn will be suppressed during this calculation.
```

Another interesting situation occurs if we fix x_0 and let c vary in $f(x) = x + cx(1 - x)$.

With the following we set $x_0 = 1.2$ and adjust the definition of f so that f depends on c: $f(x) = x + cx(1 - x)$.

```
In[663]:= Clear[f, x]

        f[c_][x_] := x + c x(1 - x)//N

        x[c_][n_] := x[c][n] = f[c][x[c][n - 1]]//N

        x[c_][0] := 1.2//N;
```

In tb, we create a list of lists of the form $\{x_c(n)|n = 200, \ldots, 300\}$ for 350 equally spaced values of c between 0 and 3.5. As before, Mathematica issues several error messages, which we ignore.

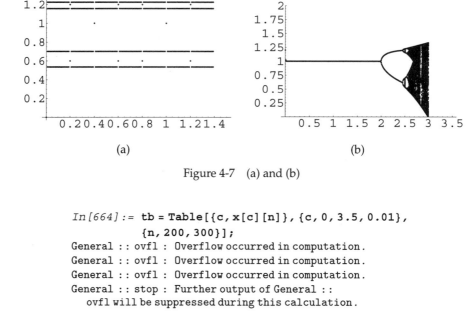

Figure 4-7   (a) and (b)

```
In[664]:= tb = Table[{c, x[c][n]}, {c, 0, 3.5, 0.01},
              {n, 200, 300}];
General :: ovfl : Overflow occurred in computation.
General :: ovfl : Overflow occurred in computation.
General :: ovfl : Overflow occurred in computation.
General :: stop : Further output of General ::
   ovfl will be suppressed during this calculation.

In[665]:= Short[tb]
Out[665]= {{{0, 1.2}, {0, 1.2}, {0, 1.2}, ≪95≫, {0, 1.2},
              {0, 1.2}, {0, 1.2}}, ≪350≫}
```

tb is then converted to a list of points with Flatten and the resulting list is plotted in Figure 4-7 (b) with ListPlot. This plot is called a **bifurcation diagram**.

```
In[666]:= tb2 = Flatten[tb, 1];

          f2 = ListPlot[tb2, PlotRange → {0, 2}]
Graphics :: gptn : Coordinate Overflow[] in {3.01, Overflow[]}
   is not a floating point number.
Graphics :: gptn : Coordinate Overflow[] in {3.01, Overflow[]}
   is not a floating point number.
Graphics :: gptn : Coordinate Overflow[] in {3.01, Overflow[]}
   is not a floating point number.
General :: stop : FurtheroutputofGraphics :: gptnwillbesuppressed
   during this calculation.

In[667]:= Show[GraphicsArray[{f1, f2}]]
```

As indicated earlier, elements of lists can be numbers, ordered pairs, functions, and even other lists. You can also use Mathematica to manipulate lists in numerous ways. Most importantly, the Map function is used to apply a function to a list:

$$\texttt{Map[f,\{x1,x2,\ldots,xn\}]}$$

returns the list $\{f(x_1),\ f(x_2),\ \ldots,\ f(x_n)\}$. We will discuss other operations that can be performed on lists in the following sections.

A function f is **listable** if f[list] and Map[f,list] are equivalent.

EXAMPLE 4.1.8 (Hermite Polynomials): The **Hermite polynomials**, $H_n(x)$, satisfy the differential equation $y'' - 2xy' + 2ny = 0$ and the orthogonality relation $\int_{-\infty}^{\infty} H_n(x)H_m(x)e^{-x^2}\,dx = \delta_{mn}2^n n!\sqrt{\pi}$. The Mathematica command HermiteH[n,x] yields the Hermite polynomial $H_n(x)$. (a) Create a table of the first five Hermite polynomials. (b) Evaluate each Hermite polynomial if $x = 1$. (c) Compute the derivative of each Hermite polynomial in the table. (d) Compute an antiderivative of each Hermite polynomial in the table. (e) Graph the five Hermite polynomials on the interval $[-1, 1]$. (f) Verify that $H_n(x)$ satisfies $y'' - 2xy' + 2ny = 0$ for $n = 1, 2 \ldots, 5$.

SOLUTION: We proceed by using HermiteH together with Table to define hermitetable to be the list consisting of the first five Hermite polynomials.

```
In[668]:= hermitetable = Table[HermiteH[n, x], {n, 1, 5}]
Out[668]= {2 x, -2 + 4 x², -12 x + 8 x³, 12 - 48 x² + 16 x⁴,
            120 x - 160 x³ + 32 x⁵}
```

We then use ReplaceAll (->) to evaluate each member of hermitetable if x is replaced by 1.

```
In[669]:= hermitetable/.x → 1
Out[669]= {2, 2, -4, -20, -8}
```

Functions like D and Integrate are listable. Thus, each of the following commands differentiate each element of hermitetable with respect to x. In the second case, we have used a *pure function*: given an argument #, D[#,x] & differentiates # with respect to x. Use the & symbol to indicate the end of a pure function.

```
In[670]:= D[hermitetable, x]
Out[670]= {2, 8x, -12+24x², -96x+64x³, 120-480x²+160x⁴}

In[671]:= Map[D[#, x]&, hermitetable]
Out[671]= {2, 8x, -12+24x², -96x+64x³, 120-480x²+160x⁴}
```

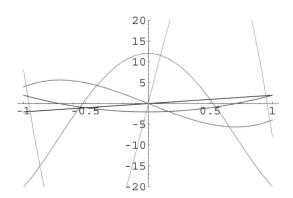

Figure 4-8   Graphs of $H_1(x)$ (in black), $H_2(x)$, $H_3(x)$, $H_4(x)$, and $H_5(x)$ (in light gray)

Similarly, we use `Integrate` to antidifferentiate each member of
`hermitetable` with respect to x. Remember that Mathematica does
not automatically include the "+C" that we include when we antidifferentiate.

In[672]:= **Integrate[hermitetable, x]**

Out[672]= $\left\{x^2, -2x + \dfrac{4x^3}{3}, -6x^2 + 2x^4, 12x - 16x^3 + \dfrac{16x^5}{5},\right.$

$\left. 60x^2 - 40x^4 + \dfrac{16x^6}{3}\right\}$

In[673]:= **Map[Integrate[#, x]&, hermitetable]**

Out[673]= $\left\{x^2, -2x + \dfrac{4x^3}{3}, -6x^2 + 2x^4, 12x - 16x^3 + \dfrac{16x^5}{5},\right.$

$\left. 60x^2 - 40x^4 + \dfrac{16x^6}{3}\right\}$

To graph the list `hermitetable`, we use `Plot` to plot each function in
the set `hermitetable` on the interval $[-2, 2]$ in Figure 4-8. Be sure to
include `hermitetable` within the `Evaluate` command as indicated.
In this case, we specify that the displayed y-values correspond to the
interval $[-20, 20]$. Note how `Table` and `GrayLevel` are used to create
a list of `GrayLevel`s in `grays`. The plots of the Hermite polynomials
are then shaded according to `grays`. The graph of $H_1(x)$ is in black and
successive plots are lighter with the graph of $H_5(x)$ the lightest gray.

> When plotting lists of functions, evaluate them first with `Evaluate` in the `Plot` command.

In[674]:= **grays = Table[GrayLevel[i], {i, 0, 0.6, 0.6/4}];**

Plot[Evaluate[hermitetable], {x, -1, 1},
PlotStyle → grays, PlotRange → {-20, 20}]

`hermitetable[[n]]` returns the nth element of `hermitetable`,
which corresponds to $H_n(x)$. Thus,

$In[675]:=$ **verifyde =**
    Table[D[hermitetable[[n]], {x, 2}] - 2x
    D[hermitetable[[n]], x] + 2 n hermitetable[[n]] //
    Simplify, {n, 1, 5}]

$Out[675]=$ $\{0, 0, 0, 0, 0\}$

computes and simplifies $H_n'' - 2xH_n' + 2nH_n$ for $n = 1, 2, \ldots, 5$. We use Table and Integrate to compute $\int_{-\infty}^{\infty} H_n(x)H_m(x)e^{-x^2}\,dx$ for $n = 1, 2, \ldots, 5$ and $m = 1, 2, \ldots, 5$.

$In[676]:=$ **verifyortho =**
    Table[Integrate[hermitetable[[n, 2]]
    hermitetable[[m, 2]] Exp[-x^2],
    {x, -∞, ∞}], {n, 1, 5}, {m, 1, 5}]

$Out[676]=$ $\left\{\left\{\frac{\sqrt{\pi}}{2}, 0, 6\sqrt{\pi}, 0, -120\sqrt{\pi}\right\}, \left\{0, 12\sqrt{\pi}, 0, -144\sqrt{\pi}, 0\right\},\right.$
$\left\{6\sqrt{\pi}, 0, 120\sqrt{\pi}, 0, -2400\sqrt{\pi}\right\},$
$\left\{0, -144\sqrt{\pi}, 0, 1728\sqrt{\pi}, 0\right\},$
$\left.\left\{-120\sqrt{\pi}, 0, -2400\sqrt{\pi}, 0, 48000\sqrt{\pi}\right\}\right\}$

To view a table in traditional row-and-column form use TableForm, as we do here illustrating the use of the TableHeadings option.

$In[677]:=$ **TableForm[verifyortho,**
    TableHeadings → {{"m = 1", "m = 2", "m = 3",
    "m = 4", "m = 5"},
    {"n = 1", "n = 2", "n = 3", "n = 4", "n = 5"}}]

	n = 1	n = 2	n = 3	n = 4	n = 5
m = 1	$\frac{\sqrt{\pi}}{2}$	0	$6\sqrt{\pi}$	0	$-120\sqrt{\pi}$
m = 2	0	$12\sqrt{\pi}$	0	$-144\sqrt{\pi}$	0
m = 3	$6\sqrt{\pi}$	0	$120\sqrt{\pi}$	0	$-2400\sqrt{\pi}$
m = 4	0	$-144\sqrt{\pi}$	0	$1728\sqrt{\pi}$	0
m = 5	$-120\sqrt{\pi}$	0	$-2400\sqrt{\pi}$	0	$48000\sqrt{\pi}$

$Out[677]=$

Be careful when using TableForm: TableForm[table] is no longer a list and cannot be manipulated like a list.

∎

4.2 Manipulating Lists: More on `Part` and `Map`

Often, Mathematica's output is given to us as a list that we need to use in subsequent calculations. Elements of a list are extracted with `Part` (`[[...]]`): `list[[i]]` returns the *i*th element of `list`; `list[[i,j]]` (or `list[[i]][[j]]`) returns the *j*th element of the *i*th element of `list`, and so on.

EXAMPLE 4.2.1: Let $f(x) = 3x^4 - 8x^3 - 30x^2 + 72x$. Locate and classify the critical points of $y = f(x)$.

SOLUTION: We begin by clearing all prior definitions of f and then defining f. The critical numbers are found by solving the equation $f'(x) = 0$. The resulting list is named `critnums`.

```
In[678]:= Clear[f]

       f[x_] = 3x^4 - 8x^3 - 30x^2 + 72x;

       critnums = Solve[f'[x] == 0]
Out[678]= {{x -> -2}, {x -> 1}, {x -> 3}}
```

`critnums` is actually a list of lists. For example, the number −2 is the second part of the first part of the second part of `critnums`.

```
In[679]:= critnums[[1]]
Out[679]= {x -> -2}

In[680]:= critnums[[1,1]]
Out[680]= x -> -2

In[681]:= critnums[[1,1,2]]
Out[681]= -2
```

Similarly, the numbers 1 and 3 are extracted with `critnums[[2,1,2]]` and `critnums[[3,1,2]]`, respectively.

```
In[682]:= critnums[[2,1,2]]

       critnums[[3,1,2]]
Out[682]= 1
Out[682]= 3
```

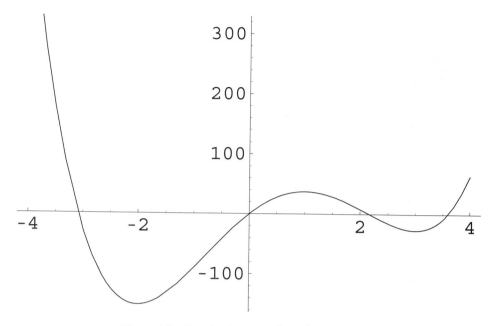

Figure 4-9   Graph of $f(x) = 3x^4 - 8x^3 - 30x^2 + 72x$

We locate and classify the points by evaluating $f(x)$ and $f''(x)$ for each of the numbers in `critnums`. `f[x]/.x->a` replaces each occurrence of x in $f(x)$ by a, so entering

> *In[683]:=* `{x, f[x], f''[x]}/.critnums`
>
> *Out[683]=* `{{-2, -152, 180}, {1, 37, -72}, {3, -27, 120}}`

replaces each x in the list $\{x, f(x), f''(x)\}$ by each of the x-values in `critnums`.

By the Second Derivative Test, we conclude that $y = f(x)$ has relative minima at the points $(-2, -152)$ and $(3, -27)$ while $f(x)$ has a relative maximum at $(1, 37)$. In fact, because $\lim_{x\to\pm\infty} = \infty$, -152 is the absolute minimum value of $f(x)$. These results are confirmed by the graph of $y = f(x)$ in Figure 4-9.

> *In[684]:=* `Plot[f[x], {x, -4, 4}]`

■

Map is a very powerful and useful function: Map [f, list] creates a list consisting of elements obtained by evaluating f for each element of list, provided that each member of list is an element of the domain of f. Note that if f is **listable**, f [list] produces the same result as Map [f, list].

To determine if f is listable, enter Attributes [f].

EXAMPLE 4.2.2: Entering

In[685] := **t1 = Table[n, {n, 1, 100}]**

Out[685]= {1, 2, 3, 4, 5, 6, 7, 8, 9, 10, 11, 12, 13, 14, 15, 16,
17, 18, 19, 20, 21, 22, 23, 24, 25, 26, 27, 28, 29,
30, 31, 32, 33, 34, 35, 36, 37, 38, 39, 40, 41, 42,
43, 44, 45, 46, 47, 48, 49, 50, 51, 52, 53, 54, 55,
56, 57, 58, 59, 60, 61, 62, 63, 64, 65, 66, 67, 68,
69, 70, 71, 72, 73, 74, 75, 76, 77, 78, 79, 80, 81,
82, 83, 84, 85, 86, 87, 88, 89, 90, 91, 92, 93, 94,
95, 96, 97, 98, 99, 100}

computes a list of the first 100 integers and names the result t1. We then define $f(x) = x^2$ and use Map to square each number in t1.

In[686] := **f[x_] = x^2**

Out[686]= x^2

In[687] := **Map[f, t1]**

Out[687]= {1, 4, 9, 16, 25, 36, 49, 64, 81, 100, 121, 144, 169,
196, 225, 256, 289, 324, 361, 400, 441, 484, 529,
576, 625, 676, 729, 84, 841, 900, 961, 1024, 1089,
1156, 1225, 1296, 1369, 1444, 1521, 1600, 1681,
1764, 1849, 1936, 2025, 2116, 2209, 2304, 2401,
2500, 2601, 2704, 2809, 2916, 3025, 3136, 3249,
3364, 3481, 3600, 3721, 3844, 3969, 4096, 4225,
4356, 4489, 4624, 4761, 4900, 5041, 5184, 5329,
5476, 5625, 5776, 5929, 6084, 6241, 6400, 6561,
6724, 6889, 7056, 7225, 7396, 7569, 7744, 7921,
8100, 8281, 8464, 8649, 8836, 8281, 8464, 8649,
8836, 9025, 9216, 9409, 9604, 9801, 10000}

The same result is accomplished by the pure function that squares its argument. Note how # denotes the argument of the pure function; the & symbol marks the end of the pure function.

In[688]:= **Map[#^2&, t1]**

Out[688]= {1, 4, 9, 16, 25, 36, 49, 64, 81, 100, 121, 144, 169, 196, 225, 256, 289, 324, 361, 400, 441, 484, 529, 576, 625, 676, 729, 784, 841, 900, 961, 1024, 1089, 1156, 1225, 1296, 1369, 1444, 1521, 1600, 1681, 1764, 1849, 1936, 2025, 2116, 2209, 2304, 2401, 2500, 2601, 2704, 2809, 2916, 3025, 3136, 3249, 3364, 3481, 3600, 3721, 3844, 3969, 4096, 4225, 4356, 4489, 4624, 4761, 4900, 5041, 5184, 5329, 5476, 5625, 5776, 5929, 6084, 6241, 6400, 6561, 6724, 6889, 7056, 7225, 7396, 7569, 7744, 7921, 8100, 8281, 8464, 8649, 8836, 9025, 9216, 9409, 9604, 9801, 10000}

On the other hand, entering

In[689]:= **t1 = Table[{a, b}, {a, 1, 5}, {b, 1, 5}]**

Out[689]= {{{1, 1}, {1, 2}, {1, 3}, {1, 4}, {1, 5}}, {{2, 1}, {2, 2}, {2, 3}, {2, 4}, {2, 5}}, {{3, 1}, {3, 2}, {3, 3}, {3, 4}, {3, 5}}, {{4, 1}, {4, 2}, {4, 3}, {4, 4}, {4, 5}}, {{5, 1}, {5, 2}, {5, 3}, {5, 4}, {5, 5}}}

is a list (of length 5) of lists (each of length 5). Use Flatten to obtain a list of 25 points, which we name t2.

In[690]:= **t2 = Flatten[t1, 1]**

Out[690]= {{1, 1}, {1, 2}, {1, 3}, {1, 4}, {1, 5}, {2, 1}, {2, 2}, {2, 3}, {2, 4}, {2, 5}, {3, 1}, {3, 2}, {3, 3}, {3, 4}, {3, 5}, {4, 1}, {4, 2}, {4, 3}, {4, 4}, {4, 5}, {5, 1}, {5, 2}, {5, 3}, {5, 4}, {5, 5}}

f is a function of two variables. Given an ordered pair (x, y), $f((x, y))$ returns the ordered triple $(x, y, x^2 + y^2)$.

In[691]:= **f[{x_, y_}] = {{x, y}, x^2 + y^2};**

We then use Map to apply f to t2.

In[692]:= **Map[f, t2]**

Out[692]= {{{1, 1}, 2}, {{1, 2}, 5}, {{1, 3}, 10}, {{1, 4}, 17},
        {{2, 4}, 20}, {{1, 5}, 26}, {{2, 1}, 5}, {{2, 2}, 8},
        {{2, 3}, 13}, {{2, 5}, 29}, {{3, 1}, 10}, {{3, 2}, 13},
        {{3, 3}, 18}, {{3, 4}, 25}, {{3, 5}, 34}, {{4, 1}, 17},
        {{4, 2}, 20}, {{4, 3}, 25}, {{4, 4}, 32}, {{4, 5}, 41},
        {{5, 1}, 26}, {{5, 2}, 29}, {{5, 3}, 34}, {{5, 4}, 41},
        {{5, 5}, 50}}

We accomplish the same result with a pure function. Observe how #[[1]] and #[[2]] are used to represent the first and second arguments: given a list of length 2, the pure function returns the list of ordered triples consisting of the first element of the list, the second element of the list, and the sum of the squares of the first and second elements.

In[693]:= **Map[{{#[[1]], #[[2]]}, #[[1]]^2 + #[[2]]^2}&, t2]**

Out[693]= {{{1, 1}, 2}, {{1, 2}, 5}, {{1, 3}, 10}, {{1, 4}, 17},
        {{1, 5}, 26}, {{2, 1}, 5}, {{2, 2}, 8}, {{2, 3}, 13},
        {{2, 4}, 20}, {{2, 5}, 29}, {{3, 1}, 10}, {{3, 2}, 13},
        {{3, 3}, 18}, {{3, 4}, 25}, {{3, 5}, 34}, {{4, 1}, 17},
        {{4, 2}, 20}, {{4, 3}, 25}, {{4, 4}, 32}, {{4, 5}, 41},
        {{5, 1}, 26}, {{5, 2}, 29}, {{5, 3}, 34}, {{5, 4}, 41},
        {{5, 5}, 50}}

EXAMPLE 4.2.3: Make a table of the values of the trigonometric functions $y = \sin x$, $y = \cos x$, and $y = \tan x$ for the principal angles.

SOLUTION: We first construct a list of the principal angles which is accomplished by defining t1 to be the list consisting of $n\pi/4$ for $n = 0$, $1, \ldots, 8$ and t2 to be the list consisting of $n\pi/6$ for $n = 0, 1, \ldots, 12$. The principal angles are obtained by taking the union of t1 and t2. Union[t1,t2] joins the lists t1 and t2, removes repeated elements, and sorts the results. If we did not wish to remove repeated elements and sort the result, the command Join[t1,t2] concatenates the lists t1 and t2.

In[694]:= **t1 = Table$\left[\dfrac{n\pi}{4}, \{n, 0, 8\}\right]$;**

t2 = Table$\left[\dfrac{n\pi}{6}, \{n, 0, 12\}\right]$;

In[695]:= **prinangles = Union[t1, t2]**

Out[695]= $\left\{0, \dfrac{\pi}{6}, \dfrac{\pi}{4}, \dfrac{\pi}{3}, \dfrac{\pi}{2}, \dfrac{2\pi}{3}, \dfrac{3\pi}{4}, \dfrac{5\pi}{6}, \pi, \right.$
$\left. \dfrac{7\pi}{6}, \dfrac{5\pi}{4}, \dfrac{4\pi}{3}, \dfrac{3\pi}{2}, \dfrac{5\pi}{3}, \dfrac{7\pi}{4}, \dfrac{11\pi}{6}, 2\pi \right\}$

We can also use the symbol ∪, which is obtained by clicking on the ∪ button on the **BasicTypesetting** palette to represent Union.

The **BasicTypesetting** palette:

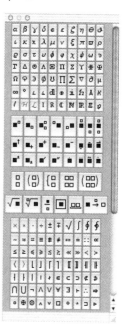

In[696]:= **prinangles = t1 ∪ t2**

Out[696]= $\left\{0, \dfrac{\pi}{6}, \dfrac{\pi}{4}, \dfrac{\pi}{3}, \dfrac{\pi}{2}, \dfrac{2\pi}{3}, \dfrac{3\pi}{4}, \dfrac{5\pi}{6}, \pi, \right.$
$\left. \dfrac{7\pi}{6}, \dfrac{5\pi}{4}, \dfrac{4\pi}{3}, \dfrac{3\pi}{2}, \dfrac{5\pi}{3}, \dfrac{7\pi}{4}, \dfrac{11\pi}{6}, 2\pi \right\}$

Next, we define $f(x)$ to be the function that returns the ordered quadruple $(x, \sin x, \cos x, \tan x)$ and compute the value of $f(x)$ for each number in prinangles with Map naming the resulting table prinvalues. prinvalues is not displayed because a semi-colon is included at the end of the command.

In[697]:= **Clear[f]**

  f[x_] = {x, Sin[x], Cos[x], Tan[x]};

In[698]:= **prinvalues = Map[f, prinangles];**

Finally, we use TableForm illustrating the use of the TableHeadings option to display prinvalues in row-and-column form; the columns are labeled x, $\sin x$, $\cos x$, and $\tan x$.

Remember that the result of using TableForm is not a list so cannot be manipulated like lists.

In[699]:= **TableForm[prinvalues,**
  TableHeadings →
  {None, {"x", "sin(x)", "cos(x)", "tan(x)"}}]

$$
\begin{array}{llll}
x & \sin(x) & \cos(x) & \tan(x) \\
0 & 0 & 1 & 0 \\
\dfrac{\pi}{6} & \dfrac{1}{2} & \dfrac{\sqrt{3}}{2} & \dfrac{1}{\sqrt{3}} \\
\dfrac{\pi}{4} & \dfrac{1}{\sqrt{2}} & \dfrac{1}{\sqrt{2}} & 1 \\
\dfrac{\pi}{3} & \dfrac{\sqrt{3}}{2} & \dfrac{1}{2} & \sqrt{3} \\
\dfrac{\pi}{2} & 1 & 0 & \text{ComplexInfinity} \\
\dfrac{2\pi}{3} & \dfrac{\sqrt{3}}{2} & -\dfrac{1}{2} & -\sqrt{3} \\
\dfrac{3\pi}{4} & \dfrac{1}{\sqrt{2}} & -\dfrac{1}{\sqrt{2}} & -1 \\
\dfrac{5\pi}{6} & \dfrac{1}{2} & -\dfrac{\sqrt{3}}{2} & -\dfrac{1}{\sqrt{3}} \\
\pi & 0 & -1 & 0 \\
\dfrac{7\pi}{6} & -\dfrac{1}{2} & -\dfrac{\sqrt{3}}{2} & \dfrac{1}{\sqrt{3}} \\
\dfrac{5\pi}{4} & -\dfrac{1}{\sqrt{2}} & -\dfrac{1}{\sqrt{2}} & 1 \\
\dfrac{4\pi}{3} & -\dfrac{\sqrt{3}}{2} & -\dfrac{1}{2} & \sqrt{3} \\
\dfrac{3\pi}{2} & -1 & 0 & \text{ComplexInfinity} \\
\dfrac{5\pi}{3} & -\dfrac{\sqrt{3}}{2} & \dfrac{1}{2} & -\sqrt{3} \\
\dfrac{7\pi}{4} & -\dfrac{1}{\sqrt{2}} & \dfrac{1}{\sqrt{2}} & -1 \\
\dfrac{11\pi}{6} & -\dfrac{1}{2} & \dfrac{\sqrt{3}}{2} & -\dfrac{1}{\sqrt{3}} \\
2\pi & 0 & 1 & 0
\end{array}
$$

Out[699]=

■

In the table, note that functions like $y = \tan x$ are undefined at certain values of x. $y = \tan x$ is undefined at odd multiples of $\pi/2$ and Mathematica appropriately returns ComplexInfinity at those values of x for which $y = \tan x$ is undefined.

Remark. The result of using TableForm is not a list (or table) and calculations on it using commands like Map cannot be performed. TableForm helps you see results in a more readable format. To avoid confusion, do not assign the results of using TableForm any name: adopting this convention avoids any possible manipulation of TableForm objects.

object=name assigns the object object the name name.

We can use Map on any list, including lists of functions and/or other lists.

Lists of functions are graphed with Plot:

    Plot[Evaluate[listoffunctions],{x,a,b}]

graphs the list of functions of x, listoffunctions, for $a \le x \le b$.

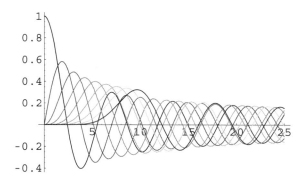

Figure 4-10   Graphs of $J_n(x)$ for $n = 0, 1, 2, \ldots, 8$

EXAMPLE 4.2.4 (Bessel Functions): The **Bessel functions of the first kind**, $J_n(x)$, are nonsingular solutions of $x^2y'' + xy' + (x^2 - n^2)y = 0$. `BesselJ[n,x]` returns $J_n(x)$. Graph $J_n(x)$ for $n = 0, 1, 2, \ldots, 8$.

SOLUTION: In `t1`, we use `Table` and `BesselJ` to create a list of $J_n(x)$ for $n = 0, 1, 2, \ldots, 8$.

```
In[700]:= t1 = Table[BesselJ[n, x], {n, 0, 8}];
```

Next, we define a list, named `grays`, consisting of `GrayLevel[i]` for 8 equally spaced values of i between 0 and 0.8. We then use `Plot` to graph each function in `t1`; the graphs in Figure 4-10 are shaded according to `grays`. In the plot, the graph of $J_0(x)$ is black. Subsequent plots are lighter; the lightest gray is the graph of $J_8(x)$.

```
In[701]:= grays = Table[GrayLevel[i], {i, 0, 0.8, 0.8/7}];

          Plot[Evaluate[t1], {x, 0, 25}, PlotStyle → grays]
```

A different effect is achieved by graphing each function separately. To do so, we define the function `pfunc`. Given a function of x, `f`, `pfunc[f]` plots the function for $0 \leq x \leq 100$. The resulting graphic is not displayed because the option `DisplayFunction->Identity` is included in the `Plot` command. We then use `Map` to apply `pfunc` to each element of `t1`. The result is a list of 9 graphics objects, which we name `t2`. A nice way to display 9 graphics is as a 3×3 array so we use `Partition` to convert `t2` from a list of length 9 to a list of lists, each with length

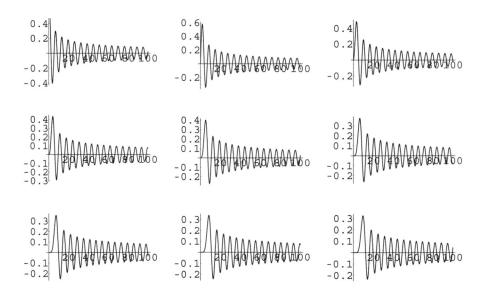

Figure 4-11   In the first row, from left to right, graphs of $J_0(x)$, $J_1(x)$, and $J_2(x)$; in the second row, from left to right, graphs of $J_3(x)$, $J_4(x)$, and $J_5(x)$; in the third row, from left to right, graphs of $J_6(x)$, $J_7(x)$, and $J_8(x)$

3–a 3 × 3 array. `Partition[list,n]` returns a list of lists obtained by partitioning `list` into n-element subsets.

Think of `Flatten` and
`Partition` as inverse
functions.

```
In[702]:= pfunc[f_] := Plot[f, {x, 0, 100},
              DisplayFunction → Identity];

          t2 = Map[pfunc, t1];

          t3 = Partition[t2, 3];
```

Instead of defining `pfunc`, you can use a pure function instead. The following accomplishes the same result. We display `t3` using `Show` together with `GraphicsArray` in Figure 4-11.

```
In[703]:= t2 = Map[Plot[#, {x, 0, 100},
              DisplayFunction → Identity]&, t1];

          t3 = Partition[t2, 3];

          Show[GraphicsArray[t3]]
```

■

EXAMPLE 4.2.5 (Dynamical Systems): Let $f_c(x) = x^2 + c$ and consider the dynamical system given by $x_0 = 0$ and $x_{n+1} = f_c(x_n)$. Generate a bifurcation diagram of f_c.

SOLUTION: First, recall that `Nest[f,x,n]` computes the repeated composition $f^n(x)$. Then, in terms of a composition,

Compare the approach used here with the approach used in Example 4.1.7.

$$x_{n+1} = f_c(x_n) = f_c^n(0).$$

We will compute $f_c^n(0)$ for various values of c and "large" values of n so we begin by defining `cvals` to be a list of 300 equally spaced values of c between -2.5 and 1.

In[704]:= **cvals = Table[c, {c, -2.5, 1., 3.5/299}];**

We then define $f_c(x) = x^2 + c$. For a given value of c, `f[c]` is a function of one variable, x, while the form `f[c_,x_]:=...` results in a function of two variables.

In[705]:= **Clear[f]**

f[c_][x_] := x^2 + c

To iterate f_c for various values of c, we define h. For a given value of c, $h(c)$ returns the list of points $\{(c, f_c^{100}(0)), (c, f_c^{101}(0)), \ldots, (c, f_c^{200}(0))\}$.

In[706]:= **h[c_] := {Table[{c, Nest[f[c], 0, n]},**
{n, 100, 200}]}

We then use `Map` to apply h to the list `cvals`. Observe that Mathematica generates several error messages when numerical precision is exceeded. We choose to disregard the error messages.

In[707]:= **t1 = Map[h, cvals];**
```
General :: ovfl : Overflow occurred in computation.
General :: ovfl : Overflow occurred in computation.
General :: ovfl : Overflow occurred in computation.
General :: stop : Further output of General ::
   ovfl will be suppressed during this calculation.
```

`t1` is a list (of length 300) of lists (each of length 101). To obtain a list of points (or, lists of length 2), we use `Flatten`. The resulting set of points is plotted with `ListPlot` in Figure 4-12. Observe that Mathematica again displays several error messages, which are not displayed here

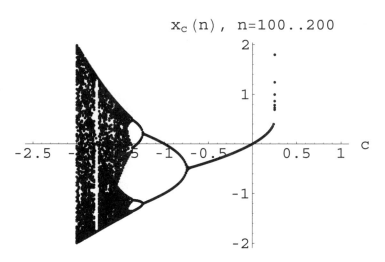

Figure 4-12   Bifurcation diagram of f_c

for length considerations, that we ignore: Mathematica only plots the points with real coordinates and ignores those containing `Overflow[]`.

```
In[708]:= t2 = Flatten[t1, 2];

        ListPlot[t2, AxesLabel → {"c", x_c(n), n = 100.0.2}]
```

■

4.2.1 More on Graphing Lists; Graphing Lists of Points Using Graphics Primitives

Include the `PlotJoined->True` option in a `ListPlot` command to connect successive points with line segments.

Using *graphics primitives* like `Point` and `Line` gives you even more flexibility. `Point[{x,y}]` represents a point at (x, y).

$$\text{Line}[\{\{x1,y1\},\{x2,y2\},\ldots,\{xn,yn\}\}]$$

represents a sequence of points $(x_1, y_1), (x_2, y_2), \ldots, (x_n, y_n)$ connected with line segments. A graphics primitive is declared to be a graphics object with `Graphics`: `Show[Graphics[Point[{x,y}]]]` displays the point (x, y). The advantage of using primitives is that each primitive is affected by the options that directly precede it.

Table 4-1   Union membership as a percentage of the labor force

Year	Union Membership as a Percentage of the Labor Force
1930	11.6
1935	13.2
1940	26.9
1945	35.5
1950	31.5
1955	33.2
1960	31.4
1965	28.4
1970	27.3
1975	25.5
1980	21.9
1985	18.0
1990	16.1

EXAMPLE 4.2.6: Table 4-1 shows the percentage of the United States labor force that belonged to unions during certain years. Graph the data represented in the table.

SOLUTION: We begin by entering the data represented in the table as dataunion:

```
In[709]:= dataunion = {{30,11.6},{35,13.2},{40,26.9},
             {45,35.5},{50,31.5},{55,33.2},{60,31.4},
             {65,28.4},{70,27.3},{75,25.5},{80,21.9},
             {85,18.},{90,16.1}};
```

the x-coordinate of each point corresponds to the year, where x is the number of years past 1900, and the y-coordinate of each point corresponds to the percentage of the United States labor force that belonged to unions in the given year. We then use ListPlot to graph the set of points represented in dataunion in lp1, lp2 (illustrating the PlotStyle option), and lp3 (illustrating the PlotJoined option). All three plots are displayed side-by-side in Figure 4-13 using Show together with GraphicsArray.

```
In[710]:= lp1 = ListPlot[dataunion]

In[711]:= lp2 = ListPlot[dataunion,
             PlotStyle → PointSize[0.03]]
```

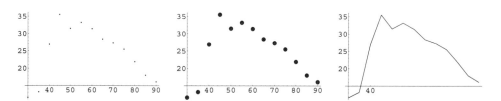

Figure 4-13   Union membership as a percentage of the labor force

In[712]:= **lp3 = ListPlot[dataunion, PlotJoined → True]**

In[713]:= **Show[GraphicsArray[{lp1, lp2, lp3}]]**

An alternative to using ListPlot is to use Show, Graphics, and Point to view the data represented in dataunion. In the following command we use Map to apply the function Point to each pair of data in dataunion. The result is not a graphics object and cannot be displayed with Show.

In[714]:= **datapts1 = Map[Point, dataunion]**

Out[714]= {Point[{30, 11.6}], Point[{35, 13.2}],
          Point[{40, 26.9}], Point[{45, 35.5}],
          Point[{50, 31.5}], Point[{55, 33.2}],
          Point[{60, 31.4}], Point[{65, 28.4}],
          Point[{70, 27.3}], Point[{75, 25.5}],
          Point[{80, 21.9}], Point[{85, 18.}],
          Point[{90, 16.1}]}

Next, we use Show and Graphics to declare the set of points Map[Point,dataunion] as graphics objects and display the resulting graphics object. The command PointSize[.03] specifies that the points be displayed as filled circles of radius 0.03 of the displayed graphics object.

In[715]:= **dp1 = Show[Graphics[{PointSize[0.03], datapts1},
          Axes → Automatic]]**

The collection of all commands contained within a Graphics command is contained in braces { . . . }. Each graphics primitive is affected by the options like PointSize, GrayLevel (or RGBColor) directly preceding it. Thus,

In[716]:= **datapts2 = Map[{GrayLevel[Random[]],
          Point[#]}&, dataunion]**

Out[716]= {{GrayLevel[0.401706], Point[{30, 11.6}]},
              {GrayLevel[0.709086], Point[{35, 13.2}]},
              {GrayLevel[0.310305], Point[{40, 26.9}]},
              {GrayLevel[0.946182], Point[{45, 35.5}]},
              {GrayLevel[0.430326], Point[{50, 31.5}]},
              {GrayLevel[0.0457745], Point[{55, 33.2}]},
              {GrayLevel[0.525196], Point[{60, 31.4}]},
              {GrayLevel[0.395095], Point[{65, 28.4}]},
              {GrayLevel[0.777691], Point[{70, 27.3}]},
              {GrayLevel[0.0661548], Point[{75, 25.5}]},
              {GrayLevel[0.378523], Point[{80, 21.9}]},
              {GrayLevel[0.0846463], Point[{85, 18.}]},
              {GrayLevel[0.519354], Point[{90, 16.1}]}}}

In[717]:= **dp2 = Show[Graphics[{PointSize[0.03], datapts2},**
             Axes → Automatic]]

displays the points in dataunion in various shades of gray and

In[718]:= **datapts3 = Map[{PointSize[Random[Real,**
             {0.008, 0.1}]], GrayLevel[Random[]],
             Point[#]}&, dataunion]

Out[718]= {{PointSize[0.0491743], GrayLevel[0.469353],
              Point[{30, 11.6}]}, {PointSize[0.0848502],
               GrayLevel[0.563721], Point[{35, 13.2}]},
              {PointSize[0.0536195], GrayLevel[0.798519],
               Point[{40, 26.9}]},
              {PointSize[0.0856063], GrayLevel[0.196485],
               Point[{45, 35.5}]},
              {PointSize[0.0278527], GrayLevel[0.189742],
               Point[{50, 31.5}]},
              {PointSize[0.0501316], GrayLevel[0.794779],
               Point[{55, 33.2}]},
              {PointSize[0.0546168], GrayLevel[0.879437],
               Point[{60, 31.4}]},
              {PointSize[0.0550828], GrayLevel[0.364453],
               Point[{65, 28.4}]},
              {PointSize[0.0504055], GrayLevel[0.354242],
               Point[{70, 27.3}]},
              {PointSize[0.0187341], GrayLevel[0.586762],
               Point[{75, 25.5}]},
              {PointSize[0.0443193], GrayLevel[0.975719],
               Point[{80, 21.9}]},
              {PointSize[0.0109466], GrayLevel[0.0674086],
               Point[{85, 18.}]},
              {PointSize[0.095145], GrayLevel[0.506366],
               Point[{90, 16.1}]}}}

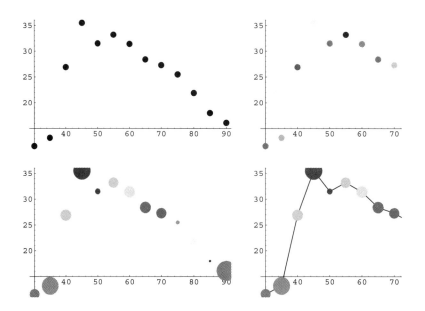

Figure 4-14   Union membership as a percentage of the labor force

$In[719]:=$ **dp3 = Show[Graphics[{datapts3}, Axes → Automatic]]**

shows the points in datunion in various sizes and in various shades of gray. We connect successive points with line segments

$In[720]:=$ **connectpts = Graphics[Line[datunion]];**

$In[721]:=$ **dp4 = Show[connectpts, dp3, Axes → Automatic]**

$In[722]:=$ **Show[GraphicsArray[{{dp1, dp2}, {dp3, dp4}}]]**

and show all four plots in Figure 4-14 using Show and GraphicsArray.

∎

With the speed of today's computers and the power of Mathematica, it is relatively easy now to carry out many calculations that required supercomputers and sophisticated programming experience just a few years ago.

EXAMPLE 4.2.7 (Julia Sets): Plot Julia sets for $f(z) = \lambda \cos z$ if $\lambda = .66i$ and $\lambda = .665i$.

SOLUTION: The sets are visualized by plotting the points (a, b) for which $|f^n(a + bi)|$ is *not* large in magnitude so we begin by forming our complex grid. Using `Table` and `Flatten`, we define `complexpts` to be a list of 62,500 points of the form $a + bi$ for 250 equally spaced real values of a between 0 and 8 and 300 equally spaced real values of b between -4 and 4 and then $f(z) = .66i \cos z$.

```
In[723]:= complexpts = Flatten[Table[a + b I,
              {a, 0., 8., 8/249}, {b, -4., 4., 6/249}], 1];
```

```
In[724]:= Clear[f]
```

```
          f[z_] = 0.66I Cos[z]
Out[724]= 0.66 i Cos[z]
```

For a given value of $c = a + bi$, $h(c)$ returns the ordered triple consisting of the real part of c, the imaginary part of c, and the value of $f^{200}(c)$.

```
In[725]:= h[c_] := {Re[c], Im[c], Nest[f, c, 200]}
```

We then use `Map` to apply h to `complexpts`. Observe that Mathematica generates several error messages. When machine precision is exceeded, we obtain an `Overflow[]` error message; numerical results smaller than machine precision results in an `Underflow[]` error message.

```
In[726]:= t1 = Map[h, complexpts]//Chop;
General :: ovfl : Overflow occurred in computation.
General :: ovfl : Overflow occurred in computation.
General :: ovfl : Overflow occurred in computation.
General :: stop : Further output of General ::
    ovfl will be suppressed during this calculation.
General :: unfl : Underflow occurred in computation.
General :: unfl : Underflow occurred in computation.
General :: unfl : Underflow occurred in computation.
General :: stop : Further output of General ::
    unfl will be suppressed during this calculation.
```

We use the error messages to our advantage. In `t2`, we select those elements of `t1` for which the third coordinate *is not* `Indeterminate`, which corresponds to the ordered triples $(a, b, f^n(a + bi))$ for which $|f^n(a + bi)|$ *is not* large in magnitude while in `t2b`, we select those elements of `t1` for which the third coordinate *is* `Indeterminate`, which corresponds to the ordered triples $(a, b, f^n(a + bi))$ for which $|f^n(a + bi)|$ *is* large in magnitude.

```
In[727]:= t2 = Select[t1, Not[#[[3]] === Indeterminate]&];
```

```
In[728]:= t2b = Select[t1, #[[3]] === Indeterminate&];
```

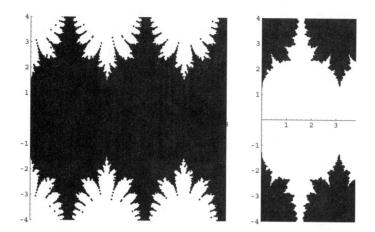

Figure 4-15   Julia set for $0.66i \cos z$

Lists of ordered pairs (a, b) are obtained in `t3` and `t3b` by applying `pt` to each list, `t2` and `t2b`, respectively,

```
In[729]:= pt[{x_, y_, z_}] := {x, y}
```

```
In[730]:= t3 = Map[pt, t2];

           t3b = Map[pt, t2b];
```

which are then graphed with `ListPlot` and shown side-by-side in Figure 4-15 using `Show` and `GraphicsArray`. As expected, the images are inversions of each other.

```
In[731]:= lp1 = ListPlot[t3, PlotRange → {{0, 8}, {-4, 4}},
              AspectRatio → Automatic,
              DisplayFunction → Identity];

           lp2 = ListPlot[t3b, PlotRange → {{0, 8}, {-4, 4}},
              AspectRatio → Automatic,
              DisplayFunction → Identity];

           Show[GraphicsArray[{lp1, lp2}]]
```

Similar error messages are encountered but we have not included them due to length considerations.

Changing λ from $0.66i$ to $0.665i$ results in a surprising difference in the plots. We proceed as before but increase the number of sample points to 120,000. See Figure 4-16.

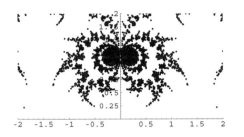

Figure 4-16   Julia set for $0.665i \cos z$

```
In[732]:= complexpts = Flatten[Table[a + b I,
                {a, -2., 2., 4/399}, {b, 0., 2., 2/299}], 1];

In[733]:= Clear[f]

            f[z_] = 0.665I Cos[z];

In[734]:= h[c_] := {Re[c], Im[c], Nest[f, c, 200]}

In[735]:= t1 = Map[h, complexpts]//Chop;

In[736]:= t2 = Select[t1, Not[#[[3]] === Indeterminate]&];

In[737]:= t2 = Select[t2, Not[#[[3]] === Overflow[]]&];

In[738]:= t2b = Select[t1, #[[3]] === Indeterminate&];

In[739]:= pt[{x_, y_, z_}] := {x, y}

In[740]:= t3 = Map[pt, t2];

            t3b = Map[pt, t2b];

In[741]:= lp1 = ListPlot[t3, PlotRange -> {{-2, 2}, {0, 2}},
                AspectRatio -> Automatic,
                DisplayFunction -> Identity];

            lp2 = ListPlot[t3b, PlotRange -> {{-2, 2}, {0, 2}},
                AspectRatio -> Automatic,
                DisplayFunction -> Identity];

            Show[GraphicsArray[{lp1, lp2}]]
```

To see detail, we take advantage of pure functions, `Map`, and graphics primitives in three different ways. In Figure 4-17, the shading of the point (a, b) is assigned according to the distance of $f^{200}(a + bi)$ from the origin. The color black indicates a distance of zero from the origin; as the distance increases, the shading of the point becomes lighter.

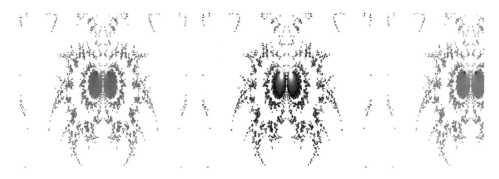

Figure 4-17   Shaded Julia sets for $0.665i \cos z$

```
In[742] := t2p = Map[{#[[1]], #[[2]],
             Min[Abs[#[[3]]], 3]}&, t2];

           t2p2 = Map[{GrayLevel[#[[3]]/3],
             Point[{#[[1]], #[[2]]}]}&, t2p];

           jp1 = Show[Graphics[t2p2],
             PlotRange → {{-2, 2}, {0, 2}}, AspectRatio → 1]


In[743] := t2p = Map[{#[[1]], #[[2]],
             Min[Abs[Re[#[[3]]]], 0.25]}&, t2];

           t2p2 = Map[{GrayLevel[#[[3]]/0.25],
             Point[{#[[1]], #[[2]]}]}&, t2p];

           jp2 = Show[Graphics[t2p2],
             PlotRange → {{-2, 2}, {0, 2}}, AspectRatio → 1]


In[744] := t2p = Map[{#[[1]], #[[2]],
             Min[Abs[Im[#[[3]]]], 2.5]}&, t2];

           t2p2 = Map[{GrayLevel[#[[3]]/2.5],
             Point[{#[[1]], #[[2]]}]}&, t2p];

           jp3 = Show[Graphics[t2p2],
             PlotRange → {{-2, 2}, {0, 2}}, AspectRatio → 1]


In[745] := Show[GraphicsArray[{jp1, jp2, jp3}]]
```

∎

4.2.2 Miscellaneous List Operations

4.2.2.1 Other List Operations

Some other Mathematica commands used with lists include:

1. `Append[list,element]`, which appends `element` to `list`;
2. `AppendTo[list,element]`, which appends `element` to `list` and names the result `list`;
3. `Drop[list,n]`, which returns the list obtained by dropping the first n elements from `list`;
4. `Drop[list,-n]`, which returns the list obtained by dropping the last n elements of `list`;
5. `Drop[list,{n,m}]`, which returns the list obtained by dropping the nth through mth elements of `list`;
6. `Drop[list,{n}]`, which returns the list obtained by dropping the nth element of `list`;
7. `Prepend[list,element]`, which prepends `element` to `list`; and
8. `PrependTo[list,element]`, which prepends `element` to `list` and names the result `list`.

4.2.2.2 Alternative Way to Evaluate Lists by Functions

Abbreviations of several of the commands discussed in this section are summarized in the following table.

`@@ Apply`	`//` (function application)	`{...} List`
`/@ Map`	`[[...]] Part`	

4.3 Mathematics of Finance

The use of lists and tables is quite useful in economic applications that deal with interest rates, annuities, and amortization. Mathematica is, therefore, of great use in these types of problems through its ability to show the results of problems in tabular form. Also, if a change is made in the problem, Mathematica can easily recompute the results.

4.3.1 Compound Interest

A common problem in economics is the determination of the amount of interest earned from an investment. If P dollars are invested for t years at an annual interest rate of $r\%$ compounded m times per year, the **compound amount**, $A(t)$, at time t is given by

$$A(t) = P\left(1 + \frac{r}{m}\right)^{mt}.$$

If P dollars are invested for t years at an annual interest rate of $r\%$ compounded continuously, the compound amount, $A(t)$, at time t is given by $A(t) = Pe^{mt}$.

EXAMPLE 4.3.1: Suppose \$12,500 is invested at an annual rate of 7% compounded daily. How much money has accumulated and how much interest has been earned at the end of each five year period for $t = 0$, 5, 10, 15, 20, 25, 30? How much money has accumulated if interest is compounded continuously instead of daily?

SOLUTION: We define ac[t] to give the total value of the investment at the end of t years and interest[t] to yield the total amount of interest earned at the end of t years. Then Table and TableForm are used to produce the table of ordered triples corresponding to the year, total value of the investment, and total interest earned.

In[746]:= **Clear[ac, interest]**

$$\mathtt{ac[t\_] = 12500\left(1 + \frac{0.07}{365}\right)^{365t};}$$

interest[t_] = ac[t] - 12500;

In[747]:= **Table[{t, ac[t], interest[t]}, {t, 0, 30, 5}]//**
          TableForm

```
          0  12500    0
          5  17737.7 5237.75
          10 25170.2 12670.2
Out[747]= 15 35717.  23217.
          20 50683.2 38183.2
          25 71920.5 59420.5
          30 102057. 89556.6
```

Next, we compute the value of the investment if interest is compounded continuously.

```
In[748]:= Clear[ac]

        ac[t_] = 12500 Exp[0.07t];

        TableForm[Table[{t, ac[t]}, {t, 0, 30, 5}]]
        0   12500
        5   17738.3
        10  25171.9
Out[748]= 15 35720.6
        20  50690.
        25  71932.5
        30  102077.
```

The problem can be redefined for arbitrary values of t, P, r, and n as follows.

```
In[749]:= Clear[ac, interest, results]

        ac[t_, P_, r_, n_] = P(1 + r/n)^(nt);

        interest[t_, P_, r_, n_] = ac[t, P, r, n] - P;

        results[{t0_, t1_, m_}, P_, r_, n_] :=
          Table[{t, ac[t, P, r, n], interest[t, P, r, n]},
                {t, t0, t1, m}]//TableForm
```

Hence, any problem of this type can be worked using these functions.

EXAMPLE 4.3.2: Suppose $10,000 is invested at an interest rate of 12% compounded daily. Create a table consisting of the total value of the investment and the interest earned at the end of 0, 5, 10, 15, 20, and 25 years. What is the total value and interest earned on an investment of $15,000 invested at an interest rate of 15% compounded daily at the end of 0, 10, 20, and 30 years?

SOLUTION: In this case, we use the function `results` defined above. Here, $t0=0$, $t1=25$, $m=5$, $P=10000$, $r=0.12$, and $n=365$.

```
In[750]:= results[{0, 25, 5}, 10000, 0.12, 365]
        0   10000    0
        5   18219.4  8219.39
        10  33194.6  23194.6
Out[750]= 15  60478.6  50478.6
        20  110188.  100188.
        25  200756.  190756.
```

If the conditions are changed to $t0=0, t1=30, m=10, P=15000, r=0.15$, and $n=365$, the desired table can be quickly calculated.

$$In[751] := \textbf{results}[\{0, 30, 10\}, 15000, 0.15, 365]$$

$$Out[751] = \begin{array}{lll} 0 & 15000 & 0 \\ 10 & 67204.6 & 52204.6 \\ 20 & 301097. & 286097. \\ 30 & 1.34901 \times 10^6 & 1.33401 \times 10^6 \end{array}$$

∎

4.3.2 Future Value

If R dollars are deposited at the end of each period for n periods in an annuity that earns interest at a rate of $j\%$ per period, the **future value** of the annuity is

$$S_{\text{future}} = R\frac{(1 + j)^n - 1}{j}.$$

EXAMPLE 4.3.3: Define a function `future` that calculates the future value of an annuity. Compute the future value of an annuity where $250 is deposited at the end of each month for 60 months at a rate of 7% per year. Make a table of the future values of the annuity where $150 is deposited at the end of each month for $12t$ months at a rate of 8% per year for $t = 1, 5, 9, 13, \ldots, 21, 25$.

SOLUTION: After defining `future`, we use `future` to calculate that the future value of an annuity where $250 is deposited at the end of each month for 60 months at a rate of 7% per year is $17,898.22.

$$In[752] := \textbf{Clear}[\textbf{r}, \textbf{n}]$$

$$\textbf{future}[\textbf{r}\_, \textbf{j}\_, \textbf{n}\_] = \frac{\textbf{r} ((1 + \textbf{j})^{\textbf{n}} - 1)}{\textbf{j}};$$

$$In[753] := \textbf{future}\left[250, \frac{0.07}{12}, 512\right]$$

$$Out[753] = 17898.2$$

For the second problem, we use `Table` and `future` to compute the future values of the annuity where $150 is deposited at the end of each month for $12t$ months at a rate of 8% per year for $t = 1, 5, 9, 13, \ldots,$ 21, 25. The first column in the following table corresponds to the time (in years) and the second column corresponds to the future value of the annuity.

$In[754]:=$ **Table** $\left[\left\{t, \text{future}\left[150, \dfrac{0.08}{12}, 12t\right]\right\}, \right.$

$\left. \{t, 1, 25, 4\}\right]$ **//TableForm**

$$
\begin{array}{ll}
1 & 1867.49 \\
5 & 11021.5 \\
9 & 23614.4 \\
Out[754]= 13 & 40938.1 \\
17 & 64769.6 \\
21 & 97553.8 \\
25 & 142654.
\end{array}
$$

∎

4.3.3 Annuity Due

If R dollars are deposited at the beginning of each period for n periods with an interest rate of $j\%$ per period, the **annuity due** is

$$S_{\text{due}} = R\left[\frac{(1 + j)^{n+1} - 1}{j} - 1\right].$$

EXAMPLE 4.3.4: Define a function due that computes the annuity due. Use due to (a) compute the annuity due of $500 deposited at the beginning of each month at an annual rate of 12% compounded monthly for three years; and (b) calculate the annuity due of 100k$ deposited at the beginning of each month at an annual rate of 9% compounded monthly for 10 years for $k = 1, 2, 3, \ldots, 10$.

SOLUTION: In the same manner as the previous example, we first define due and then use due to compute the annuity due of $500 deposited at the beginning of each month at an annual rate of 12% compounded monthly for three years.

$In[755]:=$ **due[r_, j_, n_]** $= \dfrac{r\,((1+j)^{n+1} - 1)}{j} - r;$

$In[756]:=$ **due** $\left[500, \dfrac{0.12}{12}, 312\right]$

$Out[756]=$ 21753.8

We then use `Table` and `due` to calculate the annuity due of $100k$ deposited at the beginning of each month at an annual rate of 9% compounded monthly for 10 years for $k = 1, 2, 3, \ldots, 10$. The first column corresponds to the amount deposited each month at an annual rate of 9% compounded monthly and the second column corresponds to the value of the annuity.

$$In[757] := \texttt{Table}\left[\left\{\texttt{100k, due}\left[\texttt{100k,}\, \frac{0.09}{12}, \texttt{1012}\right]\right\},\right.$$
$$\left.\{\texttt{k, 1, 10}\}\right]//\texttt{TableForm}$$

$$Out[757] = \begin{array}{ll} 100 & 19496.6 \\ 200 & 38993.1 \\ 300 & 58489.7 \\ 400 & 77986.3 \\ 500 & 97482.8 \\ 600 & 116979. \\ 700 & 136476. \\ 800 & 155973. \\ 900 & 175469. \\ 1000 & 194966. \end{array}$$

∎

EXAMPLE 4.3.5: Compare the annuity due on a $100k$ monthly investment at an annual rate of 8% compounded monthly for $t = 5, 10, 15, 20$ and $k = 1, 2, 3, 4, 5$.

SOLUTION: We use `Table` and `due` to calculate `due[100 k,0.08/12,t 12]`, corresponding to the annuity due of $100k$ deposited monthly at an annual rate of 8% compounded monthly for t years, for $k = 1, 2, 3, 4$ and $t = 5, 10, 15, 20$. Notice that the rows correspond to the annuity due on a $100, $200, $300, $400, and $500 monthly investment for 5, 10, 15, and 20 years, respectively. For example, the annuity due on $300 deposited monthly at an annual rate of 8% compounded monthly for 15 years is $104,504.

$$In[758] := \texttt{Table}\left[\texttt{due}\left[\texttt{100k,}\, \frac{0.08}{12}, \texttt{t12}\right], \{\texttt{k, 1, 5}\},\right.$$
$$\left.\{\texttt{t, 5, 20, 5}\}\right]//\texttt{TableForm}$$

$$Out[758] = \begin{array}{llll} 7396.67 & 18416.6 & 34834.5 & 59294.7 \\ 14793.3 & 36833.1 & 69669. & 118589. \\ 22190. & 55249.7 & 104504. & 177884. \\ 29586.7 & 73666.3 & 139338. & 237179. \\ 36983.4 & 92082.8 & 174173. & 296474. \end{array}$$

∎

4.3.4 Present Value

Another type of problem deals with determining the amount of money that must be invested in order to insure a particular return on the investment over a certain period of time. The **present value**, P, of an annuity of n payments of R dollars each at the end of consecutive interest periods with interest compounded at a rate of $j\%$ per period is

$$P = R\frac{1 - (1 + j)^{-n}}{j}.$$

EXAMPLE 4.3.6: Define a function present to compute the present value of an annuity. (a) Find the amount of money that would have to be invested at 7 1/2% compounded annually to provide an ordinary annuity income of \$45,000 per year for 40 years; and (b) find the amount of money that would have to be invested at 8% compounded annually to provide an ordinary annuity income of \$20, 000 + \$5000k per year for 35 years for k = 0, 1, 2, 3, 4, and 5 years.

SOLUTION: In the same manner as in the previous examples, we first define the function present which calculates the present value of an annuity. We then use present to calculate the amount of money that would have to be invested at 7 1/2% compounded annually to provide an ordinary annuity income of \$45,000 per year for 40 years.

```
In[759]:= present[r_, j_, n_] = r (1 - (1 + j)^-n) / j;
```

```
In[760]:= present[45000, 0.075, 40]
```

```
Out[760]= 566748.
```

Also, we use Table to find the amount of money that would have to be invested at 8% compounded annually to provide an ordinary annuity income of \$20, 000 + \$5000k per year for 35 years for k = 0, 1, 2, 3, 4, and 5. In the table, the first column corresponds to the annuity income and the second column corresponds to the present value of the annuity.

```
In[761]:= Table[{20000 + 5000k,
              present[20000 + 5000k, 0.08, 35]},
              {k, 0, 5}]//TableForm
```

$$Out[761] = \begin{matrix} 20000 & 233091. \\ 25000 & 291364. \\ 30000 & 349637. \\ 35000 & 407910. \\ 40000 & 466183. \\ 45000 & 524456. \end{matrix}$$

∎

4.3.5 Deferred Annuities

The present value of a **deferred annuity** of R dollars per period for n periods deferred for k periods with an interest rate of j per period is

$$P_{\text{def}} = R \left[\frac{1 - (1 + j)^{-(n+k)}}{j} - \frac{1 - (1 + j)^{-k}}{j} \right].$$

EXAMPLE 4.3.7: Define a function `def[r,n,k,j]` to compute the value of a deferred annuity where r equals the amount of the deferred annuity, n equals the number of years in which the annuity is received, k equals the number of years in which the lump sum investment is made, and j equals the rate of interest. Use `def` to compute the lump sum that would have to be invested for 30 years at a rate of 15% compounded annually to provide an ordinary annuity income of $35,000 per year for 35 years. How much money would have to be invested at the ages of 25, 35, 45, 55, and 65 at a rate of 8 1/2% compounded annually to provide an ordinary annuity income of $30,000 per year for 40 years beginning at age 65?

SOLUTION: As in the previous examples, we first define `def` and then use `def` to compute the lump sum that would have to be invested for 30 years at a rate of 15% compounded annually to provide an ordinary annuity income of $35,000 per year for 35 years. The function `def` that computes the present value of a deferred annuity where r equals the amount of the deferred annuity, n equals the number of years in which the annuity is received, k equals the number of years in which the lump sum investment is made, and j equals the rate of interest is defined.

$$In[762] := \texttt{def[r\_, n\_, k\_, j\_]} = r \left(\frac{1 - (1 + j)^{-(n+k)}}{j} - \frac{1 - (1 + j)^{-k}}{j} \right);$$

```
In[763]:= def[35000, 35, 30, 0.15]

Out[763]= 3497.58
```

To answer the second question, we note that the number of years the annuity is deferred is equal to 65 (the age at retirement) minus the age at which the money is initially invested and then use `Table` and `def` to compute the amount of money that would have to be invested at the ages of 25, 35, 45, 55, and 65 at a rate of 8 1/2% compounded annually to provide an ordinary annuity income of $30,000 per year for 40 years beginning at age 65. Note that the first column corresponds to the current age of the individual, the second column corresponds to the number of years from retirement, and the third column corresponds to the present value of the annuity.

```
In[764]:= Table[{k, 65 - k, def[30000, 40, 65 - k, 0.085]},
                {k, 25, 65, 10}]//TableForm

          25 40 12988.8
          35 30 29367.4
Out[764]= 45 20 66399.2
          55 10 150127.
          65 0  339436.
```

■

4.3.6 Amortization

A loan is **amortized** if both the principal and interest are paid by a sequence of equal periodic payments. A loan of P dollars at interest rate j per period may be amortized in n equal periodic payments of R dollars made at the end of each period, where

$$R = \frac{Pj}{1 - (1 + j)^{-n}}.$$

The function, `amort[p,j,n]`, defined next determines the monthly payment needed to amortize a loan of p dollars with an interest rate of j compounded monthly over n months. A second function, `totintpaid[p,j,n]`, calculates the total amount of interest paid to amortize a loan of p dollars with an interest rate of j% compounded monthly over n months.

```
In[765]:= amort[p_, j_, n_] = pj / (1 - (1+j)^{-n});

In[766]:= totintpaid[p_, j_, n_] = n amort[p, j, n] - p;
```

EXAMPLE 4.3.8: What is the monthly payment necessary to amortize a loan of $75,000 with an interest rate of 9.5% compounded monthly over 20 years?

SOLUTION: The first calculation uses `amort` to determine the necessary monthly payment to amortize the loan. The second calculation determines the total amount paid on a loan of $75,000 at a rate of 9.5% compounded monthly over 20 years while the third shows how much of this amount was paid towards the interest.

$$In[767] := \textbf{amort}\left[\textbf{75000}, \frac{\textbf{0.095}}{\textbf{12}}, \textbf{240}\right]$$

$$Out[767] = 699.098$$

$$In[768] := \textbf{240 amort}\left[\textbf{75000}, \frac{\textbf{0.095}}{\textbf{12}}, \textbf{240}\right]$$

$$Out[768] = 167784.$$

$$In[769] := \textbf{totintpaid}\left[\textbf{75000}, \frac{\textbf{0.095}}{\textbf{12}}, \textbf{240}\right]$$

$$Out[769] = 92783.6$$

■

EXAMPLE 4.3.9: What is the monthly payment necessary to amortize a loan of $80,000 at an annual rate of $j\%$ in 20 years for $j = 8, 8.5, 9, 9.5, 10,$ and 10.5?

SOLUTION: We use `amort` to calculate the necessary monthly payments. The first column corresponds to the annual interest rate and the second column corresponds to the monthly payment.

$$In[770] := \textbf{Table}\left[\left\{\textbf{j}, \textbf{amort}\left[\textbf{80000}, \frac{\textbf{j}}{\textbf{12}}, \textbf{240}\right]\right\},\right.$$

$$\left.\{\textbf{j}, \textbf{0.08}, \textbf{0.105}, \textbf{0.005}\}\right] // \textbf{TableForm}$$

$$Out[770] = \begin{array}{ll} 0.08 & 669.152 \\ 0.085 & 694.259 \\ 0.09 & 719.781 \\ 0.095 & 745.705 \\ 0.1 & 772.017 \\ 0.105 & 798.704 \end{array}$$

■

In many cases, the amount paid towards the principal of the loan and the total amount that remains to be paid after a certain payment need to be computed. This is easily accomplished with the functions unpaidbalance and curprinpaid defined using the function amort[p,j,n] that was previously defined.

$In[771] :=$ **unpaidbalance[p_, j_, n_, m_] =**
    present[amort[p, j, n], j, n - m]

$Out[771] = \dfrac{(1 - (1 + j)^{m-n})\, p}{1 - (1 + j)^{-n}}$

$In[772] :=$ **curprinpaid[p_, j_, n_, m_] =**
    p - unpaidbalance[p, j, n, m]

$Out[772] = p - \dfrac{(1 - (1 + j)^{m-n})\, p}{1 - (1 + j)^{-n}}$

EXAMPLE 4.3.10: What is the unpaid balance of the principal at the end of the fifth year of a loan of $60,000 with an annual interest rate of 8% scheduled to be amortized with monthly payments over a period of ten years? What is the total interest paid immediately after the 60th payment?

SOLUTION: We use the functions unpaidbalance and curprinpaid, defined above, to calculate that of the original $60,000 loan, $24,097.90 has been paid at the end of five years; $35,902.10 is still owed on the loan.

$In[773] :=$ **unpaidbalance$\left[60000, \dfrac{0.08}{12}, 120, 60\right]$**

$Out[773] = 35902.1$

$In[774] :=$ **curprinpaid$\left[60000, \dfrac{0.08}{12}, 120, 60\right]$**

$Out[774] = 24097.9$

■

Mathematica can also be used to determine the total amount of interest paid on a loan using the following function

$In[775] :=$ **curintpaid[p_, j_, n_, m_] =**
    m amort[p, j, n] - curprinpaid[p, j, n, m]

$Out[775] = -p + \dfrac{(1 - (1 + j)^{m-n})\, p}{1 - (1 + j)^{-n}} + \dfrac{j\,m\,p}{1 - (1 + j)^{-n}}$

where curintpaid[p,j,n,m] computes the interest paid on a loan of $p amortized at a rate of j per period over n periods immediately after the mth payment.

EXAMPLE 4.3.11: What is the total interest paid on a loan of $60,000 with an interest rate of 8% compounded monthly amortized over a period of ten years (120 months) immediately after the 60th payment?

SOLUTION: Using `curintpaid`, we see that the total interest paid is $19,580.10.

$$In[776] := \mathtt{curintpaid}\left[60000, \frac{0.08}{12}, 120, 60\right]$$

$$Out[776] = 19580.1$$

■

Using the functions defined above, amortization tables can be created that show a breakdown of the payments made on a loan.

EXAMPLE 4.3.12: What is the monthly payment necessary to amortize a loan of $45,000 with an interest rate of 7% compounded monthly over a period of 15 years (180 months)? What is the total principal and interest paid after 0, 3, 6, 9, 12, and 15 years?

SOLUTION: We first use `amort` to calculate the monthly payment necessary to amortize the loan.

$$In[777] := \mathtt{amort}\left[45000, \frac{0.07}{12}, 1512\right]$$

$$Out[777] = 404.473$$

Next, we use `Table`, `curprinpaid`, and `curintpaid` to determine the interest and principal paid at the end of 0, 3, 6, 9, 12, and 15 years.

$$In[778] := \mathtt{Table}\left[\left\{t, \mathtt{curprinpaid}\left[45000, \frac{0.07}{12}, 1512, 12t\right],\right.\right.$$

$$\left.\mathtt{curintpaid}\left[45000, \frac{0.07}{12}, 1512, 12t\right]\right\},$$

$$\left.\{t, 0, 15, 3\}\right]//\mathtt{TableForm}$$

$$Out[778] =$$

0	0.	0.
3	5668.99	8892.03
6	12658.4	16463.6
9	21275.9	22407.2
12	31900.6	26343.5
15	45000	27805.1

Note that the first column represents the number of years, the second column represents the principal paid, and the third column represents the interest paid. Thus, at the end of 12 years, $31,900.60 of the principal has been paid and $26,343.50 has been paid in interest.

■

Because `curintpaid[p,j,n,y]` computes the interest paid on a loan of $p amortized at a rate of j per period over n periods immediately after the yth payment, and `curintpaid[p,j,n,y-12]` computes the interest paid on a loan of $p amortized at a rate of j per period over n periods immediately after the (y – 12)th payment,

$$\texttt{curintpaid[p,j,n,y]-curintpaid[p,j,n,y-12]}$$

yields the amount of interest paid on a loan of $p amortized at a rate of j per period over n periods between the (y – 12)th and yth payment. Consequently, the interest paid and the amount of principal paid over a year can also be computed.

EXAMPLE 4.3.13: Suppose that a loan of $45,000 with interest rate of 7% compounded monthly is amortized over a period of 15 years (180 months). What is the principal and interest paid during each of the first five years of the loan?

SOLUTION: We begin by defining the functions `annualintpaid` and `annualprinpaid` that calculate the interest and principal paid during the yth year on a loan of $p amortized at a rate of j per period over n periods.

```
In[779]:= annualintpaid[p_,j_,n_,y_] :=
              curintpaid[p,j,n,y]-
                curintpaid[p,j,n,y-12];

          annualprinpaid[p_,j_,n_,y_] :=
              curprinpaid[p,j,n,y]-
                curprinpaid[p,j,n,y-12];
```

We then use these functions along with `Table` to calculate the principal and interest paid during the first five years of the loan. Note that the first column represents the number of years the loan has been held, the second column represents the interest paid on the loan during the year, and the third column represents the amount of the principal that has been paid.

$$In[780] := \texttt{Table}\Big[\Big\{\texttt{t, annualintpaid}\Big[45000, \frac{0.07}{12}, 1512,$$
$$12\texttt{t}\Big], \texttt{annualprinpaid}\Big[45000, \frac{0.07}{12},$$
$$1512, 12\texttt{t}\Big]\Big\}, \{\texttt{t, 1, 5, 1}\}\Big]//\texttt{TableForm}$$

$$Out[780] = \begin{array}{lll} 1 & 3094.26 & 1759.41 \\ 2 & 2967.08 & 1886.6 \\ 3 & 2830.69 & 2022.98 \\ 4 & 2684.45 & 2169.22 \\ 5 & 2527.64 & 2326.03 \end{array}$$

For example, we see that during the third year of the loan, $2830.69 was paid in interest and $2022.98 what paid on the principal.

∎

4.3.7 More on Financial Planning

We can use many of the functions defined above to help make decisions about financial planning.

EXAMPLE 4.3.14: Suppose a retiree has $1,200,000. If she can invest this sum at 7%, compounded annually, what level payment can she withdraw annually for a period of 40 years?

SOLUTION: The answer to the question is the same as the monthly payment necessary to amortize a loan of $1,200,000 at a rate of 7% compounded annually over a period of 40 years. Thus, we use amort to see that she can withdraw $90,011 annually for 40 years.

$$In[781] := \texttt{amort[1200000, 0.07, 40]}$$
$$Out[781] = 90011.$$

∎

EXAMPLE 4.3.15: Suppose an investor begins investing at a rate of d dollars per year at an annual rate of j%. Each year the investor increases the amount invested by i%. How much has the investor accumulated after m years?

SOLUTION: The following table illustrates the amount invested each year and the value of the annual investment after m years.

Year	Rate of Increase	Annual Interest	Amount Invested	Value after m Years
0	$j\%$	d	$(1 + j\%)^m d$	
1	$i\%$	$j\%$	$(1 + i\%)d$	$(1 + i\%)(1 + j\%)^{m-1}d$
2	$i\%$	$j\%$	$(1 + i\%)^2 d$	$(1 + i\%)^2(1 + j\%)^{m-2}d$
3	$i\%$	$j\%$	$(1 + i\%)^3 d$	$(1 + i\%)^3(1 + j\%)^{m-3}d$
k	$i\%$	$j\%$	$(1 + i\%)^k d$	$(1 + i\%)^k(1 + j\%)^{m-k}d$
m	$i\%$	$j\%$	$(1 + i\%)^m d$	$(1 + i\%)^m d$

It follows that the total value of the amount invested for the first k years after m years is given by:

Year	Total Investment
0	$(1 + j\%)^m d$
1	$(1 + j\%)^m d + (1 + i\%)(1 + j\%)^{m-1}d$
2	$(1 + j\%)^m d + (1 + i\%)(1 + j\%)^{m-1}d + (1 + i\%)^2(1 + j\%)^{m-2}d$
3	$\sum_{n=0}^{3}(1 + i\%)^n(1 + j\%)^{m-n}d$
k	$\sum_{n=0}^{k}(1 + i\%)^n(1 + j\%)^{m-n}d$
m	$\sum_{n=0}^{m}(1 + i\%)^n(1 + j\%)^{m-n}d$

The command Sum can be used to find a closed form of the sums $\sum_{n=0}^{k}(1 + i\%)^n(1 + j\%)^{m-n}d$ and $\sum_{n=0}^{m}(1 + i\%)^n(1 + j\%)^{m-n}d$. We use Sum to find the sum $\sum_{n=0}^{k}(1 + i\%)^n(1 + j\%)^{m-n}d$ and name the result closedone. We then use Factor and PowerExpand to first write closedone as a single fraction and then factor the numerator.

$In[782]:=$ **closedone = Simplify$\left[\sum_{n=0}^{k} (1 + i)^n (1 + j)^{m-n} d \right]$**

$Out[782]= \dfrac{d\,(1 + j)^m \left(-1 - j + \left(\frac{1+i}{1+j}\right)^k + i\left(\frac{1+i}{1+j}\right)^k \right)}{i - j}$

$In[783]:=$ **Factor[PowerExpand[closedone]]**

$Out[783]= \dfrac{d\,(1 + j)^{-k+m} \left((1 + i)^k + i\,(1 + i)^k - (1 + j)^k - j\,(1 + j)^k \right)}{i - j}$

In the same way, Sum is used to find a closed form of $\sum_{n=0}^{m}(1 + i\%)^n(1 + j\%)^{m-n}d$, naming the result closedtwo.

$In[784]:=$ **closedtwo $= \sum_{n=0}^{m} (1 + i)^n (1 + j)^{m-n} d$**

$Out[784]= \dfrac{d\,(1 + j)^m \left(-1 - j + \left(\frac{1+i}{1+j}\right)^m + i\left(\frac{1+i}{1+j}\right)^m \right)}{i - j}$

In[785]:= **Factor[PowerExpand[closedtwo]]**

$$Out[785]= -\frac{d\,(-(1+i)^m - i\,(1+i)^m + (1+j)^m + j\,(1+j)^m)}{i-j}$$

These results are used to define the functions investment[{d,i,j}, {k,m}] and investmenttot[{d,i,j},m] that return the value of the investment after *k* and *m* years, respectively. In each case, notice that output cells can be edited like any other input or text cell. Consequently, we use editing features to copy and paste the result when we define these functions.

In[786]:= **investment[{d_, i_, j_}, {k_, m_}] =**

$$\frac{1}{-i+j}$$
$$\left(d(1+j)^m\right.$$
$$\left(1+j-(1+i)^k\,(1+j)^{-k}-i\,(1+i)^k\,(1+j)^{-k}\right)\right);$$

In[787]:= **investmenttot[{d_, i_, j_}, m_] =**

$$\frac{d\,(-(1+i)^m - i\,(1+i)^m + (1+j)^m + j\,(1+j)^m)}{-i+j};$$

Finally, investment and investmenttot are used to illustrate various financial scenarios. In the first example, investment is used to compute the value after 25 years of investing \$6500 the first year and then increasing the amount invested 5% per year for 5, 10, 15, 20, and 25 years assuming a 15% rate of interest on the amount invested. The built-in function AccountingForm is used to convert numbers expressed in exponential notation to ordinary notation. In the second example, investmenttot is used to compute the value after 25 years of investing \$6500 the first year and then increasing the amount invested 5% per year for 25 years assuming various rates of interest. The results are named scenes and are displayed in AccountingForm.

In[788]:= **results =**
Table[{t, investment[{6500, 0.05, 0.15},
{t, 25}]},
{t, 5, 25, 5}]//TableForm

$$
\begin{array}{ll}
5 & 1.03506 \times 10^6 \\
10 & 1.55608 \times 10^6 \\
Out[788]= \quad 15 & 1.88668 \times 10^6 \\
20 & 2.09646 \times 10^6 \\
25 & 2.22957 \times 10^6
\end{array}
$$

```
In[789]:= TableForm[AccountingForm[results]]
              5  1.03507 10⁶
              10 1.55608 10⁶
Out[789]= 15 1.88668 10⁶
              20 2.09646 10⁶
              25 2.22957 10⁶

In[790]:= scenes =
              Table[{i, investmenttot[{6500, 0.05, i}, 25]},
                {i, 0.08, 0.2, 0.02}];

          AccountingForm[TableForm[scenes]]
          0.08 832147.
          0.1  1.08713 10⁶
          0.12 1.43784 10⁶
Out[790]= 0.14 1.9219 10⁶
          0.16 2.59164 10⁶
          0.18 3.51967 10⁶
          0.2  4.80652 10⁶
```

■

Another interesting investment problem is discussed in the following example. In this case, Mathematica is useful in solving a recurrence equation that occurs in the problem. The command

$$\text{RSolve}[\{\text{equations}\}, \text{a}[\text{n}], \text{n}]$$

attempts to solve the recurrence equations `equations` for the variable `a[n]` with no dependence on `a[j]`, $j \le n - 1$.

EXAMPLE 4.3.16: I am 50 years old and I have $500,000 that I can invest at a rate of 7% annually. Furthermore, I wish to receive a payment of $50,000 the first year. Future annual payments should include cost-of-living adjustments at a rate of 3% annually. Is $500,000 enough to guarantee this amount of annual income if I live to be 80 years old?

SOLUTION: Instead of directly solving the above problem, let's solve a more general problem. Let a denote the amount invested and p the first-year payment. Let a_n denote the balance of the principal at the end of year n. Then, the amount of the nth payment, the interest earned on the principal, the decrease in principal, and the principal balance at the end of year n are shown in the table for various values of n. Observe that if $(1 + j)^{n-1} > (1 + j)a_{n-1}$, then the procedure terminates and the amount received in year n is $(1 + j)a_{n-1}$.

Year	Amount	Interest Principal	From Balance	Principal
1	p	ia	$p - ia$	$a_1 = (1 + i)a - p$
2	$(1 + j)p$	ia_1	$(1 + j)p - ia_1$	$a_2 = (1 + i)a_1 - (1 + j)p$
3	$(1 + j)^2 p$	ia_2	$(1 + j)^2 p - ia_2$	$a_3 = (1 + i)a_2 - (1 + j)^2 p$
4	$(1 + j)^3 p$	ia_3	$(1 + j)^3 p - ia_3$	$a_4 = (1 + i)a_3 - (1 + j)^3 p$
n	$(1 + j)^{n-1} p$	ia_{n-1}	$(1 + j)^{n-1} p - ia_{n-1}$	$a_n = (1 + i)a_{n-1} - (1 + j)^{n-1} p$

The recurrence equation $a_n = (1 + i)a_{n-1} - (1 + j)^{n-1} p$ is solved for a_n with no dependence on a_{n-1}. After clearing several definitions of variable names, we use RSolve to solve the recurrence equation given above where the initial balance is represented by amount. Hence, a_n is given by the expression found in bigstep.

$In[791]:= $ **eq1 = a[1] == (1 + i) amount - p;**

eq2 = a[n] == (1 + i) a[n - 1] - (1 + j)$^{n-1}$ p;

bigstep = RSolve[{eq1, eq2}, a[n], n]

$Out[791]= \left\{\left\{a[n] \rightarrow \dfrac{(1 + i)^n \left(\text{amount } i - \text{amount } j - p + \left(\frac{1+j}{1+i}\right)^n p\right)}{i - j}\right\}\right\}$

We then define am[n, amount, i, p, j] to be the explicit solution found in bigstep. Last we compute am[n, a, i, p, j] which corresponds to the balance of the principal of a dollars invested under the above conditions at the end of the nth year.

$In[792]:= $ **am[n_, amount_, i_, p_, j_] =**

$$-\dfrac{(-1 - i) (1 + i)^{-1+n} (\text{amount } i - \text{amount } j - p)}{i - j} -$$

$$\dfrac{(-1 - j) (1 + j)^{-1+n} p}{i - j} //\textbf{Together;}$$

$In[793]:= $ **am[n, a, i, p, j]**

$Out[793]= \dfrac{a\, i\, (1 + i)^n - a\, (1 + i)^n\, j - (1 + i)^n p + (1 + j)^n p}{i - j}$

To answer the question, we first define annuitytable in the following. For given a, i, p, j, and m, annuitytable[a, i, p, j, m] returns an ordered triple corresponding to the year, amount of income received in that year, and principal balance at the end of the year for m years.

In[794] := **annuitytable[a_, i_, p_, j_, m_] :=**
      TableForm[
        Table[{k, (1 + j)$^{k-1}$ p, am[k, a, i, p, j]},
        {k, 1, m}]]

Then we compute annuitytable[500000,.07,50000,.03,15]. In this case, we see that the desired level of income is only guaranteed for 13 years which corresponds to an age of 63 because the principal balance is negative after 13 years.

In[795] := **annuitytable[500000, 0.07, 50000, 0.03, 15]**

1	50000	485000.
2	51500.	467450.
3	53045.	447126.
4	54636.4	423789.
5	56275.4	397179.
6	57963.7	367018.
7	59702.6	333006.
Out[795]= 8	61493.7	294823.
9	63338.5	252122.
10	65238.7	204532.
11	67195.8	151653.
12	69211.7	93057.4
13	71288.	28283.4
14	73426.7	-43163.5
15	75629.5	-121814.

An alternative method of defining annuitytable is presented next. Here we use For and ++

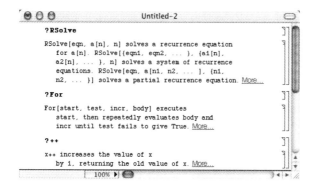

to define annuitytable so that the procedure terminates when the principal is negative or after 50 years.

```
In[796] := Clear[annuitytable]

        annuitytable[a_, i_, p_, j_] := Module[{},
             For[k = 1, am[k, a, i, p, j] ≥ 0 && k ≤ 50,
                k + +,
                Print[{k, (1 + j)^(k-1) p, am[k, a, i, p, j]}]];
             Print[{k, am[k - 1, a, i, p, j], 0}]]
```

We see that if the first year payment is \$29,000, 3% increases can occur annually for 30 years.

```
In[797] := annuitytable[500000, 0.07, 29000, 0.03]
{1, 29000, 506000.}
{2, 29870., 511550.}
{3, 30766.1, 516592.}
{4, 31689.1, 521065.}
{5, 32639.8, 524900.}
{6, 33618.9, 528024.}
{7, 34627.5, 530358.}
{8, 35666.3, 531816.}
{9, 36736.3, 532307.}
{10, 37838.4, 531730.}
{11, 38973.6, 529978.}
{12, 40142.8, 526934.}
{13, 41347.1, 522472.}
{14, 42587.5, 516457.}
{15, 43865.1, 508744.}
{16, 45181.1, 499175.}
{17, 46536.5, 487581.}
{18, 47932.6, 473779.}
{19, 49370.6, 457573.}
{20, 50851.7, 438752.}
{21, 52377.2, 417087.}
{22, 53948.5, 392335.}
{23, 55567., 364231.}
{24, 57234., 332493.}
{25, 58951., 296817.}
{26, 60719.6, 256874.}
{27, 62541.1, 212314.}
{28, 64417.4, 162759.}
{29, 66349.9, 107802.}
{30, 68340.4, 47007.9}
{31, 47007.9, 0}
```

We can also investigate other problems. For example, a 30-year mort-gage of \$80,000 with an annual interest rate of 8.125% requires monthly payments of approximately \$600 (\$7200 annually) to amortize the loan in 30 years. However, using annuitytable, we see that if the amount of the payments is increased by 3% each year, the 30-year mortgage is amortized in 17 years. In the following result, the first column corresponds to the year of the loan, the second column the annual payment, and the third column the principal balance.

```
In[798]:= annuitytable[80000, 0.08125, 7200, 0.03]
{1, 7200, 79300.}
{2, 7416., 78327.1}
{3, 7638.48, 77052.7}
{4, 7867.63, 75445.6}
{5, 8103.66, 73471.9}
{6, 8346.77, 71094.7}
{7, 8597.18, 68274.}
{8, 8855.09, 64966.2}
{9, 9120.74, 61123.9}
{10, 9394.37, 56695.9}
{11, 9676.2, 51626.2}
{12, 9966.48, 45854.4}
{13, 10265.5, 39314.6}
{14, 10573.4, 31935.4}
{15, 10890.6, 23639.5}
{16, 11217.4, 14342.9}
{17, 11553.9, 3954.36}
{18, 3954.36, 0}
```

■

4.4 Other Applications

We now present several other applications that we find interesting and require the manipulation of lists. The examples also illustrate (and combine) many of the skills that were demonstrated in the earlier chapters.

4.4.1 Approximating Lists with Functions

Another interesting application of lists is that of curve-fitting. The commands

1. Fit[data,functionset,variables] fits the list of data points data using the functions in functionset by the method of least-squares. The functions in functionset are functions of the variables listed in variables; and

2. `InterpolatingPolynomial[data,x]` fits the list of n data points `data` with an $n - 1$ degree polynomial in the variable x.

EXAMPLE 4.4.1: Define `datalist` to be the list of numbers consisting of 1.14479, 1.5767, 2.68572, 2.5199, 3.58019, 3.84176, 4.09957, 5.09166, 5.98085, 6.49449, and 6.12113. (a) Find a quadratic approximation of the points in `datalist`. (b) Find a fourth degree polynomial approximation of the points in `datalist`.

SOLUTION: The approximating function obtained via the least-squares method with `Fit` is plotted along with the data points in Figure 4-18. Notice that many of the data points are not very close to the approximating function. A better approximation is obtained using a polynomial of higher degree (4).

```
In[799]:= Clear[datalist]

          datalist = {1.14479, 1.5767, 2.68572,
             2.5199, 3.58019, 3.84176,
             4.09957, 5.09166, 5.98085, 6.49449,
             6.12113};

In[800]:= p1 = ListPlot[datalist,
             DisplayFunction → Identity];

In[801]:= Clear[y]
```

$$y[x_-] = Fit\left[datalist, \left\{1, x, x^2\right\}, x\right]$$

$Out[801] = 0.508266 + 0.608688\,x - 0.00519281\,x^2$

```
In[802]:= p2 = Plot[y[x], {x, -1, 11},
             DisplayFunction → Identity];

          Show[p1, p2, DisplayFunction →
             $DisplayFunction]

In[803]:= Clear[y]
```

$$y[x_-] = Fit\left[datalist, \left\{1, x, x^2, x^3, x^4\right\}, x\right]$$

$Out[803] = -0.54133 + 2.02744\,x - 0.532282\,x^2 +$
$\qquad 0.0709201\,x^3 - 0.00310985\,x^4$

To check its accuracy, the second approximation is graphed simultaneously with the data points in Figure 4-19.

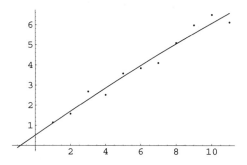

Figure 4-18   The graph of a quadratic fit shown with the data points

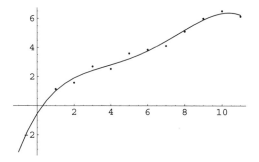

Figure 4-19   The graph of a quartic fit shown with the data points

```
In[804]:= p3 = Plot[y[x], {x, -1, 11},
            DisplayFunction → Identity];

        Show[p1, p3, DisplayFunction →
            $DisplayFunction]
```

■

Next, consider a list of data points made up of ordered pairs.

EXAMPLE 4.4.2: Table 4-2 shows the average percentage of petroleum products imported to the United States for certain years. (a) Graph the points corresponding to the data in the table and connect the consecutive points with line segments. (b) Use `InterpolatingPolynomial` to find a function that approximates the data in the table. (c) Find a fourth degree polynomial approximation of the data in the table. (d) Find a trigonometric approximation of the data in the table.

Year	Percent	Year	Percent
1973	34.8105	1983	28.3107
1974	35.381	1984	29.9822
1975	35.8167	1985	27.2542
1976	40.6048	1986	33.407
1977	47.0132	1987	35.4875
1978	42.4577	1988	38.1126
1979	43.1319	1989	41.57
1980	37.3182	1990	42.1533
1981	33.6343	1991	39.5108
1982	28.0988		

Table 4-2   Petroleum products imported to the United States for certain years

SOLUTION: We begin by defining data to be the set of ordered pairs represented in the table: the *x*-coordinate of each point represents the number of years past 1900 and the *y*-coordinate represents the percentage of petroleum products imported to the United States.

```
In[805]:= data = {{73., 34.8105}, {74., 35.381},
            {75., 35.8167}, {76., 40.6048},
            {77., 47.0132}, {78., 42.4577},
            {79., 43.1319}, {80., 37.3182},
            {81., 33.6343}, {82., 28.0988},
            {83., 28.3107}, {84., 29.9822},
          {85., 27.2542}, {86., 33.407},
            {87., 35.4875}, {88., 38.1126},
            {89., 41.57}, {90., 42.1533},
            {91., 39.5108}};
```

We use ListPlot to graph the ordered pairs in data. Note that because the option PlotStyle->PointSize[.03] is included within the ListPlot command, the points are larger than they would normally be. We also use ListPlot with the option PlotJoined->True to graph the set of points data and connect consecutive points with line segments. Then we use Show to display lp1 and lp2 together in Figure 4-20. Note that in the result, the points are easy to distinguish because of their larger size.

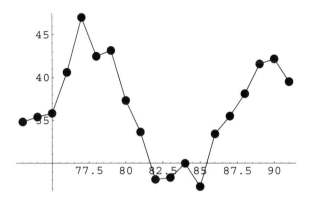

Figure 4-20   The points in Table 4-2 connected by line segments

```
In[806]:= lp1 = ListPlot[data,
                PlotStyle → PointSize[0.03],
                DisplayFunction → Identity];

          lp2 = ListPlot[data, PlotJoined → True,
                DisplayFunction → Identity];

          Show[lp1, lp2,
                DisplayFunction → $DisplayFunction]
```

Next, we use `InterpolatingPolynomial` to find a polynomial approximation, p, of the data in the table. Note that the result is lengthy, so `Short` is used to display an abbreviated form of p. We then graph p and show the graph of p along with the data in the table for the years corresponding to 1971 to 1993 in Figure 4-21. Although the interpolating polynomial agrees with the data exactly, the interpolating polynomial oscillates wildly.

```
In[807]:= p = InterpolatingPolynomial[data, x];

          Short[p, 3]
Out[807]= 34.8105+
                (0.5705 + (-0.0674 + (≪1≫) (-75. + x)) (-74. + x))
                (-73. + x)

In[808]:= plotp = Plot[p, {x, 71, 93},
                DisplayFunction → Identity];

          Show[plotp, lp1, PlotRange → {0, 50},
                DisplayFunction → $DisplayFunction]
```

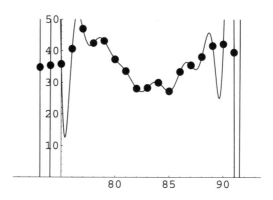

Figure 4-21   Even though interpolating polynomials agree with the data exactly, they may have extreme oscillations, even for relatively small data sets

To find a polynomial that approximates the data but does not oscillate wildly, we use $\texttt{Fit}$. Again, we graph the fit and display the graph of the fit and the data simultaneously. In this case, the fit does not identically agree with the data but does not oscillate wildly as illustrated in Figure 4-22.

$In[809]:= $ **Clear[p]**

$$p = \texttt{Fit}\Big[\texttt{data}, \{1, x, x^2, x^3, x^4\}, x\Big]$$
$Out[809]= -198884. + 9597.83\,x-$
$$173.196\,x^2 + 1.38539\,x^3 - 0.00414481\,x^4$$

$In[810]:= $ **plotp = Plot[p, {x, 71, 93},**
$$\textbf{DisplayFunction} \rightarrow \textbf{Identity]};$$

$$\textbf{Show[plotp, lp1, PlotRange} \rightarrow \{0, 50\},$$
$$\textbf{DisplayFunction} \rightarrow \textbf{\$DisplayFunction]}$$

In addition to curve-fitting with polynomials, Mathematica can also fit the data with trigonometric functions. In this case, we use $\texttt{Fit}$ to find an approximation of the data of the form $p = c_1 + c_2 \sin x + c_3 \sin (x/2) + c_4 \cos x + c_5 \cos (x/2)$. As in the previous two cases, we graph the fit and display the graph of the fit and the data simultaneously; the results are shown in Figure 4-23.

See texts like Abell, Braselton, and Rafter's *Statistics with Mathematica* [3] for a more sophisticated discussion of curve-fitting and related statistical applications.

$In[811]:= $ **Clear[p]**

$$p = \texttt{Fit}\Big[\texttt{data}, \Big\{1, \texttt{Sin[x]}, \texttt{Sin}\Big[\frac{x}{2}\Big], \texttt{Cos[x]},$$
$$\texttt{Cos}\Big[\frac{x}{2}\Big]\Big\}, x\Big]$$

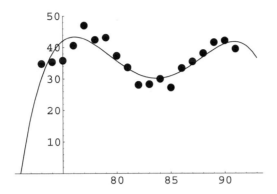

Figure 4-22 Even though the fit does not agree with the data exactly, the oscillations seen in Figure 4-21 do not occur

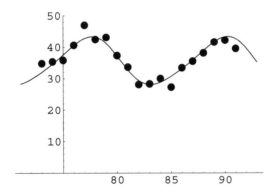

Figure 4-23 You can use Fit to approximate data by a variety of functions

$$Out[811] = 35.4237 + 4.25768 \, \text{Cos}\left[\frac{x}{2}\right] - 0.941862 \, \text{Cos}[x] +$$
$$6.06609 \, \text{Sin}\left[\frac{x}{2}\right] + 0.0272062 \, \text{Sin}[x]$$

```
In[812]:= plotp = Plot[p, {x, 71, 93},
            DisplayFunction → Identity];

        Show[plotp, lp1, PlotRange → {0, 50},
            DisplayFunction → $DisplayFunction]
```

■

4.4.2 Introduction to Fourier Series

Many problems in applied mathematics are solved through the use of Fourier series. Mathematica assists in the computation of these series in several ways. Suppose that $y = f(x)$ is defined on $-p < x < p$. Then the Fourier series for $f(x)$ is

$$\frac{1}{2}a_0 + \sum_{n=1}^{\infty} \left(a_n \cos \frac{n\pi x}{p} + b_n \sin \frac{n\pi x}{p} \right) \tag{4.1}$$

where

$$a_0 = \frac{1}{p} \int_{-p}^{p} f(x)\,dx$$

$$a_n = \frac{1}{p} \int_{-p}^{p} f(x) \cos \frac{n\pi x}{p}\,dx \quad n = 1,\ 2\ \ldots \tag{4.2}$$

$$b_n = \frac{1}{p} \int_{-p}^{p} f(x) \sin \frac{n\pi x}{p}\,dx \quad n = 1,\ 2\ \ldots$$

The **kth term of the Fourier series** (4.1) is

$$a_n \cos \frac{n\pi x}{p} + b_n \sin \frac{n\pi x}{p}. \tag{4.3}$$

The **kth partial sum of the Fourier series** (4.1) is

$$\frac{1}{2}a_0 + \sum_{n=1}^{k} \left(a_n \cos \frac{n\pi x}{p} + b_n \sin \frac{n\pi x}{p} \right). \tag{4.4}$$

It is a well-known theorem that if $y = f(x)$ is a periodic function with period $2p$ and $f'(x)$ is continuous on $[-p,\ p]$ except at finitely many points, then at each point x the Fourier series for $f(x)$ converges and

$$\frac{1}{2}a_0 + \sum_{n=1}^{\infty} \left(a_n \cos \frac{n\pi x}{p} + b_n \sin \frac{n\pi x}{p} \right) = \frac{1}{2} \left(\lim_{z \to x^+} f(z) + \lim_{z \to x^-} f(z) \right).$$

In fact, if the series $\sum_{n=1}^{\infty} (|a_n| + |b_n|)$ converges, then the Fourier series converges uniformly on $(-\infty, \infty)$.

EXAMPLE 4.4.3: Let $f(x) = \begin{cases} -x, & -1 \le x < 0 \\ 1, & 0 \le x < 1 \\ f(x-2), & x \ge 1 \end{cases}$. Compute and graph the

first few partial sums of the Fourier series for $f(x)$.

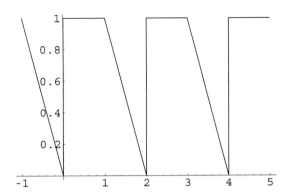

Figure 4-24   Plot of a few periods of $f(x)$

SOLUTION: We begin by clearing all prior definitions of f. We then define the piecewise function $f(x)$ and graph $f(x)$ on the interval $[-1, 5]$ in Figure 4-24.

```
In[813]:= Clear[f]

         f[x_] := 1/; 0 ≤ x < 1

         f[x_] := -x/; -1 ≤ x < 0

         f[x_] := f[x - 2]/; x ≥ 1

In[814]:= graphf = Plot[f[x], {x, -1, 5}]
```

The Fourier series coefficients are computed with the integral formulas in equation (**??**). Executing the following commands defines p to be 1, a[0] to be an approximation of the integral $a_0 = \frac{1}{p}\int_{-p}^{p} f(x)\,dx$, a[n] to be an approximation of the integral $a_n = \frac{1}{p}\int_{-p}^{p} f(x)\cos\frac{n\pi x}{p}\,dx$, and b[n] to be an approximation of the integral $b_n = \frac{1}{p}\int_{-p}^{p} f(x)\sin\frac{n\pi x}{p}\,dx$.

```
In[815]:= Clear[a, b, fs, p]

         p = 1;

         a[0] = NIntegrate[f[x], {x, -p, p}]
                ─────────────────────────────
                            2L

Out[815]= 0.75
```

$$In[816] := a[n\_] :=$$
$$\frac{NIntegrate\left[f[x] \ Cos\left[\frac{n\pi x}{p}\right], \{x, -p, p\}\right]}{L}$$

$$b[n\_] :=$$
$$\frac{NIntegrate\left[f[x] \ Sin\left[\frac{n\pi x}{p}\right], \{x, -p, p\}\right]}{L}$$

A table of the coefficients a[i] and b[i] for $i = 1, 2, 3, \ldots, 10$ is generated with Table and named coeffs. Several error messages are generated because of the discontinuities but the resulting approximations are satisfactory for our purposes. The elements in the first column of the table represent the a_i's and the second column represents the b_i's. Notice how the elements of the table are extracted using double brackets with coeffs.

$$In[817] := coeffs = Table[\{a[i], b[i]\}, \{i, 1, 10\}];$$

NIntegrate :: ncvb :
   NIntegrate failed to converge to prescribed accuracy
      after 7 recursive bisections in x near $x = -1$..

NIntegrate :: ncvb :
   NIntegrate failed to converge to prescribed accuracy
      after 7 recursive bisections in x near $x = -1$..

NIntegrate :: ncvb :
   NIntegrate failed to converge to prescribed accuracy
      after 7 recursive bisections in x near $x = -1$..

General :: stop : Further output of NIntegrate :: ncvb
   will be suppressed during this calculation.

NIntegrate :: ploss :
   Numerical integration stopping due to loss of
      precision. Achieved neither the requested
      PrecisionGoal nor AccuracyGoal; suspect one of
      the following : highly oscillatory integrand
      or the true value of the integral is 0. If
      your integrand is oscillatory try using the
      option Method- > Oscillatory in NIntegrate.

```
In[818]:= TableForm[coeffs]
          -0.202642        0.31831
          -3.42608 × 10⁻¹⁷ 0.159155
          -0.0225158       0.106103
          -4.51028 × 10⁻¹⁷ 0.0795775
          -0.00810569      0.063662
Out[818]= -5.0307 × 10⁻¹⁷  0.0530516
          -0.00413556      0.0454728
          -1.18178 × 10⁻¹⁶ 0.0397887
          -0.00250176      0.0353678
          -1.47451 × 10⁻¹⁷ 0.031831
```

The first element of the list is extracted with `coeffs[[1]]`.

```
In[819]:= coeffs[[1]]
Out[819]= {-0.202642, 0.31831}
```

The first element of the second element of `coeffs` and the second element of the third element of `coeffs` are extracted with `coeffs[[2,1]]` and `coeffs[[3,2]]`, respectively.

```
In[820]:= coeffs[[2,1]]
Out[820]= -3.42608 × 10⁻¹⁷

In[821]:= coeffs[[3,2]]
Out[821]= 0.106103
```

After the coefficients are calculated, the nth partial sum of the Fourier series is obtained with `Sum`. The kth term of the Fourier series, $a_k \cos(k\pi x) + b_k \sin(k\pi x)$, is defined in `fs`. Hence, the nth partial sum of the series is given by

$$a_0 + \sum_{k=1}^{n} [a_k \cos(k\pi x) + b_k \sin(k\pi x)] = \text{a[0]} + \sum_{k=1}^{n} \text{fs[k, x]},$$

which is defined in `fourier` using `Sum`. We illustrate the use of `fourier` by finding `fourier[2,x]` and `fourier[3,x]`.

```
In[822]:= fs[k_, x_] := coeffs[[k, 1]] Cos[k π x] +
              coeffs[[k, 2]] Sin[k π x]

In[823]:= fourier[n_, x_] := a[0] + ∑(k=1 to n) fs[k, x]

In[824]:= fourier[2, x]
Out[824]= 0.75 - 0.202642 Cos[π x] - 3.42608 × 10⁻¹⁷ Cos[2 π x] +
          0.31831 Sin[π x] + 0.159155 Sin[2 π x]
```

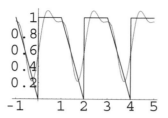 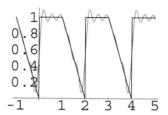

Figure 4-25   The first few terms of a Fourier series for a periodic function plotted with the function

```
In[825]:= fourier[3,x]
Out[825]= 0.75 - 0.202642 Cos[πx] - 3.42608 × 10⁻¹⁷ Cos[2 πx] -
          0.0225158 Cos[3 πx] + 0.31831 Sin[πx] +
          0.159155 Sin[2 πx] + 0.106103 Sin[3 πx]
```

To see how the Fourier series approximates the periodic function, we plot the function simultaneously with the Fourier approximation for $n = 2$ and $n = 5$. The results are displayed together using GraphicsArray in Figure 4-25.

```
In[826]:= graphtwo = Plot[fourier[2,x],{x,-1,5},
              PlotStyle → GrayLevel[0.4],
              DisplayFunction → Identity];

          bothtwo = Show[graphtwo,graphf];

In[827]:= graphfive = Plot[fourier[5,x],{x,-1,5},
              PlotStyle → GrayLevel[0.4],
              DisplayFunction → Identity];

          bothfive = Show[graphfive,graphf];

          Show[GraphicsArray[{bothtwo,bothfive}]]
```

Application: The One-Dimensional Heat Equation

A typical problem in applied mathematics that involves the use of Fourier series is that of the **one-dimensional heat equation**. The boundary value problem that describes the temperature in a uniform rod with insulated surface is

$$k\frac{\partial^2 u}{\partial x^2} = \frac{\partial u}{\partial t}, \ 0 < x < a, \ t > 0,$$

$$u(0, t) = T_0, \ t > 0,$$

$$u(a, t) = T_a, \ t > 0, \ \text{and} \qquad\qquad (4.5)$$

$$u(x, 0) = f(x), \ 0 < x < a.$$

In this case, the rod has "fixed end temperatures" at $x = 0$ and $x = a$. $f(x)$ is the initial temperature distribution. The solution to the problem is

$$u(x, t) = \underbrace{T_0 + \frac{1}{a}(T_a - T_0)x}_{v(x)} + \sum_{n=1}^{\infty} b_n \sin(\lambda_n x) e^{-\lambda_n^2 kt}, \qquad (4.6)$$

where
$$\lambda_n = n\pi/a \qquad \text{and} \qquad b_n = \frac{2}{a}\int_0^a (f(x) - v(x)) \sin\frac{n\pi x}{a}\, dx,$$

and is obtained through separation of variables techniques. The coefficient b_n in the solution equation (4.6) is the Fourier series coefficient b_n of the function $f(x) - v(x)$, where $v(x)$ is the **steady-state temperature**.

EXAMPLE 4.4.4: Solve $\begin{cases} \dfrac{\partial^2 u}{\partial x^2} = \dfrac{\partial u}{\partial t}, \ 0 < x < 1, \ t > 0, \\ u(0, t) = 10, \ u(1, t) = 10, \ t > 0, \\ u(x, 0) = 10 + 20\sin^2 \pi x. \end{cases}$

SOLUTION: In this case, $a = 1$ and $k = 1$. The fixed end temperatures are $T_0 = T_a = 10$, and the initial heat distribution is $f(x) = 10 + 20\sin^2 \pi x$. The steady-state temperature is $v(x) = 10$. The function $f(x)$ is defined and plotted in Figure 4-26. Also, the steady-state temperature, $v(x)$, and the eigenvalue are defined. Finally, Integrate is used to define a function that will be used to calculate the coefficients of the solution.

```
In[828]:= Clear[f]

        f[x_] := 10 + 20Sin[π x]²

        Plot[f[x], {x, 0, 1}, PlotRange → {0, 30}]

In[829]:= v[x_] := 10

        λ[n_] := n π
                 ───
                  4

        b[n_] := b[n] = ∫₀⁴ (f[x] - v[x]) Sin[n π x / 4] dx
```

Notice that **b[n]** is defined using the form **b[n_]:=b[n]=...** so that Mathematica "remembers" the values of **b[n]** computed and thus

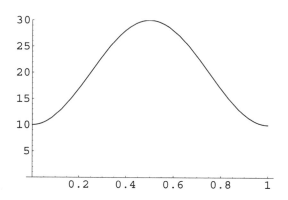

Figure 4-26   Graph of $f(x) = 10 + 20 \sin^2 \pi x$

avoids recomputing previously computed values. In the following table, we compute exact and approximate values of b[1],..., b[10].

```
In[830]:= Table[{n,b[n],b[n]//N}, {n,1,10}]//
             TableForm
```

$$
Out[830] =
\begin{array}{lll}
1 & \dfrac{5120}{63\,\pi} & 25.869 \\[2mm]
2 & 0 & 0. \\[2mm]
3 & \dfrac{1024}{33\,\pi} & 9.87725 \\[2mm]
4 & 0 & 0. \\[2mm]
5 & \dfrac{1024}{39\,\pi} & 8.35767 \\[2mm]
6 & 0 & 0. \\[2mm]
7 & \dfrac{1024}{21\,\pi} & 15.5214 \\[2mm]
8 & 0 & 0. \\[2mm]
9 & -\dfrac{5120}{153\,\pi} & -10.6519 \\[2mm]
10 & 0 & 0.
\end{array}
$$

Let $S_m = b_m \sin(\lambda_m x)\, e^{-\lambda_m^2 t}$. Then, the desired solution, $u(x, t)$, is given by

$$
u(x, t) = v(x) + \sum_{m=1}^{\infty} S_m.
$$

Let $u(x, t, n) = v(x) + \sum_{m=1}^{n} S_m$. Notice that $u(x, t, n) = u(x, t, n-1) + S_n$. Consequently, approximations of the solution to the heat equation are obtained recursively taking advantage of Mathematica's ability to compute recursively. The solution is first defined for $n = 1$ by u[x,t,1]. Subsequent partial sums, u[x,t,n], are obtained by adding the nth term of the series, S_n, to u[x,t,n-1].

$$In[831] := u[x\_, t\_, 1] :=$$
$$v[x] + b[1] \, \text{Sin}[\lambda[1] \, x] \, \text{Exp}\left[-\lambda[1]^2 \, t\right]$$

$$u[x\_, t\_, n\_] :=$$
$$u[x, t, n-1] + b[n] \, \text{Sin}[\lambda[n] \, x]$$
$$\text{Exp}\left[-\lambda[n]^2 \, t\right]$$

By defining the solution in this manner a table can be created that includes the partial sums of the solution. In the following table, we compute the first, fourth, and seventh partial sums of the solution to the problem.

$$In[832] := \text{Table}[u[x, t, n], \{n, 1, 7, 3\}]$$

$$Out[832] = \left\{ 10 + \frac{5120 \, e^{-\frac{\pi^2 t}{16}} \, \text{Sin}\left[\frac{\pi x}{4}\right]}{63 \, \pi}, \right.$$

$$10 + \frac{5120 \, e^{-\frac{\pi^2 t}{16}} \, \text{Sin}\left[\frac{\pi x}{4}\right]}{63 \, \pi} + \frac{1024 \, e^{-\frac{9\pi^2 t}{16}} \, \text{Sin}\left[\frac{3\pi x}{4}\right]}{33 \, \pi},$$

$$10 + \frac{5120 \, e^{-\frac{\pi^2 t}{16}} \, \text{Sin}\left[\frac{\pi x}{4}\right]}{63 \, \pi} + \frac{1024 \, e^{-\frac{9\pi^2 t}{16}} \, \text{Sin}\left[\frac{3\pi x}{4}\right]}{33 \, \pi} +$$

$$\left. \frac{1024 \, e^{-\frac{25\pi^2 t}{16}} \, \text{Sin}\left[\frac{5\pi x}{4}\right]}{39 \, \pi} + \frac{1024 \, e^{-\frac{49\pi^2 t}{16}} \, \text{Sin}\left[\frac{7\pi x}{4}\right]}{21 \, \pi} \right\}$$

To generate graphics that can be animated, we use a Do loop. The 10th partial sum of the solution is plotted for $t = 0$ to $t = 1$ using a step-size in t of 1/24. Remember that u[x,t,n] is determined with a Table command so that Evaluate must be used in the Do command so that Mathematica first computes the solution u and then evaluates u at the particular values of x. Otherwise, u is recalculated for each value of x. The plots of the solution obtained can be animated as indicated in the following screen shot.

$$In[833] := \text{Do}\left[\text{Plot}[\text{Evaluate}[u[x, t, 10]], \{x, 0, 1\}, \right.$$
$$\left. \text{PlotRange} \rightarrow \{0, 60\}], \left\{t, 0, 1, \frac{1}{24}\right\}\right]$$

Alternatively, we may generate several graphics and display the resulting set of graphics as a GraphicsArray. We plot the 10th partial sum of the solution for $t = 0$ to $t = 1$ using a step-size of 1/15. The resulting 16 graphs are named graphs which are then partitioned into four element subsets with Partition and named toshow. We then use Show and GraphicsArray to display toshow in Figure 4-27.

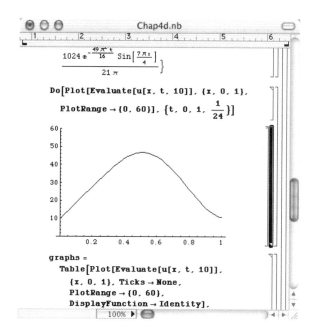

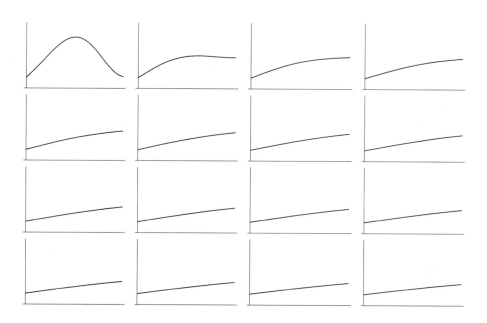

Figure 4-27   Temperature distribution in a uniform rod with insulated surface

```
In[834]:= graphs =
            Table[Plot[Evaluate[u[x, t, 10]],
                {x, 0, 1}, Ticks → None,
                PlotRange → {0, 60},
                DisplayFunction → Identity],
            {t, 0, 1, 1/15}];

          toshow = Partition[graphs, 4];

          Show[GraphicsArray[toshow]]
```

∎

Fourier series and generalized Fourier series arise in too many applications to list. Examples using them illustrate Mathematica's power to manipulate lists, symbolics, and graphics.

Application: The Wave Equation on a Circular Plate

For a classic approach to the subject see Graff's Wave Motion in Elastic Solids, *[10].*

The vibrations of a circular plate satisfy the equation

$$D \nabla^4 w(r, \theta, t) + \rho h \frac{\partial^2 w(r, \theta, t)}{\partial t^2} = q(r, \theta, t), \qquad (4.7)$$

where $\nabla^4 w = \nabla^2 \nabla^2 w$ and ∇^2 is the **Laplacian in polar coordinates,** which is defined by

$$\nabla^2 = \frac{1}{r} \frac{\partial}{\partial r} \left(r \frac{\partial}{\partial r} \right) + \frac{1}{r^2} \frac{\partial^2}{\partial \theta^2} = \frac{\partial^2}{\partial r^2} + \frac{1}{r} \frac{\partial}{\partial r} + \frac{1}{r^2} \frac{\partial^2}{\partial \theta^2}.$$

Assuming no forcing so that $q(r, \theta, t) = 0$ and $w(r, \theta, t) = W(r, \theta)e^{-i\omega t}$, equation (4.7) can be written as

$$\nabla^4 W(r, \theta) - \beta^4 W(r, \theta) = 0, \qquad \beta^4 = \omega^2 \rho h/D. \qquad (4.8)$$

For a clamped plate, the boundary conditions are $W(a, \theta) = \partial W(a, \theta)/\partial r = 0$ and after *much work* (see [10]) the **normal modes** are found to be

$$W_{nm}(r, \theta) = \left[J_n(\beta_{nm} r) - \frac{J_n(\beta_{nm} a)}{I_n(\beta_{nm} a)} I_n(\beta_{nm} r) \right] \binom{\sin n\theta}{\cos n\theta}. \qquad (4.9)$$

In equation (4.9), $\beta_{nm} = \lambda_{nm}/a$ where λ_{nm} is the *m*th solution of

$$I_n(x)J_n'(x) - J_n(x)I_n'(x) = 0, \qquad (4.10)$$

where $J_n(x)$ is the Bessel function of the first kind of order *n* and $I_n(x)$ is the **modified Bessel function of the first kind** of order *n*, related to $J_n(x)$ by $i^n I_n(x) = J_n(ix)$.

See Example 4.2.4.

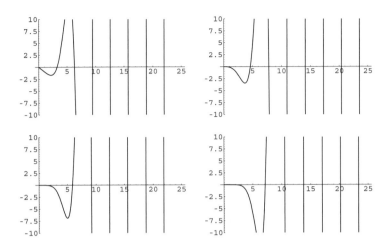

Figure 4-28   Plot of $I_n(x)J_n'(x) - J_n(x)I_n'(x)$ for $n = 0$ and 1 in the first row; $n = 2$ and 3 in the second row

The Mathematica command BesselI [n, x] returns $I_n(x)$.

EXAMPLE 4.4.5: Graph the first few normal modes of the clamped circular plate.

SOLUTION: We must determine the value of λ_{nm} for several values of n and m so we begin by defining eqn [n] [x] to be $I_n(x)J_n'(x) - J_n(x)I_n'(x)$. The mth solution of equation (4.10) corresponds to the mth zero of the graph of eqn [n] [x] so we graph eqn [n] [x] for $n = 0$, 1, 2, and 3 with Plot in Figure 4-28.

```
In[835] := eqn[n_][x_] := BesselI[n, x]D[BesselJ[n, x], x]
                        - BesselJ[n, x]D[BesselI[n, x], x]
```

The result of the Table and Plot command is a list of length four

```
In[836] := p1 = Table[Plot[eqn[n][x], {x, 0, 25},
              PlotRange → {-10, 10},
              DisplayFunction → Identity], {n, 0, 3}]
Out[836] = {-Graphics-, -Graphics-, -Graphics-, -Graphics-}
```

so we use Partition to create a 2×2 array of graphics which is displayed using Show and GraphicsArray.

```
In[837] := p2 = Show[GraphicsArray[Partition[p1, 2]]]
```

To determine λ_{nm} we use `FindRoot`. Recall that to use `FindRoot` to solve an equation an initial approximation of the solution must be given. For example,

```
In[838]:= lambda01 = FindRoot[eqn[0][x] == 0, {x, 3.04}]
Out[838]= {x → 3.19622}
```

approximates λ_{01}, the first solution of equation (4.10) if $n = 0$. However, the result of `FindRoot` is a list. The specific value of the solution is the second part of the first part of the list, `lambda01`, extracted from the list with `Part ([[...]])`.

```
In[839]:= lambda01[[1, 2]]
Out[839]= 3.19622
```

Thus,

We use the graphs in Figure 4-28 to obtain initial approximations of each solution.

```
In[840]:= λ0s = Map[FindRoot[eqn[0][x] == 0, {x, #}][[1, 2]]&,
              {3.04, 6.2, 9.36, 12.5, 15.7}]
Out[840]= {3.19622, 6.30644, 9.4395, 12.5771, 15.7164}
```

approximates the first five solutions of equation (4.10) if $n = 0$ and then returns the specific value of each solution. We use the same steps to approximate the first five solutions of equation (4.10) if $n = 1, 2$, and 3.

```
In[841]:= λ1s = Map[FindRoot[eqn[1][x] == 0, {x, #}][[1, 2]]&,
                  {4.59, 7.75, 10.9, 14.1, 17.2}]
Out[841]= {4.6109, 7.79927, 10.9581, 14.1086, 17.2557}
In[842]:= λ2s = Map[FindRoot[eqn[2][x] == 0, {x, #}][[1, 2]]&,
                  {5.78, 9.19, 12.4, 15.5, 18.7}]
Out[842]= {5.90568, 9.19688, 12.4022, 15.5795, 18.744}
In[843]:= λ3s = Map[FindRoot[eqn[3][x] == 0, {x, #}][[1, 2]]&,
                  {7.14, 10.5, 13.8, 17, 20.2}]
Out[843]= {7.14353, 10.5367, 13.7951, 17.0053, 20.1923}
```

All four lists are combined together in λs.

```
In[844]:= λs = {λ0s, λ1s, λ2s, λ3s}
Out[844]= {{3.19622, 6.30644, 9.4395, 12.5771, 15.7164},
            {4.6109, 7.79927, 10.9581, 14.1086, 17.2557},
            {5.90568, 9.19688, 12.4022, 15.5795, 18.744},
            {7.14353, 10.5367, 13.7951, 17.0053, 20.1923}}
```

For $n = 0, 1, 2$, and 3 and $m = 1, 2, 3, 4$, and 5, λ_{nm} is the mth part of the $(n + 1)$st part of λs.

Observe that the value of a does not affect the shape of the graphs of the normal modes so we use $a = 1$ and then define β_{nm}.

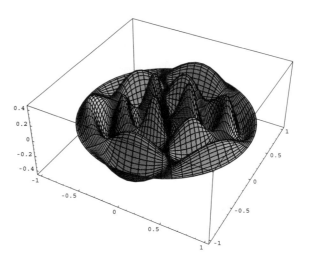

Figure 4-29   The sine part of $W_{34}(r, \theta)$

In[845]:= **a = 1;**

In[846]:= **β[n_, m_] := λs[[n + 1, m]]/a**

ws is defined to be the sine part of equation (4.9)

In[847]:= **ws[n_, m_][r, θ] :=**
        (BesselJ[n, β[n, m] r]
        -BesselJ[n, β[n, m] a]/BesselI[n, β[n, m] a]
         BesselI[n, β[n, m] r]) Sin[nθ]

and wc to be the cosine part.

In[848]:= **wc[n_, m_][r, θ] :=**
        -(BesselJ[n, β[n, m] r]
        BesselJ[n, β[n, m] a]/BesselI[n, β[n, m] a]
        BesselI[n, β[n, m] r]) Cos[nθ]

We use `ParametricPlot3D` to plot ws and wc. For example,

In[849]:= **ParametricPlot3D[{r Cos[θ],**
     r Sin[θ], ws[3, 4]
      [r, θ]}, {r, 0, 1}, {θ, -π, π}, PlotPoints $\rightarrow$ 60]

graphs the sine part of $W_{34}(r, \theta)$ shown in Figure 4-29. We use `Table` together with `ParametricPlot3D` followed by `Show` and `GraphicsArray` to graph the sine part of $W_{nm}(r, \theta)$ for $n = 0$, 1, 2, and 3 and $m = 1, 2, 3,$ and 4 shown in Figure 4.30.

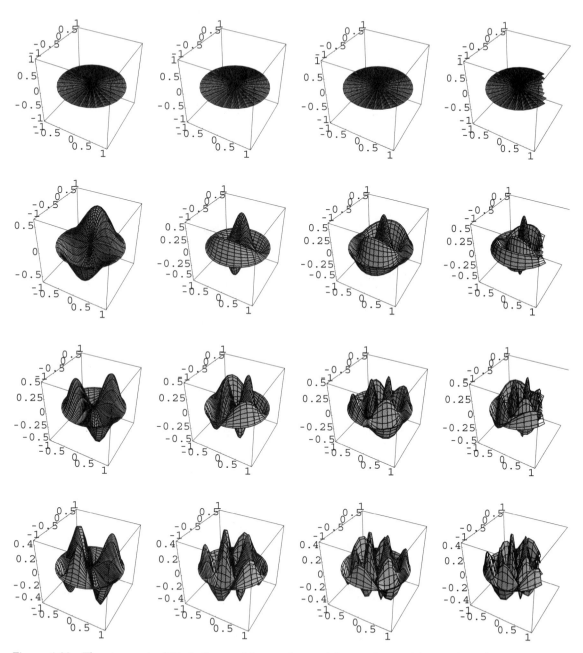

Figure 4-30   The sine part of $W_{nm}(r, \theta)$: $n = 0$ in row 1, $n = 1$ in row 2, $n = 2$ in row 3, and $n = 3$ in row 4 ($m = 1$ to 4 from left to right in each row)

```
In[850]:= ms = Table[ParametricPlot3D[{r Cos[θ],
              r Sin[θ], ws[n, m][r, θ]}, {r, 0, 1}, {θ, -π, π},
              DisplayFunction → Identity, PlotPoints → 30,
              BoxRatios → {1, 1, 1}], {n, 0, 3}, {m, 1, 4}]
Out[850]= {{-Graphics3D-, -Graphics3D-, -Graphics3D-,
            -Graphics3D-},
           {-Graphics3D-, -Graphics3D-, -Graphics3D-,
            -Graphics3D-},
           {-Graphics3D-, -Graphics3D-, -Graphics3D-,
            -Graphics3D-},
           {-Graphics3D-, -Graphics3D-, -Graphics3D-,
            -Graphics3D-}}

In[851]:= Show[GraphicsArray[ms]]
```

Identical steps are followed to graph the cosine part shown in Figure 4-31.

```
In[852]:= mc = Table[ParametricPlot3D[{r Cos[θ], r Sin[θ],
              wc[n, m][r, θ]}, {r, 0, 1}, {θ, -π, π},
              DisplayFunction → Identity, PlotPoints →
              30, BoxRatios → {1, 1, 1}], {n, 0, 3}, {m, 1, 4}]
Out[852]= {{-Graphics3D-, -Graphics3D-, -Graphics3D-,
            -Graphics3D-},
           {-Graphics3D-, -Graphics3D-, -Graphics3D-,
            -Graphics3D-},
           {-Graphics3D-, -Graphics3D-, -Graphics3D-,
            -Graphics3D-},
           {-Graphics3D-, -Graphics3D-, -Graphics3D-,
            -Graphics3D-}}

In[853]:= Show[GraphicsArray[mc]]
```

∎

4.4.3 The Mandelbrot Set and Julia Sets

See references like Barnsley's *Fractals Everywhere* [4] or Devaney and Keen's *Chaos and Fractals* [6] for detailed discussions regarding many of the topics briefly described in this section.

$f_c(x) = x^2 + c$ is the special case of $p = 2$ for $f_{p,c}(x) = x^p + c$.

In Examples 4.1.7, 4.2.5, and 4.2.7 we illustrated several techniques for plotting bifurcation diagrams and Julia sets.

Let $f_c(x) = x^2 + c$. In Example 4.2.5, we generated the c-values when plotting the bifurcation diagram of f_c. Depending upon how you think, some approaches may be easier to understand than others. With the exception of very serious calculations, the differences in the time needed to carry out the computations may be minimal so we encourage you to follow the approach that you understand. Learn new techniques as needed.

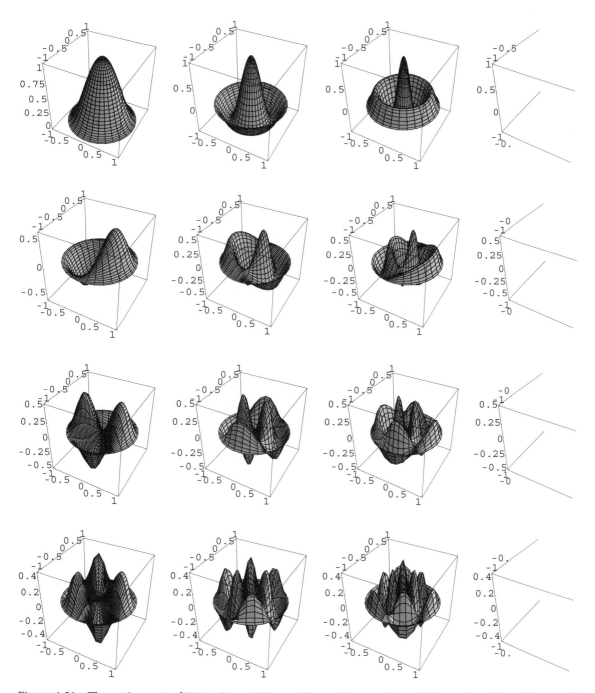

Figure 4-31   The cosine part of $W_{nm}(r, \theta)$: $n = 0$ in row 1, $n = 1$ in row 2, $n = 2$ in row 3, and $n = 3$ in row 4 ($m = 1$ to 4 from left to right in each row)

Compare the approach here
with the approach used in
Example 4.2.5.

EXAMPLE 4.4.6 (Dynamical Systems): For example, entering

In[854]:= **Clear[f, c]**

f[c_][x_] := x^2 + c//N;

defines $f_c(x) = x^2 + c$ so

In[855]:= **Nest[f[-1], x, 3]**

Out[855]= $-1. + \left(-1. + (-1. + x^2)^2\right)^2$

computes $f_{-1}{}^3(x)$ and

In[856]:= **Table[Nest[f[1/4], 0, n], {n, 101, 200}]**

Out[856]= {0.490693, 0.490779, 0.490864, 0.490948, 0.49103,
           0.49111, 0.491189, 0.491267, 0.491343, 0.491418,
           0.491492, 0.491564, 0.491635, 0.491705, 0.491774,
           0.491842, 0.491908, 0.491974, 0.492038, 0.492101,
           0.492164, 0.492225, 0.492286, 0.492345, 0.492404,
           0.492461, 0.492518, 0.492574, 0.492629, 0.492684,
           0.492737, 0.49279, 0.492842, 0.492893, 0.492944,
           0.492994, 0.493043, 0.493091, 0.493139, 0.493186,
           0.493232, 0.493278, 0.493323, 0.493368, 0.493412,
           0.493455, 0.493498, 0.49354, 0.493582, 0.493623,
           0.493664, 0.493704, 0.493744, 0.493783, 0.493821,
           0.49386, 0.493897, 0.493935, 0.493971, 0.494008,
           0.494044, 0.494079, 0.494114, 0.494149, 0.494183,
           0.494217, 0.49425, 0.494283, 0.494316, 0.494348,
           0.49438, 0.494412, 0.494443, 0.494474, 0.494505,
           0.494535, 0.494565, 0.494594, 0.494623, 0.494652,
           0.494681, 0.494709, 0.494737, 0.494765, 0.494792,
           0.494819, 0.494846, 0.494873, 0.494899, 0.494925,
           0.494951, 0.494976, 0.495002, 0.495027, 0.495051,
           0.495076, 0.4951, 0.495124, 0.495148, 0.495171}

returns a list of $f_{1/4}{}^n(0)$ for $n = 101, 102, \ldots, 200$. Thus,

In[857]:= **lgtable = Table[{c, Nest[f[c], 0, n]},**
            {c, -2, 1/4, 9/(4 * 299)}, {n, 101, 200}];

In[858]:= **Length[lgtable]**

returns a list of lists of $f_c{}^n(0)$ for $n = 101, 102, \ldots, 200$ for 300 equally
spaced values of c between -2 and 1. The list lgtable is converted to a
list of points with Flatten and plotted with ListPlot. See

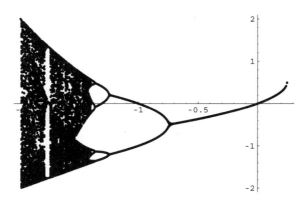

Figure 4-32   Another bifurcation diagram for f_c

Figure 4-32 and compare this result to the result obtained in Example 4.2.5.

```
In[859]:= toplot = Flatten[lgtable, 1];
```

```
In[860]:= ListPlot[toplot]
```

For a given complex number c the **Julia set**, J_c, of $f_c(x) = x^2 + c$ is the set of complex numbers, $z = a + bi$, a, b real, for which the sequence z, $f_c(z) = z^2 + c$, $f_c(f_c(z)) = (z^2 + c)^2 + c, \ldots, f_c{}^n(z), \ldots$, does *not* tend to ∞ as $n \to \infty$:

We use the notation $f^n(x)$ to represent the composition $\underbrace{(f \circ f \circ \cdots \circ f)(x)}_{n}$.

$$J_c = \left\{ z \in \mathbb{C} | z, \, z^2 + c \, \left(z^2 + c \right)^2 + c, \, \cdots \nrightarrow \infty \right\}.$$

Using a dynamical system, setting $z = z_0$ and computing $z_{n+1} = f_c(z_n)$ for large n can help us determine if z is an element of J_c. In terms of a composition, computing $f_c{}^n(z)$ for large n can help us determine if z is an element of J_c.

As before, all error messages have been deleted.

EXAMPLE 4.4.7 (Julia Sets): Plot the Julia set of $f_c(x) = x^2 + c$ if $c = -0.122561 + 0.744862i$.

SOLUTION: After defining $f_c(x) = x^2 + c$, we use `Table` together with `Nest` to compute ordered triples of the form $\left(x, y, f_{-0.122561+0.744862i}{}^{200} (x + iy) \right)$ for 150 equally spaced values of x between $-3/2$ and $3/2$ and 150 equally spaced values of y between $-3/2$ and $3/2$.

You do not need to redefine $f_c(x)$ if you have already defined it during your current Mathematica session.

```
In[861]:= Clear[f, c]
```

```
f[c_][x_] := x^2 + c//N;
```

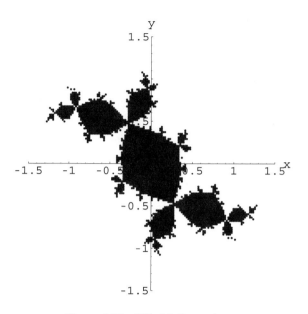

Figure 4-33   Filled Julia set for f_c

```
g1 = Table[{x, y, Nest[f[-0.122561 + 0.744862 i], x + i y, 200]},
    {x, -3/2, 3/2, 3/149}, {y, -3/2, 3/2, 3/149}];
```

```
g2 = Flatten[g1, 1];
```

We remove those elements of g2 for which the third coordinate is Overflow[]
with Select,

```
In[862]:= g3 = Select[g2, Not[#[[3]] === Overflow[]]&];
```

extract a list of the first two coordinates, (x, y), from the elements of g3,

```
In[863]:= g4 = Map[{#[[1]], #[[2]]}&, g3];
```

and plot the resulting list of points in Figure 4-33 using ListPlot.

```
In[864]:= lp1 = ListPlot[g4, PlotRange → {{-3/2, 3/2},
                {-3/2, 3/2}}, AxesLabel → {"x",
            AspectRatio → Automatic]
```

We can invert the image as well with the following commands. In the
end result, we show the Julia set and its inverted image in Figure 4-34

```
In[865]:= g3b = Select[g2, #[[3]] === Overflow[]&];
```

```
In[866]:= g4b = Map[{#[[1]], #[[2]]}&, g3b];
```

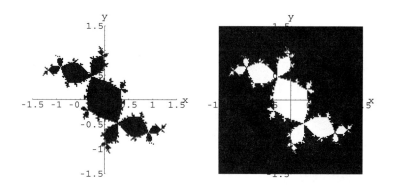

Figure 4-34   Filled Julia set for f_c on the left; the inverted set on the right

```
In[867] := lp2 = ListPlot[g4b,
               PlotRange → {{-3/2, 3/2}, {-3/2, 3/2}},
               AxesLabel → {"x", "y"}, AspectRatio → Automatic,
               DisplayFunction → Identity];

           j1 = Show[GraphicsArray[{lp1, lp2}]]
```

■

Of course, one can consider functions other than $f_c(x) = x^2 + c$ as well as rearrange the order in which we carry out the computations.

EXAMPLE 4.4.8 (Julia Sets): Plot the Julia set for $f(z) = .36e^z$.

SOLUTION: For this example, we begin by forming our complex grid first in complexpts using Table followed by Flatten. The result is a list of numbers of the form $a + bi$ for 200 equally spaced values of a between 0 and 5 and 200 equally spaced values of b between -2.5 and 2.5.

```
In[868] := complexpts = Flatten[Table[a + b I,
                {a, 0., 5., 5/199}, {b, -2.5, 2.5, 5/199}], 1];
```

After defining $f(z)$, we define $h(c)$. Given c, $h(c)$ returns the ordered triple $(\mathrm{Re}(c), \mathrm{Im}(c), f^{200}(c))$.

```
In[869] := Clear[f]

           f[z_] = 0.36 Exp[z]
```

Out[869]= 0.36 e^z

In[870]:= **h[c_] := {Re[c], Im[c], Nest[f, c, 200]}**

We then apply *h* to complexpts with Map. We use Chop to replace numbers very close to 0 with 0.

As in Examples 4.1.7, 4.2.5, and 4.2.7 Mathematica displays several Overflow[] and Underflow[] error messages that we are able to ignore. They are not shown here for length considerations.

In[871]:= **t1 = Map[h, complexpts]//Chop;**

We then use Select to extract those elements of t1 for which the third coordinate *is not* indeterminate (that is, not complex ∞) in t2 and those elements for which the third coordinate *is* indeterminate in t2b.

In[872]:= **t2 = Select[t1, Not[#[[3]] === Indeterminate]&];**

In[873]:= **t2b = Select[t1, #[[3]] === Indeterminate&];**

Applying pt to t2 and t2b results in two lists of ordered pairs that are plotted with ListPlot and shown side-by-side using Show together with GraphicsArray in Figure 4-35.

In[874]:= **pt[{x_, y_, z_}] := {x, y}**

In[875]:= **t3 = Map[pt, t2];**

            t3b = Map[pt, t2b];

In[876]:= **lp1 = ListPlot[t3, PlotRange → {{0, 5}, {-2.5, 2.5}},**
                    AspectRatio → Automatic,
                    DisplayFunction → Identity];

            lp2 = ListPlot[t3b, PlotRange → {{0, 5}, {-2.5, 2.5}},
                    AspectRatio → Automatic,
                    DisplayFunction → Identity];

            Show[GraphicsArray[{lp1, lp2}]]

∎

You have even greater control over your graphics if you use graphics primitives like Point.

As before, all error messages have been deleted.

EXAMPLE 4.4.9 (Julia Sets): Plot the Julia set for $f_c(z) = z^2 - cz$ if $c = 0.737369 + 0.67549i$.

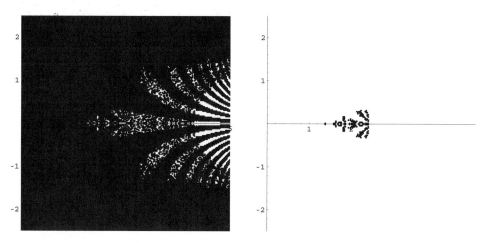

Figure 4-35 Two different views of the Julia set of $f(z) = .36e^z$: on the left, the black points (a, b) are the points for which $f^{200}(a + bi)$ is finite; on the right, the black points (a, b) are the ones for which $f^{200}(a + bi)$ is not finite

SOLUTION: We proceed as in Example 4.4.7.

```
In[877]:= Clear[f, c]

         f[c_][z_] := z^2 - c z//N;

In[878]:= g1 = Table[{x, y, Nest[f[0.737369 + 0.67549 I],
            x + I y, 200]}, {x, -1.2, 1.75, 2.95/199},
            {y, -0.7, 1.4, 2.1/199}];

         g2 = Flatten[g1, 1];

In[879]:= g3 = Select[g2, Not[#[[3]] === Overflow[]]&];
```

After removing the points that result in an `Overflow[]` error message, we code the remaining ones according to their distance from the origin.

```
In[880]:= h[{x_, y_, z_}] := {x, y, Min[Abs[z], 0.5]}

         g4 = Map[h, g3];

In[881]:= g5 = Table[{PointSize[0.005],
            GrayLevel[g4[[i, 3]]/0.5],
            Point[{g4[[i, 1]], g4[[i, 2]]}]},
            {i, 1, Length[g4]}];
```

The results are shown in Figure 4-36.

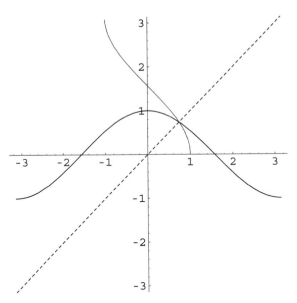

Figure 4-36   The lightest points (a, b) are the ones for which $\left|f_{0.737369+0.67549i}^{200}(z)\right|$ is the largest

```
In[882]:= Show[Graphics[g5],
             PlotRange → {{-1.2, 1.75}, {-0.7, 1.4}},
             AspectRatio → Automatic]
```

■

EXAMPLE 4.4.10 (The Ikeda Map): The **Ikeda map** is defined by

$$\mathbf{F}(x, y) = \langle \gamma + \beta\,(x \cos \tau - y \sin \tau), \beta\,(x \sin \tau + y \cos \tau)\rangle, \qquad (4.11)$$

where $\tau = \mu - \alpha/\left(1 + x^2 + y^2\right)$. If $\beta = .9$, $\mu = .4$, and $\alpha = 4.0$, plot the *basins of attraction* for F if $\gamma = .92$ and $\gamma = 1.0$.

SOLUTION: The *basins of attraction* for F are the set of points (x, y) for which $\|\mathbf{F}^n(x, y)\| \nrightarrow \infty$ as $n \to \infty$.

After defining $f[\gamma][x, y]$ to be equation (4.11) and then $\beta = .9$, $\mu = .4$, and $\alpha = 4.0$, we use Table followed by Flatten to define pts to be the list of 40,000 ordered pairs (x, y) for 200 equally spaced values of x between -2.3 and 1.3 and 200 equally spaced values of y between -2.8 and $.8$.

```
In[883]:= f[γ_][{x_, y_}] := {γ + β (x Cos[μ - α/(1 + x^2 + y^2)]
              -y Sin[μ - α/(1 + x^2 + y^2)]), β (x Sin[μ
              -α/(1 + x^2 + y^2)] + y Cos[μ - α/(1 + x^2 + y^2)])}
In[884]:= β = 0.9; μ = 0.4; α = 4.;

In[885]:= pts = Flatten[Table[{x, y}, {x, -2.3, 1.3, 3.6/199},
              {y, -2.8, 0.8, 3.6/199}], 1];
```

In l1, we use Map to compute $\left(x, y, \mathbf{F}_{.92}^{200}(x, y)\right)$ for each (x, y) in pts. In pts2, we use the graphics primitive Point and shade the points according to the maximum value of $\left\|\mathbf{F}^{200}(x, y)\right\|$ – those (x, y) for which $\mathbf{F}^{200}(x, y)$ is closest to the origin are darkest; the point (x, y) is shaded lighter as the distance of $\mathbf{F}^{200}(x, y)$ from the origin increases.

```
In[886]:= l1 = Map[{#[[1]], #[[2]], Nest[f[0.92],
              {#[[1]], #[[2]]}, 200]}&, pts];

In[887]:= g[{x_, y_, z_}] := {x, y, Sqrt[z[[1]]^2 + z[[2]]^2]};

          l2 = Map[g, l1];
In[888]:= maxl2 = Table[l2[[i, 3]], {i, 1, Length[l2]}] //Max
Out[888]= 4.33321

In[889]:= pts2 = Table[{GrayLevel[l2[[i, 3]]/(maxl2)],
              Point[{l2[[i, 1]], l2[[i, 2]]}]},
              {i, 1, Length[l2]}];

In[890]:= ik1 = Show[Graphics[pts2], AspectRatio → 1]
```

For $\gamma = 1.0$, we proceed in the same way.

```
In[891]:= l1 = Map[{#[[1]], #[[2]], Nest[f[1.], {#[[1]],
              #[[2]]}, 200]}&, pts];

In[892]:= l2 = Map[g, l1];

In[893]:= maxl2 = Table[l2[[i, 3]], {i, 1, Length[l2]}] //Max
Out[893]= 4.48421

In[894]:= pts2 = Table[{GrayLevel[l2[[i, 3]]/maxl2],
              Point[{l2[[i, 1]], l2[[i, 2]]}]},
              {i, 1, Length[l2]}];

In[895]:= ik2 = Show[Graphics[pts2], AspectRatio → 1]

In[896]:= Show[GraphicsArray[{ik1, ik2}]]
```

∎

Figure 4-37   Basins of attraction for **F** if γ = .92 (on the left) and γ = 1.0 (on the right)

The **Mandelbrot set**, M, is the set of complex numbers, $z = a+bi$, a, b real, for which the sequence z, $f_z(z) = z^2 + z$, $f_z(f_z(z)) = (z^2 + z)^2 + z$, ..., $f_z^n(z)$, ..., does *not* tend to ∞ as $n \to \infty$:

$$M = \left\{ z \in \mathbf{C} | z,\ z^2 + z \left(z^2 + z \right)^2 + z,\ \cdots \nrightarrow \infty \right\}.$$

Using a dynamical system, setting $z = z_0$ and computing $z_{n+1} = f_{z_0}(z_n)$ for large n can help us determine if z is an element of M. In terms of a composition, computing $f_z^n(z)$ for large n can help us determine if z is an element of M.

As before, all error messages have been deleted.

EXAMPLE 4.4.11 (Mandelbrot Set): Plot the Mandelbrot set.

SOLUTION: We proceed as in Example 4.4.7 except that instead of iterating $f_c(z)$ for fixed c we iterate $f_z(z)$.

```
In[897]:= Clear[f, c]

         f[c_][x_] := x^2 + c//N;

In[898]:= g1 = Table[{x, y, Nest[f[x + I y], x + I y, 200]},
             {x, -3/2, 1, 5/(2 * 149)}, {y, -1, 1, 2/149}];

         g2 = Flatten[g1, 1];

In[899]:= g3 = Select[g2, Not[#[[3]] === Overflow[]]&];

In[900]:= g4 = Map[{#[[1]], #[[2]]}&, g3];
```

The following gives us the image on the left in Figure 4-38.

```
In[901]:= lp1 = ListPlot[g4, PlotRange → {{-3/2, 1}, {-1, 1}},
             Axes → None, AspectRatio → Automatic,
             PlotStyle → PointSize[0.005]]
```

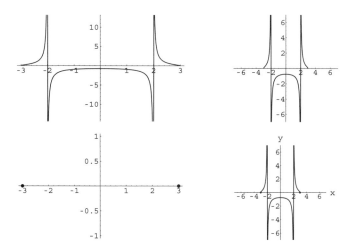

Figure 4-38   Two different views of the Mandelbrot set: on the left, the black points (a, b) are the points for which $f_{a+bi}^{200}(a + bi)$ is finite; on the right, the black points (a, b) are the ones for which $f_{a+bi}^{200}(a + bi)$ is not finite

To invert the image, we use the following to obtain the result on the right in Figure 4-38.

```
In[902]:= g3b = Select[g2, #[[3]] === Overflow[]&];
```

```
In[903]:= g4b = Map[{#[[1]], #[[2]]}&, g3b];
```

```
In[904]:= lp2 = ListPlot[g4b, PlotRange → {{-3/2, 1}, {-1, 1}},
              Axes → None, AspectRatio → Automatic,
              PlotStyle → PointSize[0.005]]
```

```
In[905]:= m1 = Show[GraphicsArray[{lp1, lp2}]]
```

In Example 4.4.11, the Mandelbrot set is obtained (or more precisely, approximated) by repeatedly composing $f_z(z)$ for a grid of z-values and then deleting those for which the values exceed machine precision. Those values greater than `$MaxNumber` result in an `Overflow[]` message; computations with `Overflow[]` result in an `Indeterminate` message.

We can generalize by considering exponents other than 2 by letting $f_{p,c} = x^p + c$. The **generalized Mandelbrot set**, M_p, is the set of complex numbers, $z = a + bi$, a, b real, for which the sequence z, $f_{p,z}(z) = z^p + z$, $f_{p,z}\left(f_{p,z}(z)\right) = (z^p + z)^p + z, \ldots, f_{p,z}{}^n(z)$, $\ldots$, does *not* tend to ∞ as $n \to \infty$:

$$M_p = \left\{ z \in \mathbb{C} \mid z, \; z^p + z, \; \left(z^p + z\right)^p + z, \; \cdots \not\to \infty \right\}.$$

Using a dynamical system, setting $z = z_0$ and computing $z_{n+1} = f_p(z_n)$ for large n can help us determine if z is an element of M_p. In terms of a composition, computing $f_p{}^n(z)$ for large n can help us determine if z is an element of M_p.

As before, all error messages have been omitted.

EXAMPLE 4.4.12 (Generalized Mandelbrot Set): After defining $f_{p,c} = x^p + c$, we use Table, Abs, and Nest to compute a list of ordered triples of the form $(x, y, |f_{p,x+iy}{}^{100}(x + iy)|)$ for p-values from 1.625 to 2.625 spaced by equal values of 1/8 and 200 values of x (y) values equally spaced between −2 and 2, resulting in 40,000 sample points of the form $x + iy$.

```
In[906]:= Clear[f,p]

          f[p_,c_][x_] := x^p + c//N;

In[907]:= g1 = Table[{x, y, Abs[Nest[f[p, x + I y],
              x + I y, 100]]}//N, {p, 1.625, 2.625, 1/8},
              {x, -2., 2., 4/199}, {y, -2., 2., 4/199}];

In[908]:= g2 = Map[Flatten[#, 1]&, g1];
```

Next, we extract those points for which the third coordinate is Indeterminate with Select, ordered pairs of the first two coordinates are obtained in g4. The resulting list of points is plotted with ListPlot in Figure 4-39.

```
In[909]:= g3 = Table[Select[g2[[i]],
              Not[#[[3]] === Indeterminate]&],
              {i, 1, Length[g2]}];

In[910]:= h[{x_, y_, z_}] := {x, y};

In[911]:= g4 = Map[h, g3, {2}];

In[912]:= t1 = Table[ListPlot[g4[[i]],
              PlotRange → {{-2, 2}, {-2, 2}},
              AspectRatio → Automatic,
              DisplayFunction → Identity], {i, 1, 9}];
          Show[GraphicsArray[Partition[t1, 3]]]
```

More detail is observed if you use the graphics primitive Point as shown in Figure 4-40. In this case, those points (x, y) for which $|f_{p,x+iy}{}^{100}(x+iy)|$ is small are shaded according to a darker GrayLevel than those points for which $|f_{p,x+iy}{}^{100}(x + iy)|$ is large.

```
In[913]:= h2[{x_, y_, z_}] := {GrayLevel[Min[{z, 0.25}]],
              Point[{x, y}]};

In[914]:= g5 = Map[h2, g3, {2}];
```

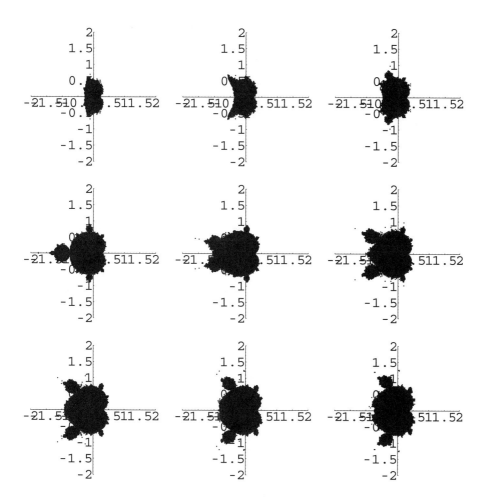

Figure 4-39   The generalized Mandelbrot set for 9 equally spaced values of *p* between 1.625
and 2.625

```
In[915]:= t1 = Table[Show[Graphics[g5[[i]]],
              PlotRange → {{-2, 2}, {-2, 2}},
              AspectRatio → Automatic,
            DisplayFunction → Identity], {i, 1, 9}];
            Show[GraphicsArray[Partition[t1, 3]]]
```

Throughout these examples, we have typically computed the iteration $f^n(z)$ for
"large" *n* like values of *n* between 100 and 200. To indicate why we have selected
those values of *n*, we revisit the Mandelbrot set plotted in Example 4.4.11.

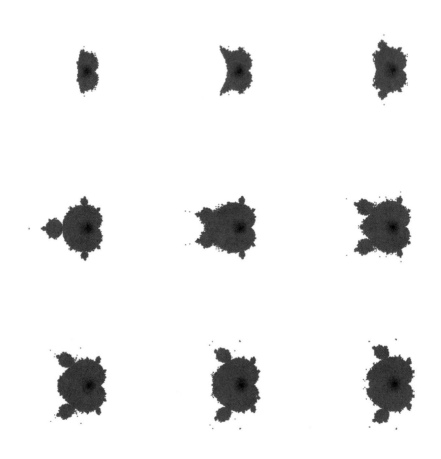

Figure 4-40   The generalized Mandelbrot set for 9 equally spaced values of p between 1.625 and 2.625 – the points (x, y) for which $\left|f_{p,x+iy}^{100}(x + iy)\right|$ is large are shaded lighter than those for which $\left|f_{p,x+iy}^{100}(x + iy)\right|$ is small

As before, all error messages
have been deleted.

EXAMPLE 4.4.13 (Mandelbrot Set): We proceed in essentially the same way as in the previous examples. After defining $f_{p,c} = x^p + c$,

```
In[916]:= Clear[f, p]

        f[p_, c_][x_] := x^p + c//N;
```

we use `Table` followed by `Map` to create a nested list. For each $n = 5$, 10,

15, 25, 50, and 100, a nested list is formed for 200 equally spaced values
of y between -1 and 1 and then 200 equally spaced values of x between
-1.5 and 1. At the bottom level of each nested list, the elements are of
the form $\left(x, y, \left|f_{2,x+iy}{}^n(x + iy)\right|\right)$.

```
In[917]:= g1 = Map[Table[{x,y,Abs[Nest[f[2,x+Iy],
              x+Iy,#]]}//N,{x,-1.5,1.,5/(2*199)},
           {y,-1.,1.,2/199}]&,
              {5,10,15,25,50,100}];
```

For each value of n, the corresponding list of ordered triples
$\left(x, y, \left|f_{2,x+iy}{}^n(x + iy)\right|\right)$ is obtained using Flatten.

```
In[918]:= g2 = Map[Flatten[#,1]&,g1];
```

We then remove those points for which the third coordinate,
$\left|f_{2,x+iy}{}^n(x + iy)\right|$, is Overflow[] (corresponding to ∞),

```
In[919]:= g3 = Table[Select[g2[[i]],Not[#[[3]] ===
              Overflow[]]&],{i,1,Length[g2]}];
```

extract (x, y) from the remaining ordered triples,

```
In[920]:= h[{x_,y_,z_}] := {x,y};
```

```
In[921]:= g4 = Map[h,g3,{2}];
```

and graph the resulting sets of points using ListPlot in Figure 4-41. As shown in Figure 4-41, we see that Mathematica's numerical
precision (and consequently decent plots) are obtained when $n = 50$ or
$n = 100$.

Fundamentally, we generated
the previous plots by
exceeding Mathematica's
numerical precision.

```
In[922]:= t1 = Table[ListPlot[g4[[i]],
           PlotRange → {{-3/2,1},{-1,1}},
              AspectRatio → Automatic,
           DisplayFunction → Identity],{i,1,6}];
```

```
Show[GraphicsArray[Partition[t1,3]]]
```

If instead, we use graphics primitives like Point and then shade
each point (x, y) according to $\left|f_{2,x+iy}{}^n(x + iy)\right|$ detail emerges quickly as
shown in Figure 4-42.

```
In[923]:= h2[{x_,y_,z_}] := {GrayLevel[Min[{z,1}]],
           Point[{x,y}]};
```

```
In[924]:= g5 = Map[h2,g3,{2}];
```

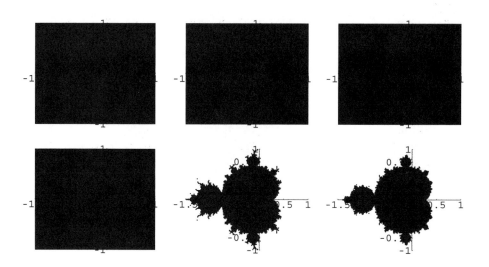

Figure 4-41   Without shading the points, the effects of iteration are difficult to see until the number of iterations is "large"

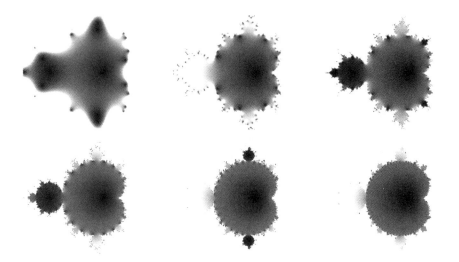

Figure 4-42   Using graphics primitives and shading, we see that we can use a relatively small number of iterations to visualize the Mandelbrot set

```
In[925]:= t1 = Table[Show[Graphics[g5[[i]]],
              PlotRange → {{-3/2, 1}, {-1, 1}},
                AspectRatio → Automatic,
            DisplayFunction → Identity], {i, 1, 6}];

      Show[GraphicsArray[Partition[t1, 3]]]
```

Thus, Figures 4-41 and 4-42 indicate that for examples like the ones illustrated here similar results could have been accomplished using far smaller values of n than $n = 100$ or $n = 200$. With fast machines, the differences in the time needed to perform the calculations is minimal; $n = 100$ and $n = 200$ appear to be a "safe" large value of n for well-studied examples like these.

Not even 10 years ago calculations like these required the use of a supercomputer and sophisticated computer programming. Now, they are accessible to virtually anyone working on a relatively new machine with just a few lines of Mathematica code. Quite amazing!

Matrices and Vectors: Topics from Linear Algebra and Vector Calculus

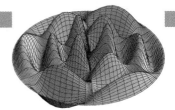

5

Chapter 5 discusses operations on matrices and vectors, including topics from linear algebra, linear programming, and vector calculus.

5.1 Nested Lists: Introduction to Matrices, Vectors, and Matrix Operations

5.1.1 Defining Nested Lists, Matrices, and Vectors

In Mathematica, a **matrix** is a list of lists where each list represents a row of the matrix. Therefore, the $m \times n$ matrix

$$\mathbf{A} = \begin{pmatrix} a_{11} & a_{12} & a_{13} & \cdots & a_{1n} \\ a_{21} & a_{22} & a_{23} & \cdots & a_{2n} \\ a_{31} & a_{32} & a_{33} & \cdots & a_{3n} \\ \vdots & \vdots & \vdots & & \vdots \\ a_{m1} & a_{m2} & a_{m3} & \cdots & a_{mn} \end{pmatrix}$$

is entered with

```
A={{a11,a12,...,a1n},{a21,a22,...,a2n},...,{am1,am2,...amn}}.
```

For example, to use Mathematica to define m to be the matrix $\mathbf{A} = \begin{pmatrix} a_{11} & a_{12} \\ a_{21} & a_{22} \end{pmatrix}$ enter the command

$$m=\{\{a11,a12\},\{a21,a22\}\}.$$

The command m=Array[a,{2,2}] produces a result equivalent to this. Once a matrix A has been entered, it can be viewed in the traditional row-and-column form using the command MatrixForm[A]. You can quickly construct 2×2 matrices by clicking on the 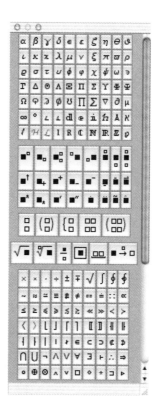 button from the **BasicTypesetting** palette, which is accessed by going to **File** under the Mathematica menu, followed by **Palettes** and then **BasicTypesetting**.

As when using TableForm, the result of using MatrixForm is no longer a list that can be manipulated using Mathematica commands. Use MatrixForm to view a matrix in traditional row-and-column form. Do not attempt to perform matrix operations on a MatrixForm object.

Alternatively, you can construct matrices of any dimension by going to the Mathematica menu under **Input** and selecting **Create Table/Matrix/Palette...**

Get Graphics Coordinates...
3D ViewPoint Selector...   ⇧⌘V
Color Selector...
Record Sound...
Get File Path...

Create Table/Matrix/Palette...   ⇧⌘C
Create Button   ▶
Edit Button...

Create Hyperlink...
Create Automatic Numbering Object...
Create Value Display Object...
Convert Automatic Objects to Literal

Copy Input from Above   ⌘L
Copy Output from Above   ⇧⌘L
Start New Cell Below   ⌥↵

Complete Selection   ⌘K
Make Template   ⇧⌘K

The resulting pop-up window allows you to create tables, matrices, and palettes. To create a matrix, select **Matrix**, enter the number of rows and columns of the matrix, and select any other options. Pressing the **OK** button places the desired matrix at the position of the cursor in the Mathematica notebook.

Create
⊙ Table (plain GridBox)   Number of rows: 3
○ Matrix
○ Palette   Number of columns: 3

Options
☐ Draw lines between rows   ☐ Fill with: 0
☐ Draw lines between columns
☐ Draw frame   ☐ Fill diagonal: 1

⊘   Cancel   OK

EXAMPLE 5.1.1: Use Mathematica to define the matrices $\begin{pmatrix} a_{11} & a_{12} & a_{13} \\ a_{22} & a_{22} & a_{23} \\ a_{31} & a_{32} & a_{33} \end{pmatrix}$ and $\begin{pmatrix} b_{11} & b_{12} & b_{13} & b_{14} \\ b_{21} & b_{22} & b_{23} & b_{24} \end{pmatrix}$.

SOLUTION: In this case, both `Table[a`$_{i,j}$`, {i, 1, 3}, {j, 1, 3}]` and `Array[a, {3, 3}]` produce equivalent results when we define `matrixa` to be the matrix

$$\begin{pmatrix} a_{11} & a_{12} & a_{13} \\ a_{22} & a_{22} & a_{23} \\ a_{31} & a_{32} & a_{33} \end{pmatrix}.$$

The commands `MatrixForm` or `TableForm` are used to display the results in traditional matrix form.

```
In[926]:= Clear[a, b, matrixa, matrixb]
General :: spell1 :
  Possible spelling error : new symbol name \"
    matrixb\" is similar to existing symbol \"
    matrixa\".

In[927]:= matrixa = Table[a_{i,j}, {i, 1, 3}, {j, 1, 3}]
Out[927]= {{a_{1,1}, a_{1,2}, a_{1,3}},
            {a_{2,1}, a_{2,2}, a_{2,3}}, {a_{3,1}, a_{3,2}, a_{3,3}}}

In[928]:= MatrixForm[matrixa]
Out[928]= \begin{pmatrix} a_{1,1} & a_{1,2} & a_{1,3} \\ a_{2,1} & a_{2,2} & a_{2,3} \\ a_{3,1} & a_{3,2} & a_{3,3} \end{pmatrix}

In[929]:= matrixa = Array[a, {3, 3}]
Out[929]= {{a[1, 1], a[1, 2], a[1, 3]},
            {a[2, 1], a[2, 2], a[2, 3]},
            {a[3, 1], a[3, 2], a[3, 3]}}

In[930]:= MatrixForm[matrixa]
Out[930]= \begin{pmatrix} a[1, 1] & a[1, 2] & a[1, 3] \\ a[2, 1] & a[2, 2] & a[2, 3] \\ a[3, 1] & a[3, 2] & a[3, 3] \end{pmatrix}
```

We may also use Mathematica to define non-square matrices.

```
In[931]:= matrixb = Array[b, {2, 4}]
Out[931]= {{b[1, 1], b[1, 2], b[1, 3], b[1, 4]},
            {b[2, 1], b[2, 2], b[2, 3], b[2, 4]}}

In[932]:= MatrixForm[matrixb]
Out[932]= \begin{pmatrix} b[1, 1] & b[1, 2] & b[1, 3] & b[1, 4] \\ b[2, 1] & b[2, 2] & b[2, 3] & b[2, 4] \end{pmatrix}
```

Equivalent results would have been obtained by entering `Table[b`$_{i,j}$`, {i, 1, 2}, {j, 1, 4}]`.

■

More generally the commands `Table[f[i,j],{i,imax},{j,jmax}]` and `Array[f,{imax,jmax}]` yield nested lists corresponding to the imax × jmax matrix

$$\begin{pmatrix} f(1,1) & f(1,2) & \cdots & f(1,\text{jmax}) \\ f(2,1) & f(2,2) & \cdots & f(2,\text{jmax}) \\ \vdots & \vdots & \vdots & \vdots \\ f(\text{imax},1) & f(\text{imax},2) & \cdots & f(\text{imax},\text{jmax}) \end{pmatrix}.$$

`Table[f[i,j],{i,imin,imax,istep},{j,jmin,jmax,jstep}]` returns the list of lists

```
{{f[imin,jmin],f[imin,jmin+jstep],...,f[imin,jmax]},
    {f[imin+istep,jmin],...,f[imin+istep,jmax]},
        ...,{f[imax,jmin],...,f[imax,jmax]}}
```

and the command

```
Table[f[i,j,k,...],{i,imin,imax,istep},{j,jmin,jmax,jstep},
    {k,kmin,kmax,kstep},...]
```

calculates a nested list; the list associated with *i* is outermost. If `istep` is omitted, the stepsize is one.

EXAMPLE 5.1.2: Define **C** to be the 3×4 matrix (c_{ij}), where c_{ij}, the entry in the *i*th row and *j*th column of **C**, is the numerical value of $\cos\left(j^2 - i^2\right)\sin\left(i^2 - j^2\right)$.

SOLUTION: After clearing all prior definitions of c, if any, we define `c[i,j]` to be the numerical value of $\cos\left(j^2 - i^2\right)\sin\left(i^2 - j^2\right)$ and then use `Array` to compute the 3×4 matrix `matrixc`.

$In[933] := $ **`Clear[c,matrixc]`**

`c[i_,j_] = N[ Cos [j² - i²] Sin [i² - j²]]`

```
General :: spell :
  Possible spelling error : new symbol name \"
    matrixc\" is similar to existing symbols
    {matrixa, matrixb}.
```

$Out[933] = $ `Cos[i² - 1. j²] Sin[i² - 1. j²]`

$In[934] := $ **`matrixc = Array[c, {3, 4}]`**

$Out[934] = $ `{{0., 0.139708, 0.143952, 0.494016},`
            `{-0.139708, 0., 0.272011, 0.452789},`
            `{-0.143952, -0.272011, 0., -0.495304}}`

In[935]:= **MatrixForm[matrixc]**

$$Out[935]= \begin{pmatrix} 0. & 0.139708 & 0.143952 & 0.494016 \\ -0.139708 & 0. & 0.272011 & 0.452789 \\ -0.143952 & -0.272011 & 0. & -0.495304 \end{pmatrix}$$

∎

EXAMPLE 5.1.3: Define the matrix $I_3 = \begin{pmatrix} 1 & 0 & 0 \\ 0 & 1 & 0 \\ 0 & 0 & 1 \end{pmatrix}$.

SOLUTION: The matrix I_3 is the 3×3 **identity matrix**. Generally, the $n \times n$ matrix with 1's on the diagonal and 0's elsewhere is the $n \times n$ identity matrix. The command `IdentityMatrix[n]` returns the $n \times n$ identity matrix.

In[936]:= **IdentityMatrix[3]**

Out[936]= {{1, 0, 0}, {0, 1, 0}, {0, 0, 1}}

The same result is obtained by going to **Input** under the Mathematica menu and selecting **Create Table/Matrix/Palette...** We then check **Matrix, Fill with**: 0 and **Fill diagonal**: 1.

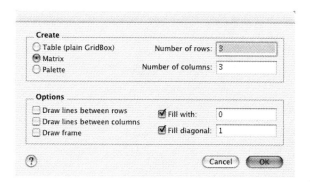

Pressing the **OK** button inserts the 3×3 identity matrix at the location of the cursor.

In[937]:= $\begin{pmatrix} 1 & 0 & 0 \\ 0 & 1 & 0 \\ 0 & 0 & 1 \end{pmatrix}$

Out[937]= {{1, 0, 0}, {0, 1, 0}, {0, 0, 1}}

∎

In Mathematica, a **vector** is a list of numbers and, thus, is entered in the same manner as lists. For example, to use Mathematica to define the row vector vectorv to be $\begin{pmatrix} v_1 & v_2 & v_3 \end{pmatrix}$ enter vectorv={v1,v2,v3}. Similarly, to define the column vector vectorv to be $\begin{pmatrix} v_1 \\ v_2 \\ v_3 \end{pmatrix}$ enter vectorv={v1,v2,v3} or vectorv={{v1}, {v2},{v3}}.

For a 2×1 vector, you can use the ⊞ button on the **Basic Typesetting** palette. Generally, with Mathematica you do not need to distinguish between row and column vectors: Mathematica performs computations with vectors and matrices correctly as long as the computations are well-defined.

With Mathematica, you do not need to distinguish between row and column vectors. Provided that computations are well-defined, Mathematica carries them out correctly. Mathematica warns of any ambiguities when they (rarely) occur.

EXAMPLE 5.1.4: Define the vector $\mathbf{w} = \begin{pmatrix} -4 \\ -5 \\ 2 \end{pmatrix}$, vectorv to be the vector $\begin{pmatrix} v_1 & v_2 & v_3 & v_4 \end{pmatrix}$ and zerovec to be the vector $\begin{pmatrix} 0 & 0 & 0 & 0 & 0 \end{pmatrix}$.

SOLUTION: To define **w**, we enter

```
In[938]:= w = {-4, -5, 2}
Out[938]= {-4, -5, 2}
```

or

```
In[939]:= w = {{-4}, {-5}, {2}};

            MatrixForm[w]
```
$$Out[939]= \begin{pmatrix} -4 \\ -5 \\ 2 \end{pmatrix}$$

To define vectorv, we use Array.

```
In[940]:= vectorv = Array[v, 4]
Out[940]= {v[1], v[2], v[3], v[4]}
```

Equivalent results would have been obtained by entering Table[v_i, {i, 1, 4}]. To define zerovec, we use Table.

```
In[941]:= zerovec = Table[0, {5}]
Out[941]= {0, 0, 0, 0, 0}
```

The same result is obtained by going to **Input** under the Mathematica menu and selecting **Create Table/Matrix/Palette...**

```
In[942]:= (0   0   0   0   0)

Out[942]= {{0, 0, 0, 0, 0}}
```

■

5.1.2 Extracting Elements of Matrices

For the 2×2 matrix m = {{$a_{1,1}$, $a_{1,2}$}, {$a_{2,1}$, $a_{2,2}$}} defined earlier, m[[1]] yields the first element of matrix m which is the list {$a_{1,1}$, $a_{1,2}$} or the first row of m; m[[2,1]] yields the first element of the second element of matrix m which is $a_{2,1}$. In general, if **m** is an $i \times j$ matrix, m[[i,j]] or Part[m,i,j] returns the unique element in the ith row and jth column of **m**. More specifically, m[[i,j]] yields the jth part of the ith part of **m**; list[[i]] or Part[list,i] yields the ith part of list; list[[i,j]] or Part[list,i,j] yields the jth part of the ith part of list, and so on.

EXAMPLE 5.1.5: Define mb to be the matrix $\begin{pmatrix} 10 & -6 & -9 \\ 6 & -5 & -7 \\ -10 & 9 & 12 \end{pmatrix}$.

(a) Extract the third row of mb. (b) Extract the element in the first row and third column of mb. (c) Display mb in traditional matrix form.

SOLUTION: We begin by defining mb. mb[[i,j]] yields the (unique) number in the ith row and jth column of mb. Observe how various components of mb (rows and elements) can be extracted and how mb is placed in MatrixForm.

```
In[943]:= mb = {{10, -6, -9}, {6, -5, -7},
                {-10, 9, 12}};

In[944]:= MatrixForm[mb]
                    10    -6    -9
Out[944]=            6    -5    -7
                   -10     9    12

In[945]:= mb[[3]]
Out[945]= {-10, 9, 12}

In[946]:= mb[[1,3]]
Out[946]= -9
```

■

If m is a matrix, the *i*th row of m is extracted with m[[i]]. The command Transpose[m] yields the transpose of the matrix m, the matrix obtained by interchanging the rows and columns of m. We extract columns of m by computing Transpose[m] and then using Part to extract rows from the transpose. Namely, if m is a matrix, Transpose[m][[i]] extracts the *i*th row from the transpose of m which is the same as the *i*th column of m.

EXAMPLE 5.1.6: Extract the second and third columns from **A** if **A** =
$\begin{pmatrix} 0 & -2 & 2 \\ -1 & 1 & -3 \\ 2 & -4 & 1 \end{pmatrix}$.

SOLUTION: We first define matrixa and then use Transpose to compute the transpose of matrixa, naming the result ta, and then displaying ta in MatrixForm.

```
In[947]:= matrixa = {{0, -2, 2}, {-1, 1, -3},
             {2, -4, 1}};
```

```
In[948]:= ta = Transpose[matrixa];
```

```
          MatrixForm[ta]
```

$Out[948]= \begin{pmatrix} 0 & -1 & 2 \\ -2 & 1 & -4 \\ 2 & -3 & 1 \end{pmatrix}$

Next, we extract the second column of matrixa using Transpose together with Part ([[...]]). Because we have already defined ta to be the transpose of matrixa, entering ta[[2]] would produce the same result.

```
In[949]:= Transpose[matrixa][[2]]
```

$Out[949]= \{-2, 1, -4\}$

To extract the third column, we take advantage of the fact that we have already defined ta to be the transpose of matrixa. Entering Transpose[matrixa][[3]] produces the same result.

```
In[950]:= ta[[3]]
Out[950]= {2,-3,1}
```

■

Other commands that can be used to manipulate matrices are included in the **Matrix Manipulation** package that is contained in the **Linear Algebra** folder (or directory).

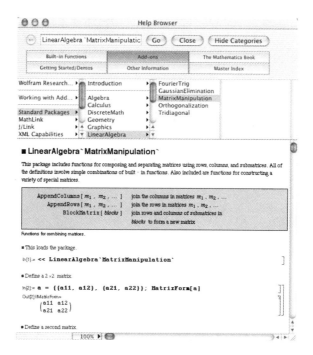

After this package has been loaded,

```
In[951]:= << LinearAlgebra`MatrixManipulation`
```

we can use commands like TakeColumns and TakeRows to extract columns and rows from a given matrix. For example, entering

```
In[952]:= TakeColumns[matrixa,{2}]//MatrixForm
Out[952]= ⎛-2⎞
         ⎜ 1⎟
         ⎝-4⎠
```

extracts the second column of the matrix **A** defined in the previous example and displays the result in MatrixForm while entering

$In[953]:=$ **TakeColumns[matrixa, {2,3}]//**
            MatrixForm

$Out[953]=$ $\begin{pmatrix} -2 & 2 \\ 1 & -3 \\ -4 & 1 \end{pmatrix}$

extracts the second and third columns of **A** and displays the result in `MatrixForm`.

5.1.3 Basic Computations with Matrices

Mathematica performs all of the usual operations on matrices. Matrix addition (**A** + **B**), scalar multiplication (k**A**), matrix multiplication (when defined) (**AB**), and combinations of these operations are all possible. The **transpose** of **A**, **A'**, is obtained by interchanging the rows and columns of **A** and is computed with the command `Transpose[A]`. If **A** is a square matrix, the determinant of **A** is obtained with `Det[A]`.

If **A** and **B** are $n \times n$ matrices satisfying **AB** = **BA** = **I**, where **I** is the $n \times n$ matrix with 1's on the diagonal and 0's elsewhere (the $n \times n$ identity matrix), **B** is called the **inverse** of **A** and is denoted by **A**$^{-1}$. If the inverse of a matrix **A** exists, the inverse is found with `Inverse[A]`. Thus, assuming that $\begin{pmatrix} a & b \\ c & d \end{pmatrix}$ has an inverse ($ad - bc \neq 0$), the inverse is

$In[954]:=$ **Inverse[{{a,b},{c,d}}]**

$Out[954]=$ $\left\{\left\{\dfrac{d}{-b\,c+a\,d}, -\dfrac{b}{-b\,c+a\,d}\right\}, \left\{-\dfrac{c}{-b\,c+a\,d}, \dfrac{a}{-b\,c+a\,d}\right\}\right\}$

EXAMPLE 5.1.7: Let **A** $= \begin{pmatrix} 3 & -4 & 5 \\ 8 & 0 & -3 \\ 5 & 2 & 1 \end{pmatrix}$ and **B** $= \begin{pmatrix} 10 & -6 & -9 \\ 6 & -5 & -7 \\ -10 & 9 & 12 \end{pmatrix}$.

Compute
    (a) **A** + **B**; (b) **B** − 4**A**; (c) the inverse of **AB**; (d) the transpose of (**A** − 2**B**)**B**; and (e) det **A** = |**A**|.

SOLUTION: We enter ma (corresponding to **A**) and mb (corresponding to **B**) as nested lists where each element corresponds to a row of the matrix. We suppress the output by ending each command with a semicolon.

$$In\,[955]\,:=\,\mathbf{ma\,=\,\{\{3\,,\,-4\,,\,5\}\,,\,\{8\,,\,0\,,\,-3\}\,,\,\{5\,,\,2\,,\,1\}\}\,;}$$

$$\mathbf{mb\,=\,\{\{10\,,\,-6\,,\,-9\}\,,\,\{6\,,\,-5\,,\,-7\}\,,}$$
$$\mathbf{\{-10\,,\,9\,,\,12\}\}\,;}$$

Entering

$$In\,[956]\,:=\,\mathbf{ma\,+\,mb//MatrixForm}$$

$$Out\,[956]=\begin{pmatrix} 13 & -10 & -4 \\ 14 & -5 & -10 \\ -5 & 11 & 13 \end{pmatrix}$$

adds matrix ma to mb and expresses the result in traditional matrix form. Entering

$$In\,[957]\,:=\,\mathbf{mb\,-\,4ma//MatrixForm}$$

$$Out\,[957]=\begin{pmatrix} -2 & 10 & -29 \\ -26 & -5 & 5 \\ -30 & 1 & 8 \end{pmatrix}$$

subtracts four times matrix ma from mb and expresses the result in traditional matrix form. Entering

$$In\,[958]\,:=\,\mathbf{Inverse[ma.mb]//MatrixForm}$$

$$Out\,[958]=\begin{pmatrix} \dfrac{59}{380} & \dfrac{53}{190} & -\dfrac{167}{380} \\ -\dfrac{223}{570} & -\dfrac{92}{95} & \dfrac{979}{570} \\ \dfrac{49}{114} & \dfrac{18}{19} & -\dfrac{187}{114} \end{pmatrix}$$

Matrix products, when defined, are computed by placing a period (.) between the matrices being multiplied. Note that a period is also used to compute the dot product of two vectors, when the dot product is defined.

computes the inverse of the matrix product **AB**. Similarly, entering

$$In\,[959]\,:=\,\mathbf{Transpose[\,(ma\,-\,2mb)\,.mb]//MatrixForm}$$

$$Out\,[959]=\begin{pmatrix} -352 & -90 & 384 \\ 269 & 73 & -277 \\ 373 & 98 & -389 \end{pmatrix}$$

computes the transpose of $(\mathbf{A} - 2\mathbf{B})\,\mathbf{B}$ and entering

$$In\,[960]\,:=\,\mathbf{Det[ma]}$$

$$Out\,[960]=\,190$$

computes the determinant of ma.

∎

EXAMPLE 5.1.8: Compute $\mathbf{AB}$ and $\mathbf{BA}$ if $\mathbf{A} = \begin{pmatrix} -1 & -5 & -5 & -4 \\ -3 & 5 & 3 & -2 \\ -4 & 4 & 2 & -3 \end{pmatrix}$

and $\mathbf{B} = \begin{pmatrix} 1 & -2 \\ -4 & 3 \\ 4 & -4 \\ -5 & -3 \end{pmatrix}$.

SOLUTION: Because $\mathbf{A}$ is a 3×4 matrix and $\mathbf{B}$ is a 4×2 matrix, $\mathbf{AB}$ is defined and is a 3×2 matrix. We define `matrixa` and `matrixb` with the following commands.

Remember that you can also define matrices by going to **Input** under the Mathematica menu and selecting **Create Table/Matrix/Palette....** After entering the desired number of rows and columns and pressing the **OK** button, a matrix template is placed at the location of the cursor that you can fill in.

$In[961] := $ **matrixa** $= \begin{pmatrix} -1 & -5 & -5 & -4 \\ -3 & 5 & 3 & -2 \\ -4 & 4 & 2 & -3 \end{pmatrix}$;

$In[962] := $ **matrixb** $= \begin{pmatrix} 1 & -2 \\ -4 & 3 \\ 4 & -4 \\ -5 & -3 \end{pmatrix}$;

We then compute the product, naming the result ab, and display ab in MatrixForm.

$In[963] := $ **ab = matrixa.matrixb;**

MatrixForm[ab]

$Out[963] = \begin{pmatrix} 19 & 19 \\ -1 & 15 \\ 3 & 21 \end{pmatrix}$

However, the matrix product $\mathbf{BA}$ is not defined and Mathematica produces error messages when we attempt to compute it.

$In[964] := $ **matrixb.matrixa**

Dot :: dotsh :
  Tensors {{1, -2}, {-4, 3}, {4, -4}, {-5, -3}}
    and {{-1, -5, -5, -4}, {-3, 5, 3, -2},
    {-4, 4, 2, -3}} have incompatible shapes .
  $Out[964] = $ {{1, -2}, {-4, 3}, {4, -4}, {-5, -3}} .
              {{-1, -5, -5, -4},
              {-3, 5, 3, -2}, {-4, 4, 2, -3}}

■

Special attention must be given to the notation that must be used in taking the product of a square matrix with itself. The following example illustrates how Mathematica interprets the expression (matrixb)^n. The command (matrixb)^n raises each element of the matrix matrixb to the nth power. The command MatrixPower is used to compute powers of matrices.

EXAMPLE 5.1.9: Let $\mathbf{B} = \begin{pmatrix} -2 & 3 & 4 & 0 \\ -2 & 0 & 1 & 3 \\ -1 & 4 & -6 & 5 \\ 4 & 8 & 11 & -4 \end{pmatrix}$. (a) Compute $\mathbf{B}^2$ and $\mathbf{B}^3$.

(b) Cube each entry of $\mathbf{B}$.

SOLUTION: After defining $\mathbf{B}$, we compute $\mathbf{B}^2$. The same results would have been obtained by entering MatrixPower[matrixb,2].

In[965]:= **matrixb = {{-2, 3, 4, 0}, {-2, 0, 1, 3},**
 **   {-1, 4, -6, 5}, {4, 8, 11, -4}};**

In[966]:= **MatrixForm[matrixb.matrixb]**

$$Out[966]= \begin{pmatrix} -6 & 10 & -29 & 29 \\ 15 & 22 & 19 & -7 \\ 20 & 13 & 91 & -38 \\ -51 & 24 & -86 & 95 \end{pmatrix}$$

Next, we use MatrixPower to compute $\mathbf{B}^3$. The same results would be obtained by entering matrixb.matrixb.matrixb.

In[967]:= **MatrixForm[MatrixPower[matrixb, 3]]**

$$Out[967]= \begin{pmatrix} 137 & 98 & 479 & -231 \\ -121 & 65 & -109 & 189 \\ -309 & 120 & -871 & 646 \\ 520 & 263 & 1381 & -738 \end{pmatrix}$$

Last, we cube each entry of $\mathbf{B}$ with ^.

In[968]:= **MatrixForm[matrixb³]**

$$Out[968]= \begin{pmatrix} -8 & 27 & 64 & 0 \\ -8 & 0 & 1 & 27 \\ -1 & 64 & -216 & 125 \\ 64 & 512 & 1331 & -64 \end{pmatrix}$$

∎

If $|\mathbf{A}| \neq 0$, the inverse of $\mathbf{A}$ can be computed using the formula

$$\mathbf{A}^{-1} = \frac{1}{|\mathbf{A}|}\mathbf{A}^a, \tag{5.1}$$

where $\mathbf{A}^a$ is the *transpose of the cofactor matrix.*

If $\mathbf{A}$ has an inverse, reducing the matrix $(\mathbf{A}|\mathbf{I})$ to reduced row echelon form results in $(\mathbf{I}|\mathbf{A}^{-1})$. This method is often easier to implement than (5.1).

The **cofactor matrix**, $\mathbf{A}^c$, of $\mathbf{A}$ is the matrix obtained by replacing each element of $\mathbf{A}$ by its cofactor.

EXAMPLE 5.1.10: Calculate $\mathbf{A}^{-1}$ if $\mathbf{A} = \begin{pmatrix} 2 & -2 & 1 \\ 0 & -2 & 2 \\ -2 & -1 & -1 \end{pmatrix}$.

SOLUTION: After defining $\mathbf{A}$ and $\mathbf{I} = \begin{pmatrix} 1 & 0 & 0 \\ 0 & 1 & 0 \\ 0 & 0 & 1 \end{pmatrix}$, we compute $|\mathbf{A}| = 12$, so $\mathbf{A}^{-1}$ exists.

```
In[969]:= << LinearAlgebra`MatrixManipulation`;
          capa = {{2, -2, 1}, {0, -2, 2}, {-2, -1, -1}};
          i3 = {{1, 0, 0}, {0, 1, 0}, {0, 0, 1}};

In[970]:= Det[capa]
Out[970]= 12
```

We use `AppendRows` to form the matrix $(\mathbf{A}|\mathbf{I})$

`AppendRows` is contained in the **MatrixManipulation** package that is located in the **LinearAlgebra** folder (or directory).

```
In[971]:= ai3 = AppendRows[capa, i3];
          MatrixForm[ai3]
```

$$Out[971]= \begin{pmatrix} 2 & -2 & 1 & 1 & 0 & 0 \\ 0 & -2 & 2 & 0 & 1 & 0 \\ -2 & -1 & -1 & 0 & 0 & 1 \end{pmatrix}$$

and then use `RowReduce` to reduce $(\mathbf{A}|\mathbf{I})$ to row echelon form.

`RowReduce[A]` reduces $\mathbf{A}$ to **reduced row echelon form**.

```
In[972]:= RowReduce[ai3]
```

$$Out[972]= \left\{\left\{1, 0, 0, \frac{1}{3}, -\frac{1}{4}, -\frac{1}{6}\right\}, \left\{0, 1, 0, -\frac{1}{3}, 0, -\frac{1}{3}\right\}, \left\{0, 0, 1, -\frac{1}{3}, \frac{1}{2}, -\frac{1}{3}\right\}\right\}$$

The result indicates that $\mathbf{A}^{-1} = \begin{pmatrix} 1/3 & -1/4 & -1/6 \\ -1/3 & 0 & -1/3 \\ -1/3 & 1/2 & -1/3 \end{pmatrix}$. We check this result with `Inverse`.

$In[973] :=$ **Inverse[capa]**

$Out[973] = \left\{\left\{\frac{1}{3}, -\frac{1}{4}, -\frac{1}{6}\right\}, \left\{-\frac{1}{3}, 0, -\frac{1}{3}\right\}, \left\{-\frac{1}{3}, \frac{1}{2}, -\frac{1}{3}\right\}\right\}$

■

5.1.4 Basic Computations with Vectors

5.1.4.1 Basic Operations on Vectors

Computations with vectors are performed in the same way as computations with matrices.

EXAMPLE 5.1.11: Let $\mathbf{v} = \begin{pmatrix} 0 \\ 5 \\ 1 \\ 2 \end{pmatrix}$ and $\mathbf{w} = \begin{pmatrix} 3 \\ 0 \\ 4 \\ -2 \end{pmatrix}$. (a) Calculate $\mathbf{v} - 2\mathbf{w}$ and

$\mathbf{v} \cdot \mathbf{w}$. (b) Find a unit vector with the same direction as $\mathbf{v}$ and a unit vector with the same direction as $\mathbf{w}$.

SOLUTION: We begin by defining $\mathbf{v}$ and $\mathbf{w}$ and then compute $\mathbf{v} - 2\mathbf{w}$ and $\mathbf{v} \cdot \mathbf{w}$.

$In[974] :=$ **v = {0, 5, 1, 2};**

                w = {3, 0, 4, -2};

$In[975] :=$ **v - 2w**

$Out[975] =$ **{-6, 5, -7, 6}**

$In[976] :=$ **v.w**

$Out[976] =$ **0**

The **norm** of the vector $\mathbf{v} = \begin{pmatrix} v_1 \\ v_2 \\ \vdots \\ v_n \end{pmatrix}$ is

$$\|\mathbf{v}\| = \sqrt{v_1^2 + v_2^2 + \cdots + v_n^2} = \sqrt{\mathbf{v} \cdot \mathbf{v}}.$$

If k is a scalar, the direction of $k\mathbf{v}$ is the same as the direction of $\mathbf{v}$. Thus, if $\mathbf{v}$ is a nonzero vector, the vector $\dfrac{1}{\|\mathbf{v}\|}\mathbf{v}$ has the same direction as $\mathbf{v}$ and because $\left\|\dfrac{1}{\|\mathbf{v}\|}\mathbf{v}\right\| = \dfrac{1}{\|\mathbf{v}\|}\|\mathbf{v}\| = 1$, $\dfrac{1}{\|\mathbf{v}\|}\mathbf{v}$ is a unit vector. We define the function $\texttt{norm}$ which, given a vector $\mathbf{v}$, computes $\|\mathbf{v}\|$. We then compute $\dfrac{1}{\|\mathbf{v}\|}\mathbf{v}$, calling the result $\texttt{uv}$, and $\dfrac{1}{\|\mathbf{w}\|}\mathbf{w}$. The results correspond to unit vectors with the same direction as $\mathbf{v}$ and $\mathbf{w}$, respectively.

```
In[977]:= norm[v_] := √v.v
```

```
In[978]:= uv = v / norm[v]
```

$$Out[978]= \left\{0, \sqrt{\frac{5}{6}}, \frac{1}{\sqrt{30}}, \sqrt{\frac{2}{15}}\right\}$$

```
In[979]:= norm[uv]
```

```
Out[979]= 1
```

```
In[980]:= w / norm[w]
```

$$Out[980]= \left\{\frac{3}{\sqrt{29}}, 0, \frac{4}{\sqrt{29}}, -\frac{2}{\sqrt{29}}\right\}$$

■

5.1.4.2 Basic Operations on Vectors in 3-Space

We review the elementary properties of vectors in 3-space. Let

$$\mathbf{u} = \langle u_1, u_2, u_3 \rangle = u_1\mathbf{i} + u_2\mathbf{j} + u_3\mathbf{k}$$

and

$$\mathbf{v} = \langle v_1, v_2, v_3 \rangle = v_1\mathbf{i} + v_2\mathbf{j} + v_3\mathbf{k}$$

be vectors in space.

1. $\mathbf{u}$ and $\mathbf{v}$ are **equal** if and only if their components are equal:

$$\mathbf{u} = \mathbf{v} \Longleftrightarrow u_1 = v_1, u_2 = v_2, \text{ and } u_3 = v_3.$$

2. The **length** (or **norm**) of $\mathbf{u}$ is

$$\|\mathbf{u}\| = \sqrt{u_1{}^2 + u_2{}^2 + u_3{}^2}.$$

3. If c is a scalar (number),

$$c\mathbf{u} = \langle cu_1, cu_2, cu_3 \rangle.$$

Vector calculus is discussed in Section 5.5.

In space, the **standard unit vectors** are $\mathbf{i} = \langle 1, 0, 0 \rangle$, $\mathbf{j} = \langle 0, 1, 0 \rangle$, and $\mathbf{k} = \langle 0, 0, 1 \rangle$. With the exception of the cross product, the vector operations discussed here are performed in the same way for vectors in the plane as they are in space. In the plane, the **standard unit vectors** are $\mathbf{i} = \langle 1, 0 \rangle$ and $\mathbf{j} = \langle 0, 1 \rangle$.

4. The **sum** of **u** and **v** is defined to be the vector

$$\mathbf{u} + \mathbf{v} = \langle u_1 + v_1, u_2 + v_2, u_3 + v_3 \rangle.$$

5. If $\mathbf{u} \neq \mathbf{0}$, a unit vector with the same direction as **u** is

A **unit vector** is a vector
with length 1.

$$\frac{1}{\|\mathbf{u}\|}\mathbf{u} = \frac{1}{\sqrt{u_1{}^2 + u_2{}^2 + u_3{}^2}} \langle u_1, u_2, u_3 \rangle.$$

6. **u** and **v** are **parallel** if there is a scalar c so that $\mathbf{u} = c\mathbf{v}$.
7. The **dot product** of **u** and **v** is

$$\mathbf{u} \cdot \mathbf{v} = u_1 v_1 + u_2 v_2 + u_3 v_3.$$

If θ is the angle between **u** and **v**,

$$\cos\theta = \frac{\mathbf{u} \cdot \mathbf{v}}{\|\mathbf{u}\| \, \|\mathbf{v}\|}.$$

Consequently, **u** and **v** are orthogonal if $\mathbf{u} \cdot \mathbf{v} = 0$.
8. The **cross product** of **u** and **v** is

$$\mathbf{u} \times \mathbf{v} = \begin{vmatrix} \mathbf{i} & \mathbf{j} & \mathbf{k} \\ u_1 & u_2 & u_3 \\ v_1 & v_2 & v_3 \end{vmatrix}$$

$$= (u_2 v_3 - u_3 v_2)\,\mathbf{i} - (u_1 v_3 - u_3 v_1)\,\mathbf{j} + (u_1 v_2 - u_2 v_1)\,\mathbf{k}.$$

You should verify that $\mathbf{u} \cdot (\mathbf{u} \times \mathbf{v}) = 0$ and $\mathbf{v} \cdot (\mathbf{u} \times \mathbf{v}) = 0$. Hence, $\mathbf{u} \times \mathbf{v}$ is orthogonal to both **u** and **v**.

Topics from linear algebra (including determinants) are discussed in more detail in the next sections. For now, we illustrate several of the basic operations listed above. In Mathematica, many vector calculations take advantage of functions contained in the **VectorAnalysis** package located in the **Calculus** directory. Use Mathematica's help facility to obtain general help regarding the **VectorAnalysis** package.

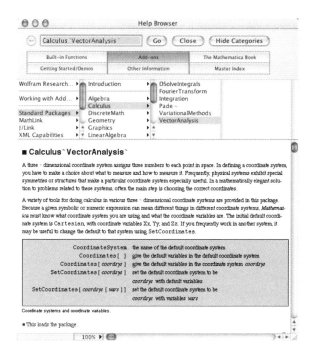

EXAMPLE 5.1.12: Let $\mathbf{u} = \langle 3, 4, 1 \rangle$ and $\mathbf{v} = \langle -4, 3, -2 \rangle$. Calculate (a) $\mathbf{u} \cdot \mathbf{v}$, (b) $\mathbf{u} \times \mathbf{v}$, (c) $\|\mathbf{u}\|$, and (d) $\|\mathbf{v}\|$. (e) Find the angle between $\mathbf{u}$ and $\mathbf{v}$. (f) Find unit vectors with the same direction as $\mathbf{u}$, $\mathbf{v}$, and $\mathbf{u} \times \mathbf{v}$.

SOLUTION: After loading the `VectorAnalysis` package, we define $\mathbf{u} = \langle 3, 4, 1 \rangle$ and $\mathbf{v} = \langle -4, 3, -2 \rangle$. Notice that to define $\mathbf{u} = \langle u_1, u_2, u_3 \rangle$ with Mathematica, we use the form

$$u=\{u1,u2,u3\}.$$

We illustrate the use of `DotProduct` and `CrossProduct`, both of which are contained in the `VectorAnalysis` package, to calculate (a)–(d).

Similarly, to define $\mathbf{u} = \langle u_1, u_2 \rangle$, we use the form u={u1,u2}.

Remark. Generally, `u.v` returns the same result as `DotProduct[u,b]`.

```
In[981]:= << Calculus`VectorAnalysis`

In[982]:= u = {3, 4, 1};
          v = {-4, 3, -2};
```

In[983]:= **udv = DotProduct[u, v]**

Out[983]= -2

In[984]:= **ucv = CrossProduct[u, v]**

Out[984]= {-11, 2, 25}

In[985]:= **v = Sqrt[u.u]**

Out[985]= $\sqrt{26}$

In[986]:= **nv = Sqrt[v.v]**

Out[986]= $\sqrt{29}$

We use the formula $\theta = \cos^{-1}\left(\frac{u \cdot v}{\|u\| \|v\|}\right)$ to find the angle θ between u and v.

In[987]:= **ArcCos[u.v/(v nv)]**

   N[%]

Out[987]= $\text{ArcCos}\left[-\sqrt{\dfrac{2}{377}}\right]$

Out[987]= 1.6437

Unit vectors with the same direction as **u**, **v**, and **u** × **v** are found next.

In[988]:= **normu = u/v**

   normv = v/nv

   nucrossv = ucv/Sqrt[ucv.ucv]

Out[988]= $\left\{\dfrac{3}{\sqrt{26}}, 2\sqrt{\dfrac{2}{13}}, \dfrac{1}{\sqrt{26}}\right\}$

Out[988]= $\left\{-\dfrac{4}{\sqrt{29}}, \dfrac{3}{\sqrt{29}}, -\dfrac{2}{\sqrt{29}}\right\}$

Out[988]= $\left\{-\dfrac{11}{5\sqrt{30}}, \dfrac{\sqrt{\frac{2}{15}}}{5}, \sqrt{\dfrac{5}{6}}\right\}$

We can graphically confirm that these three vectors are orthogonal by graphing all three vectors with the ListPlotVectorField3D function, which is contained in the PlotField3D package. After loading the PlotField3D package, the command

ListPlotVectorField3D[listofvectors]

graphs the list of vectors listofvectors. Each element of listof vectors is of the form {{u1,u2,u3},{v1,v2,v3}} where (u_1, u_2, u_3) and (v_1, v_2, v_3) are the initial and terminal points of each vector. We show the vectors in Figure 5-1.

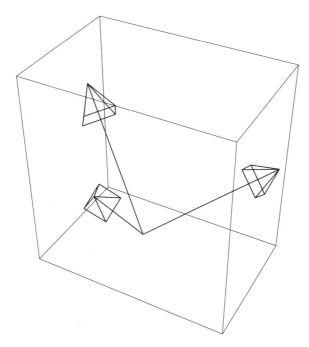

Figure 5-1   Orthogonal vectors

$In[989]:=$ **<< Graphics`PlotField3D`**

$In[990]:=$ **ListPlotVectorField3D[{{{0, 0, 0}, normu},**
                {{0, 0, 0}, normv}, {{0, 0, 0}, nucrossv}},
                VectorHeads- > True]

In the plot, the vectors do appear to be orthogonal as expected.
■

With the exception of the cross product, the calculations described above can also be performed on vectors in the plane.

EXAMPLE 5.1.13: If **u** and **v** are nonzero vectors, the **projection** of **u** onto **v** is

$$\text{proj}_{\mathbf{v}} \mathbf{u} = \frac{\mathbf{u} \cdot \mathbf{v}}{\|\mathbf{v}\|^2} \mathbf{v}.$$

Find $\text{proj}_{\mathbf{v}} \mathbf{u}$ if $\mathbf{u} = \langle -1, 4 \rangle$ and $\mathbf{v} = \langle 2, 6 \rangle$.

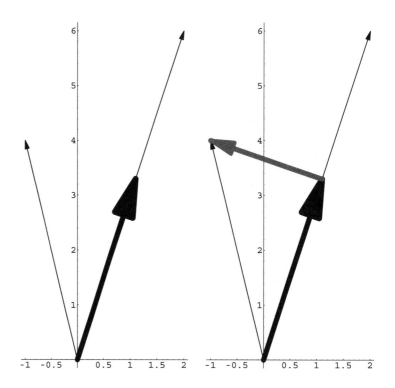

Figure 5-2   Projection of a vector

SOLUTION: We define $\mathbf{u} = \langle -1, 4 \rangle$ and $\mathbf{v} = \langle 2, 6 \rangle$ and then compute $\text{proj}_{\mathbf{v}}\,\mathbf{u}$.

```
In[991]:= u = {-1, 4};
          v = {2, 6};
          projvu = u.v v/v.v
```
$$Out[991] = \left\{ \frac{11}{10}, \frac{33}{10} \right\}$$

Finally, we graph $\mathbf{u}$, $\mathbf{v}$, and $\text{proj}_{\mathbf{v}}\,\mathbf{u}$ together using Arrow and Show in Figure 5-2.

```
In[992]:= << Graphics`Arrow`
```

```
In[993]:= ?Arrow
```

"Arrow[start, finish, (opts)]isagraphics
    primitiverepresentinganarrowstartingat
    startandendingatfinish."

```
In[994]:= p1 = Show[Graphics[
                {Arrow[{0, 0}, u], Arrow[{0, 0}, v],
                  Thickness[0.03], Arrow[{0, 0}, projvu,
                  HeadScaling- > Relative]}],
              Axes- > Automatic, AspectRatio- > Automatic,
              DisplayFunction- > Identity];

In[995]:= p2 = Show[Graphics[{Arrow[{0, 0}, u],
                Arrow[{0, 0}, v], Thickness[0.03],
                  Arrow[{0, 0}, projvu,
                  HeadScaling- > Relative],
                GrayLevel[0.4], Arrow[projvu, u,
                  HeadScaling- > Relative]}],
              Axes- > Automatic, AspectRatio- > Automatic,
              DisplayFunction- > Identity];

In[996]:= Show[GraphicsArray[{p1, p2}]]
```

In the graph, notice that $\mathbf{u} = \text{proj}_\mathbf{v}\mathbf{u} + (\mathbf{u} - \text{proj}_\mathbf{v}\mathbf{u})$ and the vector $\mathbf{u} - \text{proj}_\mathbf{v}\mathbf{u}$ is perpendicular to $\mathbf{v}$.

■

5.2 Linear Systems of Equations

5.2.1 Calculating Solutions of Linear Systems of Equations

To solve the system of linear equations $\mathbf{Ax} = \mathbf{b}$, where $\mathbf{A}$ is the coefficient matrix, $\mathbf{b}$ is the known vector and $\mathbf{x}$ is the unknown vector, we often proceed as follows: if $\mathbf{A}^{-1}$ exists, then $\mathbf{AA}^{-1}\mathbf{x} = \mathbf{A}^{-1}\mathbf{b}$ so $\mathbf{x} = \mathbf{A}^{-1}\mathbf{b}$.

EXAMPLE 5.2.1: Solve the matrix equation $\begin{pmatrix} 3 & 0 & 2 \\ -3 & 2 & 2 \\ 2 & -3 & 3 \end{pmatrix}\begin{pmatrix} x \\ y \\ z \end{pmatrix} = \begin{pmatrix} 3 \\ -1 \\ 4 \end{pmatrix}$.

SOLUTION: The solution is given by $\begin{pmatrix} x \\ y \\ z \end{pmatrix} = \begin{pmatrix} 3 & 0 & 2 \\ -3 & 2 & 2 \\ 2 & -3 & 3 \end{pmatrix}^{-1}\begin{pmatrix} 3 \\ -1 \\ 4 \end{pmatrix}$. We proceed by defining `matrixa` and `b` and then using `Inverse` to calculate `Inverse[matrixa].b` naming the resulting output $\{x, y, z\}$.

```
In[997]:= matrixa = {{3, 0, 2}, {-3, 2, 2},
               {2, -3, 3}};

          b = {3, -1, 4};
In[998]:= {x, y, z} = Inverse[matrixa].b
Out[998]= {13/23, -7/23, 15/23}
```

We verify that the result is the desired solution by calculating `matrixa.{x,y,z}`. Because the result of this procedure is $\begin{pmatrix} 3 \\ -1 \\ 4 \end{pmatrix}$, we conclude that the solution to the system is $\begin{pmatrix} x \\ y \\ z \end{pmatrix} = \begin{pmatrix} 13/23 \\ -7/23 \\ 15/23 \end{pmatrix}$.

```
In[999]:= matrixa.{x, y, z}
Out[999]= {3, -1, 4}
```

We note that this matrix equation is equivalent to the system of equations

$$3x + 2z = 3$$
$$-3x + 2y + 2z = -1,$$
$$2x - 3y + 3z = 4$$

which we are able to solve with `Solve`.

```
In[1000]:= Clear[x, y, z]

           sys =
             Thread[matrixa.{x, y, z} == {3, -1, 4}]
Out[1000]= {3 x + 2 z == 3,
             -3 x + 2 y + 2 z == -1, 2 x - 3 y + 3 z == 4}

In[1001]:= Solve[sys]
Out[1001]= {{x → 13/23, z → 15/23, y → -7/23}}
```

∎

Mathematica offers several commands for solving systems of linear equations, however, that do not depend on the computation of the inverse of **A**. The command

```
Solve[{eqn1, eqn2, ..., eqnm}, {var1, var2, ..., varn}]
```

solves an $m \times n$ system of linear equations (m equations and n unknown variables). Note that both the equations as well as the variables are entered as lists. If one wishes to solve for all variables that appear in a system, the command `Solve[{eqn1,eqn2,...eqnn}]` attempts to solve `eqn1, eqn2, ..., eqnn` for all variables that appear in them. (Remember that a double equals sign (`==`) must be placed between the left and right-hand sides of each equation.)

EXAMPLE 5.2.2: Solve the system $\begin{cases} x - 2y + z = -4 \\ 3x + 2y - z = 8 \\ -x + 3y + 5z = 0 \end{cases}$ for x, y, and z.

SOLUTION: In this case, entering either

$$\texttt{Solve[\{x-2y+z==-4,3x+2y-z==8,-x+3y+5z==0\}]}$$

or

$$\texttt{Solve[\{x-2y+z,3x+2y-z,-x+3y+5z\}==\{-4,8,0\}]}$$

gives the same result.

```
In[1002]:= Solve[{x - 2 y + z == -4, 3 x + 2 y - z == 8,
              -x + 3 y + 5 z == 0}, {x, y, z}]
Out[1002]= {{x → 1, y → 2, z → -1}}
```

Another way to solve systems of equations is based on the matrix form of the system of equations, $\mathbf{Ax} = \mathbf{b}$. This system of equations is equivalent to the matrix equation

$$\begin{pmatrix} 1 & -2 & 1 \\ 3 & 2 & -1 \\ -1 & 3 & 5 \end{pmatrix} \begin{pmatrix} x \\ y \\ z \end{pmatrix} = \begin{pmatrix} -4 \\ 8 \\ 0 \end{pmatrix}.$$

The matrix of coefficients in the previous example is entered as `matrixa` along with the vector of right-hand side values `vectorb`. After defining the vector of variables, `vectorx`, the system $\mathbf{Ax} = \mathbf{b}$ is solved explicitly with the command `Solve`.

```
In[1003]:= matrixa = {{1, -2, 1}, {3, 2, -1},
              {-1, 3, 5}}; vectorb = {-4, 8, 0};
           vectorx = {x1, y1, z1};
```

In[1004]:= **Solve[matrixa.vectorx == vectorb,**
                vectorx]

Out[1004]= $\{\{x1 \to 1, y1 \to 2, z1 \to -1\}\}$

∎

In addition to using `Solve` to solve a system of linear equations, the command

LinearSolve [A,b]

calculates the solution vector **x** of the system $\mathbf{Ax} = \mathbf{b}$. `LinearSolve` generally solves a system more quickly than does `Solve` as we see from the comments in the **Help Browser**.

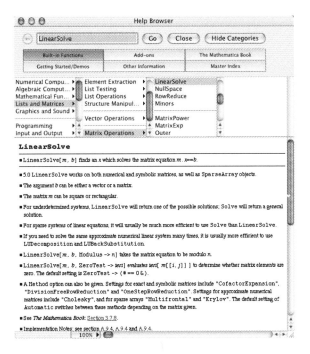

EXAMPLE 5.2.3: Solve the system $\begin{cases} 2x - 4y + z = -1 \\ 3x + y - 2z = 3 \\ -5x + y - 2z = 4 \end{cases}$. Verify that the

result returned satisfies the system.

SOLUTION: To solve the system using `Solve`, we define `eqs` to be the set of three equations to be solved and `vars` to be the variables x, y, and z and then use `Solve` to solve the set of equations `eqs` for the variables in `vars`. The resulting output is named `sols`.

$In[1005]:=$ **eqs = {2 x - 4 y + z == -1, 3 x + y - 2 z == 3,**
-5 x + y - 2 z == 4}; vars = {x, y, z};

sols = Solve[eqs, vars]

$Out[1005]=$ $\left\{\left\{x \rightarrow -\dfrac{1}{8}, y \rightarrow -\dfrac{15}{56}, z \rightarrow -\dfrac{51}{28}\right\}\right\}$

To verify that the result given in `sols` is the desired solution, we replace each occurrence of x, y, and z in `eqs` by the values found in `sols` using `ReplaceAll` (`/.`). Because the result indicates each of the three equations is satisfied, we conclude that the values given in `sols` are the components of the desired solution.

$In[1006]:=$ **eqs /. sols**

$Out[1006]=$ **{{True, True, True}}**

To solve the system using `LinearSolve`, we note that the system is equivalent to the matrix equation $\begin{pmatrix} 2 & -4 & 1 \\ 3 & 1 & -2 \\ -5 & 1 & -2 \end{pmatrix}\begin{pmatrix} x \\ y \\ z \end{pmatrix} = \begin{pmatrix} -1 \\ 3 \\ 4 \end{pmatrix}$, define `matrixa` and `vectorb`, and use `LinearSolve` to solve this matrix equation.

$In[1007]:=$ **matrixa = {{2, -4, 1}, {3, 1, -2},**
{-5, 1, -2}};

vectorb = {-1, 3, 4};

solvector =
LinearSolve[matrixa, vectorb]

$Out[1007]=$ $\left\{-\dfrac{1}{8}, -\dfrac{15}{56}, -\dfrac{51}{28}\right\}$

To verify that the results are correct, we compute `matrixa.solvector`. Because the result is $\begin{pmatrix} -1 \\ 3 \\ 4 \end{pmatrix}$, we conclude that the solution to the system is $\begin{pmatrix} x \\ y \\ z \end{pmatrix} = \begin{pmatrix} -1/8 \\ -15/36 \\ -51/28 \end{pmatrix}$.

In[1008]:= **matrixa.solvector**

Out[1008]= {-1, 3, 4}

■

EXAMPLE   5.2.4:   Solve   the   system   of   equations
$$\begin{cases} 4x_1 + 5x_2 - 5x_3 - 8x_4 - 2x_5 = 5 \\ 7x_1 + 2x_2 - 10x_3 - x_4 - 6x_5 = -4 \\ 6x_1 + 2x_2 + 10x_3 - 10x_4 + 7x_5 = -7. \\ -8x_1 - x_2 - 4x_3 + 3x_5 = 5 \\ 8x_1 - 7x_2 - 3x_3 + 10x_4 + 5x_5 = 7 \end{cases}$$

SOLUTION: We solve the system in two ways. First, we use `Solve` to solve the system. Note that in this case, we enter the equations in the form

```
set of left-hand sides==set of right-hand sides.
```

In[1009]:= **Solve[**
    {4 x[1] + 5 x[2] - 5 x[3] - 8 x[4] - 2 x[5],
        7 x[1] + 2 x[2] - 10 x[3] - x[4] - 6 x[5],
        6 x[1] + 2 x[2] + 10 x[3] - 10 x[4]+
            7 x[5], -8 x[1] - x[2] - 4 x[3] + 3 x[5],
        8 x[1] - 7 x[2] - 3 x[3] + 10 x[4]+
            5 x[5]} == {5, -4, -7, 5, 7}]

$$Out[1009]= \left\{\left\{x[1] \to \frac{1245}{6626},\right.\right.$$
$$x[2] \to \frac{113174}{9939}, x[3] \to -\frac{7457}{9939},$$
$$\left.\left. x[4] \to \frac{38523}{6626}, x[5] \to \frac{49327}{9939}\right\}\right\}$$

We also use `LinearSolve` after defining `matrixa` and `t2`. As expected, in each case, the results are the same.

$$In[1010] := \text{Clear[matrixa]}$$

$$\text{matrixa} = \{\{4, 5, -5, -8, -2\},$$
$$\{7, 2, -10, -1, -6\},$$
$$\{6, 2, 10, -10, 7\},$$
$$\{-8, -1, -4, 0, 3\},$$
$$\{8, -7, -3, 10, 5\}\};$$

$$\text{t2} = \{5, -4, -7, 5, 7\};$$

$$In[1011] := \text{LinearSolve[matrixa, t2]}$$

$$Out[1011] = \left\{ \frac{1245}{6626}, \frac{113174}{9939}, -\frac{7457}{9939}, \frac{38523}{6626}, \frac{49327}{9939} \right\}$$

■

5.2.2 Gauss–Jordan Elimination

Given the matrix equation $\mathbf{Ax} = \mathbf{b}$, where

$$\mathbf{A} = \begin{pmatrix} a_{11} & a_{12} & \cdots & a_{1n} \\ a_{21} & a_{22} & \cdots & a_{2n} \\ \vdots & \vdots & \ddots & \vdots \\ a_{m1} & a_{m2} & \cdots & a_{mn} \end{pmatrix}, \quad , \mathbf{x} = \begin{pmatrix} x_1 \\ x_2 \\ \vdots \\ x_n \end{pmatrix} \quad \text{and} \quad \mathbf{b} = \begin{pmatrix} b_1 \\ b_2 \\ \vdots \\ b_m \end{pmatrix},$$

the $m \times n$ matrix $\mathbf{A}$ is called the **coefficient matrix** for the matrix equation $\mathbf{Ax} = \mathbf{b}$ and the $m \times (n + 1)$ matrix

$$\begin{pmatrix} a_{11} & a_{12} & \cdots & a_{1n} & b_1 \\ a_{21} & a_{22} & \cdots & a_{2n} & b_2 \\ \vdots & \vdots & \ddots & \vdots & \vdots \\ a_{m1} & a_{m2} & \cdots & a_{mn} & b_m \end{pmatrix}$$

is called the **augmented** (or **associated**) **matrix** for the matrix equation. We may enter the augmented matrix associated with a linear system of equations directly or we can use commands contained in the package **MatrixManipulation** contained in the **Linear Algebra** folder (or directory) to help us construct the augmented matrix.

EXAMPLE 5.2.5: Solve the system $\begin{cases} -2x + y - 2x = 4 \\ 2x - 4y - 2z = -4 \\ x - 4y - 2z = 3 \end{cases}$ using Gauss–Jordan elimination.

SOLUTION: The system is equivalent to the matrix equation

$$\begin{pmatrix} -2 & 1 & -2 \\ 2 & -4 & -2 \\ 1 & -4 & -2 \end{pmatrix} \begin{pmatrix} x \\ y \\ z \end{pmatrix} = \begin{pmatrix} 4 \\ -4 \\ 3 \end{pmatrix}.$$

The augmented matrix associated with this system is

$$\begin{pmatrix} -2 & 1 & -2 & 4 \\ 2 & -4 & -2 & -4 \\ 1 & -4 & -2 & 3 \end{pmatrix}$$

which we construct using the command AppendRows contained in the **MatrixManipulation** package. We proceed by loading the **MatrixManipulation** package, defining matrixa and b, and then using AppendRows to construct the augmented matrix which we name augm and display in MatrixForm.

> *In[1012]:=* **<< LinearAlgebra`MatrixManipulation`**

> *In[1013]:=* **matrixa = {{-2, 1, -2}, {2, -4, -2},**
>   **{1, -4, -2}};**
>
>   **b = {{4}, {-4}, {3}};**

> *In[1014]:=* **augm = AppendRows[matrixa, b];**
>
>   **MatrixForm[augm]**

$$Out[1014] = \begin{pmatrix} -2 & 1 & -2 & 4 \\ 2 & -4 & -2 & -4 \\ 1 & -4 & -2 & 3 \end{pmatrix}$$

We calculate the solution by row-reducing augm using the built-in command RowReduce. Generally, RowReduce[A] reduces **A** to **reduced row echelon form**.

> *In[1015]:=* **RowReduce[augm]//MatrixForm**

$$Out[1015] = \begin{pmatrix} 1 & 0 & 0 & -7 \\ 0 & 1 & 0 & -4 \\ 0 & 0 & 1 & 3 \end{pmatrix}$$

From this result, we see that the solution is

$$\begin{pmatrix} x \\ y \\ z \end{pmatrix} = \begin{pmatrix} -7 \\ -4 \\ 3 \end{pmatrix}.$$

We verify this by replacing each occurrence of x, y, and z on the left-hand side of the equations by -7, -4, and 3, respectively, and noting that the components of the result are equal to the right-hand side of each equation.

$In[1016] :=$ **Clear[x, y, z]**

$$\{-2\,x + y - 2\,z,\ 2\,x - 4\,y - 2\,z,$$
$$x - 4\,y - 2\,z\}\,/.\,\{x \to -7,\, y \to -4,\, z \to 3\}$$

$Out[1016] = \{4, -4, 3\}$

∎

EXAMPLE 5.2.6: Solve

$$-3x + 2y - 2z = -10$$
$$3x - y + 2z = 7$$
$$2x - y + z = 6.$$

SOLUTION: The associated matrix is $\mathbf{A} = \begin{pmatrix} -3 & 2 & -2 & -10 \\ 3 & -1 & 2 & 7 \\ 2 & -1 & 1 & 6 \end{pmatrix}$,

defined in capa, and then displayed in traditional row-and-column form with MatrixForm.

$In[1017] :=$ **Clear[capa]**

$In[1018] :=$ **capa = {{-3, 2, -2, -10}, {3, -1, 2, 7},**
{2, -1, 1, 6}}; MatrixForm[capa]

$Out[1018] = \begin{pmatrix} -3 & 2 & -2 & -10 \\ 3 & -1 & 2 & 7 \\ 2 & -1 & 1 & 6 \end{pmatrix}$

We eliminate methodically. First, we multiply row 1 by $-1/3$ so that the first entry in the first column is 1.

$In[1019] :=$ **capa = {-1/3capa[[1]], capa[[2]], capa[[3]]}**

$Out[1019] = \left\{\left\{1, -\dfrac{2}{3}, \dfrac{2}{3}, \dfrac{10}{3}\right\}, \{3, -1, 2, 7\}, \{2, -1, 1, 6\}\right\}$

We now eliminate below. First, we multiply row 1 by -3 and add it to row 2 and then we multiply row 1 by -2 and add it to row 3.

$In[1020]:=$ **capa = {capa[[1]], -3capa[[1]] + capa[[2]],**
            -2capa[[1]] + capa[[3]]}

$Out[1020]=$ $\left\{\left\{1, -\frac{2}{3}, \frac{2}{3}, \frac{10}{3}\right\}, \{0, 1, 0, -3\}, \left\{0, \frac{1}{3}, -\frac{1}{3}, -\frac{2}{3}\right\}\right\}$

Observe that the first nonzero entry in the second row is 1. We eliminate below this entry by adding $-1/3$ times row 2 to row 3.

$In[1021]:=$ **capa = {capa[[1]], capa[[2]],**
            -1/3 capa[[2]] + capa[[3]]}

$Out[1021]=$ $\left\{\left\{1, -\frac{2}{3}, \frac{2}{3}, \frac{10}{3}\right\}, \{0, 1, 0, -3\}, \left\{0, 0, -\frac{1}{3}, \frac{1}{3}\right\}\right\}$

We multiply the third row by -3 so that the first nonzero entry is 1.

$In[1022]:=$ **capa = {capa[[1]], capa[[2]], -3capa[[3]]}**

            MatrixForm[capa]

$Out[1022]=$ $\left\{\left\{1, -\frac{2}{3}, \frac{2}{3}, \frac{10}{3}\right\}, \{0, 1, 0, -3\}, \{0, 0, 1, -1\}\right\}$

$Out[1022]=$ $\begin{pmatrix} 1 & -\frac{2}{3} & \frac{2}{3} & \frac{10}{3} \\ 0 & 1 & 0 & -3 \\ 0 & 0 & 1 & -1 \end{pmatrix}$

This matrix is equivalent to the system

$$x - \frac{2}{3}y + \frac{2}{3}z = \frac{10}{3}$$
$$y = -3$$
$$z = -1,$$

which shows us that the solution is $x = 2, y = -3, z = -1$.

Working backwards confirms this. Multiplying row 2 by $2/3$ and adding to row 1 and then multiplying row 3 by $-2/3$ and adding to row **1** results in

$In[1023]:=$ **capa = {2/3 capa[[2]] + capa[[1]], capa[[2]],**
            capa[[3]]}; capa = {-2/3 capa[[3]] + capa[[1]],
            capa[[2]], capa[[3]]}; MatrixForm[capa]

$Out[1023]=$ $\begin{pmatrix} 1 & 0 & 0 & 2 \\ 0 & 1 & 0 & -3 \\ 0 & 0 & 1 & -1 \end{pmatrix}$

which is equivalent to the system $x = 2, y = -3, z = -1$.

Equivalent results are obtained with RowReduce.

$In[1024]:=$ **capa = {{-3, 2, -2, -10}, {3, -1, 2, 7},**
            {2, -1, 1, 6}}; capa = RowReduce[capa]

            MatrixForm[capa]

> *Out[1024]=* {{1, 0, 0, 2}, {0, 1, 0, -3}, {0, 0, 1, -1}}

$$Out[1024] = \begin{pmatrix} 1 & 0 & 0 & 2 \\ 0 & 1 & 0 & -3 \\ 0 & 0 & 1 & -1 \end{pmatrix}$$

Finally, we confirm the result directly with `Solve`.

> *In[1025]:=* `Solve[{-3x + 2y - 2z == -10, 3x - y + 2z == 7, 2x -`
> `y + z == 6}]`

> *Out[1025]=* {{x → 2, y → -3, z → -1}}

∎

EXAMPLE 5.2.7: Solve

$$-3x_1 + 2x_2 + 5x_3 = -12$$
$$3x_1 - x_2 - 4x_3 = 9$$
$$2x_1 - x_2 - 3x_3 = 7.$$

SOLUTION: The associated matrix is $\mathbf{A} = \begin{pmatrix} -3 & 2 & 5 & -12 \\ 3 & -1 & -4 & 9 \\ 2 & -1 & -3 & 7 \end{pmatrix}$, which

is reduced to row echelon form with `RowReduce`.

> *In[1026]:=* `capa = {{-3, 2, 5, -12}, {3, -1, -4, 9},`
> `             {2, -1, -3, 7}}; rrcapa = RowReduce[capa];`
> `             MatrixForm[rrcapa]`

$$Out[1026] = \begin{pmatrix} 1 & 0 & -1 & 2 \\ 0 & 1 & 1 & -3 \\ 0 & 0 & 0 & 0 \end{pmatrix}$$

The result shows that the original system is equivalent to

$$\begin{matrix} x_1 - x_3 = 2 \\ x_2 + x_3 = -3 \end{matrix} \quad \text{or} \quad \begin{matrix} x_1 = 2 + x_3 \\ x_2 = -3 - x_3 \end{matrix}$$

so x_3 is *free*. That is, for any real number t, a solution to the system is

$$\begin{pmatrix} x_1 \\ x_2 \\ x_3 \end{pmatrix} = \begin{pmatrix} 2 + t \\ -3 - t \\ t \end{pmatrix} = \begin{pmatrix} 2 \\ -3 \\ 0 \end{pmatrix} + t \begin{pmatrix} 1 \\ -1 \\ 1 \end{pmatrix}.$$

The system has infinitely many solutions.

Equivalent results are obtained with `Solve`.

```
In[1027]:= Solve[{-3x1 + 2x2 + 5x3 == -12, 3x1 - x2 - 4x3 == 9,
              2x1 - x2 - 3x3 == 7}]
Solve :: "svars" :  "Equationsmaynotgivesolutions
     forallšolvevariables."
Out[1027]= {{x1 → 2 + x3, x2 → -3 - x3}}

In[1028]:= Solve[{-3x1 + 2x2 + 5x3 == -12,
              3x1 - x2 - 4x3 == 9, 2x1 - x2 - 3x3 == 7},
              {x1, x2}]
Out[1028]= {{x1 → 2 + x3, x2 → -3 - x3}}
```

■

EXAMPLE 5.2.8: Solve

$$-3x_1 + 2x_2 + 5x_3 = -14$$
$$3x_1 - x_2 - 4x_3 = 11$$
$$2x_1 - x_2 - 3x_3 = 8.$$

SOLUTION: The associated matrix is $\mathbf{A} = \begin{pmatrix} -3 & 2 & 5 & -14 \\ 3 & -1 & -4 & 11 \\ 2 & -1 & -3 & 8 \end{pmatrix}$, which

is reduced to row echelon form with `RowReduce`.

```
In[1029]:= capa = {{-3, 2, 5, -14}, {3, -1, -4, 11},
              {2, -1, -3, 8}}; RowReduce[capa]//MatrixForm
Out[1029]= ⎛1  0  -1  0⎞
           ⎜0  1   1  0⎟
           ⎝0  0   0  1⎠
```

The result shows that the original system is equivalent to

$$x_1 - x_3 = 0$$
$$x_2 + x_3 = 0$$
$$0 = 1.$$

Of course, 0 is not equal to 1: the last equation is false. The system has no solutions.

We check the calculation with `Solve`. In this case, the results indicate that `Solve` cannot find any solutions to the system.

$In[1030] :=$ **Solve[{-3x1 + 2x2 + 5x3 == -14, 3x1 - x2 - 4x3 == 11,**
 2x1 - x2 - 3x3 == 8}]
$Out[1030] =$ **{}**

Generally, if Mathematica returns nothing, the result means either that there is no solution or that Mathematica cannot solve the problem. In such a situation, we must always check using another method.

∎

EXAMPLE 5.2.9: The **nullspace** of **A** is the set of solutions to the system of equations $\mathbf{Ax} = \mathbf{0}$. Find the nullspace of $\mathbf{A} = \begin{pmatrix} 3 & 2 & 1 & 1 & -2 \\ 3 & 3 & 1 & 2 & -1 \\ 2 & 2 & 1 & 1 & -1 \\ -1 & -1 & 0 & -1 & 0 \\ 5 & 4 & 2 & 2 & -3 \end{pmatrix}$.

SOLUTION: Observe that row reducing $(\mathbf{A}|\mathbf{0})$ is equivalent to row reducing **A**. After defining **A**, we use RowReduce to row reduce **A**.

$In[1031] :=$ **capa = {{3, 2, 1, 1, -2}, {3, 3, 1, 2, -1},**
 {2, 2, 1, 1, -1}, {-1, -1, 0, -1, 0},
 {5, 4, 2, 2, -3}}; RowReduce[capa]//MatrixForm
$Out[1031] =$ $\begin{pmatrix} 1 & 0 & 0 & 0 & -1 \\ 0 & 1 & 0 & 1 & 1 \\ 0 & 0 & 1 & -1 & -1 \\ 0 & 0 & 0 & 0 & 0 \\ 0 & 0 & 0 & 0 & 0 \end{pmatrix}$

The result indicates that the solutions of $\mathbf{Ax} = \mathbf{0}$ are

$$\mathbf{x} = \begin{pmatrix} x_1 \\ x_2 \\ x_3 \\ x_4 \\ x_5 \end{pmatrix} = \begin{pmatrix} t \\ -s-t \\ s+t \\ s \\ t \end{pmatrix} = s \begin{pmatrix} 0 \\ -1 \\ 1 \\ 1 \\ 0 \end{pmatrix} + t \begin{pmatrix} 1 \\ -1 \\ 1 \\ 0 \\ 1 \end{pmatrix},$$

where s and t are any real numbers. The dimension of the nullspace, the **nullity**, is 2; a basis for the nullspace is

$$\left\{ \begin{pmatrix} 0 \\ -1 \\ 1 \\ 1 \\ 0 \end{pmatrix}, \begin{pmatrix} 1 \\ -1 \\ 1 \\ 0 \\ 1 \end{pmatrix} \right\}.$$

You can use the command `NullSpace[A]` to find a basis of the nullspace of a matrix **A** directly.

```
In[1032]:= NullSpace[capa]
Out[1032]= {{1, -1, 1, 0, 1}, {0, -1, 1, 1, 0}}
```

■

5.3 Selected Topics from Linear Algebra

5.3.1 Fundamental Subspaces Associated with Matrices

Let $A = (a_{ij})$ be an $n \times m$ matrix with entry a_{ij} in the ith row and jth column. The **row space** of **A**, row(**A**), is the spanning set of the rows of **A**; the **column space** of **A**, col(**A**), is the spanning set of the columns of **A**. If **A** is any matrix, then the dimension of the column space of **A** is equal to the dimension of the row space of **A**. The dimension of the row space (column space) of a matrix **A** is called the **rank** of **A**. The **nullspace** of **A** is the set of solutions to the system of equations $Ax = 0$. The nullspace of **A** is a subspace and its dimension is called the **nullity** of **A**. The rank of **A** is equal to the number of nonzero rows in the row echelon form of **A**, the nullity of **A** is equal to the number of zero rows in the row echelon form of **A**. Thus, if **A** is a square matrix, the sum of the rank of **A** and the nullity of **A** is equal to the number of rows (columns) of **A**.

1. `NullSpace[A]` returns a list of vectors which form a basis for the nullspace (or kernel) of the matrix **A**.
2. `RowReduce[A]` yields the reduced row echelon form of the matrix **A**.

EXAMPLE 5.3.1: Place the matrix

$$A = \begin{pmatrix} -1 & -1 & 2 & 0 & -1 \\ -2 & 2 & 0 & 0 & -2 \\ 2 & -1 & -1 & 0 & 1 \\ -1 & -1 & 1 & 2 & 2 \\ 1 & -2 & 2 & -2 & 0 \end{pmatrix}$$

in reduced row echelon form. What is the rank of **A**? Find a basis for the nullspace of **A**.

SOLUTION: We begin by defining the matrix `matrixa`. Then, `RowReduce` is used to place `matrixa` in reduced row echelon form.

```
In[1033]:= matrixa = {{-1, -1, 2, 0, -1}, {-2, 2, 0, 0, -2},
                       {2, -1, -1, 0, 1}, {-1, -1, 1, 2, 2},
                       {1, -2, 2, -2, 0}};
```

```
In[1034]:= RowReduce[matrixa]//MatrixForm
```

$$Out[1034]= \begin{pmatrix} 1 & 0 & 0 & -2 & 0 \\ 0 & 1 & 0 & -2 & 0 \\ 0 & 0 & 1 & -2 & 0 \\ 0 & 0 & 0 & 0 & 1 \\ 0 & 0 & 0 & 0 & 0 \end{pmatrix}$$

Because the row-reduced form of `matrixa` contains four nonzero rows, the rank of **A** is 4 and thus the nullity is 1. We obtain a basis for the nullspace with `NullSpace`.

```
In[1035]:= NullSpace[matrixa]
```

```
Out[1035]= {{2, 2, 2, 1, 0}}
```

As expected, because the nullity is 1, a basis for the nullspace contains one vector.

∎

EXAMPLE 5.3.2: Find a basis for the column space of

$$\mathbf{B} = \begin{pmatrix} 1 & -2 & 2 & 1 & -2 \\ 1 & 1 & 2 & -2 & -2 \\ 1 & 0 & 0 & 2 & -1 \\ 0 & 0 & 0 & -2 & 0 \\ -2 & 1 & 0 & 1 & 2 \end{pmatrix}.$$

SOLUTION: A basis for the column space of **B** is the same as a basis for the row space of the transpose of **B**. We begin by defining `matrixb` and then using `Transpose` to compute the transpose of `matrixb`, naming the resulting output `tb`.

```
In[1036]:= matrixb = {{1, -2, 2, 1, -2}, {1, 1, 2, -2, -2},
                       {1, 0, 0, 2, -1}, {0, 0, 0, -2, 0},
                       {-2, 1, 0, 1, 2}};
```

In[1037]:= **tb = Transpose[matrixb]**

Out[1037]= {{1, 1, 1, 0, -2}, {-2, 1, 0, 0, 1}, {2, 2, 0, 0, 0},
{1, -2, 2, -2, 1}, {-2, -2, -1, 0, 2}}

Next, we use RowReduce to row reduce tb and name the result rrtb.
A basis for the column space consists of the first four elements of rrtb.
We also use Transpose to show that the first four elements of rrtb are
the same as the first four columns of the transpose of rrtb. Thus, the
*j*th column of a matrix **A** can be extracted from **A** with Transpose
[A][[j]].

In[1038]:= **rrtb = RowReduce[tb];**

Transpose[rrtb]//MatrixForm

$$
Out[1038]=
\begin{pmatrix}
1 & 0 & 0 & 0 & 0 \\
0 & 1 & 0 & 0 & 0 \\
0 & 0 & 1 & 0 & 0 \\
0 & 0 & 0 & 1 & 0 \\
-\dfrac{1}{3} & \dfrac{1}{3} & -2 & -3 & 0
\end{pmatrix}
$$

We extract the first four elements of rrtb with Take. The results
correspond to a basis for the column space of **B**.

In[1039]:= **Take[rrtb, 4]**

$$
Out[1039]= \left\{ \left\{1, 0, 0, 0, -\frac{1}{3}\right\}, \left\{0, 1, 0, 0, \frac{1}{3}\right\}, \right.
$$
$$
\left. \{0, 0, 1, 0, -2\}, \{0, 0, 0, 1, -3\}\right\}
$$

∎

5.3.2 The Gram–Schmidt Process

A set of vectors $\{v_1, v_2, \ldots, v_n\}$ is **orthonormal** means that $\|v_i\| = 1$ for all values
of i and $v_i \cdot v_j = 0$ for $i \neq j$. Given a set of linearly independent vectors $S = \{v_1, v_2, \ldots, v_n\}$, the set of all linear combinations of the elements of S, $V = \text{span } S$,
is a vector space. Note that if S is an orthonormal set and $u \in \text{span } S$, then $u = (u \cdot v_1) v_1 + (u \cdot v_2) v_2 + \cdots + (u \cdot v_n) v_n$. Thus, we may easily express u as a linear
combination of the vectors in S. Consequently, if we are given any vector space, V,
it is frequently convenient to be able to find an orthonormal basis of V. We may
use the **Gram–Schmidt process** to find an orthonormal basis of the vector space
$V = \text{span } \{v_1, v_2, \ldots, v_n\}$.

We summarize the algorithm of the Gram–Schmidt process so that given a set of n linearly independent vectors $S = \{v_1, v_2, \ldots, v_n\}$, where $V = \text{span} \{v_1, v_2, \ldots, v_n\}$, we can construct a set of orthonormal vectors $\{u_1, u_2, \ldots, u_n\}$ so that $V = \text{span} \{u_1, u_2, \ldots, u_n\}$.

1. Let $u_1 = \dfrac{1}{\|v\|} v$;
2. Compute $\text{proj}_{\{u_1\}} v_2 = (u_1 \cdot v_2) u_1$, $v_2 - \text{proj}_{\{u_1\}} v_2$, and let

$$u_2 = \frac{1}{\left\| v_2 - \text{proj}_{\{u_1\}} v_2 \right\|} \left(v_2 - \text{proj}_{\{u_1\}} v_2 \right).$$

Then, span $\{u_1, u_2\} = \text{span} \{v_1, v_2\}$ and span $\{u_1, u_2, v_3, \ldots, v_n\}$ = span $\{v_1, v_1, \ldots, v_n\}$;

3. Generally, for $3 \le i \le n$, compute

$$\text{proj}_{\{u_1, u_2, \ldots, u_n\}} v_i = (u_1 \cdot v_i) u_1 + (u_2 \cdot v_i) u_2 + \cdots + (u_{i-1} \cdot v_i) u_{i-1},$$

$v_i - \text{proj}_{\{u_1, u_2, \ldots, u_n\}} v_i$, and let

$$u_1 = \frac{1}{\left\| \text{proj}_{\{u_1, u_2, \ldots, u_n\}} v_i \right\|} \left(\text{proj}_{\{u_1, u_2, \ldots, u_n\}} v_i \right).$$

Then, span $\{u_1, u_2, \ldots, u_i\} = \text{span} \{v_1, v_2, \ldots, v_i\}$ and

$$\text{span} \{u_1, u_2, \ldots, u_i, v_{i+1}, \ldots, v_n\} = \text{span} \{v_1, v_2, v_3, \ldots, v_n\};$$

and

4. Because span $\{u_1, u_2, \ldots, u_n\} = \text{span} \{v_1, v_2, \ldots, v_n\}$ and $\{u_1, u_2, \ldots, u_n\}$ is an orthonormal set, $\{u_1, u_2, \ldots, u_n\}$ is an orthonormal basis of V.

The Gram–Schmidt procedure is well-suited to computer arithmetic. The following code performs each step of the Gram–Schmidt process on a set of n linearly independent vectors $\{v_1, v_1, \ldots, v_n\}$. At the completion of each step of the procedure, gramschmidt [vecs] prints the list of vectors corresponding to

$\{u_1, u_2, \ldots, u_i, v_{i+1}, \ldots, v_n\}$ and returns the list of vectors $\{u_1, u_2, \ldots, u_n\}$. Note how comments are inserted into the code using $(* \ldots *)$.

```
In[1040]:= gramschmidt[vecs_] :=
              Module[{n, proj, u, capw},
                (*n represents the number of
                  vectors in the list vecs*)
                n = Length[vecs];
                (*proj[v, capw] computes the
                  projection of v onto capw*)
                proj[v_, capw_] :=
                  Length[capw]
                   ∑     capw[[i]].v capw[[i]];
                  i=1

                u[1] = ──────vecs[[1]]──────;
                       √vecs[[1]].vecs[[1]]
                capw = {};
                u[i_] := u[i] = Module[{stepone},
                  stepone =
                    vecs[[i]] - proj[vecs[[i]], capw];
                  Together[
                         stepone
                    ────────────────────]];
                    √stepone.stepone
                Do[
                  u[i];
                  AppendTo[capw, u[i]];
                  Print[Join[capw, Drop[vecs, i]]],
                  {i, 1, n - 1}];
                u[n];
              AppendTo[capw, u[n]]]
```

EXAMPLE 5.3.3: Use the Gram–Schmidt process to transform the basis

$$S = \left\{ \begin{pmatrix} -2 \\ -1 \\ -2 \end{pmatrix}, \begin{pmatrix} 0 \\ -1 \\ 2 \end{pmatrix}, \begin{pmatrix} 1 \\ 3 \\ -2 \end{pmatrix} \right\} \text{ of } \mathbf{R}^3 \text{ into an orthonormal basis.}$$

SOLUTION: We proceed by defining v1, v2, and v3 to be the vectors in the basis S and using gramschmidt[{v1,v2,v3}] to find an orthonormal basis.

$$In[1041] := \mathbf{v1} = \{-2, -1, -2\};$$

$$\mathbf{v2} = \{0, -1, 2\};$$

$$\mathbf{v3} = \{1, 3, -2\};$$

$$\mathbf{gramschmidt}[\{\mathbf{v1}, \mathbf{v2}, \mathbf{v3}\}]$$

$$\left\{\left\{-\tfrac{2}{3}, -\tfrac{1}{3}, -\tfrac{2}{3}\right\}, \{0, -1, 2\}, \{1, 3, -2\}\right\}$$

$$\left\{\left\{-\tfrac{2}{3}, -\tfrac{1}{3}, -\tfrac{2}{3}\right\}, \left\{-\tfrac{1}{3}, -\tfrac{2}{3}, \tfrac{2}{3}\right\}, \{1, 3, -2\}\right\}$$

$$Out[1041] = \left\{\left\{-\tfrac{2}{3}, -\tfrac{1}{3}, -\tfrac{2}{3}\right\}, \left\{-\tfrac{1}{3}, -\tfrac{2}{3}, \tfrac{2}{3}\right\}, \left\{-\tfrac{2}{3}, \tfrac{2}{3}, \tfrac{1}{3}\right\}\right\}$$

On the first line of output, the result $\{\mathbf{u}_1, \mathbf{v}_2, \mathbf{v}_3\}$ is given; $\{\mathbf{u}_1, \mathbf{u}_2, \mathbf{v}_3\}$ appears on the second line; $\{\mathbf{u}_1, \mathbf{u}_2, \mathbf{u}_3\}$ follows on the third.

■

EXAMPLE 5.3.4: Compute an orthonormal basis for the subspace of $\mathbf{R}^4$ spanned by the vectors $\begin{pmatrix} 2 \\ 4 \\ 4 \\ 1 \end{pmatrix}$, $\begin{pmatrix} -4 \\ 1 \\ -3 \\ 2 \end{pmatrix}$, and $\begin{pmatrix} 1 \\ 4 \\ 4 \\ -1 \end{pmatrix}$. Also, verify that the basis vectors are orthogonal and have norm 1.

SOLUTION: With gramschmidt, we compute the orthonormal basis vectors. Note that Mathematica names oset the last result returned by gramschmidt. The orthogonality of these vectors is then verified. Notice that Together is used to simplify the result in the case of oset[[2]].oset[[3]]. The norm of each vector is then found to be 1.

$$In[1042] := \mathbf{oset} = \mathbf{gramschmidt}[$$
$$\{\{2, 4, 4, 1\}, \{-4, 1, -3, 2\}, \{1, 4, 4, -1\}\}]$$

$$\left\{\left\{\tfrac{2}{\sqrt{37}}, \tfrac{4}{\sqrt{37}}, \tfrac{4}{\sqrt{37}}, \tfrac{1}{\sqrt{37}}\right\},\right.$$

$$\{-4, 1, -3, 2\}, \{1, 4, 4, -1\}\}$$

$$\left\{\left\{\tfrac{2}{\sqrt{37}}, \tfrac{4}{\sqrt{37}}, \tfrac{4}{\sqrt{37}}, \tfrac{1}{\sqrt{37}}\right\},\right.$$

$$\left\{-60\sqrt{\tfrac{2}{16909}}, \tfrac{93}{\sqrt{33818}}, -\tfrac{55}{\sqrt{33818}}, 44\sqrt{\tfrac{2}{16909}}\right\},$$

$$\{1, 4, 4, -1\}\}$$

$$Out[1042] = \left\{\left\{\frac{2}{\sqrt{37}}, \frac{4}{\sqrt{37}}, \frac{4}{\sqrt{37}}, \frac{1}{\sqrt{37}}\right\},\right.$$

$$\left\{-60\sqrt{\frac{2}{16909}}, \frac{93}{\sqrt{33818}}, -\frac{55}{\sqrt{33818}}, 44\sqrt{\frac{2}{16909}}\right\},$$

$$\left.\left\{-\frac{449}{\sqrt{934565}}, \frac{268}{\sqrt{934565}}, \frac{156}{\sqrt{934565}}, -\frac{798}{\sqrt{934565}}\right\}\right\}$$

The three vectors are extracted with oset using oset[[1]], oset[[2]], and oset[[3]].

```
In[1043]:= oset[[1]].oset[[2]]

            oset[[1]].oset[[3]]

            oset[[2]].oset[[3]]
Out[1043]= 0

Out[1043]= 0

Out[1043]= 0

In[1044]:= Sqrt[oset[[1]].oset[[1]]]

            Sqrt[oset[[2]].oset[[2]]]

            Sqrt[oset[[3]].oset[[3]]]
Out[1044]= 1

Out[1044]= 1

Out[1044]= 1
```

■

The package **Orthogonalization** in the **LinearAlgebra** folder (or directory) contains several useful commands.

1. GramSchmidt[{v1,v2,...}] returns an orthonormal set of vectors given the set of vectors $\{v_1, v_2, \ldots, v_n\}$. Note that this command does not illustrate each step of the Gram–Schmidt procedure as the gramschmidt function defined above.

2. Normalize[v] returns $\dfrac{1}{\|v\|}v$ given the nonzero vector v.

3. Projection[v1,v2] returns the projection of v_1 onto v_2: $\text{proj}_{v_2} v_1 = \dfrac{v_1 \cdot v_2}{\|v_2\|^2} v_2$.

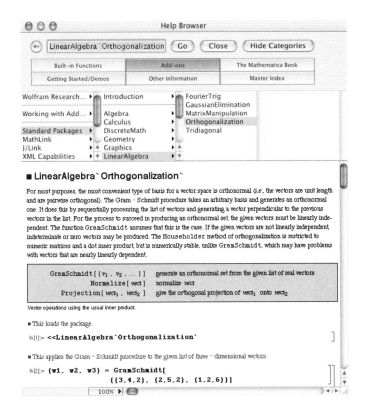

Thus, after loading the **Orthogonalization** package

$In[1045] :=$ `<< LinearAlgebra`Orthogonalization``

the command

$In[1046] :=$ `GramSchmidt[{{2,4,4,1},{-4,1,-3,2},`
                `{1,4,4,-1}}]`

$$Out[1046] = \left\{\left\{\frac{2}{\sqrt{37}}, \frac{4}{\sqrt{37}}, \frac{4}{\sqrt{37}}, \frac{1}{\sqrt{37}}\right\},\right.$$

$$\left\{-60\sqrt{\frac{2}{16909}}, \frac{93}{\sqrt{33818}}, -\frac{55}{\sqrt{33818}}, 44\sqrt{\frac{2}{16909}}\right\},$$

$$\left.\left\{-\frac{449}{\sqrt{934565}}, \frac{268}{\sqrt{934565}}, \frac{156}{\sqrt{934565}}, -\frac{798}{\sqrt{934565}}\right\}\right\}$$

returns an orthonormal basis for the subspace of $\mathbf{R}^4$ spanned by the vectors $\begin{pmatrix} 2 \\ 4 \\ 4 \\ 1 \end{pmatrix}$,

$\begin{pmatrix} -4 \\ 1 \\ -3 \\ 2 \end{pmatrix}$, and $\begin{pmatrix} 1 \\ 4 \\ 4 \\ -1 \end{pmatrix}$. The command

$In[1047]:=$ **Normalize[{2,4,4,1}]**

$Out[1047]=$ $\{ \dfrac{2}{\sqrt{37}}, \dfrac{4}{\sqrt{37}}, \dfrac{4}{\sqrt{37}}, \dfrac{1}{\sqrt{37}} \}$

finds a unit vector with the same direction as the vector $\mathbf{v} = \begin{pmatrix} 2 \\ 4 \\ 4 \\ 1 \end{pmatrix}$. Entering

$In[1048]:=$ **Projection[{2,4,4,1},{-4,1,-3,2}]**

$Out[1048]=$ $\{ \dfrac{28}{15}, -\dfrac{7}{15}, \dfrac{7}{5}, -\dfrac{14}{15} \}$

finds the projection of $\mathbf{v} = \begin{pmatrix} 2 \\ 4 \\ 4 \\ 1 \end{pmatrix}$ onto $\mathbf{w} = \begin{pmatrix} -4 \\ 1 \\ -3 \\ 2 \end{pmatrix}$.

5.3.3 Linear Transformations

A function $T : \mathbf{R}^n \longrightarrow \mathbf{R}^m$ is a **linear transformation** means that T satisfies the properties $T(\mathbf{u}+\mathbf{v}) = T(\mathbf{u}) + T(\mathbf{v})$ and $T(c\mathbf{u}) = cT(\mathbf{u})$ for all vectors $\mathbf{u}$ and $\mathbf{v}$ in $\mathbf{R}^n$ and all real numbers c. Let $T : \mathbf{R}^n \longrightarrow \mathbf{R}^m$ be a linear transformation and suppose $T(\mathbf{e}_1) = \mathbf{v}_1, T(\mathbf{e}_2) = \mathbf{v}_2, \ldots, T(\mathbf{e}_n) = \mathbf{v}_n$ where $\{\mathbf{e}_1, \mathbf{e}_2, \ldots, \mathbf{e}_n\}$ represents the standard basis of $\mathbf{R}^n$ and $\mathbf{v}_1, \mathbf{v}_2, \ldots, \mathbf{v}_n$ are (column) vectors in $\mathbf{R}^m$. The **associated matrix** of T is the $m \times n$ matrix $\mathbf{A} = \begin{pmatrix} \mathbf{v}_1 & \mathbf{v}_2 & \cdots & \mathbf{v}_n \end{pmatrix}$:

$$\text{if } \mathbf{x} = \begin{pmatrix} x_1 \\ x_2 \\ \vdots \\ x_n \end{pmatrix}, \quad T(\mathbf{x}) = T\left(\begin{pmatrix} x_1 \\ x_2 \\ \vdots \\ x_n \end{pmatrix} \right) = \mathbf{A}\mathbf{x} = \begin{pmatrix} \mathbf{v}_1 & \mathbf{v}_2 & \cdots & \mathbf{v}_n \end{pmatrix} \begin{pmatrix} x_1 \\ x_2 \\ \vdots \\ x_n \end{pmatrix}$$

Moreover, if $\mathbf{A}$ is any $m \times n$ matrix, then $\mathbf{A}$ is the associated matrix of the linear transformation defined by $T(\mathbf{x}) = \mathbf{A}\mathbf{x}$. In fact, a linear transformation T is completely determined by its action on any basis.

The **kernel** of the linear transformation T, ker (T), is the set of all vectors $\mathbf{x}$ in $\mathbf{R}^n$ such that $T(\mathbf{x}) = \mathbf{0}$: ker $(T) = \{\mathbf{x} \in \mathbf{R}^n | T(\mathbf{x}) = \mathbf{0}\}$. The kernel of T is a subspace of $\mathbf{R}^n$. Because $T(\mathbf{x}) = \mathbf{A}\mathbf{x}$ for all $\mathbf{x}$ in $\mathbf{R}^n$, ker $(T) = \{\mathbf{x} \in \mathbf{R}^n | T(\mathbf{x}) = \mathbf{0}\} = \{\mathbf{x} \in \mathbf{R}^n | \mathbf{A}\mathbf{x} = \mathbf{0}\}$ so the kernel of T is the same as the nullspace of $\mathbf{A}$.

EXAMPLE 5.3.5: Let $T : \mathbf{R}^5 \longrightarrow \mathbf{R}^3$ be the linear transformation defined by $T(\mathbf{x}) = \begin{pmatrix} 0 & -3 & -1 & -3 & -1 \\ -3 & 3 & -3 & -3 & -1 \\ 2 & 2 & -1 & 1 & 2 \end{pmatrix} \mathbf{x}$. (a) Calculate a basis for the kernel of the linear transformation. (b) Determine which of the vectors $\begin{pmatrix} 4 \\ 2 \\ 0 \\ 0 \\ -6 \end{pmatrix}$ and $\begin{pmatrix} 1 \\ 2 \\ -1 \\ -2 \\ 3 \end{pmatrix}$ is in the kernel of T.

SOLUTION: We begin by defining `matrixa` to be the matrix $\mathbf{A} = \begin{pmatrix} 0 & -3 & -1 & -3 & -1 \\ -3 & 3 & -3 & -3 & -1 \\ 2 & 2 & -1 & 1 & 2 \end{pmatrix}$ and then defining `t`. A basis for the kernel of T is the same as a basis for the nullspace of $\mathbf{A}$ found with `NullSpace`.

```
In[1049]:= Clear[t, x, matrixa]

            matrixa = {{0, -3, -1, -3, -1},
                       {-3, 3, -3, -3, -1}, {2, 2, -1, 1, 2}};

            t[x_] = matrixa.x;

In[1050]:= NullSpace[matrixa]

Out[1050]= {{-2, -1, 0, 0, 3}, {-6, -8, -15, 13, 0}}
```

Because $\begin{pmatrix} 4 \\ 2 \\ 0 \\ 0 \\ -6 \end{pmatrix}$ is a linear combination of the vectors that form a basis for the kernel, $\begin{pmatrix} 4 \\ 2 \\ 0 \\ 0 \\ -6 \end{pmatrix}$ is in the kernel while $\begin{pmatrix} 1 \\ 2 \\ -1 \\ -2 \\ 3 \end{pmatrix}$ is not. These results are verified more easily by evaluating t for each vector.

```
In[1051]:= t[{4,2,0,0,-6}]
Out[1051]= {0,0,0}

In[1052]:= t[{1,2,-1,-2,3}]
Out[1052]= {-2,9,11}
```

■

Application: Rotations

Let $\mathbf{x} = \begin{pmatrix} x_1 \\ x_2 \end{pmatrix}$ be a vector in $\mathbf{R}^2$ and θ an angle. Then, there are numbers r and ϕ given by $r = \sqrt{x_1^2 + x_2^2}$ and $\phi = \tan^{-1}(x_2/x_1)$ so that $x_1 = r\cos\phi$ and $x_2 = r\sin\phi$. When we rotate $\mathbf{x} = \begin{pmatrix} x_1 \\ x_2 \end{pmatrix} = \begin{pmatrix} r\cos\phi \\ r\sin\phi \end{pmatrix}$ through the angle θ, we obtain the vector $\mathbf{x}' = \begin{pmatrix} r\cos(\theta+\phi) \\ r\sin(\theta+\phi) \end{pmatrix}$. Using the trigonometric identities $\sin(\theta\pm\phi) = \sin\theta\cos\phi \pm \sin\phi\cos\theta$ and $\cos(\theta\pm\phi) = \cos\theta\cos\phi \mp \sin\theta\sin\phi$ we rewrite

$$\mathbf{x}' = \begin{pmatrix} r\cos(\theta+\phi) \\ r\sin(\theta+\phi) \end{pmatrix} = \begin{pmatrix} r\cos\theta\cos\phi - r\sin\theta\sin\phi \\ r\sin\theta\cos\phi + r\sin\phi\cos\theta \end{pmatrix} = \begin{pmatrix} \cos\theta & -\sin\theta \\ \sin\theta & \cos\theta \end{pmatrix} \begin{pmatrix} r\cos\phi \\ r\sin\phi \end{pmatrix}$$

$$= \begin{pmatrix} \cos\theta & -\sin\theta \\ \sin\theta & \cos\theta \end{pmatrix} \begin{pmatrix} x_1 \\ x_2 \end{pmatrix}.$$

Thus, the vector $\mathbf{x}'$ is obtained from $\mathbf{x}$ by computing $\begin{pmatrix} \cos\theta & -\sin\theta \\ \sin\theta & \cos\theta \end{pmatrix}\mathbf{x}$. Generally, if θ represents an angle, the linear transformation $T : \mathbf{R}^2 \longrightarrow \mathbf{R}^2$ defined by $T(\mathbf{x}) = \begin{pmatrix} \cos\theta & -\sin\theta \\ \sin\theta & \cos\theta \end{pmatrix}\mathbf{x}$ is called the **rotation of $\mathbf{R}^2$ through the angle θ**. We write code to rotate a polygon through an angle θ. The procedure `rotate` uses a list of n points and the rotation matrix defined in r to produce a new list of points that are joined using the `Line` graphics directive. Entering

```
Line[{{x1,y1},{x2,y2},....,{xn,yn}}]
```

represents the graphics primitive for a line in two dimensions that connects the points listed in `{{x1,y1},{x2,y2},....,{xn,yn}}`. Entering

```
Show[Graphics[Line[{{x1,y1},{x2,y2},...,{xn,yn}}]]]
```

displays the line. This rotation can be determined for one value of θ. However, a more interesting result is obtained by creating a list of rotations for a sequence of angles and then displaying the graphics objects. This is done for $\theta = 0$ to $\theta = \pi/2$

using increments of $\pi/16$. Hence, a list of nine graphs is given for the square with vertices $(-1, 1)$, $(1, 1)$, $(1, -1)$, and $(-1, -1)$ and displayed in Figure 5-3.

```
In[1053]:= r[θ_] = (Cos[θ]   -Sin[θ]);
                    (Sin[θ]    Cos[θ] )
```

```
In[1054]:= rotate[pts_, angle_] := Module[{newpts},
             newpts =
               Table[r[angle].pts[[i]],
                 {i, 1, Length[pts]}];
             newpts = AppendTo[newpts, newpts[[1]]];
             figure = Line[newpts];
             Show[Graphics[figure],
               AspectRatio → 1,
               PlotRange → {{-1.5, 1.5}, {-1.5, 1.5}},
               DisplayFunction → Identity]]
```

```
In[1055]:= graphs =
             Table[
               rotate[{{-1, 1}, {1, 1}, {1, -1}, {-1, -1}},
                 t], {t, 0, π/2, π/16}];
             array = Partition[graphs, 3];

           Show[GraphicsArray[array]]
```

5.3.4 Eigenvalues and Eigenvectors

Let $\mathbf{A}$ be an $n \times n$ matrix. λ is an **eigenvalue** of $\mathbf{A}$ if there is a *nonzero* vector, $\mathbf{v}$, called an **eigenvector**, satisfying

$$\mathbf{A}\mathbf{v} = \lambda\mathbf{v}. \tag{5.2}$$

We find the eigenvalues of $\mathbf{A}$ by solving the **characteristic polynomial**

$$|\mathbf{A} - \lambda\mathbf{I}| = 0 \tag{5.3}$$

for λ. Once we find the eigenvalues, the corresponding eigenvectors are found by solving

$$(\mathbf{A} - \lambda\mathbf{I})\,\mathbf{v} = \mathbf{0} \tag{5.4}$$

for $\mathbf{v}$.

If $\mathbf{A}$ is a square matrix,

```
Eigenvalues[A]
```

Figure 5-3   A rotated square

finds the eigenvalues of **A**,

$$\text{Eigenvectors[A]}$$

finds the eigenvectors, and

$$\text{Eigensystem[A]}$$

finds the eigenvalues and corresponding eigenvectors.

$$\text{CharacteristicPolynomial[A,lambda]}$$

finds the characteristic polynomial of **A** as a function of λ.

EXAMPLE 5.3.6: Find the eigenvalues and corresponding eigenvectors for each of the following matrices. (a) $A = \begin{pmatrix} -3 & 2 \\ 2 & -3 \end{pmatrix}$, (b) $A = \begin{pmatrix} 1 & -1 \\ 1 & 3 \end{pmatrix}$, (c) $A = \begin{pmatrix} 0 & 1 & 1 \\ 1 & 0 & 1 \\ 1 & 1 & 0 \end{pmatrix}$, (d) $A = \begin{pmatrix} -1/4 & 2 \\ -8 & -1/4 \end{pmatrix}$.

SOLUTION: (a) We begin by finding the eigenvalues. Solving

$$|A - \lambda I| = \begin{vmatrix} -3 - \lambda & 2 \\ 2 & -3 - \lambda \end{vmatrix} = \lambda^2 + 6\lambda + 5 = 0$$

gives us $\lambda_1 = -5$ and $\lambda_2 = -1$.

Observe that the same results are obtained using CharacteristicPolynomial and Eigenvalues.

```
In[1056]:= capa = {{-3,2},{2,-3}};
           CharacteristicPolynomial[capa,λ]//Factor

           e1 = Eigenvalues[capa]
Out[1056]= (1 + λ) (5 + λ)
Out[1056]= {-5, -1}
```

We now find the corresponding eigenvectors. Let $v_1 = \begin{pmatrix} x_1 \\ y_1 \end{pmatrix}$ be an eigenvector corresponding to λ_1, then

$$(A - \lambda_1 I) v_1 = 0$$

$$\left[\begin{pmatrix} -3 & 2 \\ 2 & -3 \end{pmatrix} - (-5)\begin{pmatrix} 1 & 0 \\ 0 & 1 \end{pmatrix}\right]\begin{pmatrix} x_1 \\ y_1 \end{pmatrix} = \begin{pmatrix} 0 \\ 0 \end{pmatrix}$$

$$\begin{pmatrix} 2 & 2 \\ 2 & 2 \end{pmatrix}\begin{pmatrix} x_1 \\ y_1 \end{pmatrix} = \begin{pmatrix} 0 \\ 0 \end{pmatrix},$$

which row reduces to

$$\begin{pmatrix} 1 & 1 \\ 0 & 0 \end{pmatrix}\begin{pmatrix} x_1 \\ y_1 \end{pmatrix} = \begin{pmatrix} 0 \\ 0 \end{pmatrix}.$$

That is, $x_1 + y_1 = 0$ or $x_1 = -y_1$. Hence, for any value of $y_1 \neq 0$,

$$v_1 = \begin{pmatrix} x_1 \\ y_1 \end{pmatrix} = \begin{pmatrix} -y_1 \\ y_1 \end{pmatrix} = y_1\begin{pmatrix} -1 \\ 1 \end{pmatrix}$$

is an eigenvector corresponding to λ_1. Of course, this represents infinitely many vectors. But, they are all linearly dependent. Choosing $y_1 = 1$ yields $v_1 = \begin{pmatrix} -1 \\ 1 \end{pmatrix}$. Note that you might have chosen $y_1 = -1$

and obtained $v_1 = \begin{pmatrix} 1 \\ -1 \end{pmatrix}$. However, both of our results are "correct" because these vectors are linearly dependent.

Similarly, letting $v_2 = \begin{pmatrix} x_2 \\ y_2 \end{pmatrix}$ be an eigenvector corresponding to λ_2 we solve $(\mathbf{A} - \lambda_2 \mathbf{I}) v_1 = 0$:

$$\begin{pmatrix} -2 & 2 \\ 2 & -2 \end{pmatrix} \begin{pmatrix} x_2 \\ y_2 \end{pmatrix} = \begin{pmatrix} 0 \\ 0 \end{pmatrix} \quad \text{or} \quad \begin{pmatrix} 1 & -1 \\ 0 & 0 \end{pmatrix} \begin{pmatrix} x_2 \\ y_2 \end{pmatrix} = \begin{pmatrix} 0 \\ 0 \end{pmatrix}.$$

Thus, $x_2 - y_2 = 0$ or $x_2 = y_2$. Hence, for any value of $y_2 \neq 0$,

$$v_2 = \begin{pmatrix} x_2 \\ y_2 \end{pmatrix} = \begin{pmatrix} y_2 \\ y_2 \end{pmatrix} = y_2 \begin{pmatrix} 1 \\ 1 \end{pmatrix}$$

is an eigenvector corresponding to λ_2. Choosing $y_2 = 1$ yields $v_2 = \begin{pmatrix} 1 \\ 1 \end{pmatrix}$.

We confirm these results using RowReduce.

```
In[1057]:= i2 = {{1, 0}, {0, 1}};
           ev1 = capa - e1[[1]]  i2
Out[1057]= {{2, 2}, {2, 2}}

In[1058]:= RowReduce[ev1]
Out[1058]= {{1, 1}, {0, 0}}

In[1059]:= ev2 = capa - e1[[2]]  i2

           RowReduce[ev2]
Out[1059]= {{-2, 2}, {2, -2}}
Out[1059]= {{1, -1}, {0, 0}}
```

We obtain the same results using Eigenvectors and Eigensystem.

```
In[1060]:= Eigenvectors[capa]

           Eigensystem[capa]
Out[1060]= {{-1, 1}, {1, 1}}
Out[1060]= {{-5, -1}, {{-1, 1}, {1, 1}}}
```

(b) In this case, we see that $\lambda = 2$ has multiplicity 2. There is only one linearly independent eigenvector, $v = \begin{pmatrix} -1 \\ 1 \end{pmatrix}$, corresponding to λ.

```
In[1061]:= capa = {{1, -1}, {1, 3}};
           Factor[CharacteristicPolynomial[capa, λ]]

           Eigenvectors[capa]

           Eigensystem[capa]
```

Out [1061]= $(-2 + \lambda)^2$

Out [1061]= $\{\{-1, 1\}, \{0, 0\}\}$

Out [1061]= $\{\{2, 2\}, \{\{-1, 1\}, \{0, 0\}\}\}$

(c) The eigenvalue $\lambda_1 = 2$ has corresponding eigenvector $\mathbf{v}_1 = \begin{pmatrix} 1 \\ 1 \\ 1 \end{pmatrix}$.

The eigenvalue $\lambda_{2,3} = -1$ has multiplicity 2. In this case, there are two linearly independent eigenvectors corresponding to this eigenvalue:

$$\mathbf{v}_2 = \begin{pmatrix} -1 \\ 0 \\ 1 \end{pmatrix} \text{ and } \mathbf{v}_3 = \begin{pmatrix} -1 \\ 1 \\ 0 \end{pmatrix}.$$

*In[1062]:= **capa = {{0, 1, 1}, {1, 0, 1}, {1, 1, 0}};***
           Factor[CharacteristicPolynomial[capa, λ]]

           Eigenvectors[capa]

           Eigensystem[capa]

Out [1062]= $-(-2 + \lambda) (1 + \lambda)^2$

Out [1062]= $\{\{-1, 0, 1\}, \{-1, 1, 0\}, \{1, 1, 1\}\}$

Out [1062]= $\{\{-1, -1, 2\}, \{\{-1, 0, 1\}, \{-1, 1, 0\}, \{1, 1, 1\}\}\}$

(d) In this case, the eigenvalues $\lambda_{1,2} = -\frac{1}{4} \pm 4i$ are complex conjugates. We see that the eigenvectors $\mathbf{v}_{1,2} = \begin{pmatrix} 0 \\ 2 \end{pmatrix} \pm \begin{pmatrix} 1 \\ 0 \end{pmatrix} i$ are complex conjugates as well.

*In[1063]:= **capa = {{-1/4, 2}, {-8, -1/4}};***
           Eigenvectors[capa]

           Eigensystem[capa]

Out [1063]= $\{\{i, 2\}, \{-i, 2\}\}$

Out [1063]= $\left\{\left\{-\frac{1}{4} - 4\ i, -\frac{1}{4} + 4\ i\right\}, \{\{i, 2\}, \{-i, 2\}\}\right\}$

■

5.3.5 Jordan Canonical Form

Let $\mathbf{N}_k = (n_{ij}) = \begin{cases} 1, & j = i + 1 \\ 0, & \text{otherwise} \end{cases}$ represent a $k \times k$ matrix with the indicated elements. The $k \times k$ **Jordan block matrix** is given by $\mathbf{B}(\lambda) = \lambda \mathbf{I} + \mathbf{N}_k$ where λ is a constant:

$$
\mathbf{N}_l = \begin{pmatrix} 0 & 1 & 0 & \cdots & 0 \\ 0 & 0 & 1 & \cdots & 0 \\ \vdots & \vdots & \vdots & & \vdots \\ 0 & 0 & 0 & \cdots & 1 \\ 0 & 0 & 0 & \cdots & 0 \end{pmatrix} \quad \text{and} \quad \mathbf{B}(\lambda) = \lambda\mathbf{I} + \mathbf{N}_k = \begin{pmatrix} \lambda & 1 & 0 & \cdots & 0 \\ 0 & \lambda & 1 & \cdots & 0 \\ \vdots & \vdots & \vdots & & \vdots \\ 0 & 0 & 0 & \cdots & 1 \\ 0 & 0 & 0 & \cdots & \lambda \end{pmatrix}.
$$

Hence, $\mathbf{B}(\lambda)$ can be defined as $\mathbf{B}(\lambda) = (b_{ij}) = \begin{cases} \lambda, & i = j \\ 1, & j = i+1 \\ 0, & \text{otherwise} \end{cases}$. A **Jordan matrix** has

the form

$$
\mathbf{J} = \begin{pmatrix} \mathbf{B}_1(\lambda) & 0 & \cdots & 0 \\ 0 & \mathbf{B}_2(\lambda) & \cdots & 0 \\ \vdots & \vdots & & \vdots \\ 0 & 0 & \cdots & \mathbf{B}_n(\lambda) \end{pmatrix}
$$

where the entries $\mathbf{B}_j(\lambda)$, $j = 1, 2, \ldots, n$ represent Jordan block matrices.

Suppose that $\mathbf{A}$ is an $n \times n$ matrix. Then there is an invertible $n \times n$ matrix $\mathbf{C}$ such that $\mathbf{C}^{-1}\mathbf{AC} = \mathbf{J}$ where $\mathbf{J}$ is a Jordan matrix with the eigenvalues of $\mathbf{A}$ as diagonal elements. The matrix $\mathbf{J}$ is called the **Jordan canonical form** of $\mathbf{J}$. The command

```
JordanDecomposition[m]
```

yields a list of matrices {s, j} such that m=s.j.Inverse[s] and j is the Jordan canonical form of the matrix m.

For a given matrix $\mathbf{A}$, the unique monic polynomial q of least degree satisfying $q(\mathbf{A}) = 0$ is called the **minimal polynomial of A**. Let p denote the characteristic polynomial of $\mathbf{A}$. Because $p(\mathbf{A}) = 0$, it follows that q divides p. We can use the Jordan canonical form of a matrix to determine its minimal polynomial.

EXAMPLE 5.3.7: Find the Jordan canonical form, $\mathbf{J_A}$, of $\mathbf{A} = \begin{pmatrix} 2 & 9 & -9 \\ 0 & 8 & -6 \\ 0 & 9 & -7 \end{pmatrix}$.

SOLUTION: After defining matrixa, we use JordanDecomposition to find the Jordan canonical form of a and name the resulting output ja.

```
In[1064]:= matrixa = {{2, 9, -9}, {0, 8, -6}, {0, 9, -7}};

           ja = JordanDecomposition[matrixa]
```

Out[1064]= {{{3, 0, 1}, {2, 1, 0}, {3, 1, 0}},
　　　　　　{{-1, 0, 0}, {0, 2, 0}, {0, 0, 2}}}

The Jordan matrix corresponds to the second element of `ja` extracted with `ja[[2]]` and displayed in `MatrixForm`.

In[1065]:= **ja[[2]]//MatrixForm**

$$Out[1065]= \begin{pmatrix} -1 & 0 & 0 \\ 0 & 2 & 0 \\ 0 & 0 & 2 \end{pmatrix}$$

We also verify that the matrices `ja[[1]]` and `ja[[2]]` satisfy

　　matrixa=ja[[1]].ja[[2]].Inverse[ja[[1]]].

In[1066]:= **ja[[1]].ja[[2]].Inverse[ja[[1]]]**

Out[1066]= {{2, 9, -9}, {0, 8, -6}, {0, 9, -7}}

Next, we use `CharacteristicPolynomial` to find the characteristic polynomial of `matrixa` and then verify that `matrixa` satisfies its characteristic polynomial.

In[1067]:= **p = CharacteristicPolynomial[matrixa, x]**

Out[1067]= $-4 + 3 x^2 - x^3$

In[1068]:= **-4 IdentityMatrix[3]+**
　　　　　　3 MatrixPower[matrixa, 2]-
　　　　　　MatrixPower[matrixa, 3]

Out[1068]= {{0, 0, 0}, {0, 0, 0}, {0, 0, 0}}

From the Jordan form, we see that the minimal polynomial of **A** is $(x + 1)(x - 2)$. We define the minimal polynomial to be q and then verify that `matrixa` satisfies its minimal polynomial.

In[1069]:= **q = Expand[(x + 1)(x - 2)]**

Out[1069]= $-2 - x + x^2$

In[1070]:= **-2 IdentityMatrix[3] - matrixa+**
　　　　　　MatrixPower[matrixa, 2]

Out[1070]= {{0, 0, 0}, {0, 0, 0}, {0, 0, 0}}

As expected, q divides p.

In[1071]:= **Cancel[p/q]**

Out[1071]= $2 - x$

■

EXAMPLE 5.3.8: If $\mathbf{A} = \begin{pmatrix} 3 & 8 & 6 & -1 \\ -3 & 2 & 0 & 3 \\ 3 & -3 & -1 & -3 \\ 4 & 8 & 6 & -2 \end{pmatrix}$, find the characteristic

and minimal polynomials of **A**.

SOLUTION: As in the previous example, we first define matrixa and then use JordanDecomposition to find the Jordan canonical form of **A**.

```
In[1072]:= matrixa = {{3, 8, 6, -1}, {-3, 2, 0, 3},
               {3, -3, -1, -3}, {4, 8, 6, -2}};
           ja = JordanDecomposition[matrixa]
```

$Out[1072]= \left\{\left\{\{3, -1, 1, 0\}, \left\{-1, -1, 0, \frac{1}{2}\right\},\right.\right.$

$\left.\left\{0, 2, 0, -\frac{1}{2}\right\}, \{4, 0, 1, 0\}\right\}, \{\{-1, 0, 0, 0\},$

$\left.\{0, -1, 0, 0\}, \{0, 0, 2, 1\}, \{0, 0, 0, 2\}\}\right\}$

The Jordan canonical form of **A** is the second element of ja, extracted with ja[[2]] and displayed in MatrixForm.

```
In[1073]:= ja[[2]]//MatrixForm
```

$Out[1073]= \begin{pmatrix} -1 & 0 & 0 & 0 \\ 0 & -1 & 0 & 0 \\ 0 & 0 & 2 & 1 \\ 0 & 0 & 0 & 2 \end{pmatrix}$

From this result, we see that the minimal polynomial of **A** is $(x + 1)(x - 2)^2$. We define q to be the minimal polynomial of **A** and then verify that matrixa satisfies q.

```
In[1074]:= q = Expand[(x - 2)² (x + 1)]
```

$Out[1074]= 4 - 3 x^2 + x^3$

```
In[1075]:= 4 IdentityMatrix[4] -
               3 MatrixPower[matrixa, 2] +
               MatrixPower[matrixa, 3]
```

$Out[1075]= \{\{0, 0, 0, 0\}, \{0, 0, 0, 0\},$
$\{0, 0, 0, 0\}, \{0, 0, 0, 0\}\}$

The characteristic polynomial is obtained next and named p. As expected, q divides p, verified with Cancel.

In[1076]:= **p = CharacteristicPolynomial[matrixa, x]**

Out[1076]= $4 + 4 x - 3 x^2 - 2 x^3 + x^4$

In[1077]:= **Cancel[p/q]**

Out[1077]= $1 + x$

∎

5.3.6 The QR Method

The **conjugate transpose** (or **Hermitian adjoint matrix**) of the $m \times n$ complex matrix **A** which is denoted by $\mathbf{A}^*$ is the transpose of the complex conjugate of **A**. Symbolically, we have $\mathbf{A}^* = (\bar{\mathbf{A}})^t$. A complex matrix **A** is **unitary** if $\mathbf{A}^* = \mathbf{A}^{-1}$. Given a matrix **A**, there is a unitary matrix **Q** and an upper triangular matrix **R** such that $\mathbf{A} = \mathbf{QR}$. The product matrix **QR** is called the **QR factorization of A**. The command

QRDecomposition[N[m]]

determines the QR decomposition of the matrix m by returning the list {q,r}, where q is an orthogonal matrix, r is an upper triangular matrix and m=Transpose[q].r.

EXAMPLE 5.3.9: Find the QR factorization of the matrix $\mathbf{A} = \begin{pmatrix} 4 & -1 & 1 \\ -1 & 4 & 1 \\ 1 & 1 & 4 \end{pmatrix}$.

SOLUTION: We define matrixa and then use QRDecomposition to find the QR decomposition of matrixa, naming the resulting output qrm.

In[1078]:= **matrixa = {{4, -1, 1}, {-1, 4, 1}, {1, 1, 4}};**

In[1079]:= **qrm = QRDecomposition[N[matrixa]]**

Out[1079]= {{{-0.942809, 0.235702, -0.235702},
      {-0.142134, -0.92387, -0.355335},
      {-0.301511, -0.301511, 0.904534}},
    {{-4.24264, 1.64992, -1.64992},
      {0., -3.90868, -2.48734}, {0., 0., 3.01511}}}

The first matrix in qrm is extracted with qrm[[1]] and the second with qrm[[2]].

In[1080]:= **qrm[[1]]//MatrixForm**

$$Out[1080]= \begin{pmatrix} -0.942809 & 0.235702 & -0.235702 \\ -0.142134 & -0.92387 & -0.355335 \\ -0.301511 & -0.301511 & 0.904534 \end{pmatrix}$$

In[1081]:= **qrm[[2]]//MatrixForm**

$$Out[1081]= \begin{pmatrix} -4.24264 & 1.64992 & -1.64992 \\ 0. & -3.90868 & -2.48734 \\ 0. & 0. & 3.01511 \end{pmatrix}$$

We verify that the results returned are the QR decomposition of **A**.

In[1082]:= **Transpose[qrm[[1]]].qrm[[2]]//MatrixForm**

$$Out[1082]= \begin{pmatrix} 4. & -1. & 1. \\ -1. & 4. & 1. \\ 1. & 1. & 4. \end{pmatrix}$$

∎

One of the most efficient and most widely used methods for numerically calculating the eigenvalues of a matrix is the QR Method. Given a matrix **A**, then there is a Hermitian matrix **Q** and an upper triangular matrix **R** such that $\mathbf{A} = \mathbf{QR}$. If we define a sequence of matrices $\mathbf{A}_1 = \mathbf{A}$, factored as $\mathbf{A}_1 = \mathbf{Q}_1\mathbf{R}_1$; $\mathbf{A}_2 = \mathbf{R}_1\mathbf{Q}_1$, factored as $\mathbf{A}_2 = \mathbf{R}_2\mathbf{Q}_2$; $\mathbf{A}_3 = \mathbf{R}_2\mathbf{Q}_2$, factored as $\mathbf{A}_2 = \mathbf{R}_3\mathbf{Q}_3$; and in general, $\mathbf{A}_k = \mathbf{R}_{k+1}\mathbf{Q}_{k+1}$, $k = 1, 2, \ldots$ then the sequence $\{\mathbf{A}_n\}$ converges to a triangular matrix with the eigenvalues of **A** along the diagonal or to a nearly triangular matrix from which the eigenvalues of **A** can be calculated rather easily.

EXAMPLE 5.3.10: Consider the 3×3 matrix $\mathbf{A} = \begin{pmatrix} 4 & -1 & 1 \\ -1 & 4 & 1 \\ 1 & 1 & 4 \end{pmatrix}$.

Approximate the eigenvalues of **A** with the QR Method.

SOLUTION: We define the sequence a and qr recursively. We define a using the form a[n_]:=a[n]=... and qr using the form qr[n_]:=qr[n]=... so that Mathematica "remembers" the values of a and qr computed, and thus Mathematica avoids recomputing values previously computed. This is of particular advantage when computing a[n] and qr[n] for large values of *n*.

```
In[1083]:= matrixa = {{4, -1, 1}, {-1, 4, 1}, {1, 1, 4}};

            a[1] = N[matrixa];

            qr[1] = QRDecomposition[a[1]];

In[1084]:= a[n_] :=
              a[n] = qr[n - 1][[2]].
                 Transpose[qr[n - 1][[1]]];

            qr[n_] := qr[n] = QRDecomposition[a[n]];
```

We illustrate `a[n]` and `qr[n]` by computing `qr[9]` and `a[10]`. Note that computing `a[10]` requires the computation of `qr[9]`. From the results, we suspect that the eigenvalues of **A** are 5 and 2.

```
In[1085]:= qr[9]
Out[1085]= {{{-1., 2.23173 × 10^-7, -0.000278046},
              {-8.92692 × 10^-8, -1., -0.000481589},
              {-0.000278046, -0.000481589, 1.}},
            {{-5., 1.56221 × 10^-6, -0.00194632},
              {0., -5., -0.00337112}, {0., 0., 2.}}}

In[1086]:= a[10]//MatrixForm
```

Out[1086]=

$$\begin{pmatrix} 5. & -1.78538\times 10^{-7} & -0.000556091 \\ -1.78538\times 10^{-7} & 5. & -0.000963178 \\ -0.000556091 & -0.000963178 & 2. \end{pmatrix}$$

Next, we compute `a[n]` for $n = 5$, 10, and 15, displaying the result in `TableForm`. We obtain further evidence that the eigenvalues of **A** are 5 and 2.

```
In[1087]:= Table[a[n]//MatrixForm, {n, 5, 15, 5}]//
             TableForm
```

Out[1087]=

$$\begin{pmatrix} 4.99902 & -0.001701 & 0.0542614 \\ -0.001701 & 4.99706 & 0.0939219 \\ 0.0542614 & 0.0939219 & 2.00393 \end{pmatrix}$$

$$\begin{pmatrix} 5. & -1.78538\times 10^{-7} & -0.000556091 \\ -1.78538\times 10^{-7} & 5. & -0.000963178 \\ -0.000556091 & -0.000963178 & 2. \end{pmatrix}$$

$$\begin{pmatrix} 5. & -1.87212\times 10^{-11} & 5.69438\times 10^{-6} \\ -1.87213\times 10^{-11} & 5. & 9.86295\times 10^{-6} \\ 5.69438\times 10^{-6} & 9.86295\times 10^{-6} & 2. \end{pmatrix}$$

We verify that the eigenvalues of **A** are indeed 5 and 2 with `Eigenvalues`.

```
In[1088]:= Eigenvalues[matrixa]
Out[1088]= {2, 5, 5}
```

■

5.4 Maxima and Minima Using Linear Programming

5.4.1 The Standard Form of a Linear Programming Problem

We call the linear programming problem of the following form the **standard form** of the linear programming problem:

$$\text{Minimize } Z = \underbrace{c_1 x_1 + c_2 x_2 + \cdots + c_n x_n}_{\text{function}}, \text{ subject to the restrictions}$$

$$\begin{cases} a_{11} x_1 + a_{12} x_2 + \cdots + a_{1n} x_n \geq b_1 \\ a_{21} x_1 + a_{22} x_2 + \cdots + a_{2n} x_n \geq b_2 \\ \vdots \\ a_{m1} x_1 + a_{m2} x_2 + \cdots + a_{mn} x_n \geq b_m \end{cases}$$

and $x_1 \geq 0, x_2 \geq 0, \ldots, x_n \geq 0$.

(5.5)

The command

```
ConstrainedMin[function,{inequalities},{variables}]
```

solves the standard form of the linear programming problem. Similarly, the command

```
ConstrainedMax[function,{inequalities},{variables}]
```

solves the linear programming problem: Maximize $Z = \underbrace{c_1 x_1 + c_2 x_2 + \cdots + c_n x_n}_{\text{function}}$, subject to the restrictions

$$\begin{cases} a_{11} x_1 + a_{12} x_2 + \cdots + a_{1n} x_n \geq b_1 \\ a_{21} x_1 + a_{22} x_2 + \cdots + a_{2n} x_n \geq b_2 \\ \vdots \\ a_{m1} x_1 + a_{m2} x_2 + \cdots + a_{mn} x_n \geq b_m \end{cases}$$

and $x_1 \geq 0, x_2 \geq 0, \ldots, x_n \geq 0$.

EXAMPLE 5.4.1: Maximize $Z(x_1, x_2, x_3) = 4x_1 - 3x_2 + 2x_3$ subject to the constraints $3x_1 - 5x_2 + 2x_3 \leq 60$, $x_1 - x_2 + 2x_3 \leq 10$, $x_1 + x_2 - x_3 \leq 20$, and x_1, x_2, x_3 all nonnegative.

SOLUTION: In order to solve a linear programming problem with Mathematica, the variables {x1,x2,x3} and objective function z[x1,x2,x3] are first defined. In an effort to limit the amount of typing required to complete the problem, the set of inequalities is assigned the name ineqs while the set of variables is called vars. The symbol "<=", obtained by typing the "<" key and then the "=" key, represents "less than or equal to" and is used in ineqs. Hence, the maximization problem is solved with the command

> ConstrainedMax[z[x1,x2,x3],ineqs,vars].

In[1089]:= **Clear[x1, x2, x3, z, ineqs, vars]**

vars = {x1, x2, x3};

z[x1_, x2_, x3_] = 4x1 - 3x2 + 2x3;

In[1090]:= **ineqs = {3 x1 - 5 x2 + x3 ≤ 60, x1 - x2 + 2 x3 ≤ 10,
x1 + x2 - x3 ≤ 20};**

In[1091]:= **ConstrainedMax[z[x1, x2, x3], ineqs,
vars]**

Out[1091]= {45, {x1 → 15, x2 → 5, x3 → 0}}

The solution gives the maximum value of z subject to the given constraints as well as the values of x1, x2, and x3 that maximize z. Thus, we see that the maximum value of Z is 45 if $x_1 = 15$, $x_2 = 5$, and $x_3 = 0$.
∎

We demonstrate the use of ConstrainedMin in the following example.

EXAMPLE 5.4.2: Minimize $Z(x, y, z) = 4x - 3y + 2z$ subject to the constraints $3x - 5y + z \leq 60$, $x - y + 2z \leq 10$, $x + y - z \leq 20$, and x, y, z all nonnegative.

SOLUTION: After clearing all previously used names of functions and variable values, the variables, objective function, and set of constraints for this problem are defined and entered as they were in the first example. By using

```
ConstrainedMin[z[x1,x2,x3],ineqs,vars]
```

the minimum value of the objective function is obtained as well as the variable values that give this minimum.

```
In[1092]:= Clear[x1, x2, x3, z, ineqs, vars]

           vars = {x1, x2, x3};

           z[x1_, x2_, x3_] = 4x1 - 3x2 + 2x3;

In[1093]:= ineqs = {3 x1 - 5 x2 + x3 ≤ 60, x1 - x2 + 2 x3 ≤ 10,
                    x1 + x2 - x3 ≤ 20};

In[1094]:= ConstrainedMin[z[x1, x2, x3], ineqs,
               vars]
Out[1094]= {-90, {x1 → 0, x2 → 50, x3 → 30}}
```

We conclude that the minimum value is -90 and occurs if $x_1 = 0$, $x_2 = 50$, and $x_3 = 30$.

∎

5.4.2 The Dual Problem

Given the standard form of the linear programming problem in equations (5.5), the **dual problem** is as follows: "Maximize $Y = \sum_{i=1}^{m} b_i y_i$ subject to the constraints $\sum_{i=1}^{m} a_{ij} y_i \leq c_{ij}$ for $j = 1, 2, \ldots, n$ and $y_i \geq 0$ for $i = 1, 2, \ldots, m$." Similarly, for the problem: "Maximize $Z = \sum_{j=1}^{n} c_j x_j$ subject to the constraints $\sum_{j=1}^{n} a_{ij} x_j \leq b_i$ for $i = 1, 2, \ldots, m$ and $x_j \geq 0$ for $j = 1, 2, \ldots, n$," the dual problem is as follows: "Minimize $Y = \sum_{i=1}^{m} b_i y_i$ subject to the constraints $\sum_{i=1}^{m} a_{ij} y_i \geq c_j$ for $j = 1, 2, \ldots, n$ and $y_i \geq 0$ for $i = 1, 2, \ldots, m$."

EXAMPLE 5.4.3: Maximize $Z = 6x + 8y$ subject to the constraints $5x + 2y \leq 20$, $x + 2y \leq 10$, $x \geq 0$, and $y \geq 0$. State the dual problem and find its solution.

SOLUTION: First, the original (or *primal*) problem is solved. The objective function for this problem is represented by zx. Finally, the set of inequalities for the primal is defined to be ineqsx. Using the command

ConstrainedMax[zx,ineqsx,{x[1],x[2]}],

the maximum value of zx is found to be 45.

```
In[1095]:= Clear[zx, zy, x, y, valsx, valsy, ineqsx,
              ineqsy]
```

```
In[1096]:= zx = 6 x[1] + 8 x[2];
           ineqsx = {5 x[1] + 2 x[2] ≤ 20, x[1] + 2 x[2] ≤ 10};
```

```
In[1097]:= ConstrainedMax[zx, ineqsx, {x[1], x[2]}]
```

$$Out[1097]= \left\{45, \left\{x[1] \to \frac{5}{2}, x[2] \to \frac{15}{4}\right\}\right\}$$

Because in this problem we have $c_1 = 6$, $c_2 = 8$, $b_1 = 20$, and $b_2 = 10$, the dual problem is as follows: Minimize $Z = 20y_1 + 10y_2$ subject to the constraints $5y_1 + y_2 \geq 6$, $2y_1 + 2y_2 \geq 8$, $y_1 \geq 0$, and $y_2 \geq 0$. The dual is solved in a similar fashion by defining the objective function zy and the collection of inequalities ineqsy. The minimum value obtained by zy subject to the constraints ineqsy is 45, which agrees with the result of the primal and is found with

ConstrainedMin[zy,ineqsy,{y[1],y[2]}].

```
In[1098]:= zy = 20 y[1] + 10 y[2];
           ineqsy = {5 y[1] + y[2] ≥ 6, 2 y[1] + 2 y[2] ≥ 8};
```

```
In[1099]:= ConstrainedMin[zy, ineqsy, {y[1], y[2]}]
```

$$Out[1099]= \left\{45, \left\{y[1] \to \frac{1}{2}, y[2] \to \frac{7}{2}\right\}\right\}$$

■

Of course, linear programming models can involve numerous variables. Consider the following: given the standard form linear programming problem in equations

(5.5), let $\mathbf{x} = \begin{pmatrix} x_1 \\ x_2 \\ \vdots \\ x_n \end{pmatrix}$, $\mathbf{b} = \begin{pmatrix} b_1 \\ b_2 \\ \vdots \\ b_m \end{pmatrix}$, $\mathbf{c} = \begin{pmatrix} c_1 & c_2 & \cdots & c_n \end{pmatrix}$, and $\mathbf{A}$ denote the $m \times n$ matrix

$\mathbf{A} = \begin{pmatrix} a_{11} & a_{12} & \cdots & a_{1n} \\ a_{21} & a_{22} & \cdots & a_{2n} \\ \vdots & \vdots & & \vdots \\ a_{m1} & a_{m2} & \cdots & a_{mn} \end{pmatrix}$. Then the standard form of the linear programming

problem is equivalent to finding the vector x that maximizes $Z = \mathbf{c} \cdot \mathbf{x}$ subject

to the restrictions $\mathbf{A}\mathbf{x} \geq \mathbf{b}$ and $x_1 \geq 0$, $x_2 \geq 0$, ..., $x_n \geq 0$. The dual problem is: "Minimize $Y = \mathbf{y} \cdot \mathbf{b}$ where $\mathbf{y} = \begin{pmatrix} y_1 & y_2 & \cdots & y_m \end{pmatrix}$ subject to the restrictions $\mathbf{y}\mathbf{A} \leq \mathbf{c}$ (componentwise) and $y_1 \geq 0$, $y_2 \geq 0$, ..., $y_m \geq 0$."
The command

$$\texttt{LinearProgramming[c,A,b]}$$

finds the vector x that minimizes the quantity Z=c.x subject to the restrictions A.x>=b and x>=0. LinearProgramming does not yield the minimum value of Z as did ConstrainedMin and ConstrainedMax and the value must be determined from the resulting vector.

EXAMPLE 5.4.4: Maximize $Z = 5x_1 - 7x_2 + 7x_3 + 5x_4 + 6x_5$ subject to the constraints $2x_1 + 3x_2 + 3x_3 + 2x_4 + 2x_5 \geq 10$, $6x_1 + 5x_2 + 4x_3 + x_4 + 4x_5 \geq 30$, $-3x_1 - 2x_2 - 3x_3 - 4x_4 \geq -5$, $-x_1 - x_2 - x_4 \geq -10$, and $x_1 \geq 0$ for $i = 1, 2, 3, 4$, and 5. State the dual problem. What is its solution?

SOLUTION: For this problem, $\mathbf{x} = \begin{pmatrix} x_1 \\ x_2 \\ x_3 \\ x_4 \\ x_5 \end{pmatrix}$, $\mathbf{b} = \begin{pmatrix} 10 \\ 30 \\ -5 \\ -10 \end{pmatrix}$, $\mathbf{c} = \begin{pmatrix} 5 & -7 & 7 & 5 & 6 \end{pmatrix}$,

and $\mathbf{A} = \begin{pmatrix} 2 & 3 & 3 & 2 & 2 \\ 6 & 5 & 4 & 1 & 4 \\ -3 & -2 & -3 & -4 & 0 \\ -1 & -1 & 0 & -1 & 0 \end{pmatrix}$. First, the vectors c and b are entered

and then matrix **A** is entered and named matrixa.

    In[1100]:= **Clear[matrixa, x, y, c, b]**

                  c = {5, -7, 7, 5, 6}; b = {10, 30, -5, -10};

                matrixa = {{2, 3, 3, 2, 2}, {6, 5, 4, 1, 4},
                        {-3, -2, -3, -4, 0}, {-1, -1, 0, -1, 0}};

Next, we use Array[x,5] to create the list of five elements {x[1],x[2],...,x[5]} named xvec. The command Table[x[i], {i,1,5}] returns the same list. These variables must be defined before attempting to solve this linear programming problem.

    In[1101]:= **xvec = Array[x, 5]**
    Out[1101]= **{x[1], x[2], x[3], x[4], x[5]}**

After entering the objective function coefficients with the vector c, the matrix of coefficients from the inequalities with matrixa, and the right-hand side values found in b; the problem is solved with

$$\text{LinearProgramming}[c, matrixa, b].$$

The solution is called xvec. Hence, the maximum value of the objective function is obtained by evaluating the objective function at the variable values that yield a maximum. Because these values are found in xvec, the maximum is determined with the dot product of the vector c and the vector xvec. (Recall that this product is entered as c.xvec.) This value is found to be 35/4.

> *In[1102]:=* **xvec = LinearProgramming[c, matrixa, b]**
>
> *Out[1102]=* $\left\{0, \dfrac{5}{2}, 0, 0, \dfrac{35}{8}\right\}$

> *In[1103]:=* **c.xvec**
>
> *Out[1103]=* $\dfrac{35}{4}$

Because the dual of the problem is "Minimize the number Y=y.b subject to the restrictions y.A<c and y>0," we use Mathematica to calculate y.b and y.A. A list of the dual variables {y[1],y[2],y[3],y[4]} is created with Array[y,4]. This list includes four elements because there are four constraints in the original problem. The objective function of the dual problem is, therefore, found with yvec.b, and the left-hand sides of the set of inequalities are given with yvec.matrixa.

> *In[1104]:=* **yvec = Array[y, 4]**
>
> *Out[1104]=* {y[1],y[2],y[3],y[4]}

> *In[1105]:=* **yvec.b**
>
> *Out[1105]=* 10 y[1] + 30 y[2] − 5 y[3] − 10 y[4]

> *In[1106]:=* **yvec.matrixa**
>
> *Out[1106]=* {2 y[1] + 6 y[2] − 3 y[3] − y[4],
> 3 y[1] + 5 y[2] − 2 y[3] − y[4],
> 3 y[1] + 4 y[2] − 3 y[3],
> 2 y[1] + y[2] − 4 y[3] − y[4], 2 y[1] + 4 y[2]}

Hence, we may state the dual problem as:

Minimize $Y = 10y_1 + 30y_2 - 5y_3 - 10y_4$ subject to the constraints

$$\begin{cases} 2y_1 + 6y_2 - 3y_3 - y_4 \leq 5 \\ 3y_1 + 5y_2 - 2y_3 - y_4 \leq -7 \\ 3y_1 + 4y_2 - 3y_3 \leq 7 \\ 2y_1 + y_2 - 4y_3 - y_4 \leq 5 \\ 2y_1 + 4y_2 \leq 6 \end{cases}$$

and $y_i \geq 0$ for $i = 1, 2, 3,$ and 4.

∎

Application: A Transportation Problem

A certain company has two factories, F1 and F2, each producing two products, P1 and P2, that are to be shipped to three distribution centers, D1, D2, and D3. The following table illustrates the cost associated with shipping each product from the factory to the distribution center, the minimum number of each product each distribution center needs, and the maximum output of each factory. How much of each product should be shipped from each plant to each distribution center to minimize the total shipping costs?

	F1/P1	F1/P2	F2/P1	F2/P2	Minimum
D1/P1	$0.75		$0.80		500
D1/P2		$0.50		$0.40	400
D2/P1	$1.00		$0.90		300
D2/P2		$0.75		$1.20	500
D3/P1	$0.90		$0.85		700
D3/P2		$0.80		$0.95	300
Maximum Output	1000	400	800	900	

SOLUTION: Let x_1 denote the number of units of P1 shipped from F1 to D1; x_2 the number of units of P2 shipped from F1 to D1; x_3 the number of units of P1 shipped from F1 to D2; x_4 the number of units of P2 shipped from F1 to D2; x_5 the number of units of P1 shipped from F1 to D3; x_6 the number of units of P2 shipped from F1 to D3; x_7 the number of units of P1 shipped from F2 to D1; x_8 the number of units of P2 shipped from F2 to D1; x_9 the number of units of P1 shipped from F2 to D2; x_{10} the number of units of P2 shipped from F2 to D2; x_{11} the number of units of P1 shipped from F2 to D3; and x_{12} the number of units of P2 shipped from F2 to D3.

Then, it is necessary to minimize the number

$$Z = .75x_1 + .5x_2 + x_3 + .75x_4 + .9x_5 + .8x_6 + .8x_7$$
$$+ .4x_8 + .9x_9 + 1.2x_{10} + .85x_{11} + .95x_{12}$$

subject to the constraints $x_1 + x_3 + x_5 \leq 1000$, $x_2 + x_4 + x_6 \leq 400$, $x_7 + x_9 + x_{11} \leq 800$, $x_8 + x_{10} + x_{12} \leq 900$, $x_1 + x_7 \geq 500$, $x_3 + x_9 \geq 500$, $x_5 + x_{11} \geq 700$, $x_2 + x_8 \geq 400$, $x_4 + x_{10} \geq 500$, $x_6 + x_{12} \geq 300$, and x_i nonnegative for $i = 1, 2, \ldots, 12$. In order to solve this linear programming problem, the objective function which computes the total cost, the 12 variables, and the set of inequalities must be entered. The coefficients of the objective function are given in the vector c. Using the command `Array[x,12]` illustrated in the previous example to define the list of 12 variables $\{x[1],x[2],\ldots,x[12]\}$, the objective function is given by the product z=xvec.c, where xvec is the name assigned to the list of variables.

```
In[1107]:= Clear[xvec, z, constraints, vars, c]

            c = {0.75, 0.5, 1, 0.75, 0.9, 0.8, 0.8,
                 0.4, 0.9, 1.2, 0.85, 0.95};

In[1108]:= xvec = Array[x, 12]
Out[1108]= {x[1], x[2], x[3], x[4], x[5], x[6],
            x[7], x[8], x[9], x[10], x[11], x[12]}

In[1109]:= z = xvec.c
Out[1109]= 0.75 x[1] + 0.5 x[2] + x[3] + 0.75 x[4] +
           0.9 x[5] + 0.8 x[6] + 0.8 x[7] + 0.4 x[8] +
           0.9 x[9] + 1.2 x[10] + 0.85 x[11] + 0.95 x[12]
```

The set of constraints are then entered and named `constraints` for easier use. Therefore, the minimum cost and the value of each variable which yields this minimum cost are found with the command

```
ConstrainedMin[z,constraints,xvec].
```

```
In[1110]:= constraints = {x[1] + x[3] + x[5] ≤ 1000,
              x[2] + x[4] + x[6] ≤ 400,
              x[7] + x[9] + x[11] ≤ 800,
              x[8] + x[10] + x[12] ≤ 900, x[1] + x[7] ≥ 500,
              x[3] + x[9] ≥ 300, x[5] + x[11] ≥ 700,
              x[2] + x[8] ≥ 400, x[4] + x[10] > 500,
              x[6] + x[12] > 300};
```

In[1111]:= **values = ConstrainedMin[z, constraints, xvec]**
Out[1111]= {2115., {x[1] → 500., x[2] → 0., x[3] → 0.,
                          x[4] → 400., x[5] → 200., x[6] → 0.,
                          x[7] → 0., x[8] → 400., x[9] → 300.,
                          x[10] → 100., x[11] → 500., x[12] → 300.}}

Notice that values is a list consisting of two elements: the minimum
value of the cost function, 2115, and the list of the variable values
{x[1]->500,x[2]->0, ...}. Hence, the minimum cost is obtained
with the command values[[1]] and the list of variable values that
yield the minimum cost is extracted with values[[2]].

In[1112]:= **values[[1]]**
Out[1112]= 2115.

In[1113]:= **values[[2]]**
Out[1113]= {x[1] → 500., x[2] → 0., x[3] → 0.,
                          x[4] → 400., x[5] → 200., x[6] → 0.,
                          x[7] → 0., x[8] → 400., x[9] → 300.,
                          x[10] → 100., x[11] → 500., x[12] → 300.}

Using these extraction techniques, the number of units produced by
each factory can be computed. Because x_1 denotes the number of units
of P1 shipped from F1 to D1, x_3 the number of units of P1 shipped
from F1 to D2, and x_5 the number of units of P1 shipped from F1 to
D3, the total number of units of Product 1 produced by Factory 1 is
given by the command x[1]+x[3]+x[5] /. values[[2]] which
evaluates this sum at the values of x[1], x[3], and x[5] given in the
list values[[2]].

In[1114]:= **x[1] + x[3] + x[5] /. values[[2]]**
Out[1114]= 700.

Also, the number of units of Products 1 and 2 received by each distribution center can be computed. The command x[3]+x[9] /.
values[[2]] gives the total amount of P1 received at D1 because
x[3]=amount of P1 received by D2 from F1 and x[9]= amount of
P1 received by D2 from F2. Notice that this amount is the minimum
number of units (300) of P1 requested by D1.

In[1115]:= **x[3] + x[9] /. values[[2]]**
Out[1115]= 300.

The number of units of each product that each factory produces can be
calculated and the amount of P1 and P2 received at each distribution
center is calculated in a similar manner.

```
In[1116]:= {x[1] + x[3] + x[5], x[2] + x[4] + x[6],
             x[7] + x[9] + x[11],
           x[8] + x[10] + x[12], x[1] + x[7],
             x[3] + x[9], x[5] + x[11], x[2] + x[8],
           x[4] + x[10], x[6] + x[12]} /.
             values[[2]]//TableForm
            700.
            400.
            800.
            800.
            500.
Out[1116]=  300.
            700.
            400.
            500.
            300.
```

From these results, we see that F1 produces 700 units of P1, F1 produces 400 units of P2, F2 produces 800 units of P1, F2 produces 800 units of P2, and each distribution center receives exactly the minimum number of each product it requests.

■

5.5 Selected Topics from Vector Calculus

5.5.1 Vector-Valued Functions

Basic operations on two and three-dimensional vectors are discussed in Section 5.1.4.2.

We now turn our attention to vector-valued functions. In particular, we consider vector-valued functions of the following forms.

Plane curves:	$\mathbf{r}(t) = x(t)\mathbf{i} + y(t)\mathbf{j}$	(5.6)
Space curves:	$\mathbf{r}(t) = x(t)\mathbf{i} + y(t)\mathbf{j} + z(t)\mathbf{k}$	(5.7)
Parametric surfaces:	$\mathbf{r}(s, t) = x(s, t)\mathbf{i} + y(s, t)\mathbf{j} + z(s, t)\mathbf{k}$	(5.8)
Vector fields in the plane:	$\mathbf{F}(x, y) = P(x, y)\mathbf{i} + Q(x, y)\mathbf{j}$	(5.9)
Vector fields in space:	$\mathbf{F}(x, y, z) = P(x, y, z)\mathbf{i} + Q(x, y, z)\mathbf{j} + R(x, y, z)\mathbf{k}$	(5.10)

For the vector-valued functions (5.6) and (5.7), differentiation and integration are carried out term-by-term, provided that all the terms are differentiable and integrable. Suppose that C is a smooth curve defined by $\mathbf{r}(t)$, $a \leq t \leq b$.

1. If $r'(t) \neq 0$, the **unit tangent vector**, $T(t)$, is

$$T(t) = \frac{r'(t)}{\|r'(t)\|}.$$

2. If $T'(t) \neq 0$, the **principal unit normal vector**, $N(t)$, is

$$N(t) = \frac{T'(t)}{\|T'(t)\|}.$$

3. The **arc length function**, $s(t)$, is

$$s(t) = \int_a^t \|r'(u)\| \, du.$$

In particular, the length of C on the interval $[a, b]$ is $\int_a^b \|r'(t)\| \, dt$.

It is a good exercise to show that the curvature of a circle of radius r is $1/r$.

4. The **curvature**, κ, of C is

$$\kappa = \frac{\|T'(t)\|}{\|r'(t)\|} = \frac{a(t) \cdot N(t)}{\|v(t)\|^2} = \frac{\|r'(t) \times r''(t)\|}{\|r'(t)\|^3},$$

where $v(t) = r'(t)$ and $a(t) = r''(t)$.

EXAMPLE 5.5.1 (Folium of Descartes): Consider the **folium of Descartes**,

$$r(t) = \frac{3at}{1 + t^3}i + \frac{3at^2}{1 + t^3}j$$

for $t \neq -1$, if $a = 1$. (a) Find $r'(t)$, $r''(t)$ and $\int r(t) \, dt$. (b) Find $T(t)$ and $N(t)$. (c) Find the curvature, κ. (d) Find the length of the loop of the folium.

SOLUTION: (a) After defining $r(t)$,

```
In[1117] := r[t_] = {3 a t/(1 + t^3), 3 a t^2/(1 + t^3)};
            a = 1;
```

we compute $r'(t)$ and $\int r(t) \, dt$ with ', '' and Integrate, respectively. We name $r'(t)$ dr, $r''(t)$ dr2, and $\int r(t) \, dt$ ir.

```
In[1118] := dr = Simplify[r'[t]]

            dr2 = Simplify[r''[t]]

            ir = Integrate[r[t], t]
```

$$Out[1118] = \left\{ \frac{3 - 6\ t^3}{(1 + t^3)^2}, -\frac{3\ t\ (-2 + t^3)}{(1 + t^3)^2} \right\}$$

$$Out[1118] = \left\{ \frac{18\ t^2\ (-2 + t^3)}{(1 + t^3)^3}, \frac{6\ (1 - 7\ t^3 + t^6)}{(1 + t^3)^3} \right\}$$

$$Out[1118] = \left\{ \sqrt{3}\ \text{ArcTan}\left[\frac{-1 + 2\ t}{\sqrt{3}} \right] - \text{Log}[1 + t] + \right.$$
$$\left. \frac{1}{2}\ \text{Log}\left[1 - t + t^2\right], \text{Log}\left[1 + t^3\right] \right\}$$

(b) Mathematica does not automatically make assumptions regarding the value of t, so it does not algebraically simplify $\|\mathbf{r}'(t)\|$ as we might typically do unless we use PowerExpand

PowerExpand[Sqrt[x^2]] returns x.

$$In[1119] := \mathbf{nr = PowerExpand[Sqrt[dr.dr]]//Simplify}$$

$$Out[1119] = \frac{3\ \sqrt{1 + 4\ t^2 - 4\ t^3 - 4\ t^5 + 4\ t^6 + t^8}}{(1 + t^3)^2}$$

The unit tangent vector, $\mathbf{T}(t)$ is formed in ut.

$$In[1120] := \mathbf{ut = 1/nr\ dr//Simplify}$$

$$Out[1120] = \left\{ \frac{1 - 2\ t^3}{\sqrt{1 + 4\ t^2 - 4\ t^3 - 4\ t^5 + 4\ t^6 + t^8}}, \right.$$
$$\left. -\frac{t\ (-2 + t^3)}{\sqrt{1 + 4\ t^2 - 4\ t^3 - 4\ t^5 + 4\ t^6 + t^8}} \right\}$$

We perform the same steps to compute the unit normal vector, $\mathbf{N}(t)$. In particular, note that dutb = $\|\mathbf{T}'(t)\|$.

$$In[1121] := \mathbf{dut = D[ut, t]//Simplify}$$

$$Out[1121] = \left\{ \frac{2\ t\ (-2 + t^3)\ (1 + t^3)^2}{(1 + 4\ t^2 - 4\ t^3 - 4\ t^5 + 4\ t^6 + t^8)^{3/2}}, \right.$$
$$\left. -\frac{2\ (-1 + 3\ t^6 + 2\ t^9)}{(1 + 4\ t^2 - 4\ t^3 - 4\ t^5 + 4\ t^6 + t^8)^{3/2}} \right\}$$

$$In[1122] := \mathbf{duta = dut.dut//Simplify}$$

$$Out[1122] = \frac{4\ (1 + t^3)^4}{(1 + 4\ t^2 - 4\ t^3 - 4\ t^5 + 4\ t^6 + t^8)^2}$$

$$In[1123] := \mathbf{dutb = PowerExpand[Sqrt[duta]]}$$

$$Out[1123] = \frac{2\ (1 + t^3)^2}{1 + 4\ t^2 - 4\ t^3 - 4\ t^5 + 4\ t^6 + t^8}$$

$$In[1124] := \mathbf{nt = 1/dutb\ dut//Simplify}$$

$$Out[1124] = \left\{ \frac{t\ (-2 + t^3)}{\sqrt{1 + 4\ t^2 - 4\ t^3 - 4\ t^5 + 4\ t^6 + t^8}}, \right.$$
$$\left. \frac{1 - 2\ t^3}{\sqrt{1 + 4\ t^2 - 4\ t^3 - 4\ t^5 + 4\ t^6 + t^8}} \right\}$$

(c) We use the formula $\kappa = \dfrac{\|\mathbf{T}'(t)\|}{\|\mathbf{r}'(t)\|}$ to determine the curvature in `curvature`.

```
In[1125]:= curvature = Simplify[dutb/nr]
```
$$Out[1125] = \frac{2\ (1+t^3)^4}{3\ (1+4\ t^2-4\ t^3-4\ t^5+4\ t^6+t^8)^{3/2}}$$

We graphically illustrate the unit tangent and normal vectors at $\mathbf{r}(1) = \langle 3/2,\ 3/2 \rangle$. First, we compute the unit tangent and normal vectors if $t = 1$ using `/..`

```
In[1126]:= ut1 = ut/.t- >1
```
$$Out[1126] = \left\{ -\frac{1}{\sqrt{2}},\ \frac{1}{\sqrt{2}} \right\}$$

```
In[1127]:= nt1 = nt/.t- >1
```
$$Out[1127] = \left\{ -\frac{1}{\sqrt{2}},\ -\frac{1}{\sqrt{2}} \right\}$$

We then compute the curvature if $t = 1$ in `smallk`. The center of the osculating circle at $\mathbf{r}(1)$ is found in `x0` and `y0`.

The radius of the osculating circle is $1/\kappa$; the position vector of the center is $\mathbf{r} + \dfrac{1}{\kappa}\mathbf{N}$.

```
In[1128]:= smallk = curvature/.t- >1

           N[smallk]

           N[1/smallk]

           x0 = r[t][[1]] - dr.dr
                   r[[2]]/(dr[[1]]dr2[[2]]-
                   dr2[[1]]dr[[2]])/.t- >1

           y0 = r[t][[2]]-
                   r.dr dr[[2]]/(dr[[1]]dr2[[2]]-
                   dr2[[1]]dr[[2]])/.t
                   - >1
```
$$Out[1128] = \frac{8\ \sqrt{2}}{3}$$
$$Out[1128] = 3.77124$$
$$Out[1128] = 0.265165$$
$$Out[1128] = \frac{21}{16}$$
$$Out[1128] = \frac{21}{16}$$

`Graphics[Circle[{x0, y0}, r]]` is a two-dimensional graphics object that represents a circle of radius r centered at the point (x_0, y_0). Use `Show` to display the graph.

We now load the `Arrow` package and graph $\mathbf{r}(t)$ with `ParametricPlot`. The unit tangent and normal vectors at $\mathbf{r}(1)$ are graphed with `arrow` in `a1` and `a2`. The osculating circle at $\mathbf{r}(1)$ is graphed with `Circle` in `c1`. All four graphs are displayed together with `Show` in Figure 5-4.

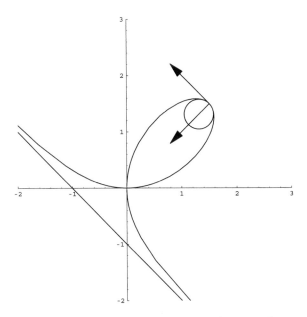

Figure 5-4   The folium with an osculating circle

```
In[1129]:= << Graphics`Arrow`
```

```
In[1130]:= p1 = ParametricPlot[Evaluate[r[t]],
             {t, -100, 100}, PlotRange- > {{-2, 3}, {-2, 3}},
             PlotPoints- > 200, AspectRatio- > 1,
             DisplayFunction- > Identity];
         p2 = Show[Graphics[{Circle[{x0, y0}, 1/smallk],
             Arrow[r[1], r[1] + ut1],
             Arrow[r[1], r[1] + nt1]}],
           DisplayFunction- > Identity];
         Show[p1, p2, DisplayFunction- > $DisplayFunction]
```

(d) The loop is formed by graphing $r(t)$ for $t \geq 0$. Hence, the length of the loop is given by the improper integral $\int_0^\infty \|r(t)\| \, dt$, which we compute with NIntegrate.

```
In[1131]:= NIntegrate[nr, {t, 0, ∞}]
```

```
Out[1131]= 4.91749
```

■

Recall that the **gradient** of $z = f(x, y)$ is the vector-valued function $\nabla f(x, y) = \langle f_x(x, y), f_y(x, y) \rangle$. Similarly, we define the **gradient** of $w = f(x, y, z)$ to be

$$\nabla f(x, y, z) = \langle f_x(x, y, z), f_y(x, y, z), f_z(x, y, z) \rangle = \frac{\partial f}{\partial x}\mathbf{i} + \frac{\partial f}{\partial y}\mathbf{j} + \frac{\partial f}{\partial z}\mathbf{k}. \qquad (5.11)$$

A vector field $\mathbf{F}$ is **conservative** if there is a function f, called a **potential function**, satisfying $\nabla f = \mathbf{F}$. In the special case that $\mathbf{F}(x, y) = P(x, y)\mathbf{i} + Q(x, y)\mathbf{j}$, $\mathbf{F}$ is conservative if and only if

$$\frac{\partial P}{\partial y} = \frac{\partial Q}{\partial x}.$$

The **divergence** of the vector field $\mathbf{F}(x, y, z) = P(x, y, z)\mathbf{i} + Q(x, y, z)\mathbf{j} + R(x, y, z)\mathbf{k}$ is the scalar field

$$\text{div } \mathbf{F} = \nabla \cdot \mathbf{F} = \frac{\partial P}{\partial x} + \frac{\partial Q}{\partial y} + \frac{\partial R}{\partial z}. \qquad (5.12)$$

The `Div` command, which is contained in the `VectorAnalysis` package, can be used to find the divergence of a vector field:

```
Div[{P(x,y,z),Q(x,y,z),R(x,y,z)},Cartesian[x,y,z]]
```

computes the divergence of $\mathbf{F}(x, y, z) = P(x, y, z)\mathbf{i} + Q(x, y, z)\mathbf{j} + R(x, y, z)\mathbf{k}$. The **laplacian** of the scalar field $w = f(x, y, z)$ is defined to be

$$\text{div}\,(\nabla f) = \nabla \cdot (\nabla f) = \nabla^2 f = \frac{\partial^2 f}{\partial x^2} + \frac{\partial^2 f}{\partial y^2} + \frac{\partial^2 f}{\partial z^2} = \Delta f. \qquad (5.13)$$

In the same way that `Div` computes the divergence of a vector field, `Laplacian`, which is also contained in the `VectorAnalysis` package, computes the laplacian of a scalar field.

The **curl** of the vector field $\mathbf{F}(x, y, z) = P(x, y, z)\mathbf{i} + Q(x, y, z)\mathbf{j} + R(x, y, z)\mathbf{k}$ is

$$\text{curl } \mathbf{F}(x, y, z) = \nabla \times \mathbf{F}(x, y, z)$$

$$= \begin{vmatrix} \mathbf{i} & \mathbf{j} & \mathbf{k} \\ \frac{\partial}{\partial x} & \frac{\partial}{\partial y} & \frac{\partial}{\partial z} \\ P(x, y, z) & Q(x, y, z) & R(x, y, z) \end{vmatrix} \qquad (5.14)$$

$$= \left(\frac{\partial R}{\partial y} - \frac{\partial Q}{\partial z}\right)\mathbf{i} - \left(\frac{\partial R}{\partial x} - \frac{\partial P}{\partial z}\right)\mathbf{j} + \left(\frac{\partial Q}{\partial x} - \frac{\partial P}{\partial y}\right)\mathbf{k}.$$

If $\mathbf{F}(x, y, z) = P(x, y, z)\mathbf{i} + Q(x, y, z)\mathbf{j} + R(x, y, z)\mathbf{k}$, $\mathbf{F}$ is conservative if and only if curl $\mathbf{F}(x, y, z) = \mathbf{0}$, in which case $\mathbf{F}$ is said to be **irrotational**.

EXAMPLE 5.5.2: Determine if

$$F(x, y) = \left(1 - 2x^2\right) ye^{-x^2-y^2}\mathbf{i} + \left(1 - 2y^2\right) xe^{-x^2-y^2}\mathbf{j}$$

is conservative. If **F** is conservative find a potential function for **F**.

SOLUTION: We define $P(x, y) = \left(1 - 2x^2\right) ye^{-x^2-y^2}$ and $Q(x, y) = \left(1 - 2y^2\right) xe^{-x^2-y^2}$. Then we use D and Simplify to see that $P_y(x, y) = Q_x(x, y)$. Hence, **F** is conservative.

```
In[1132]:= p[x_,y_] = (1 - 2x^2)y Exp[-x^2 - y^2];
           q[x_,y_] = (1 - 2y^2)x Exp[-x^2 - y^2];

In[1133]:= Simplify[D[p[x,y],y]]

           Simplify[D[q[x,y],x]]
```
$Out[1133]= e^{-x^2-y^2} \left(-1 + 2\ x^2\right) \left(-1 + 2\ y^2\right)$
$Out[1133]= e^{-x^2-y^2} \left(-1 + 2\ x^2\right) \left(-1 + 2\ y^2\right)$

We use Integrate to find f satisfying $\nabla f = \mathbf{F}$.

```
In[1134]:= i1 = Integrate[p[x,y],x] + g[y]
```
$Out[1134]= e^{-x^2-y^2}\ x\ y + g[y]$

```
In[1135]:= Solve[D[i1,y] == q[x,y],g'[y]]
```
$Out[1135]= BoxData(\{\{g'[y] \to 0\}\})$

Therefore, $g(y) = C$, where C is an arbitrary constant. Letting $C = 0$ gives us the following potential function.

```
In[1136]:= f = i1/.g[y]- > 0
```
$Out[1136]= e^{-x^2-y^2}\ x\ y$

Remember that the vectors **F** are perpendicular to the level curves of f. To see this, we normalize **F** in uv.

```
In[1137]:= uv = {p[x,y],q[x,y]}/
               Sqrt[{p[x,y],q[x,y]}.{p[x,y],q[x,y]}]//
               Simplify
```

$$Out[1137]= \left\{ - \frac{e^{-x^2-y^2}\ \left(-1 + 2\ x^2\right)\ y}{\sqrt{e^{-2\ \left(x^2+y^2\right)}\ \left(y^2 + 4\ x^4\ y^2 + x^2\ \left(1 - 8\ y^2 + 4\ y^4\right)\right)}}, \right.$$

$$\left. - \frac{e^{-x^2-y^2}\ x\ \left(-1 + 2\ y^2\right)}{\sqrt{e^{-2\ \left(x^2+y^2\right)}\ \left(y^2 + 4\ x^4\ y^2 + x^2\ \left(1 - 8\ y^2 + 4\ y^4\right)\right)}} \right\}$$

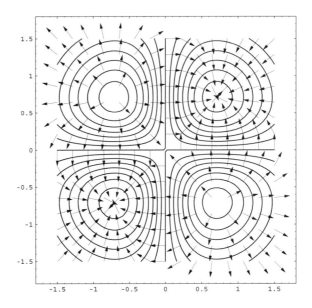

Figure 5-5   The vectors **F** are perpendicular to the level curves of f

We then graph several level curves of f in cp with ContourPlot and several vectors of uv with PlotVectorField, which is contained in the PlotField package, in fp. We show the graphs together with Show in Figure 5-5.

```
In[1138]:= << Graphics`PlotField`

            cp = ContourPlot[f, {x, -3/2, 3/2}, {y, -3/2, 3/2},
              ontours->15, ContourShading->False,
              PlotPoints->60, D
              isplayFunction->Identity];

In[1139]:= fp = PlotVectorField[uv, {x, -3/2, 3/2},
              {y, -3/2, 3/2}, DisplayFunction->Identity];
Power :: "infy" :
  "Infiniteexpression10encountered."
Power :: "infy" :
  "Infiniteexpression10encountered."

In[1140]:= Show[cp, fp,
              DisplayFunction->$DisplayFunction]
```

Note that we can use PlotGradientField, which is contained in the PlotField package, to graph several vectors of ∇f. However, the vectors are scaled and it can be difficult to see that the vectors are perpendicular to the level curves of f. The advantage of proceeding this way is that by graphing unit vectors, it is easier to see that the vectors are perpendicular to the level curves of f in the resulting plot.

∎

EXAMPLE 5.5.3: (a) Show that

$$\mathbf{F}(x, y, z) = -10xy^2\mathbf{i} + (3z^3 - 10x^2y)\mathbf{j} + 9yz^2\mathbf{k}$$

is irrotational. (b) Find f satisfying $\nabla f = \mathbf{F}$. (c) Compute div $\mathbf{F}$ and $\nabla^2 f$.

SOLUTION: (a) After defining $\mathbf{F}(x, y, z)$, we use `Curl`, which is contained in the `VectorAnalysis` package, to see that curl $\mathbf{F}(x, y, z) = \mathbf{0}$.

```
In[1141]:= << Calculus`VectorAnalysis`
```

```
In[1142]:= BoxData({Clear[f],
            f[x_, y_, z_] = {-10x y^2,
            3z^3 - 10x^2 y, 9 y z ^2}})
Out[1142]= {-10 x y², -10 x² y + 3 z³, 9 y z²}
```

```
In[1143]:= Curl[f[x, y, z]]
Out[1143]= {0, 0, 0}
```

(b) We then use `Integrate` to find $w = f(x, y, z)$ satisfying $\nabla f = \mathbf{F}$.

```
In[1144]:= i1 = Integrate[f[x, y, z][[1]], x] + g[y, z]
Out[1144]= -5 x² y² + g[y, z]
```

```
In[1145]:= i2 = D[i1, y]
Out[1145]= BoxData(-10 x² y + g^(1,0) [y, z])
```

```
In[1146]:= BoxData(Solve[i2 == f[x, y, z][[2]], g^(1,0) [y, z]])
Out[1146]= BoxData({{g^(1,0) [y, z] → 3 z³}})
```

```
In[1147]:= i3 = Integrate[3z^3, y] + h[z]
Out[1147]= 3 y z³ + h[z]
```

```
In[1148]:= i4 = i1/.g[y, z]->i3
Out[1148]= -5 x² y² + 3 y z³ + h[z]
```

```
In[1149]:= Solve[D[i4, z] == f[x, y, z][[3]]]
Out[1149]= BoxData({{h'[z] → 0}})
```

With $h(z) = C$ and $C = 0$ we have $f(x, y, z) = -5x^2y^2 + 3yz^3$.

```
In[1150]:= lf = -5 x² y² + 3 y z³;
```

∇f is orthogonal to the level surfaces of f. To illustrate this, we use ContourPlot3D, which is contained in the ContourPlot3D package, to graph the level surface of $w = f(x, y, z)$ corresponding to $w = -1$ for $-2 \leq x \leq 2$, $-2 \leq y \leq 2$, and $-2 \leq z \leq 2$ in pf. We then use PlotGradientField3D, which is contained in the PlotField3D package, to graph several vectors in the gradient field of f over the same domain in gradf. The two plots are shown together with Show in Figure 5-6. In the plot, notice that the vectors appear to be perpendicular to the surface.

```
In[1151]:= << Graphics`PlotField3D`

           << Graphics`ContourPlot3D`

In[1152]:= pf = ContourPlot3D[lf, {x, -2, 2}, {y, -2, 2},
              {z, -2, 2}, PlotPoints- > {5, 7},
              DisplayFunction- > Identity];

In[1153]:= gf = PlotGradientField3D[lf, {x, -2, 2},
              {y, -2, 2}, {z, -2, 2},
              DisplayFunction- > Identity];

In[1154]:= Show[pf, gf,
              DisplayFunction- > $DisplayFunction]
```

For (c), we take advantage of Div and Laplacian. As expected, the results are the same.

```
In[1155]:= Div[f[x, y, z], Cartesian[x, y, z]]
Out[1155]= -10 x² - 10 y² + 18 y z

In[1156]:= Laplacian[lf, Cartesian[x, y, z]]
Out[1156]= -10 x² - 10 y² + 18 y z
```

■

5.5.2 Line Integrals

If **F** is continuous on the smooth curve C with parametrization $\mathbf{r}(t)$, $a \leq t \leq b$, the **line integral** of **F** on C is

$$\int_C \mathbf{F} \cdot d\mathbf{r} = \int_a^b \mathbf{F} \cdot \mathbf{r}'(t)\, dt \tag{5.15}$$

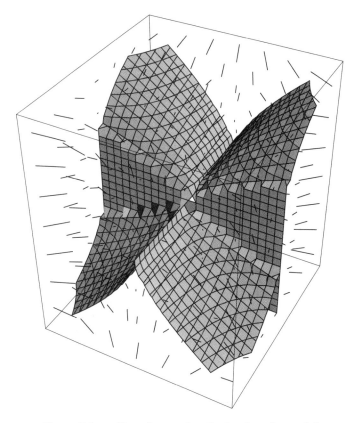

Figure 5-6 ∇f is orthogonal to the level surfaces of f

If $\mathbf{F}$ is conservative and C is piecewise smooth, line integrals can be evaluated using the *Fundamental Theorem of Line Integrals*.

Theorem 19 (Fundamental Theorem of Line Integrals). *If $\mathbf{F}$ is conservative and the curve C defined by $\mathbf{r}(t)$, $a \leq t \leq b$ is piecewise smooth,*

$$\int_C \mathbf{F} \cdot d\mathbf{r} = f\left(\mathbf{r}(b)\right) - f\left(\mathbf{r}(a)\right) \qquad (5.16)$$

where $\mathbf{F} = \nabla f$.

EXAMPLE 5.5.4: Find $\int_C \mathbf{F} \cdot d\mathbf{r}$ where $\mathbf{F}(x, y) = (e^{-y} - ye^{-x})\,\mathbf{i} + (e^{-x} - xe^{-y})\,\mathbf{j}$ and C is defined by $\mathbf{r}(t) = \cos t\,\mathbf{i} + \ln(2t/\pi)\,\mathbf{j}$, $\pi/2 \leq t \leq 4\pi$.

SOLUTION: We see that $\mathbf{F}$ is conservative with D and find that $f(x, y) = xe^{-y} + ye^{-x}$ satisfies $\nabla f = \mathbf{F}$ with `Integrate`.

```
In[1157]:= f[x_,y_] = {Exp[-y] -y Exp[-x],
           Exp[-x] -x Exp[-y]}; r[t_] = {Cos[t],
           Log[2t/π]};
```

```
In[1158]:= BoxData({D[f[x,y][[1]],y]//Simplify,
           D[f[x,y][[2]],x]//Simplify })
Out[1158]= -e^-x - e^-y
Out[1158]= -e^-x - e^-y
```

```
In[1159]:= lf = Integrate[f[x,y][[1]],x]
Out[1159]= e^-y x + e^-x y
```

Hence, using (5.16),

$$\int_C \mathbf{F} \cdot d\mathbf{r} = \left(xe^{-y} + ye^{-x}\right)\Big]_{x=0,y=0}^{x=1,y=\ln 8} = \frac{3\ln 2}{e} + \frac{1}{8} \approx 0.890.$$

```
In[1160]:= xr[t_] = Cos[t];
           yr[t_] = Log[2 t/π];
           {xr[π/2],yr[π/2]}

           {xr[4π],yr[4π]}
Out[1160]= {0, 0}
Out[1160]= {1, Log[8]}
```

```
In[1161]:= Simplify[lf/.{x- >1, y- > Log[8]}]

           N[%]
Out[1161]= 1/8 + Log[8]/e

Out[1161]= 0.889984
```

∎

If C is a piecewise smooth simple closed curve and $P(x, y)$ and $Q(x, y)$ have continuous partial derivatives, *Green's Theorem* relates the line integral $\oint_C (P(x, y)\,dx + Q(x, y)\,dy)$ to a double integral.

Theorem 20 (Green's Theorem). *Let C be a piecewise smooth simple closed curve in the plane and R the region bounded by C. If $P(x, y)$ and $Q(x, y)$ have continuous partial derivatives on R,*

We assume that the symbol $\oint$ means to evaluate the integral in the positive (or counter-clockwise) direction.

$$\oint_C (P(x, y)\,dx + Q(x, y)\,dy) = \iint_R \left(\frac{\partial Q}{\partial x} - \frac{\partial P}{\partial y}\right) dA. \tag{5.17}$$

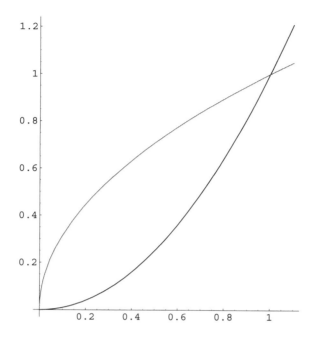

Figure 5-7 $y = x^2$ and $y = \sqrt{x}, 0 \le x \le 1$

EXAMPLE 5.5.5: Evaluate

$$\oint_C \left(e^{-x} - \sin y\right) dx + \left(\cos x - e^{-y}\right) dy$$

where C is the boundary of the region between $y = x^2$ and $x = y^2$.

SOLUTION: After defining $P(x, y) = e^{-x} - \sin y$ and $Q(x, y) = \cos x - e^{-y}$, we use Plot to determine the region R bounded by C in Figure 5-7.

```
In[1162]:= p[x_, y_] = Exp[-x] - Sin[y];
           q[x_, y_] = Cos[x] - Exp[-y];
           Plot[{x^2, Sqrt[x]}, {x, 0, 1.1},
             PlotStyle- > {GrayLevel[0], GrayLevel[0.3]},
             AspectRatio- > Automatic]
```

Using equation (5.17),

$$\oint_C \left(e^{-x} - \sin y\right) dx + \left(\cos x - e^{-y}\right) dy = \iint_R \left(\frac{\partial Q}{\partial x} - \frac{\partial P}{\partial y}\right) dA$$

$$= \iint_R (\cos y - \sin x)\, dA$$

$$= \int_0^1 \int_{x^2}^{\sqrt{x}} (\cos y - \sin x)\, dy\, dx,$$

```
In[1163]:= dqdp = Simplify[D[q[x,y],x] - D[p[x,y],y]]
```

```
Out[1163]= Cos[y] - Sin[x]
```

which we evaluate with Integrate.

```
In[1164]:= Integrate[dqdp, {x, 0, 1}, {y, x^2, Sqrt[x]}]

            N[%]
```

$$Out[1164]= -2 - \sqrt{\frac{\pi}{2}}\ \texttt{FresnelC}\left[\sqrt{\frac{2}{\pi}}\right] - \sqrt{\frac{\pi}{2}}\ \texttt{FresnelS}\left[\sqrt{\frac{2}{\pi}}\right]$$
$$+ 4\ \texttt{Sin}[1]$$

```
Out[1164]= 0.151091
```

Notice that the result is given in terms of the FresnelS and FresnelC functions, which are defined by

$$\texttt{FresnelS[x]} = \int_0^x \sin\left(\frac{\pi}{2}t^2\right) dt \quad \text{and} \quad \texttt{FresnelC[x]} = \int_0^x \cos\left(\frac{\pi}{2}t^2\right) dt.$$

A more meaningful approximation is obtained with N. We conclude that

$$\int_0^1 \int_{x^2}^{\sqrt{x}} (\cos y - \sin x)\, dy\, dx \approx 0.151.$$

■

5.5.3 Surface Integrals

Let S be the graph of $z = f(x, y)$ $(y = h(x, z), x = k(y, z))$ and let R_{xy} (R_{xz}, R_{yz}) be the projection of S onto the xy (xz, yz) plane. Then,

$$\iint_S g(x, y, z) \, dS = \iint_{R_{xy}} g(x, y, f(x, y)) \sqrt{[f_x(x, y)]^2 + [f_y(x, y)]^2 + 1} \, dA \qquad (5.18)$$

$$= \iint_{R_{xz}} g(x, h(x, z), z) \sqrt{[h_x(x, z)]^2 + [h_z(x, z)]^2 + 1} \, dA \qquad (5.19)$$

$$= \iint_{R_{yz}} g(k(y, z), y, z) \sqrt{[k_y(y, z)]^2 + [k_z(y, z)]^2 + 1} \, dA. \qquad (5.20)$$

If S is defined parametrically by

$$\mathbf{r}(s, t) = x(s, t)\mathbf{i} + y(s, t)\mathbf{j} + z(s, t)\mathbf{k}, \quad (s, t) \in R$$

the formula

$$\iint_S g(x, y, z) \, dS = \iint_R g(\mathbf{r}(s, t)) \, \|\mathbf{r}_s \times \mathbf{r}_t\| \, dA, \qquad (5.21)$$

where

$$\mathbf{r}_s = \frac{\partial x}{\partial s}\mathbf{i} + \frac{\partial y}{\partial s}\mathbf{j} + \frac{\partial z}{\partial s}\mathbf{k} \quad \text{and} \quad \mathbf{r}_t = \frac{\partial x}{\partial t}\mathbf{i} + \frac{\partial y}{\partial t}\mathbf{j} + \frac{\partial z}{\partial t}\mathbf{k},$$

is also useful.

Theorem 21 (The Divergence Theorem). *Let Q be any domain with the property that each line through any interior point of the domain cuts the boundary in exactly two points, and such that the boundary S is a piecewise smooth closed, oriented surface with unit normal $\mathbf{n}$. If $\mathbf{F}$ is a vector field that has continuous partial derivatives on Q, then*

For our purposes, a surface is **oriented** if it has two distinct sides.

$$\iiint_Q \nabla \cdot \mathbf{F} \, dV = \iiint_Q \text{div} \mathbf{F} \, dV = \iint_S \mathbf{F} \cdot \mathbf{n} \, dS \qquad (5.22)$$

In (5.22), $\iint_S \mathbf{F} \cdot \mathbf{n} \, dS$ is called the **outward flux** of the vector field $\mathbf{F}$ across the surface S. If S is a portion of the level curve $g(x, y) = C$ for some g, then a unit normal vector $\mathbf{n}$ may be taken to be either

$$\mathbf{n} = \frac{\nabla g}{\|\nabla g\|} \quad \text{or} \quad \mathbf{n} = -\frac{\nabla g}{\|\nabla g\|}.$$

If S is defined parametrically by

$$\mathbf{r}(s, t) = x(s, t)\mathbf{i} + y(s, t)\mathbf{j} + z(s, t)\mathbf{k}, \quad (s, t) \in R,$$

a unit normal vector to the surface is

$$\mathbf{n} = \frac{\mathbf{r}_s \times \mathbf{r}_t}{\|\mathbf{r}_s \times \mathbf{r}_t\|}$$

and (5.22) becomes

$$\iint_S \mathbf{F} \cdot \mathbf{n}\, dS = \iint_R \mathbf{F} \cdot (\mathbf{r}_s \times \mathbf{r}_t)\, dA.$$

EXAMPLE 5.5.6: Find the outward flux of the vector field

$$\mathbf{F}(x, y, z) = \left(xz + xyz^2\right)\mathbf{i} + \left(xy + x^2yz\right)\mathbf{j} + \left(yz + xy^2z\right)\mathbf{k}$$

through the surface of the cube cut from the first octant by the planes $x = 1, y = 1$, and $z = 1$.

SOLUTION: By the Divergence theorem,

$$\iint_{\text{cube surface}} \mathbf{F} \cdot \mathbf{n}\, dA = \iiint_{\text{cube interior}} \nabla \cdot \mathbf{F}\, dV.$$

Hence, without the Divergence theorem, calculating the outward flux would require six separate integrals, corresponding to the six faces of the cube. After defining $\mathbf{F}$, we compute $\nabla \cdot \mathbf{F}$ with `Div`.

`Div` is contained in the VectorAnalysis package. You do not need to reload the VectorAnalysis package if you have already loaded it during your *current* Mathematica session.

```
In[1165]:= << Calculus`VectorAnalysis`
```

```
In[1166]:= f[x_, y_, z_] = {x z + x y z^2, x y + x^2y z, y z
                  + x y^2 z};
```

```
In[1167]:= divf = Div[f[x, y, z], Cartesian[x, y, z]]
Out[1167]= x + y + x y^2 + z + x^2 z + y z^2
```

The outward flux is then given by

$$\iiint_{\text{cube interior}} \nabla \cdot \mathbf{F}\, dV = \int_0^1 \int_0^1 \int_0^1 \nabla \cdot \mathbf{F}\, dz\, dy\, dx = 2,$$

which we compute with `Integrate`.

```
In[1168]:= Integrate[divf, {z, 0, 1}, {y, 0, 1}, {x, 0, 1}]
Out[1168]= 2
```

∎

Theorem 22 (Stokes' Theorem). *Let S be an oriented surface with finite surface area, unit normal* $\mathbf{n}$, *and boundary C. Let* $\mathbf{F}$ *be a continuous vector field defined on S such that the components of* $\mathbf{F}$ *have continuous partial derivatives at each nonboundary point of S. Then,*

$$\oint_C \mathbf{F} \cdot d\mathbf{r} = \iint_S \text{curl } \mathbf{F} \cdot \mathbf{n}\, dS. \qquad (5.23)$$

In other words, the surface integral of the normal component of the curl of **F** taken over S equals the line integral of the tangential component of the field taken over C. In particular, if $\mathbf{F} = P(x, y, z)\mathbf{i} + Q(x, y, z)\mathbf{j} + R(x, y, z)\mathbf{k}$, then

$$\int_C (P(x, y, z)dx + Q(x, y, z)dy + R(x, y, z)dz) = \iint_S \text{curl } \mathbf{F} \cdot \mathbf{n} \, dS.$$

EXAMPLE 5.5.7: Verify Stokes' theorem for the vector field

$$\mathbf{F}(x, y, z) = \left(x^2 - y\right)\mathbf{i} + \left(y^2 - z\right)\mathbf{j} + \left(x + z^2\right)\mathbf{k}$$

and S the portion of the paraboloid $z = f(x, y) = 9 - \left(x^2 + y^2\right)$, $z \geq 0$.

SOLUTION: After loading the `VectorAnalysis` package, we define **F** and f. The curl of **F** is computed with `Curl` in `curlF`.

In[1169] := `<< Calculus`VectorAnalysis``

In[1170] := `capf[x_, y_, z_] = {x^2 - y, y^2 - z, x + z^2};`
`f[x_, y_] = 9 - (x^2 + y^2);`

In[1171] := `curlcapf = Curl[capf[x, y, z],`
`Cartesian[x, y, z]]`
Out[1171] = `{1, -1, 1}`

Next, we define the function $h(x, y, z) = z - f(x, y)$. A normal vector to the surface is given by ∇h. A unit normal vector, **n**, is then given by $\mathbf{n} = \dfrac{\nabla h}{\|\nabla h\|}$, which is computed in un.

In[1172] := `BoxData({h[x_, y_, z_] = z - f[x, y], normtosurf =`
`Grad[h[x, y, z], Cartesian[x, y, z]]})`
Out[1172] = $-9 + x^2 + y^2 + z$
Out[1172] = `{2 x, 2 y, 1}`

In[1173] := `un = Simplify`
`[normtosurf/Sqrt[normtosurf.normtosurf]]`
Out[1173] = $\left\{ \dfrac{2\ x}{\sqrt{1 + 4\ x^2 + 4\ y^2}},\ \dfrac{2\ y}{\sqrt{1 + 4\ x^2 + 4\ y^2}},\ \dfrac{1}{\sqrt{1 + 4\ x^2 + 4\ y^2}} \right\}$

The dot product curl **F** · **n** is computed in g.

In[1174] := `g = Simplify[curlcapf.un]`
Out[1174] = $\dfrac{1 + 2\ x - 2\ y}{\sqrt{1 + 4\ x^2 + 4\ y^2}}$

In this example, R, the projection of $f(x, y)$ onto the xy-plane, is the region bounded by the graph of the circle $x^2 + y^2 = 9$.

Using the surface integral evaluation formula (5.18),

$$\iint_S \operatorname{curl} \mathbf{F} \cdot \mathbf{n}\, dS = \iint_R g\left(x, y, f(x, y)\right) \sqrt{\left[f_x(x, y)\right]^2 + \left[f_y(x, y)\right]^2 + 1}\, dA$$

$$= \int_{-3}^{3} \int_{-\sqrt{9-x^2}}^{\sqrt{9-x^2}} g\left(x, y, f(x, y)\right) \sqrt{\left[f_x(x, y)\right]^2 + \left[f_y(x, y)\right]^2 + 1}\, dy\, dx$$

$$= 9\pi,$$

which we compute with `Integrate`.

```
In[1175]:= tointegrate = Simplify[(g/.z- > f[x,y])*
                Sqrt[D[f[x,y],x]^2 + D[f[x,y],y]^2 + 1]]
Out[1175]= 1 + 2 x - 2 y

In[1176]:= il = Integrate[tointegrate, {x, -3, 3},
                {y, -Sqrt[9 - x^2], Sqrt[9 - x^2]}]
Out[1176]= 9 π
```

To verify Stokes' theorem, we must compute the associated line integral. Notice that the boundary of $z = f(x, y) = 9 - (x^2 + y^2)$, $z = 0$, is the circle $x^2 + y^2 = 9$ with parametrization $x = 3\cos t$, $y = 3\sin t$, $z = 0$, $0 \le t \le 2\pi$. This parametrization is substituted into $\mathbf{F}(x, y, z)$ and named pvf.

```
In[1177]:= pvf = capf[3 Cos[t], 3 Sin[t], 0]
Out[1177]= {9 Cos[t]² - 3 Sin[t], 9 Sin[t]², 3 Cos[t]}
```

To evaluate the line integral along the circle, we next define the parametrization of the circle and calculate $d\mathbf{r}$. The dot product of pvf and dr represents the integrand of the line integral.

```
In[1178]:= r[t_] = {3 Cos[t], 3 Sin[t], 0};
                dr = r'[t]
Out[1178]= {-3 Sin[t], 3 Cos[t], 0}

In[1179]:= tointegrate = pvf.dr;
```

As before with x and y, we instruct Mathematica to assume that t is real, compute the dot product of pvf and dr, and evaluate the line integral with `Integrate`.

```
In[1180]:= Integrate[tointegrate, {t, 0, 2π}]
Out[1180]= 9 π
```

As expected, the result is 9π.

∎

5.5.4 A Note on Nonorientability

See "When is a surface *not* orientable?" by Braselton, Abell, and Braselton [5] for a detailed discussion regarding the examples in this section.

Suppose that S is the surface determined by

$$\mathbf{r}(s, t) = x(s, t)\mathbf{i} + y(s, t)\mathbf{j} + z(s, t)\mathbf{k}, \quad (s, t) \in R$$

and let

$$\mathbf{n} = \frac{\mathbf{r}_s \times \mathbf{r}_t}{\|\mathbf{r}_s \times \mathbf{r}_t\|} \quad \text{or} \quad \mathbf{n} = -\frac{\mathbf{r}_s \times \mathbf{r}_t}{\|\mathbf{r}_s \times \mathbf{r}_t\|}, \tag{5.24}$$

where

$$\mathbf{r}_s = \frac{\partial x}{\partial s}\mathbf{i} + \frac{\partial y}{\partial s}\mathbf{j} + \frac{\partial z}{\partial s}\mathbf{k} \quad \text{and} \quad \mathbf{r}_t = \frac{\partial x}{\partial t}\mathbf{i} + \frac{\partial y}{\partial t}\mathbf{j} + \frac{\partial z}{\partial t}\mathbf{k},$$

if $\|\mathbf{r}_s \times \mathbf{r}_t\| \neq 0$. If $\mathbf{n}$ is defined, $\mathbf{n}$ is orthogonal (or perpendicular) to S. We state three familiar definitions of *orientable*.

- S is **orientable** if S has a unit normal vector field, $\mathbf{n}$, that varies continuously between any two points (x_0, y_0, z_0) and (x_1, y_1, z_1) on S. (See [7].)
- S is **orientable** if S has a continuous unit normal vector field, $\mathbf{n}$. (See [7] and [19].)
- S is **orientable** if a unit vector $\mathbf{n}$ can be defined at every nonboundary point of S in such a way that the normal vectors vary continuously over the surface S. (See [14].)

A path is **order preserving** if our chosen orientation is preserved as we move along the path.

Thus, a surface like a torus is orientable.

Also see Example 2.3.18.

EXAMPLE 5.5.8 (The Torus): Using the standard parametrization of the torus, we use `ParametricPlot3D` to plot the torus if $c = 3$ and $a = 1$ in Figure 5-8.

```
In[1181]:= Clear[r]

        c = 3;

        a = 1;

        x[s_, t_] = (c + a Cos[s]) Cos[t];

        y[s_, t_] = (c + a Cos[s]) Sin[t];

        z[s_, t_] = a Sin[s];

        r[s_, t_] = {x[s, t], y[s, t], z[s, t]};
```

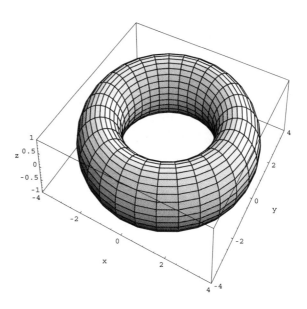

Figure 5-8   A torus

```
In[1182]:= threedplt = ParametricPlot3D[r[s,t],
              {s,-π,π},{t,-π,π},Compiled->False,
              PlotPoints->{30,30},AspectRatio->1,
              LightSources → {{{1.,0.,1.},
                GrayLevel[0.3]},{{1.,1.,1.},
              GrayLevel[0.5]},{{0.,1.,1.},
              GrayLevel[0.4]}},PlotRange->{{-4,4},
              {-4,4},{-1,1}},BoxRatios->{4,4,1},
              AxesLabel->{"x","y","z"}]
```

To plot a normal vector field on the torus, we compute $\frac{\partial}{\partial s}\mathbf{r}(s,t)$,

```
In[1183]:= rs = D[r[s,t],s]

Out[1183]= {-Cos[t] Sin[s],-Sin[s] Sin[t],Cos[s]}
```

and $\frac{\partial}{\partial t}\mathbf{r}(s,t)$.

```
In[1184]:= rt = D[r[s,t],t]

Out[1184]= {-(3+Cos[s]) Sin[t],(3+Cos[s]) Cos[t],0}
```

The cross product $\frac{\partial}{\partial s}\mathbf{r}(s,t) \times \frac{\partial}{\partial t}$ is formed in rscrossrt.

In[1185]:= **rscrossrt = Cross[rs, rt]//Simplify**

Out[1185]= {- Cos[s] (3 + Cos[s]) Cos[t],
              - Cos[s] (3 + Cos[s]) Sin[t],
              - (3 + Cos[s]) Sin[s]}

In[1186]:= **Sqrt[rscrossrt.rscrossrt]//FullSimplify**

Out[1186]= $\sqrt{(3 + \text{Cos}[s])^2}$

Using equation (5.24), we define un: given *s* and *t*, un[s,t] returns a unit normal to the torus.

In[1187]:= **Clear[un]un[s_, t_] =**
              -rscrossrt/Sqrt[rscrossrt.rscrossrt]//
              PowerExpand//FullSimplify

Out[1187]= $\left\{ \dfrac{\text{Cos}[s]\ (3 + \text{Cos}[s])\ \text{Cos}[t]}{\sqrt{(3 + \text{Cos}[s])^2}}, \right.$
              $\dfrac{\text{Cos}[s]\ (3 + \text{Cos}[s])\ \text{Sin}[t]}{\sqrt{(3 + \text{Cos}[s])^2}},$
              $\left. \dfrac{(3 + \text{Cos}[s])\ \text{Sin}[s]}{\sqrt{(3 + \text{Cos}[s])^2}} \right\}$

In[1188]:= **Map[PowerExpand, un[s, t]]**

Out[1188]= {Cos[s] Cos[t], Cos[s] Sin[t], Sin[s]}

In[1189]:= **r[s, t]**

Out[1189]= {(3+Cos[s]) Cos[t], (3+Cos[s]) Sin[t], Sin[s]}

In[1190]:= **un[s, t]**

Out[1190]= $\left\{ \dfrac{\text{Cos}[s]\ (3 + \text{Cos}[s])\ \text{Cos}[t]}{\sqrt{(3 + \text{Cos}[s])^2}}, \right.$
              $\dfrac{\text{Cos}[s]\ (3 + \text{Cos}[s])\ \text{Sin}[t]}{\sqrt{(3 + \text{Cos}[s])^2}},$
              $\left. \dfrac{(3 + \text{Cos}[s])\ \text{Sin}[s]}{\sqrt{(3 + \text{Cos}[s])^2}} \right\}$

To plot the normal vector field on the torus, we take advantage of the command ListPlotVectorField3D, which is contained in the **Plot-Field3D** package that is located in the **Graphics** folder (or directory).See Figure 5-9.

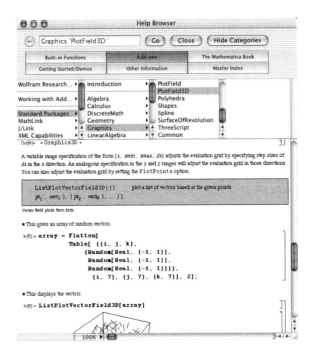

$In[1191] := \; \texttt{<< Graphics `PlotField3D`}$

$In[1192] := \; \texttt{Clear[vecs]}$

```
vecs =
Flatten[Table[{r[s, t], un[s, t]},
{s, -π, π, 2π/14}, {t, -π, π, 2π/29}], 1];
```

$In[1193] := \; \texttt{pp2 = ListPlotVectorField3D[vecs,}$
$\qquad\qquad\quad\; \texttt{VectorHeads- > True]}$

We use Show (illustrating the use of the ViewPoint option) together with GraphicsArray to see the vector field on the torus together from various angles in Figure 5-10. Regardless of the viewing angle, the figure looks the same; the torus is orientable.

$In[1194] := \; \texttt{g1 = Show[threedp1t, pp2, AspectRatio- > 1,}$
$\qquad\qquad\quad\; \texttt{PlotRange- > \{\{-5, 5\}, \{-5, 5\}, \{-2, 2\}\},}$
$\qquad\qquad\quad\; \texttt{BoxRatios- > \{4, 4, 1\},}$
$\qquad\qquad\quad\; \texttt{AxesLabel- > \{"x", "y", "z"\},}$
$\qquad\qquad\quad\; \texttt{ViewPoint- > \{2.729, -0., 2.\}]}$

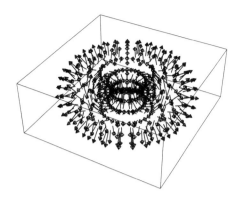

Figure 5-9   Unit normal vector field on a torus

```
In[1195]:= g2 = Show[threedp1t, pp2, AspectRatio- >1,
                 PlotRange- > {{-5, 5}, {-5, 5}, {-2, 2}},
                 BoxRatios- > {4, 4, 1},
                 AxesLabel- > {"x", "y", "z"},
                 ViewPoint- > {1.365, -2.364, 2.}]

In[1196]:= g3 = Show[threedp1t, pp2, AspectRatio- >1,
                 PlotRange- > {{-5, 5}, {-5, 5}, {-2, 2}},
                 BoxRatios- > {4, 4, 1},
                 AxesLabel- > {"x", "y", "z"},
                 ViewPoint- > {-1.365, -2.364, 2.}]

In[1197]:= g4 = Show[threedp1t, pp2, AspectRatio- >1,
                 PlotRange- > {{-5, 5}, {-5, 5}, {-2, 2}},
                 BoxRatios- > {4, 4, 1},
                 AxesLabel- > {"x", "y", "z"},
                 ViewPoint- > {-2.729, 0., 2.}]

In[1198]:= g5 = Show[threedp1t, pp2, AspectRatio- >1,
                 PlotRange- > {{-5, 5}, {-5, 5}, {-2, 2}},
                 BoxRatios- > {4, 4, 1},
                 AxesLabel- > {"x", "y", "z"},
                 ViewPoint- > {-1.365, 2.364, 2.}]

In[1199]:= g6 = Show[threedp1t, pp2, AspectRatio- >1,
                 PlotRange- > {{-5, 5}, {-5, 5}, {-2, 2}},
                 BoxRatios- > {4, 4, 1},
                 AxesLabel- > {"x", "y", "z"},
                 ViewPoint- > {1.365, 2.364, 2.}]

In[1200]:= Show[GraphicsArray[{{g1, g2}, {g3, g4},
                 {g5, g6}}]]
```

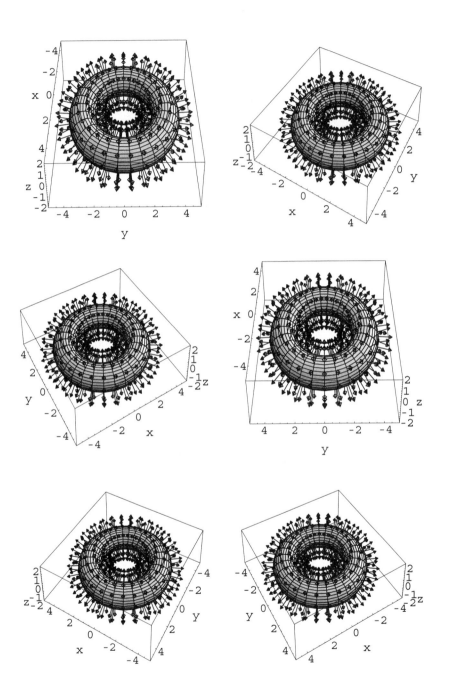

Figure 5-10   The torus is orientable

If a 2-manifold, S, has an **order reversing path** (or **not order preserving path**), S is **nonorientable** (or **not orientable**).

Determining whether a given surface S is orientable or not may be a difficult problem.

EXAMPLE 5.5.9 (The Möbius Strip): The *Möbius strip* is frequently cited as an examp of a nonorientable surface with boundary: it has one side and is physically easy to construct by hand by half twisting and taping (or pasting) together the ends of a piece of paper (for example, see [5],[7], [14], and [19]). A parametrization of the Möbius strip is $\mathbf{r}(s, t) = x(s, t)\mathbf{i} + y(s, t)\mathbf{j} + z(s, t)\mathbf{k}$, $-1 \le s \le 1$, $-\pi \le t \le \pi$, where

$$x = \left[c + s\cos\left(\frac{1}{2}t\right)\right]\cos t, \quad y = \left[c + s\cos\left(\frac{1}{2}t\right)\right]\sin t, \text{ and}$$

$$z = s\sin\left(\frac{1}{2}t\right), \quad (5.25)$$

and we assume that $c > 1$. In Figure 5-11, we graph the Möbius strip using $c = 3$.

```
In[1201]:= c = 3;

          x[s_, t_] = (c + s Cos[t/2]) Cos[t];

          y[s_, t_] = (c + s Cos[t/2]) Sin[t];

          z[s_, t_] = s Sin[t/2];

          r[s_, t_] = {x[s, t], y[s, t], z[s, t]};

In[1202]:= threedp1 = ParametricPlot3D[r[s, t], {s, -1, 1},
             {t, -π, π}, Compiled- > False,
             PlotPoints- > {30, 30}, AspectRatio- > 1,
             LightSources → {{{1., 0., 1.},
             GrayLevel[0.4]}, {{1., 1., 1.},
             GrayLevel[0.6]},
                {{0., 1., 1.}, GrayLevel[0.5]}},
             PlotRange- > {{-4, 4}, {-4, 4}, {-1, 1}},
             BoxRatios- > {4, 4, 1},
             AxesLabel- > {"x", "y", "z"}]
```

Although it is relatively easy to see in the plot that the Möbius strip has only one side, the fact that a unit vector, $\mathbf{n}$, normal to the Möbius strip at a point P reverses its direction as $\mathbf{n}$ moves around the strip to P is not obvious to the novice.

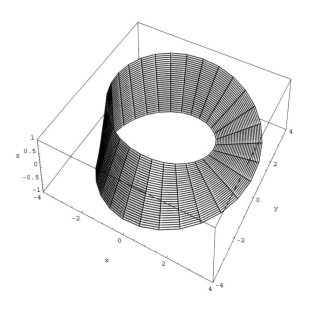

Figure 5-11   Parametric plot of equations (5.25) if $c = 3$

With Mathematica, we compute $\|\mathbf{r}_s \times \mathbf{r}_t\|$ and $\mathbf{n} = \dfrac{\mathbf{r}_s \times \mathbf{r}_t}{\|\mathbf{r}_s \times \mathbf{r}_t\|}$.

In[1203]:= **rs = D[r[s, t], s]**

Out[1203]= $\left\{ \text{Cos}\left[\dfrac{t}{2}\right] \text{Cos}[t], \text{Cos}\left[\dfrac{t}{2}\right] \text{Sin}[t], \text{Sin}\left[\dfrac{t}{2}\right] \right\}$

In[1204]:= **rt = D[r[s, t], t]**

Out[1204]= $\left\{ -\dfrac{1}{2} s \, \text{Cos}[t] \, \text{Sin}\left[\dfrac{t}{2}\right] - \left(3 + s \, \text{Cos}\left[\dfrac{t}{2}\right]\right) \text{Sin}[t], \right.$
$\left(3 + s \, \text{Cos}\left[\dfrac{t}{2}\right]\right) \text{Cos}[t] - \dfrac{1}{2} s \, \text{Sin}\left[\dfrac{t}{2}\right] \text{Sin}[t],$
$\left. \dfrac{1}{2} s \, \text{Cos}\left[\dfrac{t}{2}\right] \right\}$

In[1205]:= **rscrossrt = Cross[rs, rt]//Simplify**

Out[1205]= $\left\{ -\dfrac{1}{2} \left(-s \, \text{Cos}\left[\dfrac{t}{2}\right] + 6 \, \text{Cos}[t] + s \, \text{Cos}\left[\dfrac{3t}{2}\right]\right) \text{Sin}\left[\dfrac{t}{2}\right], \right.$
$\dfrac{1}{4} \left(-s - 6 \, \text{Cos}\left[\dfrac{t}{2}\right] - 2 s \, \text{Cos}[t] + 6 \, \text{Cos}\left[\dfrac{3t}{2}\right] + \right.$
$\left. s \, \text{Cos}[2t]\right), \text{Cos}\left[\dfrac{t}{2}\right] \left(3 + s \, \text{Cos}\left[\dfrac{t}{2}\right]\right) \right\}$

In[1206]:= **Sqrt[rscrossrt.rscrossrt]//FullSimplify**

Out[1206]= $\sqrt{9 + \dfrac{3 s^2}{4} + 6 s \, \text{Cos}\left[\dfrac{t}{2}\right] + \dfrac{1}{2} s^2 \, \text{Cos}[t]}$

In[1207]:= **Clear[un]**

un[s_, t_] = rscrossrt/Sqrt
[rscrossrt.rscrossrt]//FullSimplify

$$
Out[1207]= \left\{ \frac{s\ \text{Sin}[t] - \text{Cos}[t]\left(6\ \text{Sin}\left[\frac{t}{2}\right] + s\ \text{Sin}[t]\right)}{\sqrt{36 + 3\ s^2 + 24\ s\ \text{Cos}\left[\frac{t}{2}\right] + 2\ s^2\ \text{Cos}[t]}}, \right.
$$

$$
-\frac{3\ \text{Cos}\left[\frac{t}{2}\right] - 3\ \text{Cos}\left[\frac{3t}{2}\right] + s\ (\text{Cos}[t] + \text{Sin}[t]^2)}{\sqrt{36 + 3\ s^2 + 24\ s\ \text{Cos}\left[\frac{t}{2}\right] + 2\ s^2\ \text{Cos}[t]}},
$$

$$
\left. \frac{s + 6\ \text{Cos}\left[\frac{t}{2}\right] + s\ \text{Cos}[t]}{\sqrt{36 + 3\ s^2 + 24\ s\ \text{Cos}\left[\frac{t}{2}\right] + 2\ s^2\ \text{Cos}[t]}} \right\}
$$

Consider the path C given by $r(0, t)$, $-\pi \le t \le \pi$ that begins and ends at $\langle -3, 0, 0 \rangle$. On C, $n(0, t)$ is given by

In[1208]:= **un[0, t]**

$$
Out[1208]= \left\{ -\text{Cos}[t]\ \text{Sin}\left[\frac{t}{2}\right], \frac{1}{6}\left(-3\ \text{Cos}\left[\frac{t}{2}\right] + 3\ \text{Cos}\left[\frac{3t}{2}\right]\right), \text{Cos}\left[\frac{t}{2}\right] \right\}
$$

At $t = -\pi$, $n(0, -\pi) = \langle 1, 0, 0 \rangle$, while at $t = \pi$, $n(0, \pi) = \langle -1, 0, 0 \rangle$.

In[1209]:= **r[0, -π]**

r[0, π]

$Out[1209]= \{-3, 0, 0\}$

$Out[1209]= \{-3, 0, 0\}$

As n moves along C from $r(0, -\pi)$ to $r(0, \pi)$, the orientation of n reverses, as shown in Figure 5-12.

In[1210]:= **<< Graphics`PlotField3D`;**

vecs = Table[{r[0, t], un[0, t]},
            {t, -π, π, 2π/59}];

pp2 = ListPlotVectorField3D[vecs,
        VectorHeads- > True,
        DisplayFunction- > Identity];

In[1211]:= **Show[threedp2, pp2,**
        ViewPoint- > {-2.093, 2.124, 1.6},
        AxesLabel- > {"x", "y", "z"},
        Boxed- > False,
        DisplayFunction- > $DisplayFunction]

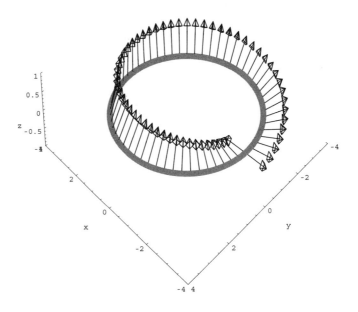

Figure 5-12   Parametric plot of equations (5.25) if $c = 3$

Several different views of Figure 5-12 on the Möbius strip shown in
Figure 5-11 are shown in Figure 5-13. C is an orientation reversing path
and we can conclude that the Möbius strip is not orientable.

An animation is particularly
striking.

```
In[1212]:= g1 = Show[threedp1, threedp2, pp2,
                ViewPoint- > {2.729, -0., 2.},
                AxesLabel- > {"x", "y", "z"}, Boxed- > False]

In[1213]:= g2 = Show[threedp1, threedp2, pp2,
                ViewPoint- > {1.365, -2.364, 2.},
                AxesLabel- > {"x", "y", "z"}, Boxed- > False]

In[1214]:= g3 = Show[threedp1, threedp2, pp2,
                ViewPoint- > {-1.365, -2.364, 2.},
                AxesLabel- > {"x", "y", "z"}, Boxed- > False]

In[1215]:= g4 = Show[threedp1, threedp2, pp2,
                ViewPoint- > {-2.729, 0., 2.},
                AxesLabel- > {"x", "y", "z"}, Boxed- > False]
```

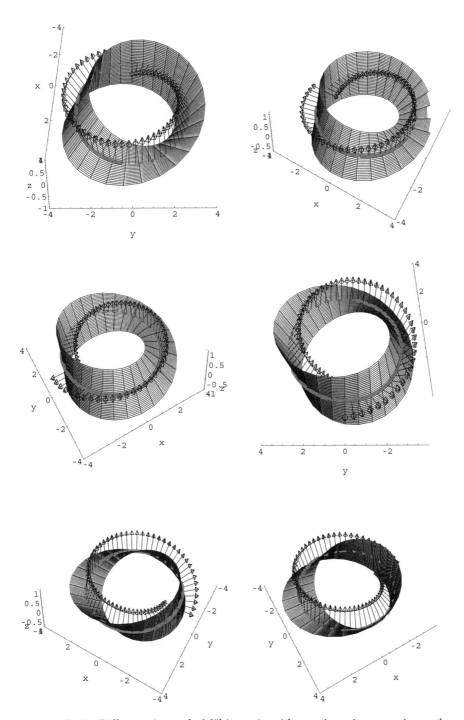

Figure 5-13   Different views of a Möbius strip with an orientation reversing path

```
In[1216]:= g5 = Show[threedp1, threedp2, pp2,
              ViewPoint- > {-1.365, 2.364, 2.},
              AxesLabel- > {"x", "y", "z"}, Boxed- > False]

In[1217]:= g6 = Show[threedp1, threedp2, pp2,
              ViewPoint- > {1.365, 2.364, 2.},
              AxesLabel- > {"x", "y", "z"}, Boxed- > False]

In[1218]:= Show[GraphicsArray[{{g1, g2}, {g3, g4},
              {g5, g6}}]]
```

EXAMPLE 5.5.10 (The Klein Bottle): The *Klein bottle* is an interesting surface with neither an inside nor an outside, which indicates to us that it is not orientable. In Figure 5-14 (a) we show the "usual" *immersion* of the Klein bottle. Although the Klein bottle does not intersect itself, it is not possible to visualize it in Euclidean 3-space without it doing so. Visualizations of 2-manifolds like the Klein bottle's "usual" rendering in Euclidean 3-space are called *immersions*. (See [11]for a nontechnical discussion of immersions.)

```
In[1219]:= r = 4 (1 - 1/2 Cos[u]);

         x1[u_, v_] = 6 (1 + Sin[u]) Cos[u]
           +r Cos[u] Cos[v];

         x2[u_, v_] = 6 (1 + Sin[u]) Cos[u] + r Cos[v + π];

         y1[u_, v_] = 16 Sin[u] + r Sin[u] Cos[v];

         y2[u_, v_] = 16 Sin[u];

         z[u_, v_] = r Sin[v];

In[1220]:= kb1a = ParametricPlot3D[{x1[s, t], y1[s, t],
           z[s, t]}, {s, 0, π}, {t, 0, 2π},
           Compiled- > False, PlotPoints- > {30, 30},
           AspectRatio- > 1, LightSources →
           {{{1., 0., 1.}, GrayLevel[0.3]}, {{1., 1., 1.},
           GrayLevel[0.5]},
             {{0., 1., 1.}, GrayLevel[0.4]}},
             AxesLabel- > {"x", "y", "z"}]
```

```
In[1221]:= kb1b = ParametricPlot3D[{x1[s,t],y1[s,t],
                z[s,t]}, {s, π, 2π}, {t, 0, 2π},
                Compiled- > False, PlotPoints- > {30, 30},
                AspectRatio- > 1,
                LightSources → {{{1.,0.,1.},
                GrayLevel[0.3]}, {{1.,1.,1.},
                GrayLevel[0.5]},
                {{0.,1.,1.}, GrayLevel[0.4]}},
                AxesLabel- > {"x", "y", "z"}]


In[1222]:= kb1 = Show[kb1a, kb1b]
```

Figure 5-14 (b) shows the *Figure-8* immersion of the Klein bottle. Notice
that it is not easy to see that the Klein bottle has neither an inside nor
an outside in Figure (5.14).

```
In[1223]:= Clear[x, y, z, r, a]

           x[u_, v_] = (a + Cos[u/2] Sin[v]
                      - Sin[u/2] Sin[2v]) Cos[u];

           y[u_, v_] = (a + Cos[u/2] Sin[v]
                      - Sin[u/2] Sin[2v]) Sin[u];

           z[u_, v_] = Sin[u/2] Sin[v] + Cos[u/2] Sin[2v];

           r[u_, v_] = {x[u, v], y[u, v], z[u, v]};


In[1224]:= r[s, t]
```

$$Out[1224] = \left\{ \cos[s] \left(a + \cos\left[\frac{s}{2}\right] \sin[t] - \sin\left[\frac{s}{2}\right] \sin[2t]\right), \right.$$
$$\sin[s] \left(a + \cos\left[\frac{s}{2}\right] \sin[t] - \sin\left[\frac{s}{2}\right] \sin[2t]\right),$$
$$\left. \sin\left[\frac{s}{2}\right] \sin[t] + \cos\left[\frac{s}{2}\right] \sin[2t]\right\}$$

```
In[1225]:=    a = 3;
           x[u_, v_] = (a + Cos[u/2] Sin[v]
            - Sin[u/2] Sin[2v]) Cos[u];
           y[u_, v_] = (a + Cos[u/2] Sin[v]
            - Sin[u/2] Sin[2v]) Sin[u];
           z[u_, v_] = Sin[u/2] Sin[v] + Cos[u/2] Sin[2v];
           r[u_, v_] = {x[u, v], y[u, v], z[u, v]};
```

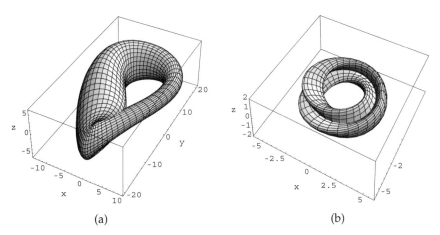

(a)   (b)

Figure 5-14   Two different immersions of the Klein bottle: (a) the "usual" immersion; (b) the Figure-8 immersion

```
In[1226]:= kb2 = ParametricPlot3D[r[s, t], {s, -π, π},
               {t, -π, π}, Compiled- > False,
               PlotPoints- > {30, 30}, AspectRatio- > 1,
               LightSources → {{{1., 0., 1.},
               GrayLevel[0.3]}, {{1., 1., 1.},
               GrayLevel[0.5]},
               {{0., 1., 1.}, GrayLevel[0.4]}},
               AxesLabel- > {"x", "y", "z"},
               PlotRange- > {{-6, 6}, {-6, 6}, {-2, 2}},
               BoxRatios- > {4, 4, 1}]

In[1227]:= Show[GraphicsArray[{kb1, kb2}]]
```

In fact, to many readers it may not be clear whether the Klein bottle is orientable or nonorientable, especially when we compare the graph to the graphs of the Möbius strip and torus in the previous examples.

A parametrization of the Figure-8 immersion of the Klein bottle (see [20]) is $\mathbf{r}(s, t) = x(s, t)\mathbf{i} + y(s, t)\mathbf{j} + z(s, t)\mathbf{k}$, $-\pi \leq s \leq \pi$, $-\pi \leq t \leq \pi$, where

$$x = \left[c + \cos\left(\frac{1}{2}s\right) \sin t - \sin\left(\frac{1}{2}s\right) \sin 2t \right] \cos s,$$

$$y = \left[c + \cos\left(\frac{1}{2}s\right) \sin t - \sin\left(\frac{1}{2}s\right) \sin 2t \right] \sin s,$$

(5.26)

and

$$z = \sin\left(\frac{1}{2}s\right) \sin t + \cos\left(\frac{1}{2}s\right) \sin 2t.$$

The plot in Figure 5-14 (b) uses equation (5.26) if $c = 3$.

Using (5.24), let

$$\mathbf{n} = \frac{\mathbf{r}_s \times \mathbf{r}_t}{\|\mathbf{r}_s \times \mathbf{r}_t\|}.$$

Let C be the path given by

$$\mathbf{r}(t, t) = x(t, t)\mathbf{i} + y(t, t)\mathbf{j} + z(t, t)\mathbf{k}, \quad -\pi \le t \le \pi \tag{5.27}$$

that begins and ends at $\mathbf{r}(-\pi, -\pi) = \mathbf{r}(\pi, \pi) = \langle -3, 0, 0 \rangle$ and where the components are given by (5.26). The components of $\mathbf{r}$ and $\mathbf{n}$ are computed with Mathematica. The final calculations are quite lengthy so we suppress the output of the last few by placing a semicolon ($;$) at the end of those commands.

```
In[1228] := rs = D[r[s, t], s]
```
$$Out[1228] = \left\{ \cos[s] \left(-\frac{1}{2} \sin\left[\frac{s}{2}\right] \sin[t] \right.\right.$$
$$\left. -\frac{1}{2} \cos\left[\frac{s}{2}\right] \sin[2t] \right) - \sin[s] \left(3 + \cos\left[\frac{s}{2}\right] \sin[t] \right.$$
$$\left. -\sin\left[\frac{s}{2}\right] \sin[2t] \right), \sin[s] \left(-\frac{1}{2} \sin\left[\frac{s}{2}\right] \sin[t] \right.$$
$$\left. -\frac{1}{2} \cos\left[\frac{s}{2}\right] \sin[2t] \right) + \cos[s] \left(3 + \cos\left[\frac{s}{2}\right] \sin[t] \right.$$
$$\left. -\sin\left[\frac{s}{2}\right] \sin[2t] \right), \frac{1}{2} \cos\left[\frac{s}{2}\right] \sin[t]$$
$$\left. -\frac{1}{2} \sin\left[\frac{s}{2}\right] \sin[2t] \right\}$$

```
In[1229] := rt = D[r[s, t], t]
```
$$Out[1229] = \left\{ \cos[s] \left(\cos\left[\frac{s}{2}\right] \cos[t] - 2 \cos[2t] \sin\left[\frac{s}{2}\right] \right), \right.$$
$$\left(\cos\left[\frac{s}{2}\right] \cos[t] - 2 \cos[2t] \sin\left[\frac{s}{2}\right] \right) \sin[s],$$
$$\left. 2 \cos\left[\frac{s}{2}\right] \cos[2t] + \cos[t] \sin\left[\frac{s}{2}\right] \right\}$$

```
In[1230] := rscrossrt = Cross[rs, rt];
```

```
In[1231] := normcross = Sqrt[rscrossrt.rscrossrt];
```

```
In[1232] := Clear[un]

           un[s_, t_] = -rscrossrt/Sqrt
           [rscrossrt.rscrossrt]
```

At $t = -\pi$, $\mathbf{n}(-\pi, -\pi) = \left\langle \frac{1}{\sqrt{5}}, 0, \frac{2}{\sqrt{5}} \right\rangle$, while at $t = \pi$, $\mathbf{n}(\pi, \pi) = \left\langle -\frac{1}{\sqrt{5}}, 0, -\frac{2}{\sqrt{5}} \right\rangle$ so as $\mathbf{n}$ moves along C from $\mathbf{r}(-\pi, -\pi)$ to $\mathbf{r}(\pi, \pi)$, the orientation of $\mathbf{n}$ reverses. Several different views of the orientation reversing path on the Klein bottle shown in Figure 5-14 (b) are shown in Figure 5-15.

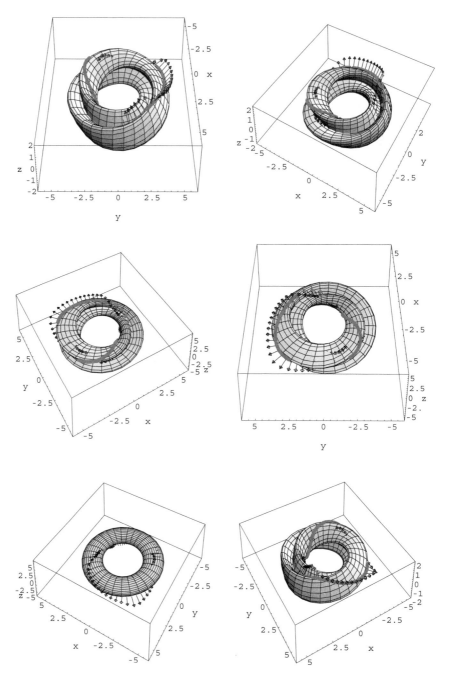

Figure 5-15   Different views of the Figure-8 immersion of the Klein bottle with an orientation reversing path

```
In[1233]:= l1 = Table[r[s, s], {s, -π, π, 2π/179}];

In[1234]:= threedp2 = Show[Graphics3D[{Thickness[0.02],
                GrayLevel[0.6], Line[l1]}],
            Axes- >Automatic,
            PlotRange- >{{-4, 4}, {-4, 4}, {-4, 4}},
            BoxRatios- >{4, 4, 1},
            AspectRatio- >1,
            DisplayFunction- >Identity];

In[1235]:= << Graphics`PlotField3D`;

            vecs = Table[{r[s, s], un[s, s]},
              {s, -π, π, 2π/59}];

            pp2 = ListPlotVectorField3D[vecs,
                VectorHeads- >True,
                DisplayFunction- >Identity];

In[1236]:= pp3 = Show[threedp2, pp2,
                AxesLabel- >{"x", "y", "z"},
                Boxed- >False,
                DisplayFunction- >$DisplayFunction]

In[1237]:= g1 = Show[kb2, threedp2, pp2,
                AspectRatio- >1,
                PlotRange- >{{-6, 6}, {-6, 6}, {-2, 2}},
                BoxRatios- >{4, 4, 1},
                AxesLabel- >{"x", "y", "z"},
                ViewPoint- >{2.729, -0., 2.}]

In[1238]:= g2 = Show[kb2, threedp2, pp2, AspectRatio- >1,
                PlotRange- >{{-6, 6}, {-6, 6}, {-2, 2}},
                BoxRatios- >{4, 4, 1},
                AxesLabel- >{"x", "y", "z"},
                ViewPoint- >{1.365, -2.364, 2.}]

In[1239]:= g3 = Show[kb2, threedp2, pp2, AspectRatio- >1,
                PlotRange- >{{-6, 6}, {-6, 6}, {-6, 6}},
                BoxRatios- >{4, 4, 1},
                AxesLabel- >{"x", "y", "z"},
                ViewPoint- >{-1.365, -2.364, 2.}]

In[1240]:= g4 = Show[kb2, threedp2, pp2, AspectRatio- >1,
                PlotRange- >{{-6, 6}, {-6, 6}, {-6, 6}},
                BoxRatios- >{4, 4, 1},
                AxesLabel- >{"x", "y", "z"},
                ViewPoint- >{-2.729, 0., 2.}]
```

```
In[1241]:= g5 = Show[threedplt, pp2, AspectRatio- > 1,
               PlotRange- > {{-6, 6}, {-6, 6}, {-6, 6}},
               BoxRatios- > {4, 4, 1},
               AxesLabel- > {"x", "y", "z"},
               ViewPoint- > {-1.365, 2.364, 2.}]

In[1242]:= g6 = Show[kb2, pp3, AspectRatio- > 1,
               PlotRange- > {{-6, 6}, {-6, 6}, {-2, 2}},
               BoxRatios- > {4, 4, 1},
               AxesLabel- > {"x", "y", "z"},
               ViewPoint- > {1.365, 2.364, 2.}]

In[1243]:= Show[GraphicsArray[{{g1, g2}, {g3, g4},
               {g5, g6}}]]
```

C is an orientation reversing path and we can conclude that the Klein
bottle is not orientable.

Applications Related to Ordinary and Partial Differential Equations

Chapter 6 discusses Mathematica's differential equations commands. The examples used to illustrate the various commands are similar to examples routinely done in a one or two-semester differential equations course.

For more detailed discussions regarding Mathematica and differential equations see references like Abell and Braselton's *Differential Equations with Mathematica*, [1].

6.1 First-Order Differential Equations

6.1.1 Separable Equations

Because they are solved by integrating, separable differential equations are usually the first introduced in the introductory differential equations course.

Definition 2 (Separable Differential Equation). *A differential equation of the form*

$$f(y)\, dy = g(t)\, dt \qquad (6.1)$$

*is called a first-order **separable differential equation**.*

We solve separable differential equations by integrating.

Remark. The command

```
DSolve[y'[t]==f[t,y[t]],y[t],t]
```

attempts to solve $y' = dy/dt = f(t, y)$ for y.

EXAMPLE 6.1.1: Solve each of the following equations: (a) $y' - y^2 \sin t = 0$; (b) $y' = \alpha y \left(1 - \frac{1}{K} y\right)$, $K, \alpha > 0$ constant.

SOLUTION: (a) The equation is separable:

$$\frac{1}{y^2} dy = \sin t \, dt$$

$$\int \frac{1}{y^2} dy = \int \sin t \, dt$$

$$-\frac{1}{y} = -\cos t + C$$

$$y = \frac{1}{\cos t + C}.$$

We check our result with DSolve.

$In[1244] :=$ **sola = DSolve[y'[t]-y[t]^2 Sin[t] == 0, y[t], t]**

$Out[1244] = \left\{\left\{y[t] \to \dfrac{1}{-C[1] + Cos[t]}\right\}\right\}$

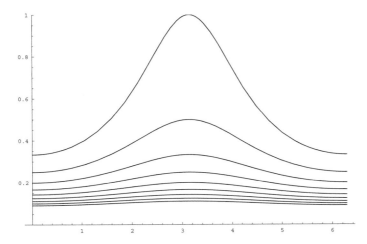

Figure 6-1   Several solutions of $y' - y^2 \sin t = 0$

Observe that the result is given as a list. The formula for the solution is the second part of the first part of the first part of `sola`.

$$In[1245] := \textbf{sola[[1, 1, 2]]}$$

$$Out[1245] = \frac{1}{-C[1] + Cos[t]}$$

We then graph the solution for various values of C with `Plot` in Figure 6-1.

$$In[1246] := \textbf{toplota = Table[sola[[1, 1, 2]]/.C[1]- > -i,}$$
$$\textbf{\{i, 2, 10\}]}$$

$$Out[1246] = \left\{ \frac{1}{2 + Cos[t]}, \frac{1}{3 + Cos[t]}, \frac{1}{4 + Cos[t]}, \frac{1}{5 + Cos[t]}, \right.$$
$$\frac{1}{6 + Cos[t]}, \frac{1}{7 + Cos[t]}, \frac{1}{8 + Cos[t]}, \frac{1}{9 + Cos[t]},$$
$$\left. \frac{1}{10 + Cos[t]} \right\}$$

expression `/.` x->y replaces all occurrences of x in expression by y. `Table[a[k],{k,n,m}]` generates the list a_n, a_{n+1}, . . . , a_{m-1}, a_m.

$$In[1247] := \textbf{Plot[Evaluate[toplota], \{t, 0, 2\pi\},}$$
$$\textbf{PlotRange- > \{0, 1\}, AxesOrigin- > \{0, 0\}]}$$

To graph the list of functions list for $a \le x \le b$, enter `Plot[Evaluate[list],{x,a,b}]`

(b) After separating variables, we use partial fractions to integrate:

$$y' = \alpha y \left(1 - \frac{1}{K} y\right)$$

$$\frac{1}{\alpha y \left(1 - \frac{1}{K} y\right)} dy = dt$$

$$\frac{1}{\alpha} \left(\frac{1}{y} + \frac{1}{K - y}\right) = dt$$

$$\frac{1}{\alpha} \left(\ln |y| - \ln |K - y|\right) = C_1 + t$$

$$\frac{y}{K - y} = Ce^{\alpha t}$$

$$y = \frac{CKe^{\alpha t}}{Ce^{\alpha t} - 1}.$$

We check the calculations with Mathematica. First, we use Apart to find the partial fraction decomposition of $\frac{1}{\alpha y \left(1 - \frac{1}{K} y\right)}$.

```
In[1248]:= s1 = Apart[1/(α y (1-1/k y)),y]
Out[1248]=  1      1
           ---- - --------
           y α    (-k+y) α
```

Then, we use Integrate to check the integration.

```
In[1249]:= s2 = Integrate[s1,y]
Out[1249]= Log[y]    Log[-k+y]
           ------  - ---------
             α           α
```

Last, we use use Solve to solve $\frac{1}{\alpha}(\ln |y| - \ln |K - y|) = ct$ for y.

```
In[1250]:= Solve[s2 == c + t,y]
                     c+t α
                    e     k
Out[1250]= {{y → --------------}}
                        c+t α
                 -1 + e
```

We can use DSolve to find a general solution of the equation

```
In[1251]:= solb = DSolve[y'[t] == α y[t] (1-1/k y[t]),
             y[t],t]
                       t α
                      e   k
Out[1251]= {{y[t] → -----------}}
                     t α    C[1]
                    e    - e
```

as well as find the solution that satisfies the initial condition $y(0) = y_0$.

```
In[1252]:= solc = DSolve[{y'[t] ==  y[t] (1- y[t]),
             y[0] == y0},y[t],t]
                        t
                       e  y0
Out[1252]= {{y[t] → ------------}}
                             t
                    1 - y0 + e  y0
```

The equation $y' = \alpha y \left(1 - \frac{1}{K} y\right)$ is called the **logistic equation** (or **Verhulst equation**) and is used to model the size of a population that is

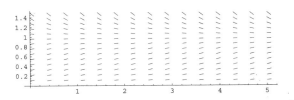

Figure 6-2   A typical direction field for the logistic equation

not allowed to grow in an unbounded manner. Assuming that $y(0) > 0$, then all solutions of the equation have the property that $\lim_{t\to\infty} y(t) = K$.

To see this, we set $\alpha = K = 1$ and use `PlotVectorField`, which is contained in the **PlotField** package that is located in the **Graphics** directory to graph the direction field associated with the equation in Figure 6-2.

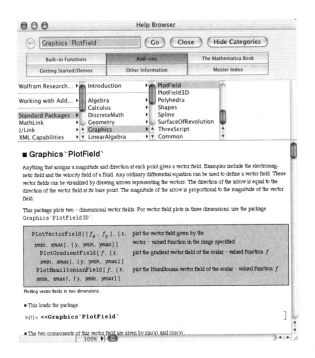

$In[1253]:=$ `<< Graphics'PlotField';`
      `pvf = PlotVectorField[{1, y(1 - y)}, {t, 0, 5},`
            `{y, 0, 5/2}, HeadLength- > 0, Axes- > Automatic]`

The property is more easily seen when we graph various solutions along with the direction field as done next in Figure 6-3.

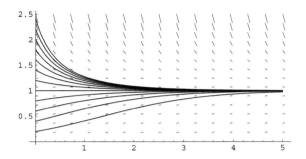

Figure 6-3   A typical direction field for the logistic equation along with several solutions

```
In[1254]:= toplot = Table[solc[[1, 1, 2]]/.y0- > i/5, {i, 1, 12}];
           sols = Plot[Evaluate[toplot],
                   {t, 0, 5}, DisplayFunction- > Identity];
           Show[pvf, sols]
```

∎

6.1.2 Linear Equations

Definition 3 (First-Order Linear Equation). *A differential equation of the form*

$$a_1(t)\frac{dy}{dt} + a_0(t)y = f(t), \tag{6.2}$$

*where $a_1(t)$ is not identically the zero function, is a first-order **linear differential equation**.*

Assuming that $a_1(t)$ is not identically the zero function, dividing equation (6.2) by $a_1(t)$ gives us the **standard form** of the first-order linear equation:

$$\frac{dy}{dt} + p(t)y = q(t). \tag{6.3}$$

If $q(t)$ is identically the zero function, we say that the equation is **homogeneous**. The **corresponding homogeneous equation** of equation (6.3) is

$$\frac{dy}{dt} + p(t)y = 0. \tag{6.4}$$

Observe that equation (6.4) is separable:

$$\frac{dy}{dt} + p(t)y = 0$$

$$\frac{1}{y} dy = -p(t) \, dt$$

$$\ln |y| = -\int p(t) \, dt + C$$

$$y = Ce^{-\int p(t) \, dt}.$$

Notice that any constant multiple of a solution to a linear homogeneous equation is also a solution. Now suppose that y is any solution of equation (6.3) and y_p is a particular solution of equation (6.3). Then,

$$\left(y - y_p\right)' + p(t)\left(y - y_p\right) = y' + p(t)y - \left(y_p' + p(t)y_p\right)$$
$$= q(t) - q(t) = 0.$$

> A **particular solution** is a specific solution to the equation that does not contain any arbitrary constants.

Thus, $y - y_p$ is a solution to the corresponding homogeneous equation of equation (6.3). Hence,

$$y - y_p = Ce^{-\int p(t) \, dt}$$

$$y = Ce^{-\int p(t) \, dt} + y_p$$

$$y = y_h + y_p,$$

where $y_h = Ce^{-\int p(t) \, dt}$. That is, a general solution of equation (6.3) is

$$y = y_h + y_p,$$

where y_p is a particular solution to the nonhomogeneous equation and y_h is a general solution to the corresponding homogeneous equation. Thus, to solve equation (6.3), we need to first find a general solution to the corresponding homogeneous equation, y_h, which we can accomplish through separation of variables, and then find a particular solution, y_p, to the nonhomogeneous equation.

If y_h is a solution to the corresponding homogeneous equation of equation (6.3) then for any constant C, Cy_h is also a solution to the corresponding homogeneous equation. Hence, it is impossible to find a particular solution to equation (6.3) of this form. Instead, we search for a particular solution of the form $y_p = u(t)y_h$, where $u(t)$ is *not* a constant function. Assuming that a particular solution, y_p, to equation (6.3) has the form $y_p = u(t)y_h$, differentiating gives us

$$y_p' = u'y_h + uy_h'$$

and substituting into equation (6.3) results in

$$y_p' + p(t)y_p = u'y_h + uy_h' + p(t)uy_h = q(t).$$

y_h is a solution to the
corresponding homogeneous
equation so $y_h' + p(t)y_h = 0$.

Because $uy_h' + p(t)uy_h = u\,[y_h' + p(t)y_h] = u \cdot 0 = 0$, we obtain

$$u'y_h = q(t)$$

$$u' = \frac{1}{y_h}q(t)$$

$$u' = e^{\int p(t)\,dt}q(t)$$

$$u = \int e^{\int p(t)\,dt}q(t)\,dt$$

so

$$y_p = u(t)\,y_h = Ce^{-\int p(t)\,dt}\int e^{\int p(t)\,dt}q(t)\,dt.$$

Because we can include an arbitrary constant of integration when evaluating $\int e^{\int p(t)\,dt}q(t)\,dt$, it follows that we can write a general solution of equation (6.3) as

$$y = e^{-\int p(t)\,dt}\int e^{\int p(t)\,dt}q(t)\,dt. \tag{6.5}$$

Alternatively, multiplying equation (6.3) by the **integrating factor** $\mu(t) = e^{\int p(t)\,dt}$ gives us the same result:

$$e^{\int p(t)\,dt}\frac{dy}{dt} + p(t)e^{\int p(t)\,dt}y = q(t)e^{\int p(t)\,dt}$$

$$\frac{d}{dt}\left(e^{\int p(t)\,dt}y\right) = q(t)e^{\int p(t)\,dt}$$

$$e^{\int p(t)\,dt}y = \int q(t)e^{\int p(t)\,dt}\,dt$$

$$y = e^{-\int p(t)\,dt}\int q(t)e^{\int p(t)\,dt}\,dt.$$

Thus, first-order linear equations can always be solved, although the resulting integrals may be difficult or impossible to evaluate exactly.

Mathematica is able to solve the general form of the first-order equation, the initial-value problem $y' + p(t)y = q(t)$, $y(0) = y_0$,

```
In[1255]:= DSolve[y'[t] +p[t]y[t] == q[t],y[t],t]
Out[1255]= {{y[t] → e^(-∫₀ᵗ p[DSolve't]dDSolve't) C[1] + e^(-∫₀ᵗ p[DSolve't]dDSolve't)
           ∫₀ᵗ e^(∫₀^DSolve't p[DSolve't]dDSolve't) q[DSolve't]dDSolve't}}

In[1256]:= DSolve[{y'[t] +p[t]y[t] == q[t],y[0] == y0},y[t],t]
Out[1256]= {{y[t] → e^(-∫₀ᵗ p[DSolve't]dDSolve't) (y0+
           ∫₀ᵗ e^(∫₀^DSolve't p[DSolve't]dDSolve't) q[DSolve't]dDSolve't)}}
```

as well as the corresponding homogeneous equation,

$$In[1257] := \texttt{DSolve[y'[t] +p[t]y[t] == 0,y[t],t]}$$
$$Out[1257] = \left\{\left\{y[t] \to e^{-\int_0^t p[\text{DSolve't}]\text{dDSolve't}} \ C[1]\right\}\right\}$$

$$In[1258] := \texttt{DSolve[\{y'[t] +p[t]y[t] == 0,y[0] == y0\},y[t],t]}$$
$$Out[1258] = \left\{\left\{y[t] \to e^{-\int_0^t p[\text{DSolve't}]\text{dDSolve't}} \ y0\right\}\right\}$$

although the results contain unevaluated integrals.

EXAMPLE 6.1.2 (Exponential Growth): Let $y = y(t)$ denote the size of a population at time t. If y grows at a rate proportional to the amount present, y satisfies

$$\frac{dy}{dt} = \alpha y, \tag{6.6}$$

where α is the **growth constant**. If $y(0) = y_0$, using equation (6.5) results in $y = y_0 e^{\alpha t}$. We use DSolve to confirm this result.

$$In[1259] := \texttt{DSolve[\{y'[t] == \alpha \ y[t],y[0] == y0\},y[t],t]}$$
$$Out[1259] = \{\{y[t] \to e^{t \ \alpha} \ y0\}\}$$

EXAMPLE 6.1.3: Solve each of the following equations: (a) $dy/dt = k(y - y_s)$, $y(0) = y_0$, k and y_s constant (b) $y' - 2ty = t$ (c) $ty' - y = 4t \cos 4t - \sin 4t$

$dy/dt = k(y - y_s)$ models Newton's Law of Cooling: the rate at which the temperature, $y(t)$, changes in a heating/cooling body is proportional to the difference between the temperature of the body and the constant temperature, y_s, of the surroundings.

SOLUTION: By hand, we rewrite the equation and obtain

$$\frac{dy}{dt} - ky = -ky_s.$$

A general solution of the corresponding homogeneous equation

$$\frac{dy}{dt} - ky = 0$$

is $y_h = e^{kt}$. Because k and $-ky_s$ are constants, we suppose that a particular solution of the nonhomogeneous equation, y_p, has the form $y_p = A$, where A is a constant.

Assuming that $y_p = A$, we have $y'_p = 0$ and substitution into the nonhomogeneous equation gives us

This will turn out to be a lucky guess. If there is not a solution of this form, we would not find one of this form.

$$\frac{dy_p}{dt} - ky_p = -KA = -ky_s \quad \text{so} \quad A = y_s.$$

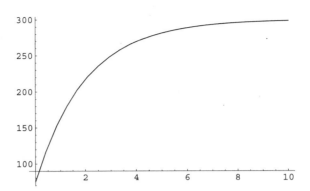

Figure 6-4   The temperature of the body approaches the temperature of its surroundings

Thus, a general solution is $y = y_h + y_p = Ce^{kt} + y_s$. Applying the initial condition $y(0) = y_0$ results in $y = y_s + (y_0 - y_s)e^{kt}$.

We obtain the same result with DSolve. We graph the solution satisfying $y(0) = 75$ assuming that $k = -1/2$ and $y_s = 300$ in Figure 6-4. Notice that $y(t) \to y_s$ as $t \to \infty$.

> In[1260]:= **sola = DSolve[{y'[t] == k(y[t] - ys),**
>   **y[0] == y0}, y[t], t]**
> Out[1260]= $\{\{y[t] \to e^{k\ t}\ (y0 - ys) + ys\}\}$

> In[1261]:= **tp = sola[[1, 1, 2]] /. {k- > -1/2, ys- > 300,**
>   **y0- > 75}; Plot[tp, {t, 0, 10}]**

(b) The equation is in standard form and we identify $p(t) = -2t$. Then, the integrating factor is $\mu(t) = e^{\int p(t)\,dt} = e^{-t^2}$. Multiplying the equation by the integrating factor, $\mu(t)$, results in

$$e^{-t^2}(y' - 2ty) = te^{-t^2} \qquad \text{or} \qquad \frac{d}{dt}\left(ye^{-t^2}\right) = te^{-t^2}.$$

Integrating gives us

$$ye^{-t^2} = -\frac{1}{2}e^{-t^2} + C \qquad \text{or} \qquad y = -\frac{1}{2} + Ce^{t^2}.$$

We confirm the result with DSolve.

> In[1262]:= **DSolve[y'[t] - 2t y[t] == t, y[t], t]**
> Out[1262]= $\left\{\left\{y[t] \to -\frac{1}{2} + e^{t^2}\ C[1]\right\}\right\}$

(c) In standard form, the equation is $y' - y/t = (4t\cos 4t - \sin 4t)/t$ so $p(t) = -1/t$. The integrating factor is $\mu(t) = e^{\int p(t)\,dt} = e^{-\ln t} = 1/t$ and multiplying the equation by the integrating factor and then integrating gives us

$$\frac{1}{t}\frac{dy}{dt} - \frac{1}{t^2}y = \frac{1}{t^2}(4t\cos 4t - \sin 4t)$$

$$\frac{d}{dt}\left(\frac{1}{t}y\right) = \frac{1}{t^2}(4t\cos 4t - \sin 4t)$$

$$\frac{1}{t}y = \frac{\sin 4t}{t} + C$$

$$y = \sin 4t + Ct,$$

where we use the `Integrate` function to evaluate $\int \frac{1}{t^2}(4t\cos 4t - \sin 4t)\,dt = \frac{\sin 4t}{t} + C.$

```
In[1263]:= Integrate[(4 t Cos[4t] - Sin[4t])/t^2, t]
Out[1263]= Sin[4 t]
           ────────
              t
```

We confirm this result with `DSolve`.

```
In[1264]:= sol = DSolve[y'[t] - y[t]/t == (4 t Cos[4t]
                  - Sin[4t])/t, y[t], t]
Out[1264]= {{y[t] → t C[1] + Sin[4 t]}}
```

In the general solution, observe that *every* solution satisfies $y(0) = 0$. That is, the initial-value problem

$$\frac{dy}{dt} - \frac{1}{t}y = \frac{1}{t^2}(4t\cos 4t - \sin 4t), \qquad y(0) = 0$$

has infinitely many solutions. We see this in the plot of several solutions that is generated with `Plot` in Figure 6-5.

```
In[1265]:= toplot = Table[sol[[1, 1, 2]]/.C[1]- >i,
                  {i, -5, 5}];
           Plot[Evaluate[toplot], {t, -2π, 2π},
              PlotRange- > {-2π, 2π}, AspectRatio- >1]
```

■

6.1.2.1 Application: Free-Falling Bodies

The motion of objects can be determined through the solution of first-order initial-value problems. We begin by explaining some of the theory that is needed to set up the differential equation that models the situation.

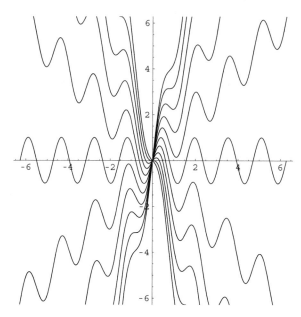

Figure 6-5   Every solution satisfies $y(0) = 0$

Newton's Second Law of Motion: The rate at which the momentum of a body changes with respect to time is equal to the resultant force acting on the body.

Because the body's momentum is defined as the product of its mass and velocity, this statement is modeled as

$$\frac{d}{dt}(mv) = F,$$

where m and v represent the body's mass and velocity, respectively, and F is the sum of the forces (the resultant force) acting on the body. Because m is constant, differentiation leads to the well-known equation

$$m\frac{dv}{dt} = F.$$

If the body is subjected only to the force due to gravity, then its velocity is determined by solving the differential equation

$$m\frac{dv}{dt} = mg \qquad \text{or} \qquad \frac{dv}{dt} = g,$$

where $g = 32\text{ft}/\text{s}^2$ (English system) and $g = 9.8\text{m}/\text{s}^2$ (metric system). This differential equation is applicable only when the resistive force due to the medium (such as air resistance) is ignored. If this offsetting resistance is considered, we must discuss all of the forces acting on the object. Mathematically, we write the equation as

$$m\frac{dv}{dt} = \sum (\text{forces acting on the object})$$

where the direction of motion is taken to be the positive direction. Because air resistance acts against the object as it falls and g acts in the same direction of the motion, we state the differential equation in the form

$$m\frac{dv}{dt} = mg + (-F_R) \qquad \text{or} \qquad m\frac{dv}{dt} = mg - F_R,$$

where F_R represents this resistive force. Note that down is assumed to be the positive direction. The resistive force is typically proportional to the body's velocity, v, or the square of its velocity, v^2. Hence, the differential equation is linear or nonlinear based on the resistance of the medium taken into account.

EXAMPLE 6.1.4: An object of mass $m = 1$ is dropped from a height of 50 feet above the surface of a small pond. While the object is in the air, the force due to air resistance is v. However, when the object is in the pond, it is subjected to a buoyancy force equivalent to $6v$. Determine how much time is required for the object to reach a depth of 25 feet in the pond.

SOLUTION: This problem must be broken into two parts: an initial-value problem for the object above the pond, and an initial-value problem for the object below the surface of the pond. The initial-value problem above the pond's surface is found to be

$$\begin{cases} dv/dt = 32 - v \\ v(0) = 0. \end{cases}$$

However, to define the initial-value problem to find the velocity of the object beneath the pond's surface, the velocity of the object when it reaches the surface must be known. Hence, the velocity of the object above the surface must be determined by solving the initial-value problem above. The equation $dv/dt = 32 - v$ is separable and solved with DSolve in d1.

```
In[1266]:= Clear[v, y]

        d1 = DSolve[{v'[t] == 32 - v[t], v[0] == 0},
             v[t], t]
```

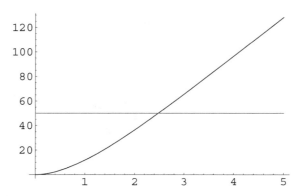

Figure 6-6   The object has traveled 50 feet when $t \approx 2.5$

Out[1266]= $\{\{v[t] \to 32\,e^{-t}\,(-1+e^{t})\}\}$

In order to find the velocity when the object hits the pond's surface we must know the time at which the distance traveled by the object (or the displacement of the object) is 50. Thus, we must find the displacement function, which is done by integrating the velocity function obtaining $s(t) = 32e^{-t} + 32t - 32$.

In[1267]:= **p1 = DSolve[{y'[t] == e^{-t} (-32 + 32 e^{t}), y[0] == 0},**
              y[t], t]
Out[1267]= $\{\{y[t] \to 32\,e^{-t}\,(1-e^{t}+e^{t}\,t)\}\}$

The displacement function is graphed with Plot in Figure 6-6. The value of t at which the object has traveled 50 feet is needed. This time appears to be approximately 2.5 seconds.

In[1268]:= **Plot[{e^{-t} (32 - 32 e^{t} + 32 e^{t} t), 50}, {t, 0, 5},**
              PlotStyle → {GrayLevel[0], GrayLevel[0.5]}]

A more accurate value of the time at which the object hits the surface is found using FindRoot. In this case, we obtain $t \approx 2.47864$. The velocity at this time is then determined by substitution into the velocity function resulting in $v(2.47864) \approx 29.3166$. Note that this value is the initial velocity of the object when it hits the surface of the pond.

In[1269]:= **t1 = FindRoot[p1[[1, 1, 2]] == 50, {t, 2.5}]**
Out[1269]= $\{t \to 2.47864\}$

In[1270]:= **v1 = d1[[1, 1, 2]] /. t1**
Out[1270]= **29.3166**

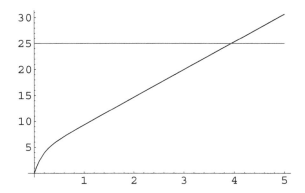

Figure 6-7   After approximately 4 seconds, the object is 25 feet below the surface of the pond

Thus, the initial-value problem that determines the velocity of the object beneath the surface of the pond is given by

$$\begin{cases} dv/dt = 32 - 6v \\ v(0) = 29.3166. \end{cases}$$

The solution of this initial-value problem is $v(t) = \frac{16}{3} + 23.9833e^{-t}$ and integrating to obtain the displacement function (the initial displacement is 0) we obtain $s(t) = 3.99722 - 3.99722e^{-6t} + \frac{16}{3}t$. These steps are carried out in d2 and p2.

```
In[1271]:= d2 = DSolve[{v'[t] == 32 - 6 v[t], v[0] == v1},
                 v[t], t]
Out[1271]= {{v[t] → e^-6t (23.9832 + 5.33333 e^6t)}}

In[1272]:= p2 = DSolve[{y'[t] == d2[[1, 1, 2]], y[0] == 0},
                 y[t], t]
Out[1272]= {{y[t] → 2.71828^-6.t (-3.99721+
                 3.99721 2.71828^6.t + 5.33333 2.71828^6.t t)}}
```

This displacement function is then plotted in Figure 6-7 to determine when the object is 25 feet beneath the surface of the pond. This time appears to be near 4 seconds.

```
In[1273]:= Plot[{p2[[1, 1, 2]], 25}, {t, 0, 5},
                 PlotStyle → {GrayLevel[0], GrayLevel[0.5]}]
```

A more accurate approximation of the time at which the object is 25 feet beneath the pond's surface is obtained with FindRoot. In this case, we obtain $t \approx 3.93802$. Finally, the time required for the object to reach the pond's surface is added to the time needed for it to travel 25 feet beneath the surface to see that approximately 6.41667 seconds are required for the object to travel from a height of 50 feet above the pond to a depth of 25 feet below the surface.

```
In[1274] := t2 = FindRoot[p2[[1, 1, 2]] == 25, {t, 4}]
Out[1274] = {t → 3.93802}

In[1275] := t1[[1, 2]] + t2[[1, 2]]
Out[1275] = 6.41667
```

∎

6.1.3 Nonlinear Equations

Mathematica can solve a variety of nonlinear first-order equations that are typically encountered in the introductory differential equations course.

EXAMPLE 6.1.5: Solve each: (a) $(\cos x + 2xe^y) dx + (\sin y + x^2 e^y - 1) dy = 0$; (b) $(y^2 + 2xy) dx - x^2 dy = 0$.

SOLUTION: (a) Notice that $(\cos x + 2xe^y) dx + (\sin y + x^2 e^y - 1) dy = 0$ can be written as $dy/dx = -(\cos x + 2xe^y)/(\sin x + x^2 e^y - 1)$. The equation is an example of an *exact equation*. A theorem tells us that the equation

$$M(x, y)dx + N(x, y)dy = 0$$

is **exact** if and only if $\partial M/\partial y = \partial N/\partial x$.

```
In[1276] := m = Cos[x] + 2 x Exp[y];
            n = Sin[y] + x^2 Exp[y] - 1;
            D[m, y]

            D[n, x]
Out[1276] = 2 e^y x
Out[1276] = 2 e^y x
```

We solve exact equations by integrating. Let $F(x, y) = C$ satisfy $(y \cos x + 2xe^y)dx + (\sin y + x^2 e^y - 1) dy = 0$. Then,

$$F(x, y) = \int (\cos x + 2xe^y) dx = \sin x + x^2 e^y + g(y),$$

where $g(y)$ is a function of y.

```
In[1277]:= f1 = Integrate[m, x]
Out[1277]= e^y x^2 + Sin[x]
```

We next find that $g'(y) = \sin y - 1$ so $g(y) = -\cos y - y$. Hence, a general solution of the equation is

$$\sin x + x^2 e^y - \cos y - y = C.$$

```
In[1278]:= f2 = D[f1, y]
Out[1278]= e^y x^2
```

```
In[1279]:= f3 = Solve[f2 + c == n, c]
Out[1279]= {{c → -1 + Sin[y]}}
```

```
In[1280]:= Integrate[f3[[1, 1, 2]], y]
Out[1280]= -y - Cos[y]
```

We confirm this result with DSolve. Notice that Mathematica warns us that it cannot solve for y explicitly and returns the same implicit solution obtained by us.

```
In[1281]:= mf = m/.y- >y[x];
           nf = n/.y- >y[x];
           sol = DSolve[mf + nf y'[x] == 0, y[x], x]
Solve :: "tdep" :  "Theequationsappeartoinvolve
     transcendentalfunctionsofthevariablesin
     anessentiallynon-algebraicway."
Out[1281]= Solve[e^y[x] x^2 - Cos[y[x]] + Sin[x] - y[x] == C[1],
                 {y[x]}]
```

Graphs of several solutions using the values of C generated in cvals are graphed with ContourPlot in Figure 6-8.

```
In[1282]:= sol[[1, 1]]
Out[1282]= e^y[x] x^2 - Cos[y[x]] + Sin[x] - y[x]
```

```
In[1283]:= sol2 = sol[[1, 1]]/.y[x]- >y
Out[1283]= e^y x^2 - y - Cos[y] + Sin[x]
```

```
In[1284]:= cvals = Table[sol2/.{x- >-3π/2, y- >i},
                   {i, 0, 6π, 6π/24}]//
                 N
```

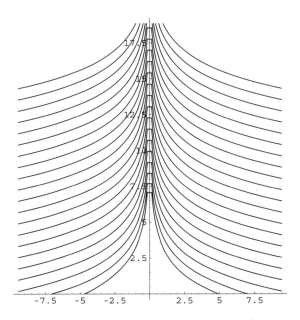

Figure 6-8   Graphs of several solutions of $\left(\cos x + 2xe^y\right)dx + \left(\sin y + x^2e^y - 1\right)dy = 0$

```
Out[1284]= {22.2066,
              48.2128, 106.254,
              233.647, 512.735,
              1124.85, 2468.28,
              5416.56, 11885.2,
              26074.5, 57196.7,
              125457., 275169.,
              603531., 1.32372 10^6,
              2.9033 10^6, 6.36776 10^6,
              1.39663 10^7, 3.0632 10^7,
              6.71846 10^7, 1.47355 10^8,
              3.2319 10^8, 7.08847 10^8,
              1.5547 10^9, 3.40989 10^9}

In[1285]:= ContourPlot[sol2, {x, -3π, 3π}, {y, 0, 6π},
              ContourShading- > False, Frame- > False,
              Axes- > Automatic, AxesOrigin- > {0, 0},
              Contours- > cvals, PlotPoints- > 60]
```

(b) We can write $\left(y^2 + 2xy\right)dx - x^2dy = 0$ as $dy/dx = \left(y^2 + 2xy\right)/x^2$. A first-order equation is **homogeneous** if it can be written in the form

$$\frac{dy}{dx} = F\left(\frac{y}{x}\right).$$

Homogeneous equations are reduced to separable equations with either the substitution $y = ux$ or $x = vy$. In this case, we have that $dy/dx = (y/x)^2 + 2(y/x)$ so the equation is homogeneous.

Let $y = ux$. Then, $dy = u\,dx + x\,du$. Substituting into $\left(y^2 + 2xy\right)dx - x^2 dy = 0$ and separating gives us

$$\left(y^2 + 2xy\right)dx - x^2 dy = 0$$
$$\left(u^2 x^2 + 2ux^2\right)dx - x^2(u\,dx + x\,du) = 0$$
$$\left(u^2 + 2u\right)dx - (u\,dx + x\,du) = 0$$
$$\left(u^2 + u\right)dx = x\,du$$
$$\frac{1}{u(u+1)}du = \frac{1}{x}dx.$$

Integrating the left and right-hand sides of this equation with `Integrate`,

```
In[1286]:= Integrate[1/(u(u+1)),u]
Out[1286]= Log[u] - Log[1+u]
```

```
In[1287]:= Integrate[1/x,x]
Out[1287]= Log[x]
```

exponentiating, resubstituting $u = y/x$, and solving for y gives us

$$\ln|u| - \ln|u+1| = \ln|x| + C$$
$$\frac{u}{u+1} = Cx$$
$$\frac{\frac{y}{x}}{\frac{y}{x}+1} = Cx$$
$$y = \frac{Cx^2}{1 - Cx}.$$

```
In[1288]:= Solve[(y/x)/(y/x+1) == c x, y]
Out[1288]= {{y -> -(cx^2)/(cx-1)}}
```

We confirm this result with `DSolve` and then graph several solutions with `Plot` in Figure 6-9.

```
In[1289]:= sol = DSolve[y[x]^2+2x y[x]-x^2y'[x] == 0, y[x], x]
Out[1289]= {{y[x] -> -(x^2 C[1])/(-1 + x C[1])}}
```

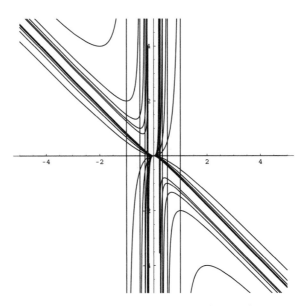

Figure 6-9   Graphs of several solutions of $\left(y^2 + 2xy\right)dx - x^2dy = 0$

```
In[1290]:= toplot = Table[sol[[1, 1, 2]]/.C[1]->i,
              {i, -5, 5}];
           Plot[Evaluate[toplot], {x, -5, 5}, PlotRange->
              {-5, 5}, AspectRatio->Automatic]
```

∎

6.1.4 Numerical Methods

If numerical results are desired, use NDSolve:

```
NDSolve[{y'[t]==f[t,y[t]],y[t0]==y0},y[t],{t,a,b}]
```

attempts to generate a numerical solution of

$$\begin{cases} dy/dt = f(t, y) \\ y(t_0) = y_0 \end{cases}$$

valid for $a \le t \le b$.

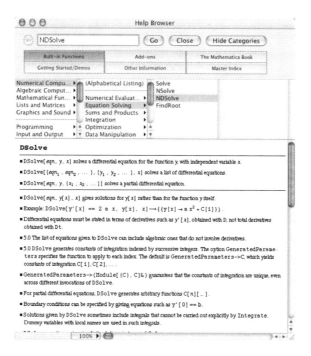

EXAMPLE 6.1.6: Consider

$$\frac{dy}{dt} = \left(t^2 - y^2\right)\sin y, \; y(0) = -1.$$

(a) Determine $y(1)$. (b) Graph $y(t)$, $-1 \le t \le 10$.

SOLUTION: We first remark that DSolve can neither exactly solve the differential equation $y' = \left(t^2 - y^2\right)\sin y$ nor find the solution that satisfies $y(0) = -1$.

```
In[1291]:=  sol = DSolve[y'[t] == (t^2 - y[t]^2) Sin[t],
                y[t], t]
```

```
Out[1291]=  BoxData(DSolve[y'[t] == Sin[t] (t² - y[t]²),
                y[t], t])
```

```
In[1292]:=  sol = DSolve[{y'[t] == (t^2 - y[t]^2) Sin[t],
                  y[0] == y0}, y[t], t]
```

```
Out[1292]=  BoxData(DSolve[{y'[t] == Sin[t] (t² - y[t]²),
                  y[0] == y0}, y[t], t])
```

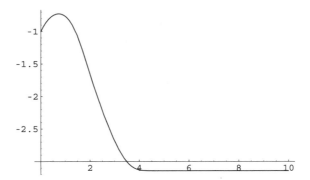

Figure 6-10   Graph of the solution to $y' = (t^2 - y^2) \sin y$, $y(0) = -1$

However, we obtain a numerical solution valid for $0 \le t \le 1000$ using the NDSolve function.

```
In[1293]:= sol = NDSolve[{y'[t] == (t^2 - y[t]^2) Sin[y[t]],
               y[0] == -1}, y[t], {t, 0, 1000}]
Out[1293]= BoxData({{y[t] → InterpolatingFunction
               [{{0., 1000.}}, " <> "][t]}})
```

Entering sol /.t->1 evaluates the numerical solution if $t = 1$.

```
In[1294]:= sol /.t- >1
Out[1294]= {{y[1] → -0.766014}}
```

The result means that $y(1) \approx -.766$. We use the Plot command to graph the solution for $0 \le t \le 10$ in Figure 6-10.

```
In[1295]:= Plot[Evaluate[y[t]/.sol], {t, 0, 10}]
```

■

EXAMPLE 6.1.7 (Logistic Equation with Predation): Incorporating predation into the **logistic equation**, $y' = \alpha y\left(1 - \dfrac{1}{K}y\right)$, results in

$$\frac{dy}{dt} = \alpha y\left(1 - \frac{1}{K}y\right) - P(y),$$

where $P(y)$ is a function of y describing the rate of predation. A typical choice for P is $P(y) = ay^2/(b^2 + y^2)$ because $P(0) = 0$ and P is bounded above: $\lim_{t\to\infty} P(y) < \infty$.

Remark. Of course, if $\lim_{t\to\infty} y(t) = Y$, then $\lim_{t\to\infty} P(y) = aY^2/(b^2 + Y^2)$. Generally, however, $\lim_{t\to\infty} P(y) \neq a$ because $\lim_{t\to\infty} y(t) \leq K \neq \infty$, for some $K \geq 0$, in the predation situation.

If $\alpha = 1$, $a = 5$, and $b = 2$, graph the direction field associated with the equation as well as various solutions if (a) $K = 19$ and (b) $K = 20$.

SOLUTION: (a) We define eqn [k] to be

$$\frac{dy}{dt} = y\left(1 - \frac{1}{K}y\right) - \frac{5y^2}{4 + y^2}.$$

In[1296]:= `<< Graphics`PlotField`

In[1297]:= `eqn[k_] = y'[t] == y[t] (1 - 1/k y[t])`
`                -5y[t]^2/(4 + y[t]^2);`

We use PlotVectorField to graph the direction field in Figure 6-11 (a) and then the direction field along with the solutions that satisfy $y(0) =$ .5, $y(0) = .2$, and $y(0) = 4$ in Figure 6-11 (b).

In[1298]:= `pvf19 = PlotVectorField[{1, y(1 - 1/19 y) -`
`                5y^2/(4 + y^2)}, {t, 0, 10}, {y, 0, 6},`
`                Axes- > Automatic, HeadLength- > 0,`
`                DisplayFunction- > Identity];`

In[1299]:= `n1 = NDSolve[{eqn[19], y[0] == 0.5}, y[t],`
`                {t, 0, 10}];`
`            n2 = NDSolve[{eqn[19], y[0] == 2}, y[t],`
`                {t, 0, 10}];`
`            n3 = NDSolve[{eqn[19], y[0] == 4}, y[t],`
`                {t, 0, 10}];`

In[1300]:= `solplot = Plot[Evaluate[y[t]/.{n1, n2, n3}],`
`                {t, 0, 10}, PlotStyle- > Thickness[0.01],`
`                DisplayFunction- > Identity];`

The same results can be obtained using Map.

In[1301]:= `numsols = Map[NDSolve[`
`                {eqn[19], y[0] == #}, y[t], {t, 0, 10}]&,`
`                {0.5, 2, 4}];`
`            solplot = Plot[Evaluate[y[t]/.numsols],`
`                {t, 0, 10}, PlotStyle- > Thickness[0.01],`
`                DisplayFunction- > Identity];`

In[1302]:= `Show[GraphicsArray[{pvf19, Show[pvf19, solplot]}]]`

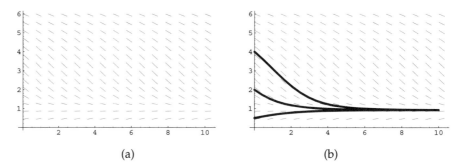

Figure 6-11   (a) Direction field and (b) direction field with three solutions

In the plot, notice that all nontrivial solutions appear to approach an equilibrium solution. We determine the equilibrium solution by solving $y' = 0$

$In[1303]:=$ **eqn[19][[2]]**

$Out[1303]=$ $\left(1 - \dfrac{y[t]}{19}\right) y[t] - \dfrac{5\ y[t]^2}{4 + y[t]^2}$

$In[1304]:=$ **Solve[eqn[19.][[2]] == 0, y[t]]**

$Out[1304]=$ $\{\{y[t] \to 0.\}, \{y[t] \to 0.923351\},$
$\{y[t] \to 9.03832 - 0.785875\ i\},$
$\{y[t] \to 9.03832 + 0.785875\ i\}\}$

to see that it is $y \approx 0.923$.

(b) We carry out similar steps for (b). First, we graph the direction field with PlotVectorField in Figure 6-12.

$In[1305]:=$ **pvf20 = PlotVectorField[{1, y(1 - 1/20 y) - 5y^2/**
(4 + y^2)}, {t, 0, 10}, {y, 0, 20}, Axes- >
Automatic, HeadLength- > 0,
AspectRatio- > 1/GoldenRatio];

We then use Map together with NDSolve to numerically find the solution satisfying $y(0) = .5i$, for $i = 1, 2, \ldots, 40$ and name the resulting list numsols. The functions contained in numsols are graphed with Plot in solplot.

$In[1306]:=$ **numsols = Map[NDSolve[{eqn[20], y[0] == #}, y[t],**
{t, 0, 10}]&, Table[0.5i, {i, 1, 40}]];
solplot = Plot[Evaluate[y[t]/.numsols], {t, 0, 10},
PlotStyle- > Thickness[0.005],
DisplayFunction- > Identity];

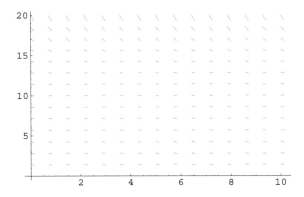

Figure 6-12   Direction field

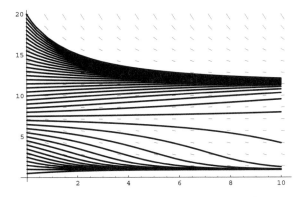

Figure 6-13   Direction field with several solutions

Last, we display the direction field along with the solution graphs in solplot using Show in Figure 6-13.

```
In[1307]:= Show[pvf20, solplot]
```

Notice that there are three nontrivial equilibrium solutions that are found by solving $y' = 0$.

```
In[1308]:= Solve[eqn[20.][[2]] == 0, y[t]]
Out[1308]= {{y[t] → 0.}, {y[t] → 0.926741},
            {y[t] → 7.38645},
            {y[t] → 11.6868}}
```

In this case, $y \approx .926$ and $y \approx 11.687$ are stable while $y \approx 7.386$ is unstable.

■

6.2 Second-Order Linear Equations

We now present a concise discussion of second-order linear equations, which are extensively discussed in the introductory differential equations course.

6.2.1 Basic Theory

The **general form** of the **second-order linear equation** is

$$a_2(t)\frac{d^2y}{dt^2} + a_1(t)\frac{dy}{dt} + a_0(t)y = f(t), \qquad (6.7)$$

where $a_2(t)$ is not identically the zero function.

The **standard form** of the second-order linear equation (6.7) is

$$\frac{d^2y}{dt^2} + p(t)\frac{dy}{dt} + q(t)y = f(t). \qquad (6.8)$$

The **corresponding homogeneous equation** of equation (6.8) is

$$\frac{d^2y}{dt^2} + p(t)\frac{dy}{dt} + q(t)y = 0. \qquad (6.9)$$

A **general solution** of equation (6.9) is $y = c_1y_1 + c_2y_2$ where

1. y_1 and y_2 are solutions of equation (6.9), and
2. y_1 and y_2 are *linearly independent*.

If y_1 and y_2 are solutions of equation (6.9), then y_1 and y_2 are **linearly independent** if and only if the **Wronskian**,

$$W\left(\{y_1, y_2\}\right) = \begin{vmatrix} y_1 & y_2 \\ y_1' & y_2' \end{vmatrix} = y_1y_2' - y_1'y_2, \qquad (6.10)$$

is not the zero function. If y_1 and y_2 are linearly independent solutions of equation (6.9), we call the set $S = \{y_1, y_2\}$ a **fundamental set of solutions** for equation (6.9).

A particular solution, y_p, is a solution that does not contain any arbitrary constants.

Let y be a general solution of equation (6.8) and y_p be a particular solution of equation (6.8). It follows that $y - y_p$ is a solution of equation (6.9) so $y - y_p = y_h$ where y_h is a general solution of equation (6.9). Hence, $y = y_h + y_p$. That is, to solve the nonhomogeneous equation, we need a general solution, y_h, of the corresponding homogeneous equation and a particular solution, y_p, of the nonhomogeneous equation.

6.2.2 Constant Coefficients

Suppose that the coefficient functions of equation (6.7) are constants: $a_2(t) = a$, $a_1(t) = b$, and $a_0(t) = c$ and that $f(t)$ is identically the zero function. In this case, equation (6.7) becomes

$$ay'' + by' + cy = 0. \tag{6.11}$$

Now suppose that $y = e^{kt}$, k constant, is a solution of equation (6.11). Then, $y' = ke^{kt}$ and $y'' = k^2 e^{kt}$. Substitution into equation (6.11) then gives us

$$ay'' + by' + cy = ak^2 e^{kt} + bke^{kt} + ce^{kt}$$
$$= e^{kt}\left(ak^2 + bk + c\right) = 0.$$

Because $e^{kt} \neq 0$, the solutions of equation (6.11) are determined by the solutions of

$$ak^2 + bk + c = 0, \tag{6.12}$$

called the **characteristic equation** of equation (6.11).

Theorem 23. *Let k_1 and k_2 be the solutions of equation (6.12).*

1. *If $k_1 \neq k_2$ are real and distinct, two linearly independent solutions of equation (6.11) are $y_1 = e^{k_1 t}$ and $y_2 = e^{k_2 t}$; a general solution of equation (6.11) is*

$$y = c_1 e^{k_1 t} + c_2 e^{k_2 t}.$$

2. *If $k_1 = k_2$, two linearly independent solutions of equation (6.11) are $y_1 = e^{k_1 t}$ and $y_2 = te^{k_1 t}$; a general solution of equation (6.11) is*

$$y = c_1 e^{k_1 t} + c_2 te^{k_1 t}.$$

3. *If $k_{1,2} = \alpha \pm \beta i$, $\beta \neq 0$, two linearly independent solutions of equation (6.11) are $y_1 = e^{\alpha t} \cos \beta t$ and $y_2 = e^{\alpha t} \sin \beta t$; a general solution of equation (6.11) is*

$$y = e^{\alpha t}(c_1 \cos \beta t + c_2 \sin \beta t).$$

EXAMPLE 6.2.1: Solve each of the following equations: (a) $6y'' + y' - 2y = 0$; (b) $y'' + 2y' + y = 0$; (c) $16y'' + 8y' + 145y = 0$.

SOLUTION: (a) The characteristic equation is $6k^2 + k - 2 = (3k+2)(2k-1) = 0$ with solutions $k = -2/3$ and $k = 1/2$. We check with either `Factor` or `Solve`.

```
In[1309]:= Factor[6k^2 + k - 2]
```

```
        Solve[6k^2 + k - 2 == 0]
Out[1309]= (-1 + 2 k) (2 + 3 k)
```

$$Out[1309]= \left\{\left\{k \to -\frac{2}{3}\right\}, \left\{k \to \frac{1}{2}\right\}\right\}$$

Then, a fundamental set of solutions is $\{e^{-2t/3}, e^{t/2}\}$ and a general solution is

$$y = c_1 e^{-2t/3} + c_2 e^{t/2}.$$

Of course, we obtain the same result with `DSolve`.

```
In[1310]:= DSolve[6y''[t] + y'[t] - 2y[t] == 0, y[t], t]
Out[1310]= {{y[t] → e^(-2 t/3) C[1] + e^(t/2) C[2]}}
```

(b) The characteristic equation is $k^2 + 2k + 1 = (k + 1)^2 = 0$ with solution $k = -1$, which has multiplicity two, so a fundamental set of solutions is $\{e^{-t}, te^{-t}\}$ and a general solution is

$$y = c_1 e^{-t} + c_2 t e^{-t}.$$

We check the calculation in the same way as in (a).

```
In[1311]:= Factor[k^2 + 2k + 1]
```

```
        Solve[k^2 + 2k + 1 == 0]
```

```
        DSolve[y''[t] + 2y'[t] + y[t] == 0, y[t], t]
Out[1311]= (1 + k)^2
Out[1311]= {{k → -1}, {k → -1}}
Out[1311]= {{y[t] → e^(-t) C[1] + e^(-t) t C[2]}}
```

(c) The characteristic equation is $16k^2 + 8k + 145 = 0$ with solutions $k_{1,2} = -\frac{1}{4} \pm 3i$ so a fundamental set of solutions is $\{e^{-t/4} \cos 3t, e^{-t/4} \sin 3t\}$ and a general solution is

$$y = e^{-t/4} (c_1 \cos 3t + c_2 \sin 3t).$$

The calculation is verified in the same way as in (a) and (b).

```
In[1312]:= Factor[16k^2 + 8k + 145, GaussianIntegers->
           True]
           Solve[16k^2 + 8k + 145 == 0]
```

```
           DSolve[16y''[t] + 8y'[t] + 145y[t] == 0, y[t], t]
Out[1312]= ((1 - 12 i) + 4 k) ((1 + 12 i) + 4 k)
```

Out[1312]= $\left\{\left\{k \rightarrow -\frac{1}{4} - 3\ i\right\}, \left\{k \rightarrow -\frac{1}{4} + 3\ i\right\}\right\}$

Out[1312]= $\left\{\left\{y[t] \rightarrow e^{-t/4}\ C[2]\ Cos[3\ t] - e^{-t/4}\ C[1]\ Sin[3\ t]\right\}\right\}$

■

EXAMPLE 6.2.2: Solve

$$64\frac{d^2y}{dt^2} + 16\frac{dy}{dt} + 1025y = 0,\ y(0) = 1,\ \frac{dy}{dt}(0) = 3.$$

SOLUTION: A general solution of $64y'' + 16y' + 1025y = 0$ is $y = e^{-t/8}(c_1 \sin 4t + c_2 \cos 4t)$.

In[1313]:= **gensol = DSolve[64y''[t] + 16y'[t] + 1025y[t] == 0,**
              y[t], t]

Out[1313]= $\left\{\left\{y[t] \rightarrow e^{-t/8}\ C[2]\ Cos[4\ t] - e^{-t/8}\ C[1]\right.\right.$
              $\left.\left. Sin[4\ t]\right\}\right\}$

Applying $y(0) = 1$ shows us that $c_2 = 1$.

In[1314]:= **e1 = y[t]/.gensol[[1]]/.t- > 0**

Out[1314]= C[2]

Computing y'

In[1315]:= **D[y[t]/.gensol[[1]], t]**

Out[1315]= $-4\ e^{-t/8}\ C[1]\ Cos[4\ t] - \frac{1}{8}\ e^{-t/8}\ C[2]\ Cos[4\ t]$

              $+ \frac{1}{8}\ e^{-t/8}\ C[1]\ Sin[4\ t] - 4\ e^{-t/8}\ C[2]\ Sin[4\ t]$

and then $y'(0)$, shows us that $-4c_1 - \frac{1}{8}c_2 = 3$.

In[1316]:= **e2 = D[y[t]/.gensol[[1]], t]/.t- > 0**

Out[1316]= $-4\ C[1] - \frac{C[2]}{8}$

Solving for c_1 and c_2 with Solve shows us that $c_1 = -25/32$ and $c_1 = 1$.

In[1317]:= **cvals = Solve[{e1 == 1, e2 == 3}]**

Out[1317]= $\left\{\left\{C[1] \rightarrow -\frac{25}{32}, C[2] \rightarrow 1\right\}\right\}$

Thus, $y = e^{-t/8}\left(\frac{-25}{32}\sin 4t + \cos 4t\right)$, which we graph with Plot in Figure 6-14.

In[1318]:= **sol = y[t]/.gensol[[1]]/.cvals[[1]]**

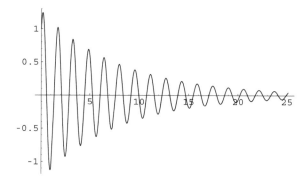

Figure 6-14   The solution to the initial-value problem tends to 0 as $t \to \infty$

$Out[1318]= \text{e}^{-\text{t}/8} \ \text{Cos}[4 \ \text{t}] + \dfrac{25}{32} \ \text{e}^{-\text{t}/8} \ \text{Sin}[4 \ \text{t}]$

$In[1319]:= \textbf{Plot[sol, \{t, 0, 8}\boldsymbol{\pi}\textbf{\}]}$

We verify the calculation with DSolve.

$In[1320]:= \textbf{DSolve[\{64y''[t] + 16y'[t] + 1025y[t] == 0,}$
$\qquad\qquad\textbf{y[0] == 1, y'[0] == 2\}, y[t], t]}$

$Out[1320]= \left\{\left\{\text{y[t]} \to \text{e}^{-\text{t}/8} \ \left(\text{Cos}[4 \ \text{t}] + \dfrac{17}{32} \ \text{Sin}[4 \ \text{t}]\right)\right\}\right\}$

■

Application: Harmonic Motion

Suppose that a mass is attached to an elastic spring that is suspended from a rigid support such as a ceiling. According to Hooke's law, the spring exerts a restoring force in the upward direction that is proportional to the displacement of the spring.

> **Hooke's Law:** $F = ks$, where $k > 0$ is the constant of proportionality or spring constant, and s is the displacement of the spring.

Using Hooke's law and assuming that $x(t)$ represents the displacement of the mass from the equilibrium position at time t, we obtain the initial-value problem

$$\begin{cases} m\dfrac{d^2x}{dt^2} + kx = 0 \\ x(0) = \alpha, \ \dfrac{dx}{dt}(0) = \beta. \end{cases}$$

Note that the initial conditions give the initial displacement and velocity, respectively. This differential equation disregards all retarding forces acting on the motion of the mass and a more realistic model which takes these forces into account is needed. Studies in mechanics reveal that resistive forces due to damping are proportional to a power of the velocity of the motion. Hence, $F_R = c \, dx/dt$ or $F_R = c \, (dx/dt)^3$, where $c > 0$, are typically used to represent the damping force. Then, we have the following initial-value problem assuming that $F_R = c \, dx/dt$:

$$\begin{cases} m\dfrac{d^2x}{dt^2} + c\dfrac{dx}{dt} + kx = 0 \\ x(0) = \alpha, \ \dfrac{dx}{dt}(0) = \beta. \end{cases}$$

Problems of this type are characterized by the value of $c^2 - 4mk$ as follows.

1. $c^2 - 4mk > 0$. This situation is said to be **overdamped** because the damping coefficient c is large in comparison to the spring constant k.
2. $c^2 - 4mk = 0$. This situation is described as **critically damped** because the resulting motion is oscillatory with a slight decrease in the damping coefficient c.
3. $c^2 - 4mk < 0$. This situation is called **underdamped** because the damping coefficient c is small in comparison with the spring constant k.

EXAMPLE 6.2.3: Classify the following differential equations as over-damped, underdamped, or critically damped. Also, solve the corresponding initial-value problem using the given initial conditions and investigate the behavior of the solutions.

(a) $\dfrac{d^2x}{dt^2} + 8\dfrac{dx}{dt} + 16x = 0$ subject to $x(0) = 0$ and $\dfrac{dx}{dt}(0) = 1$;

(b) $\dfrac{d^2x}{dt^2} + 5\dfrac{dx}{dt} + 4x = 0$ subject to $x(0) = 1$ and $\dfrac{dx}{dt}(0) = 1$; and

(c) $\dfrac{d^2x}{dt^2} + \dfrac{dx}{dt} + 16x = 0$ subject to $x(0) = 0$ and $\dfrac{dx}{dt}(0) = 1$.

SOLUTION: For (a), we identify $m = 1$, $c = 8$, and $k = 16$ so that $c^2 - 4mk = 0$, which means that the differential equation $x'' + 8x' + 16x = 0$ is critically damped. After defining del, we solve the equation subject to the initial conditions and name the resulting output sol1. We then graph the solution shown in Figure 6-15 by selecting and copying the result given in sol1 to the subsequent Plot command. If you prefer working with **InputForm**, the formula for the solution to the initial-value problem is extracted from sol1 with sol1[[1,1,2]]. Thus, entering Plot[sol[[1,1,2]],{t,0,4}] displays the same graph as that obtained with the following Plot command. Note that replacing

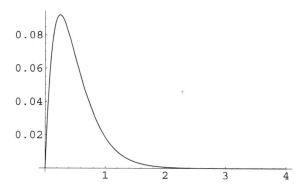

Figure 6-15   Critically damped motion

soll[[1,1,2]] with Evaluate[x[t]/.soll] in the Plot command also produces the same result.

In[1321]:= **Clear[de1, x, t]**

de1 = x''[t] + 8 x'[t] + 16 x[t] == 0;
soll = DSolve[{de1, x[0] == 0, x'[0] == 1},
        x[t], t]

Out[1321]= {{x[t] → e^{-4t} t}}

In[1322]:= **Plot[e^{-4t} t, {t, 0, 4}]**

For (b), we proceed in the same manner. We identify $m = 1$, $c = 5$,
and $k = 4$ so that $c^2 - 4mk = 9$ and the equation $x'' + 5x' + 4x = 0$ is
overdamped. We then define de2 to be the equation and the solution to
the initial-value problem obtained with DSolve, sol2 and then graph
$x(t)$ on the interval [0, 4] in Figure 6-16.

In[1323]:= **Clear[de2, x, t]**

de2 = x''[t] + 5 x'[t] + 4 x[t] == 0;
sol2 = DSolve[{de2, x[0] == 1, x'[0] == 1},
        x[t], t]

Out[1323]= $\left\{\left\{x[t] \rightarrow \frac{1}{3} e^{-4t} (-2 + 5 e^{3t})\right\}\right\}$

In[1324]:= **Plot[sol2[[1, 1, 2]], {t, 0, 4}]**

For (c), we proceed in the same manner as in (a) and (b) to show that
the equation is underdamped because the value of $c^2 - 4mk$ is −63. See
Figure 6-17.

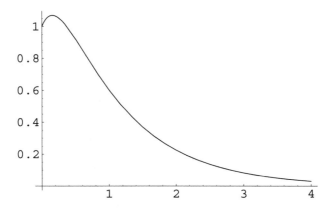

Figure 6-16   Overdamped motion

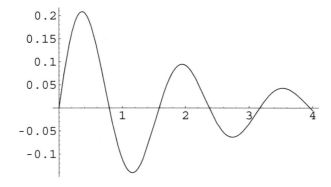

Figure 6-17   Underdamped motion

In[1325]:= **Clear[de3, x, t]**

de3 = x''[t] + x'[t] + 16 x[t] == 0;
sol3 = DSolve[{de3, x[0] == 0, x'[0] == 1},
     x[t], t]

Out[1325]= $\left\{\left\{x[t] \to \dfrac{2\, e^{-t/2}\, \text{Sin}\left[\frac{3\sqrt{7}\,t}{2}\right]}{3\sqrt{7}}\right\}\right\}$

In[1326]:= **Plot[sol3[[1, 1, 2]], {t, 0, 4}]**

6.2.3 Undetermined Coefficients

If equation (6.7) has constant coefficients and $f(t)$ is a product of terms t^n, $e^{\alpha t}$, α constant, $\cos \beta t$, and/or $\sin \beta t$, β constant, *undetermined coefficients* can often be used to find a particular solution of equation (6.7). The key to implementing the method is to *judiciously* choose the correct form of y_p.

Assume that a general solution, y_h, of the corresponding homogeneous equation has been found and that each term of $f(t)$ has the form

$$t^n e^{\alpha t} \cos \beta t \qquad \text{or} \qquad t^n e^{\alpha t} \sin \beta t.$$

For *each* term of $f(t)$, write down the *associated set*

$$F = \left\{ t^n e^{\alpha t} \cos \beta t, t^n e^{\alpha t} \sin \beta t, t^{n-1} e^{\alpha t} \cos \beta t, t^{n-1} e^{\alpha t} \sin \beta t, \ldots, e^{\alpha t} \cos \beta t, e^{\alpha t} \sin \beta t, \right\}.$$

If any element of F is a solution to the corresponding homogeneous equation, multiply each element of F by t^m, where m is the smallest positive integer so that none of the elements of $t^m F$ are solutions to the corresponding homogeneous equation. A particular solution will be a linear combination of the functions in all the F's.

EXAMPLE 6.2.4: Solve

$$4 \frac{d^2 y}{dt^2} - y = t - 2 - 5 \cos t - e^{-t/2}.$$

SOLUTION: The corresponding homogeneous equation is $4y'' - y = 0$ with general solution $y_h = c_1 e^{-t/2} + c_2 e^{t/2}$.

```
In[1327]:= DSolve[4y''[t] - y[t] == 0, y[t], t]
Out[1327]= {{y[t] → e^{-t/2} C[1] + e^{t/2} C[2]}}
```

A fundamental set of solutions for the corresponding homogeneous equation is $S = \left\{ e^{-t/2}, e^{t/2} \right\}$. The associated set of functions for $t - 2$ is $F_1 = \{1, t\}$, the associated set of functions for $-5 \cos t$ is $F_2 = \{\cos t, \sin t\}$, and the associated set of functions for $-e^{-t/2}$ is $F_3 = \left\{ e^{-t/2} \right\}$. Note that $e^{-t/2}$ is an element of S so we multiply F_3 by t resulting in $t F_3 = \left\{ t e^{-t/2} \right\}$. Then, we search for a particular solution of the form

No element of F_1 is contained in S and no element of F_2 is contained in S.

$$y_p = A + Bt + C \cos t + D \sin t + Et e^{-t/2},$$

where $A, B, C, D,$ and E are constants to be determined.

```
In[1328] := yp[t_] = a+b t+c Cos[t]+d Sin[t]+e t Exp[-t/2]
```

$$Out[1328] = a + b\ t + e\ e^{-t/2}\ t + c\ Cos[t] + d\ Sin[t]$$

Computing y_p' and y_p''

```
In[1329] := dyp = yp'[t]

           d2yp = yp''[t]
```

$$Out[1329] = b + e\ e^{-t/2} - \frac{1}{2}\ e\ e^{-t/2}\ t + d\ Cos[t] - c\ Sin[t]$$

$$Out[1329] = -e\ e^{-t/2} + \frac{1}{4}\ e\ e^{-t/2}\ t - c\ Cos[t] - d\ Sin[t]$$

and substituting into the nonhomogeneous equation results in

$$-A - Bt - 5C\cos t - 5D\sin t - 4Ee^{-t/2} = t - 2 - 5\cos t - e^{-t/2}.$$

```
In[1330] := eqn = 4 yp''[t]-yp[t] == t-2-5 Cos[t]-Exp[-t/2]
```

$$Out[1330] = -a - b\ t - e\ e^{-t/2}\ t - c\ Cos[t] - d\ Sin[t]$$
$$+ 4\left(-e\ e^{-t/2} + \frac{1}{4}\ e\ e^{-t/2}\ t - c\ Cos[t] - d\ Sin[t]\right) ==$$
$$-2 - e^{-t/2} + t - 5\ Cos[t]$$

Equating coefficients results in

$$-A = -2 \qquad -B = 1 \qquad -5C = -5 \qquad -5D = 0 \qquad -4E = -1$$

so $A = 2$, $B = -1$, $C = 1$, $D = 0$, and $E = 1/4$.

```
In[1331] := cvals = Solve[{-a == -2, -b == 1, -5c == -5, -5d == 0,
                -4e == -1}]
```

$$Out[1331] = \left\{\left\{a \rightarrow 2,\ b \rightarrow -1,\ c \rightarrow 1,\ d \rightarrow 0,\ e \rightarrow \frac{1}{4}\right\}\right\}$$

y_p is then given by $y_p = 2 - t + \cos t + \frac{1}{4}te^{-t/2}$

```
In[1332] := yp[t]/.cvals[[1]]
```

$$Out[1332] = 2 - t + \frac{1}{4}\ e^{-t/2}\ t + Cos[t]$$

and a general solution is given by

$$y = y_h + y_p = c_1 e^{-t/2} + c_2 e^{t/2} + 2 - t + \cos t + \frac{1}{4}te^{-t/2}.$$

Note that $-A - Bt - 5C\cos t - 5D\sin t - 4Ee^{-t/2} = t - 2 - 5\cos t - e^{-t/2}$ is true for *all* values of t. Evaluating for five different values of t gives us five equations that we then solve for A, B, C, D, and E, resulting in the same solutions as already obtained.

In[1333] := **e1 = eqn/.t- > 0**

Out[1333]= $-a - c + 4 \ (-c - e) == -8$

In[1334] := **e2 = eqn/.t- > π/2**

e3 = eqn/.t- > π

e4 = eqn/.t- > 1

e5 = eqn/.t- > 2

Out[1334]= $-a - d - \dfrac{b \ \pi}{2}$

$-\dfrac{1}{2} \ e \ e^{-\pi/4} \ \pi + 4 \ \left(-d - e \ e^{-\pi/4} + \dfrac{1}{8} \ e \ e^{-\pi/4} \ \pi\right) ==$

$-2 - e^{-\pi/4} + \dfrac{\pi}{2}$

Out[1334]= $-a + c - b \ \pi - e \ e^{-\pi/2} \ \pi + 4 \ \left(c - e \ e^{-\pi/2} + \dfrac{1}{4} \ e \ e^{-\pi/2} \ \pi\right) ==$

$3 - e^{-\pi/2} + \pi$

Out[1334]= $-a - b - \dfrac{e}{\sqrt{e}} - c \ Cos[1]$

$-d \ Sin[1] + 4 \ \left(-\dfrac{3 \ e}{4 \ \sqrt{e}} - c \ Cos[1] - d \ Sin[1]\right) ==$

$-1 - \dfrac{1}{\sqrt{e}} - 5 \ Cos[1]$

Out[1334]= $-a - 2 \ b - \dfrac{2 \ e}{e} - c \ Cos[2]$

$-d \ Sin[2] + 4 \ \left(-\dfrac{e}{2 \ e} - c \ Cos[2] - d \ Sin[2]\right) ==$

$-\dfrac{1}{e} - 5 \ Cos[2]$

In[1335] := **Solve[{e1, e2, e3, e4, e5}, {a, b, c, d, e}]//Simplify**

Out[1335]= $\left\{\left\{d \to 0, b \to -1, a \to 2, c \to 1, e \to \dfrac{1}{4}\right\}\right\}$

Last, we check our calculation with DSolve and simplify.

In[1336] := **sol2 = DSolve[4y''[t] - y[t] == t - 2 - 5 Cos[t]**
- Exp[-t/2], y[t], t]

Out[1336]= $\left\{\left\{y[t] \to e^{-t/2} \ C[1] + e^{t/2} \ C[2]\right.\right.$

$+ \dfrac{1}{4} \ \left(e^{-t/2} - 2 \ t + 2 \ Cos[t] - 4 \ Sin[t]\right)$

$+ e^{-t/2} \ \left(2 \ e^{t/2} + \dfrac{t}{4} - \dfrac{1}{2} \ e^{t/2} \ t + \dfrac{1}{2} \ e^{t/2} \ Cos[t]\right.$

$\left.\left.\left.+ e^{t/2} \ Sin[t]\right)\right\}\right\}$

In[1337] := **Simplify[sol2]**

Out[1337]= $\left\{\left\{y[t] \rightarrow \frac{1}{4} e^{-t/2} \left(1 + 8 e^{t/2} + t - 4 e^{t/2} t + 4 C[1]\right.\right.\right.$
$\left.\left.\left. + 4 e^t C[2]\right) + Cos[t]\right\}\right\}$

∎

EXAMPLE 6.2.5: Solve $y'' + 4y = \cos 2t$, $y(0) = 0$, $y'(0) = 0$.

SOLUTION: A general solution of the corresponding homogeneous equation is $y_h = c_1 \cos 2t + c_2 \sin 2t$. For this equation, $F = \{\cos 2t, \sin 2t\}$. Because elements of F are solutions to the corresponding homogeneous equation, we multiply each element of F by t resulting in $tF = \{t \cos 2t, t \sin 2t\}$. Therefore, we assume that a particular solution has the form

$$y_p = At \cos 2t + Bt \sin 2t,$$

where A and B are constants to be determined. Proceeding in the same manner as before, we compute y'_p and y''_p

In[1338] := **yp[t_] = a t Cos[2t] + b t Sin[2 t];**
yp'[t]

yp''[t]

Out[1338]= a Cos[2 t] + 2 b t Cos[2 t] + b Sin[2 t]
    - 2 a t Sin[2 t]

Out[1338]= 4 b Cos[2 t] - 4 a t Cos[2 t] - 4 a Sin[2 t]
    - 4 b t Sin[2 t]

and then substitute into the nonhomogeneous equation.

In[1339] := **eqn = yp''[t] + 4yp[t] == Cos[2t]**

Out[1339]= 4 b Cos[2 t] - 4 a t Cos[2 t] - 4 a Sin[2 t]
    - 4 b t Sin[2 t] + 4 (a t Cos[2 t]
    + b t Sin[2 t]) == Cos[2 t]

Equating coefficients readily yields $A = 0$ and $B = 1/4$. Alternatively, remember that $-4A \sin 2t + 4B \cos 2t = \cos 2t$ is true for *all* values of t. Evaluating for two values of t and then solving for A and B gives the same result.

In[1340] := **e1 = eqn/.t->0**

e2 = eqn/.t->π/4

cvals = Solve[{e1, e2}]

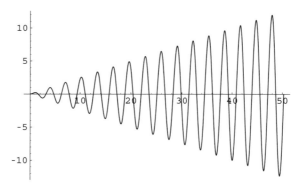

Figure 6-18   The forcing function causes the solution to become unbounded as $t \to \infty$

Out[1340]= 4 b == 1

Out[1340]= -4 a == 0

Out[1340]= $\left\{\left\{a \to 0, b \to \dfrac{1}{4}\right\}\right\}$

It follows that $y_p = \frac{1}{4}t \sin 2t$ and $y = c_1 \cos 2t + c_2 \sin 2t + \frac{1}{4}t \sin 2t$.

In[1341]:= **yp[t]/.cvals[[1]]**

Out[1341]= $\dfrac{1}{4}$ t Sin[2 t]

In[1342]:= **y[t_] = c1 Cos[2t]+c2 Sin[2t]+1/4 t Sin[2t]**

Out[1342]= c1 Cos[2 t] + c2 Sin[2 t] + $\dfrac{1}{4}$ t Sin[2 t]

Applying the initial conditions

In[1343]:= **y'[t]**

Out[1343]= 2 c2 Cos[2 t] + $\dfrac{1}{2}$ t Cos[2 t]

$+ \dfrac{1}{4}$ Sin[2 t] - 2 c1 Sin[2 t]

In[1344]:= **cvals = Solve[{y[0] == 0, y'[0] == 0}]**

Out[1344]= {{c1 → 0, c2 → 0}}

results in $y = \frac{1}{4}t \sin 2t$, which we graph with **Plot** in Figure 6-18.

In[1345]:= **y[t]/.cvals[[1]]**

Out[1345]= $\dfrac{1}{4}$ t Sin[2 t]

In[1346]:= **Plot[Evaluate[y[t]/.cvals[[1]]], {t, 0, 16π}]**

We verify the calculation with **DSolve**.

```
In[1347]:= Clear[y]

        DSolve[y''[t] + 4y[t] == Cos[2t],
            y[0] == 0, y'[0] == 0}, y[t], t]
```
$$Out[1347]= \left\{\left\{y[t] \to \frac{1}{4} \; t \; Sin[2 \; t]\right\}\right\}$$

∎

6.2.4 Variation of Parameters

Let $S = \{y_1, y_2\}$ be a fundamental set of solutions for equation (6.9). To solve the nonhomogeneous equation (6.8), we need to find a particular solution, y_p of equation (6.8). We search for a particular solution of the form

$$y_p = u_1(t)y_1(t) + u_2(t)y_2(t), \tag{6.13}$$

A particular solution, y_p, is a solution that does not contain any arbitrary constants.

where u_1 and u_2 are functions of t. Differentiating equation (6.13) gives us

$$y_p' = u_1'y_1 + u_1y_1' + u_2'y_2 + u_2y_2'.$$

Assuming that

$$y_1u_1' + y_2u_2' = 0 \tag{6.14}$$

results in $y_p' = u_1y_1' + u_2y_2'$. Computing the second derivative then yields

$$y_p'' = u_1'y_1' + u_1y_1'' + u_2'y_2' + u_2y_2''.$$

Observe that it is pointless to search for solutions of the form $y_p = c_1y_1 + c_2y_2$ where c_1 and c_2 are constants because for every choice of c_1 and c_2, $c_1y_1 + c_2y_2$ is a solution to the corresponding homogeneous equation.

Substituting y_p, y_p', and y_p'' into equation (6.8) and using the facts that

$$u_1\left(y_1'' + p\,y_1' + q\,y_1\right) = 0 \quad \text{and} \quad u_2\left(y_2'' + p\,y_2' + q\,y_2\right) = 0$$

(because y_1 and y_2 are solutions to the corresponding homogeneous equation) results in

$$\frac{d^2y_p}{dt^2} + p(t)\frac{dy_p}{dt} + q(t)y_p = u_1'y_1' + u_1y_1'' + u_2'y_2' + u_2y_2'' + p(t)\left(u_1y_1' + u_2y_2'\right)$$
$$+ q(t)\left(u_1y_1 + u_2y_2\right)$$
$$= y_1'u_1' + y_2'u_2' = f(t). \tag{6.15}$$

Observe that equation (6.14) and equation (6.15) form a system of two linear equations in the unknowns u_1' and u_2':

$$y_1u_1' + y_2u_2' = 0$$
$$y_1'u_1' + y_2'u_2' = f(t). \tag{6.16}$$

Applying Cramer's Rule gives us

$$u_1' = \frac{\begin{vmatrix} 0 & y_2 \\ f(t) & y_2' \end{vmatrix}}{\begin{vmatrix} y_1 & y_2 \\ y_1' & y_2' \end{vmatrix}} = -\frac{y_2(t)f(t)}{W(S)} \quad \text{and} \quad u_2' = \frac{\begin{vmatrix} y_1 & 0 \\ y_1' & f(t) \end{vmatrix}}{\begin{vmatrix} y_1 & y_2 \\ y_1' & y_2' \end{vmatrix}} = \frac{y_1(t)f(t)}{W(S)}, \tag{6.17}$$

where $W(S)$ is the Wronskian, $W(S) = \begin{vmatrix} y_1 & y_2 \\ y_1' & y_2' \end{vmatrix}$. After integrating to obtain u_1 and u_2, we form y_p and then a general solution, $y = y_h + y_p$.

EXAMPLE 6.2.6: Solve $y'' + 9y = \sec 3t$, $y(0) = 0$, $y'(0) = 0$, $0 \le t < \pi/6$.

SOLUTION: The corresponding homogeneous equation is $y'' + 9y = 0$ with general solution $y_h = c_1 \cos 3t + c_2 \sin 3t$. Then, a fundamental set of solutions is $S = \{\cos 3t, \sin 3t\}$ and $W(S) = 3$, as we see using Det, and Simplify.

```
In[1348]:= fs = {Cos[3t], Sin[3t]};
           wm = {fs, D[fs, t]};
           wm//MatrixForm

           wd = Simplify[Det[wm]]
Out[1348]= ( Cos[3 t]       Sin[3 t]   )
           ( -3 Sin[3 t]   3 Cos[3 t]  )
Out[1348]= 3
```

We use equation (6.17) to find $u_1 = \frac{1}{9} \ln \cos 3t$ and $u_2 = \frac{1}{3}t$.

```
In[1349]:= u1 = Integrate[-Sin[3t] Sec[3t]/3, t]

           u2 = Integrate[Cos[3t] Sec[3t]/3, t]
Out[1349]= 1
           ─ Log[Cos[3 t]]
           9
           t
Out[1349]= ─
           3
```

It follows that a particular solution of the nonhomogeneous equation is $y_p = \frac{1}{9} \cos 3t \ln \cos 3t + \frac{1}{3}t \sin 3t$ and a general solution is $y = y_h + y_p = c_1 \cos 3t + c_2 \sin 3t + \frac{1}{9} \cos 3t \ln \cos 3t + \frac{1}{3}t \sin 3t$.

```
In[1350]:= yp = u1 Cos[3t] + u2 Sin[3t]
           1                           1
Out[1350]= ─ Cos[3 t] Log[Cos[3 t]] + ─ t Sin[3 t]
           9                           3
```

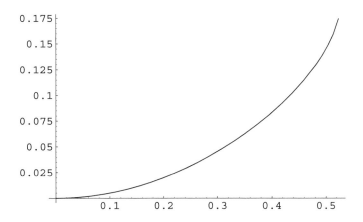

Figure 6-19 The domain of the solution is $-\pi/6 < t < \pi/6$

Identical results are obtained using DSolve.

The negative sign in the output does not affect the result because C[1] is arbitrary.

$In[1351]:=$ **DSolve[y″[t] + 9y[t] == Sec[3t], y[t], t]**

$Out[1351]=$ $\left\{\left\{y[t] \to C[2]\ Cos[3\ t]\right.\right.$

$+\dfrac{1}{9}\ Cos[3\ t]\ Log[Cos[3\ t]]$

$\left.\left.+\dfrac{1}{3}\ t\ Sin[3\ t] - C[1]\ Sin[3\ t]\right\}\right\}$

Applying the initial conditions gives us $c_1 = c_2 = 0$ so we conclude that the solution to the initial value problem is $y = \frac{1}{9}\cos 3t\ \ln\cos 3t + \frac{1}{3}t\sin 3t$.

$In[1352]:=$ **sol = DSolve[{y″[t] + 9y[t] == Sec[3t], y[0] == 0,**
        y′[0] == 0}, y[t], t]

$Out[1352]=$ $\left\{\left\{y[t] \to \dfrac{1}{9}\ (Cos[3\ t]\ Log[Cos[3\ t]]\right.\right.$

$\left.\left.+3\ t\ Sin[3\ t])\right\}\right\}$

We graph the solution with Plot in Figure 6-19.

$In[1353]:=$ **Plot[Evaluate[y[t]/.sol], {t, 0, π/6}]**

■

6.3 Higher-Order Linear Equations

6.3.1 Basic Theory

The **standard form of the nth-order linear equation** is

$$\frac{d^n y}{dt^n} + a_{n-1}(t)\frac{d^{n-1} y}{dt^{n-1}} + \cdots + a_1(t)\frac{dy}{dt} + a_0(t)y = f(t). \tag{6.18}$$

The **corresponding homogeneous equation** of equation (6.18) is

$$\frac{d^n y}{dt^n} + a_{n-1}(t)\frac{d^{n-1} y}{dt^{n-1}} + \cdots + a_1(t)\frac{dy}{dt} + a_0(t)y = 0. \tag{6.19}$$

Let $y_1, y_2, \ldots, y_n$ be n solutions of equation (6.19). The set $S = \{y_1, y_2, \ldots, y_n\}$ is **linearly independent** if and only if the **Wronskian**,

$$W(S) = \begin{vmatrix} y_1 & y_2 & y_3 & \cdots & y_n \\ y_1' & y_2' & y_3' & \cdots & y_n' \\ y_1'' & y_2'' & y_3'' & \cdots & y_n'' \\ y_1^{(3)} & y_2^{(3)} & y_3^{(3)} & \cdots & y_n^{(3)} \\ \vdots & \vdots & \vdots & \cdots & \vdots \\ y_1^{(n-1)} & y_2^{(n-1)} & y_3^{(n-1)} & \cdots & y_n^{(n-1)} \end{vmatrix} \tag{6.20}$$

is not identically the zero function. S is **linearly dependent** if S is not linearly independent.

If $y_1, y_2, \ldots, y_n$ are n linearly independent solutions of equation (6.19), we say that $S = \{y_1, y_2, \ldots, y_n\}$ is a **fundamental set** for equation (6.19) and a **general solution** of equation (6.19) is $y = c_1 y_1 + c_2 y_2 + c_3 y_3 + \cdots + c_n y_n$.

A **general solution** of equation (6.18) is $y = y_h + y_p$ where y_h is a general solution of the corresponding homogeneous equation and y_p is a particular solution of equation (6.18).

6.3.2 Constant Coefficients

If

$$\frac{d^n y}{dt^n} + a_{n-1}\frac{d^{n-1} y}{dt^{n-1}} + \cdots + a_1\frac{dy}{dt} + a_0 y = 0$$

has real constant coefficients, we assume that $y = e^{kt}$ and find that k satisfies the **characteristic equation**

$$k^n + a_{n-1}k^{n-1} + \cdots + a_1 k + a_0 = 0. \tag{6.21}$$

If a solution k of equation (6.21) has multiplicity m, m linearly independent solutions corresponding to k are

$$e^{kt}, te^{kt}, \ldots, t^{m-1}e^{kt}.$$

If a solution $k = \alpha + \beta i$, $\beta \neq 0$, of equation (6.21) has multiplicity m, $2m$ linearly independent solutions corresponding to $k = \alpha + \beta i$ (and $k = \alpha - \beta i$) are

$$e^{\alpha t}\cos\beta t, e^{\alpha t}\sin\beta t, te^{\alpha t}\cos\beta t, te^{\alpha t}\sin\beta t, \ldots, t^{m-1}e^{\alpha t}\cos\beta t, t^{m-1}e^{\alpha t}\sin\beta t.$$

EXAMPLE 6.3.1: Solve $12y''' - 5y'' - 6y' - y = 0$.

SOLUTION: The characteristic equation is

$$12k^3 - 5k^2 - 6k - 1 = (k-1)(3k+1)(4k+1) = 0$$

with solutions $k_1 = -1/3$, $k_2 = -1/4$, and $k_3 = 1$.

> `Factor[expression]` attempts to factor expression.

```
In[1354]:= Factor[12k^3 - 5k^2 - 6k - 1]
Out[1354]= (-1 + k) (1 + 3 k) (1 + 4 k)
```

Thus, three linearly independent solutions of the equation are $y_1 = e^{-t/3}$, $y_2 = e^{-t/4}$, and $y_3 = e^t$; a general solution is $y = c_1 e^{-t/3} + c_2 e^{-t/4} + c_3 e^t$. We check with DSolve.

```
In[1355]:= DSolve[12y'''[t]-5y''[t]-6y'[t]-y[t] == 0,y[t],t]
Out[1355]= {{y[t] → e^{-t/3} C[1] + e^{-t/4} C[2] + e^t C[3]}}
```

■

EXAMPLE 6.3.2: Solve $y''' + 4y' = 0$, $y(0) = 0$, $y'(0) = 1$, $y''(0) = -1$.

SOLUTION: The characteristic equation is $k^3 + 4k = k(k^2 + 4) = 0$ with solutions $k_1 = 0$ and $k_{2,3} = \pm 2i$ that are found with Solve.

> Enter ?Solve to obtain basic help regarding the Solve function.

```
In[1356]:= Solve[k^3 + 4k == 0]
Out[1356]= {{k → 0}, {k → -2 i}, {k → 2 i}}
```

Three linearly independent solutions of the equation are $y_1 = 1$, $y_2 = \cos 2t$, and $y_3 = \sin 2t$. A general solution is $y = c_1 + c_2 \sin 2t + c_3 \cos 2t$.

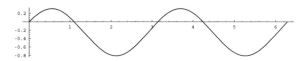

Figure 6-20　Graph of $y = -\frac{1}{4} + \frac{1}{2}\sin 2t + \frac{1}{4}\cos 2t$

$In[1357] :=$ **gensol = DSolve[y'''[t] + 4y'[t] == 0, y[t], t]**

$Out[1357] =$ $\left\{\left\{y[t] \to C[3] + \frac{1}{2} C[1] \ Cos[2 \ t] + \frac{1}{2} C[2]\right.\right.$
　　　　　　　$\left.\left.Sin[2 \ t]\right\}\right\}$

Application of the initial conditions shows us that $c_1 = -1/4$, $c_2 = 1/2$, and $c_3 = 1/4$ so the solution to the initial-value problem is $y = -\frac{1}{4} + \frac{1}{2}\sin 2t + \frac{1}{4}\cos 2t$. We verify the computation with DSolve and graph the result with Plot in Figure 6-20.

$In[1358] :=$ **e1 = y[t]/.gensol[[1]]/.t-> 0**

$Out[1358] =$ $\frac{C[1]}{2} + C[3]$

$In[1359] :=$ **e2 = D[y[t]/.gensol[[1]], t]/.t-> 0**

　　　　　　e3 = D[y[t]/.gensol[[1]], {t, 2}]/.t-> 0

$Out[1359] =$ $C[2]$

$Out[1359] =$ $-2 \ C[1]$

$In[1360] :=$ **cvals = Solve[{e1 == 0, e2 == 1, e3 == -1}]**

$Out[1360] =$ $\left\{\left\{C[1] \to \frac{1}{2}, C[2] \to 1, C[3] \to -\frac{1}{4}\right\}\right\}$

$In[1361] :=$ **partsol = DSolve[**
　　　　　　{y'''[t] + 4y'[t] == 0, y[0] == 0, y'[0] == 1,
　　　　　　y''[0] == -1}, y[t], t]

$Out[1361] =$ $\left\{\left\{y[t] \to -\frac{1}{4} + \frac{1}{4} \ Cos[2 \ t] + \frac{1}{2} \ Sin[2 \ t]\right\}\right\}$

$In[1362] :=$ **Plot[Evaluate[y[t]/.partsol], {t, 0, 2π},**
　　　　　　AspectRatio-> Automatic]

■

EXAMPLE 6.3.3: Find a differential equation with general solution $y = c_1e^{-2t/3} + c_2te^{-2t/3} + c_3t^2e^{-2t/3} + c_4\cos t + c_5\sin t + c_6t\cos t + c_7t\sin t + c_8t^2\cos t + c_9t^2\sin t$.

SOLUTION: A linear homogeneous differential equation with constant coefficients that has this general solution has fundamental set of solutions

$$S = \left\{ e^{-2t/3}, te^{-2t/3}, t^2 e^{-2t/3}, \cos t, \sin t, t \cos t, t \sin t, t^2 \cos t, t^2 \sin t \right\}.$$

Hence, in the characteristic equation $k = -2/3$ has multiplicity 3 while $k = \pm i$ has multiplicity 3. The characteristic equation is

$$27\left(k + \frac{2}{3}\right)^3 (k - i)^3 (k + i)^3 = k^9 + 2k^8 + \frac{13}{3}k^7 + \frac{170}{27}k^6 + 7k^5 + \frac{62}{9}k^4$$
$$+ 5k^3 + \frac{26}{9}k^2 + \frac{4}{3}k + \frac{8}{27},$$

where we use Mathematica to compute the multiplication with Expand.

$$In[1363] := \textbf{Expand[27(k + 2/3)^3(k^2 + 1)^3]}$$
$$Out[1363] = 8 + 36 \ k + 78 \ k^2 + 135 \ k^3 + 186 \ k^4 + 189 \ k^5$$
$$+ 170 \ k^6 + 117 \ k^7 + 54 \ k^8 + 27 \ k^9$$

Thus, a differential equation obtained after dividing by 27 with the indicated general solution is

$$\frac{d^9 y}{dt^9} + 2\frac{d^8 y}{dt^8} + \frac{13}{3}\frac{d^7 y}{dt^7} + \frac{170}{27}\frac{d^6 y}{dt^6} + 7\frac{d^5 y}{dt^5} + \frac{62}{9}\frac{d^4 y}{dt^4}$$
$$+ 5\frac{d^3 y}{dt^3} + \frac{26}{9}\frac{d^2 y}{dt^2} + \frac{4}{3}\frac{dy}{dt} + \frac{8}{27}y = 0.$$

∎

6.3.3 Undetermined Coefficients

For higher-order linear equations with constant coefficients, the method of undetermined coefficients is the same as for second-order equations discussed in Section 6.2.3 , provided that the forcing function involves appropriate terms.

EXAMPLE 6.3.4: Solve

$$\frac{d^3 y}{dt^3} + \frac{2}{3}\frac{d^2 y}{dt^2} + \frac{145}{9}\frac{dy}{dt} = e^{-t}, \ y(0) = 1, \ \frac{dy}{dt}(0) = 2, \ \frac{d^2 y}{dt^2}(0) = -1.$$

SOLUTION: The corresponding homogeneous equation, $y''' + \frac{2}{3}y'' + \frac{145}{9}y' = 0$, has general solution $y_h = c_1 + (c_2 \sin 4t + c_3 \cos 4t)\, e^{-t/3}$ and a fundamental set of solutions for the corresponding homogeneous equation is $S = \left\{1,\, e^{-t/3} \cos 4t,\, e^{-t/3} \sin 4t\right\}$.

```
In[1364]:= DSolve[y'''[t] + 2/3y''[t] + 145/9y'[t] == 0,
               y[t], t]//Simplify
```
$$Out[1364]= \left\{\left\{y[t] \to C[3] + \frac{3}{145}\; e^{-t/3}\right.\right.$$
$$((12\;C[1] - C[2])\;Cos[4\;t] + (C[1]$$
$$\left.\left.+ 12\;C[2])\;Sin[4\;t])\right\}\right\}$$

For e^{-t}, the associated set of functions is $F = \left\{e^{-t}\right\}$. Because no element of F is an element of S, we assume that $y_p = Ae^{-t}$, where A is a constant to be determined. After defining y_p, we compute the necessary derivatives

```
In[1365]:= yp[t_] = a Exp[-t];
               yp'[t]

               yp''[t]

               yp'''[t]
```
$$Out[1365]= -a\;e^{-t}$$
$$Out[1365]= a\;e^{-t}$$
$$Out[1365]= -a\;e^{-t}$$

and substitute into the nonhomogeneous equation.

```
In[1366]:= eqn = yp'''[t]+2/3yp''[t]+145/9yp'[t] == Exp[-t]
```
$$Out[1366]= -\frac{148}{9}\;a\;e^{-t} == e^{-t}$$

Equating coefficients and solving for A gives us $A = -9/148$ so $y_p = -\frac{9}{148}e^{-t}$ and a general solution is $y = y_h + y_p$.

Remark. `SolveAlways[equation,variable]` attempts to solve equation so that it is true for all values of `variable`.

```
In[1367]:= SolveAlways[eqn, t]
```
$$Out[1367]= \left\{\left\{a \to -\frac{9}{148}\right\}\right\}$$

We verify the result with `DSolve`.

```
In[1368]:= gensol = DSolve[y'''[t] + 2/3y''[t]
                    + 145/9y'[t] == Exp[-t], y[t], t]
```

$Out[1368] = \left\{\left\{y[t] \to -\dfrac{9\ e^{-t}}{148} - \left(\dfrac{3}{145} - \dfrac{36\ i}{145}\right)\ e^{\left(-\frac{1}{3} - 4\ i\right)\ t}\ C[1]\right.\right.$

$\qquad\qquad\qquad \left.\left. - \left(\dfrac{9}{290} - \dfrac{3\ i}{1160}\right)\ e^{\left(-\frac{1}{3} + 4\ i\right)\ t}\ C[2] + C[3]\right\}\right\}$

To obtain a real-valued solution, we use `ComplexExpand`:

$In[1369] :=$ `?ComplexExpand`

"ComplexExpand[expr] expands expr assuming
  that all variables are real. ComplexExpand[
  expr, x1, x2, ...] expands expr assuming
  that variables matching any of the xi are complex."

$In[1370] :=$ `s1 = ComplexExpand[y[t]/.gensol[[1]]]`

$Out[1370] = -\dfrac{9\ e^{-t}}{148} + C[3] - \left(\dfrac{3}{145} - \dfrac{36\ i}{145}\right)\ e^{-t/3}\ C[1]\ Cos[4\ t]$

$\qquad\qquad - \left(\dfrac{9}{290} - \dfrac{3\ i}{1160}\right)\ e^{-t/3}\ C[2]\ Cos[4\ t]$

$\qquad\qquad + \left(\dfrac{36}{145} + \dfrac{3\ i}{145}\right)\ e^{-t/3}\ C[1]\ Sin[4\ t]$

$\qquad\qquad - \left(\dfrac{3}{1160} + \dfrac{9\ i}{290}\right)\ e^{-t/3}\ C[2]\ Sin[4\ t]$

$In[1371] :=$ `t1 = Coefficient[s1, Exp[-t/3] Cos[4t]]`

$Out[1371] = \left(-\dfrac{3}{145} + \dfrac{36\ i}{145}\right)\ C[1] - \left(\dfrac{9}{290} - \dfrac{3\ i}{1160}\right)\ C[2]$

$In[1372] :=$ `t2 = Coefficient[s1, Exp[-t/3] Sin[4t]]`

$Out[1372] = \left(\dfrac{36}{145} + \dfrac{3\ i}{145}\right)\ C[1] - \left(\dfrac{3}{1160} + \dfrac{9\ i}{290}\right)\ C[2]$

$In[1373] :=$ `t3 = C[3]`

$Out[1373] =$ `C[3]`

$In[1374] :=$ `Clear[c1, c2, c3]`

$\qquad\qquad$ `s2 = Solve[{t1 == c1, t2 == c2, t3 == c3}, {C[1],`
$\qquad\qquad\qquad$ `C[2], C[3]}]`

$Out[1374] = \left\{\left\{C[1] \to \left(-\dfrac{1}{6} - 2\ i\right)\ (c1 + i\ c2), C[2] \to\right.\right.$

$\qquad\qquad\qquad \left.\left. \left(-16 - \dfrac{4\ i}{3}\right)\ (c1 - i\ c2), C[3] \to c3\right\}\right\}$

The result indicates that the form returned by `DSolve` is equivalent to

$In[1375] :=$ `s3 = s1/.s2[[1]]//Simplify`

$Out[1375] = c3 - \dfrac{9\ e^{-t}}{148} + c1\ e^{-t/3}\ Cos[4\ t] + c2\ e^{-t/3}\ Sin[4\ t]$

To apply the initial conditions, we compute $y(0) = 1$, $y'(0) = 2$, and $y''(0) = -1$

$In[1376]:=$ **e1 = (s3/.t- > 0) == 1**

e2 = (D[s3, t]/.t- > 0) == 2

e3 = (D[s3, {t, 2}]/.t- > 0) == -1

$Out[1376]=$ $-\dfrac{9}{148} + c1 + c3 == 1$

$Out[1376]=$ $\dfrac{9}{148} - \dfrac{c1}{3} + 4\ c2 == 2$

$Out[1376]=$ $-\dfrac{9}{148} - \dfrac{143\ c1}{9} - \dfrac{8\ c2}{3} == -1$

and solve for c_1, c_2, and c_3.

$In[1377]:=$ **cvals = Solve[{e1, e2, e3}]**

$Out[1377]=$ $\left\{\left\{c1 \rightarrow -\dfrac{471}{21460},\ c2 \rightarrow \dfrac{20729}{42920},\ c3 \rightarrow \dfrac{157}{145}\right\}\right\}$

The solution of the initial-value problem is obtained by substituting these values into the general solution.

$In[1378]:=$ **s3/.cvals[[1]]**

$Out[1378]=$ $\dfrac{157}{145} - \dfrac{9\ e^{-t}}{148} - \dfrac{471\ e^{-t/3}\ \cos[4\ t]}{21460}$
$+ \dfrac{20729\ e^{-t/3}\ \sin[4\ t]}{42920}$

We check by using DSolve to solve the initial-value problem and graph the result with Plot in Figure 6-21.

$In[1379]:=$ **sol = DSolve[{y'''[t] + 2/3y''[t]**
+ 145/9y'[t] == Exp[-t],
y[0] == 1, y'[0] == 2, y''[0] == -1}, y[t],
t]

$Out[1379]=$ $\left\{\left\{y[t] \rightarrow \dfrac{157}{145} - \dfrac{9\ e^{-t}}{148} - \left(\dfrac{471}{42920}\right.\right.\right.$
$\left.- \dfrac{20729\ i}{85840}\right)\ e^{\left(-\frac{1}{3}-4\ i\right)\ t}$
$\left.\left.\left.- \left(\dfrac{471}{42920} + \dfrac{20729\ i}{85840}\right)\ e^{\left(-\frac{1}{3}+4\ i\right)\ t}\right\}\right\}\right.$

$In[1380]:=$ **realsol = ComplexExpand[y[t]/.sol[[1]]]**

$Out[1380]=$ $\dfrac{157}{145} - \dfrac{9\ e^{-t}}{148} - \dfrac{471\ e^{-t/3}\ \cos[4\ t]}{21460}$
$+ \dfrac{20729\ e^{-t/3}\ \sin[4\ t]}{42920}$

$In[1381]:=$ **Plot[realsol, {t, 0, 2π},**
AspectRatio- > Automatic]

∎

Figure 6-21 The solution of the equation that satisfies $y(0) = 1$, $y'(0) = 2$, and $y''(0) = -1$

EXAMPLE 6.3.5: Solve

$$\frac{d^8y}{dt^8} + \frac{7}{2}\frac{d^7y}{dt^7} + \frac{73}{2}\frac{d^6y}{dt^6} + \frac{229}{2}\frac{d^5y}{dt^5} + \frac{801}{2}\frac{d^4y}{dt^4} +$$

$$976\frac{d^3y}{dt^3} + 1168\frac{d^2y}{dt^2} + 640\frac{dy}{dt} + 128y = te^{-t} + \sin 4t + t.$$

SOLUTION: Solving the characteristic equation

```
In[1382]:= Solve[k^8 + 7/2k^7 + 73/2k^6 + 229/2k^5+
               801/2k^4 + 976k^3 + 1168k^2 + 640k + 128 ==
           0]
```

$$Out[1382] = \left\{ \{k \rightarrow -1\}, \{k \rightarrow -1\}, \{k \rightarrow -1\}, \left\{k \rightarrow -\frac{1}{2}\right\}, \{k \rightarrow -4 \ i\}, \right.$$
$$\left. \{k \rightarrow -4 \ i\}, \{k \rightarrow 4 \ i\}, \{k \rightarrow 4 \ i\} \right\}$$

shows us that the solutions are $k_1 = -1/2$, $k_2 = -1$ with multiplicity 3, and $k_{3,4} = \pm 4i$, each with multiplicity 2. A fundamental set of solutions for the corresponding homogeneous equation is

$$S = \left\{ e^{-t/2}, e^{-t}, te^{-t}, t^2e^{-t}, \cos 4t, t\cos 4t, \sin 4t, t\sin 4t \right\}.$$

A general solution of the corresponding homogeneous equation is

$$y_h = c_1e^{-t/2} + \left(c_2 + c_3t + c_4t^2\right)e^{-t} + (c_5 + c_7t)\sin 4t + (c_6 + c_8t)\cos 4t.$$

```
In[1383]:= gensol = DSolve[D[y[t],{t,8}] + 7/2D[y[t],{t,7}]
               + 73/2D[y[t],{t,6}] + 229/2D[y[t],{t,5}]
               + 801/2D[y[t],{t,4}] + 976D[y[t],{t,3}]
               + 1168D[y[t],{t,2}] + 640D[y[t],t]
               + 128y[t] == 0, y[t], t]
```

$$Out[1383] = \left\{ \{y[t] \rightarrow e^{-t} \ C[1] + e^{-t} \ t \ C[2] + e^{-t} \ t^2 \ C[3] \right.$$
$$+ e^{-t/2} \ C[4] + C[6] \ \text{Cos}[4 \ t]$$
$$+ t \ C[8] \ \text{Cos}[4 \ t] - C[5] \ \text{Sin}[4 \ t]$$
$$\left. - t \ C[7] \ \text{Sin}[4 \ t]\} \right\}$$

The associated set of functions for te^{-t} is $F_1 = \left\{ e^{-t}, te^{-t} \right\}$. We multiply F_1 by t^n, where n is the smallest nonnegative integer so that no element of

$t^n F_1$ is an element of S: $t^3 F_1 = \{t^3 e^{-t}, t^4 e^{-t}\}$. The associated set of functions for $\sin 4t$ is $F_2 = \{\cos 4t, \sin 4t\}$. We multiply F_2 by t^n, where n is the smallest nonnegative integer so that no element of $t^n F_2$ is an element of S: $t^2 F_2 = \{t^2 \cos 4t, t^2 \sin 4t\}$. The associated set of functions for t is $F_3 = \{1, t\}$. No element of F_3 is an element of S.

Thus, we search for a particular solution of the form

$$y_p = A_1 t^3 e^{-t} + A_2 t^4 e^{-t} + A_3 t^2 \cos 4t + A_4 t^2 \sin 4t + A_5 + A_6 t,$$

where the A_i are constants to be determined.

After defining y_p, we compute the necessary derivatives

Remark. We have used `Table` twice for typesetting purposes. You can compute the derivatives using `Table[{n,D[yp[t],{t,n}]},{n,1,8}]`.

```
In[1384]:= yp[t_] = a[1]t^3 Exp[-t] + a[2]t^4 Exp[-t] +
              a[3]t^2 Cos[4t] + a[4]t^2 Sin[4t] + a[5] + a[6]t
Out[1384]= e^-t t^3 a[1] + e^-t t^4 a[2] + a[5] + t a[6]
            +t^2 a[3] Cos[4 t] + t^2 a[4] Sin[4 t]

In[1385]:= Table[{n, D[yp[t], {t, n}]}, {n, 1, 4}]
Out[1385]= {{1, 3 e^-t t^2 a[1] - e^-t t^3 a[1] + 4 e^-t t^3 a[2]
               -e^-t t^4 a[2] + a[6] + 2 t a[3] Cos[4 t]
               +4 t^2 a[4] Cos[4 t] - 4 t^2 a[3] Sin[4 t]
               +2 t a[4] Sin[4 t]},
             {2, 6 e^-t t a[1] - 6 e^-t t^2 a[1] + e^-t t^3 a[1]
               +12 e^-t t^2 a[2] - 8 e^-t t^3 a[2] + e^-t t^4 a[2]
               +2 a[3] Cos[4 t] - 16 t^2 a[3] Cos[4 t]
               +16 t a[4] Cos[4 t] - 16 t a[3] Sin[4 t]
               +2 a[4] Sin[4 t] - 16 t^2 a[4] Sin[4 t]},
             {3, 6 e^-t a[1] - 18 e^-t t a[1] + 9 e^-t t^2 a[1]
               -e^-t t^3 a[1] + 24 e^-t t a[2] - 36 e^-t t^2 a[2]
               +12 e^-t t^3 a[2] - e^-t t^4 a[2]
               -96 t a[3] Cos[4 t] + 24 a[4] Cos[4 t]
               -64 t^2 a[4] Cos[4 t] - 24 a[3] Sin[4 t]
               +64 t^2 a[3] Sin[4 t]
               -96 t a[4] Sin[4 t]},
             {4, -24 e^-t a[1] + 36 e^-t t a[1] - 12 e^-t t^2 a[1]
               +e^-t t^3 a[1] + 24 e^-t a[2] - 96 e^-t t a[2]
               +72 e^-t t^2 a[2] - 16 e^-t t^3 a[2]
               +e^-t t^4 a[2] - 192 a[3] Cos[4 t]
               +256 t^2 a[3] Cos[4 t] - 512 t a[4] Cos[4 t]
               +512 t a[3] Sin[4 t] - 192 a[4] Sin[4 t]
               +256 t^2 a[4] Sin[4 t]}}
```

In[1386]:= **Table[{n, D[yp[t], {t, n}]}, {n, 5, 8}]**

Out[1386]= $\{\{5, 60\ e^{-t}\ a[1] - 60\ e^{-t}\ t\ a[1] + 15\ e^{-t}\ t^2\ a[1]$
$\quad - e^{-t}\ t^3\ a[1] - 120\ e^{-t}\ a[2] + 240\ e^{-t}\ t\ a[2]$
$\quad - 120\ e^{-t}\ t^2\ a[2] + 20\ e^{-t}\ t^3\ a[2] - e^{-t}\ t^4\ a[2]$
$\quad + 2560\ t\ a[3]\ Cos[4\ t] - 1280\ a[4]\ Cos[4\ t]$
$\quad + 1024\ t^2\ a[4]\ Cos[4\ t] + 1280\ a[3]\ Sin[4\ t]$
$\quad - 1024\ t^2\ a[3]\ Sin[4\ t] + 2560\ t\ a[4]\ Sin[4\ t]\},$
$\{6, -120\ e^{-t}\ a[1] + 90\ e^{-t}\ t\ a[1] - 18\ e^{-t}\ t^2\ a[1]$
$\quad + e^{-t}\ t^3\ a[1] + 360\ e^{-t}\ a[2] - 480\ e^{-t}\ t\ a[2]$
$\quad + 180\ e^{-t}\ t^2\ a[2] - 24\ e^{-t}\ t^3\ a[2] + e^{-t}\ t^4\ a[2]$
$\quad + 7680\ a[3]\ Cos[4\ t] - 4096\ t^2\ a[3]\ Cos[4\ t]$
$\quad + 12288\ t\ a[4]\ Cos[4\ t] - 12288\ t\ a[3]\ Sin[4\ t]$
$\quad + 7680\ a[4]\ Sin[4\ t] - 4096\ t^2\ a[4]\ Sin[4\ t]\},$
$\{7, 210\ e^{-t}\ a[1] - 126\ e^{-t}\ t\ a[1] + 21\ e^{-t}\ t^2\ a[1]$
$\quad - e^{-t}\ t^3\ a[1] - 840\ e^{-t}\ a[2] + 840\ e^{-t}\ t\ a[2]$
$\quad - 252\ e^{-t}\ t^2\ a[2] + 28\ e^{-t}\ t^3\ a[2] - e^{-t}\ t^4\ a[2]$
$\quad - 57344\ t\ a[3]\ Cos[4\ t] + 43008\ a[4]\ Cos[4\ t]$
$\quad - 16384\ t^2\ a[4]\ Cos[4\ t] - 43008\ a[3]\ Sin[4\ t]$
$\quad + 16384\ t^2\ a[3]\ Sin[4\ t] - 57344\ t\ a[4]\ Sin[4\ t]\},$
$\{8, -336\ e^{-t}\ a[1]+$
$\quad 168\ e^{-t}\ t\ a[1] - 24\ e^{-t}\ t^2\ a[1] + e^{-t}\ t^3\ a[1]$
$\quad + 1680\ e^{-t}\ a[2] - 1344\ e^{-t}\ t\ a[2]$
$\quad + 336\ e^{-t}\ t^2\ a[2] - 32\ e^{-t}\ t^3\ a[2] + e^{-t}\ t^4\ a[2]$
$\quad - 229376\ a[3]\ Cos[4\ t] + 65536\ t^2\ a[3]\ Cos[4\ t]$
$\quad - 262144\ t\ a[4]\ Cos[4\ t]$
$\quad + 262144\ t\ a[3]\ Sin[4\ t] - 229376\ a[4]\ Sin[4\ t]$
$\quad + 65536\ t^2\ a[4]\ Sin[4\ t]\}\}$

and substitute into the nonhomogeneous equation, naming the result eqn. At this point we can either equate coefficients and solve for A_i or use the fact that eqn is true for *all* values of *t*.

In[1387]:= **eqn = D[yp[t], {t, 8}] + 7/2D[yp[t], {t, 7}]**
+ 73/2D[yp[t], {t, 6}] + 229/2D[yp[t],
{t, 5}] + 801/2D[yp[t], {t, 4}] + 976D[yp[t],
{t, 3}] + 1168D[yp[t], {t, 2}]
+ 640D[yp[t], t] + 128yp[t] ==
t Exp[-t] + Sin[4t] + t//
Simplify

Out[1387]= e^{-t} $(-867 \ a[1] + 7752 \ a[2] - 3468 \ t \ a[2]$
$+ 128 + e^t \ a[5] + 640 \ e^t \ a[6] + 128 \ e^t \ t \ a[6])$
$- 64 \ (369 \ a[3] - 428 \ a[4]) \ Cos[4 \ t]$
$- 64 \ (428 \ a[3] + 369 \ a[4]) \ Sin[4 \ t] ==$
$t + e^{-t} \ t + Sin[4 \ t]$

We substitute in six values of *t*

In[1388]:= **sysofeqs = Table[eqn/.t- > n//N, {n, 0, 5}]**

Out[1388]= {-867. a[1.]+
            7752. a[2.]-
            64. (369. a[3.]-
                428. a[4.])+
            128. a[5.]+
            640. a[6.] == 0,
            41.8332 (369. a[3.]-
                428. a[4.])+
            48.4354 (428. a[3.]+
                369. a[4.])+
            0.367879 (-867. a[1.]+
                4284. a[2.]+
                347.94 a[5.]+
                2087.64 a[6.]) ==
        0.611077,
            9.312 (369. a[3.]-
                428. a[4.])-
            63.3189 (428. a[3.]+
                369. a[4.])+
            0.135335 (-867. a[1.]+
                816. a[2.]+
                945.799 a[5.]+
                6620.59 a[6.]) ==
        3.26003,
            -54.0067 (369. a[3.]-
                428. a[4.])+
            34.3407 (428. a[3.]+
                369. a[4.])+
            0.0497871 (-867. a[1.]-
                2652. a[2.]+
                2570.95 a[5.]+
                20567.6 a[6.]) ==
        2.61279,

```
Out[1388]=  61.2902 (369. a[3.]-
                     428. a[4.])+
               18.4258 (428. a[3.]+
                        369. a[4.])+
               0.0183156 (-867. a[1.]-
                          6120. a[2.]+
                          6988.56 a[5.]+
                          62897.1 a[6.]) ==
             3.78536,
            -26.1173 (369. a[3.]-
                      428. a[4.])-
               58.4285 (428. a[3.]+
                        369. a[4.])+
               0.00673795
                  (-867. a[1.]-
                   9588. a[2.]+
                   18996.9 a[5.]+
                   189969. a[6.]) ==
             5.94663}
```

and then solve for A_i.

```
In[1389]:= coeffs =
             Solve[sysofeqs, {a[1.],a[2.],a[3.],a[4.],a[5.],a[6.]}]

Out[1389]= {{a[1.] → -0.00257819,
             a[2.] → -0.000288351,
             a[3.] → -0.0000209413,
             a[4.] → -0.0000180545,
             a[5.] → -0.0390625,
             a[6.] → 0.0078125}}
```

y_p is obtained by substituting the values for A_i into y_p and a general
solution is $y = y_h + y_p$. DSolve is able to find an exact solution.

```
In[1390]:= gensol = DSolve[D[y[t], {t, 8}] + 7/2D[y[t], {t, 7}]
                    + 73/2D[y[t], {t, 6}] + 229/2D[y[t], {t, 5}]
                    + 801/2D[y[t], {t, 4}] + 976D[y[t], {t, 3}]
                    + 1168D[y[t], {t, 2}] + 640D[y[t], t] + 128y[t] ==
                 t Exp[-t] + Sin[4t] + t, y[t], t]//
               Simplify
```

$$Out[1390]= \left\{\left\{y[t] \rightarrow -\frac{5}{128} - \frac{2924806\ e^{-t}}{24137569} + \frac{t}{128} - \frac{86016\ e^{-t}\ t}{1419857}\right.\right.$$

$$-\frac{1270\ e^{-t}\ t^2}{83521} - \frac{38\ e^{-t}\ t^3}{14739} - \frac{e^{-t}\ t^4}{3468} + e^{-t}\ C[1]$$

$$+ e^{-t}\ t\ C[2]$$

$$+ e^{-t}\ t^2\ C[3] + e^{-t/2}\ C[4]$$

$$+ \left(\frac{9041976373}{199643253056000} - \frac{107\ t^2}{5109520} + C[6]\right.$$

$$+ t\ \left(-\frac{1568449}{45168156800} + C[8]\right)\right)\ Cos[4\ t]$$

$$+ \left(\frac{13794625331}{798573012224000} + \frac{20406\ t}{352876225}\right.$$

$$\left.\left.\left.-\frac{369\ t^2}{20438080} - C[5] - t\ C[7]\right) Sin[4\ t]\right\}\right\}$$

∎

Variation of Parameters

In the same way as with second-order equations, we assume that a particular solution of the nth-order linear equation (6.18) has the form $y_p = u_1(t)y_1 + u_2(t)y_2 + \cdots + u_n(t)y_n$, where $S = \{y_1, y_2, \ldots, y_n\}$ is a fundamental set of solutions to the corresponding homogeneous equation (6.19). With the assumptions

$$y_p{}' = y_1 u_1{}' + y_2 u_2{}' + \cdots + y_n u_n{}' = 0$$
$$y_p{}'' = y_1{}' u_1{}' + y_2{}' u_2{}' + \cdots + y_n{}' u_n{}' = 0 \qquad (6.22)$$
$$\vdots$$
$$y_p{}^{(n-1)} = y_1{}^{(n-2)} u_1{}' + y_2{}^{(n-2)} u_2{}' + \cdots + y_n{}^{(n-2)} u_n{}' = 0$$

we obtain the equation

$$y_1{}^{(n-1)} u_1{}' + y_2{}^{(n-1)} u_2{}' + \cdots + y_n{}^{(n-1)} u_n{}' = f(t). \qquad (6.23)$$

Equations (6.22) and (6.23) form a system of n linear equations in the unknowns $u_1{}', u_2{}', \ldots, u_n{}'$. Applying Cramer's Rule,

$$u_i{}' = \frac{W_i(S)}{W(S)}, \qquad (6.24)$$

where $W(S)$ is given by equation (6.20) and $W_i(S)$ is the determinant of the matrix obtained by replacing the ith column of

$$\begin{pmatrix} y_1 & y_2 & \cdots & y_n \\ y_1{}' & y_2{}' & \cdots & y_n{}' \\ \vdots & \vdots & \cdots & \vdots \\ y_1{}^{(n-1)} & y_2{}^{(n-1)} & \cdots & y_n{}^{(n-1)} \end{pmatrix} \qquad \text{by} \qquad \begin{pmatrix} 0 \\ 0 \\ \vdots \\ f(t) \end{pmatrix}.$$

EXAMPLE 6.3.6: Solve $y^{(3)} + 4y' = \sec 2t$.

SOLUTION: A general solution of the corresponding homogeneous equation is $y_h = c_1 + c_2 \cos 2t + c_3 \sin 2t$; a fundamental set is $S = \{1, \cos 2t, \sin 2t\}$ with Wronskian $W(S) = 8$.

```
In[1391]:= yh = DSolve[y'''[t] + 4y'[t] == 0, y[t], t]
```

$$Out[1391]= \left\{\left\{y[t] \to C[3] + \frac{1}{2}\ C[1]\ Cos[2\ t] + \frac{1}{2}\ C[2]\ Sin[2\ t]\right\}\right\}$$

```
In[1392]:= s = {1, Cos[2t], Sin[2t]};
           ws = {s, D[s, t], D[s, {t, 2}]};
           MatrixForm[ws]
```

$$Out[1392]= \begin{pmatrix} 1 & Cos[2\ t] & Sin[2\ t] \\ 0 & -2\ Sin[2\ t] & 2\ Cos[2\ t] \\ 0 & -4\ Cos[2\ t] & -4\ Sin[2\ t] \end{pmatrix}$$

```
In[1393]:= dws = Simplify[Det[ws]]
```

$$Out[1393]= 8$$

Using variation of parameters to find a particular solution of the nonhomogeneous equation, we let $y_1 = 1$, $y_2 = \cos 2t$, and $y_3 = \sin 2t$ and assume that a particular solution has the form $y_p = u_1 y_1 + u_2 y_2 + u_3 y_3$. Using the variation of parameters formula, we obtain

$$u_1' = \frac{1}{8}\begin{vmatrix} 0 & \cos 2t & \sin 2t \\ 0 & -2\sin 2t & 2\cos 2t \\ \sec 2t & -4\cos 2t & -4\sin 2t \end{vmatrix} = \frac{1}{4}\sec 2t \quad \text{so} \quad u_1 = \frac{1}{8}\ln|\sec 2t + \tan 2t|,$$

$$u_2' = \frac{1}{8}\begin{vmatrix} 1 & 0 & \sin 2t \\ 0 & 0 & 2\cos 2t \\ 0 & \sec 2t & -4\sin 2t \end{vmatrix} = -\frac{1}{4} \quad \text{so} \quad u_2 = -\frac{1}{4}t,$$

and

$$u_3' = \frac{1}{8}\begin{vmatrix} 1 & \cos 2t & 0 \\ 0 & -2\sin 2t & 0 \\ 0 & -4\cos 2t & \sec 2t \end{vmatrix} = -\frac{1}{2}\tan 2t \quad \text{so} \quad u_3 = \frac{1}{8}\ln|\cos 2t|,$$

where we use Det and Integrate to evaluate the determinants and integrals. In the case of u_1, the output given by Mathematica looks different than the result we obtained by hand but using properties of logarithms ($\ln(a/b) = \ln a - \ln b$) and trigonometric identities ($\cos^2 x + \sin^2 x = 1$, $\sin 2x = 2 \sin x \cos x$, $\cos^2 x - \sin^2 x = \cos 2x$, and the reciprocal identities) shows us that

$$\frac{1}{8}(\ln|\cos t + \sin t| - \ln|\cos t + \sin t|) = \frac{1}{8}\ln\left|\frac{\cos t + \sin t}{\cos t - \sin t}\right|$$

$$= \frac{1}{8}\ln\left|\frac{\cos t + \sin t}{\cos t - \sin t} \cdot \frac{\cos t + \sin t}{\cos t + \sin t}\right|$$

$$= \frac{1}{8}\ln\left|\frac{\cos^2 t + 2\cos t \sin t + \sin^2 t}{\cos^2 t - \sin^2 t}\right|$$

$$= \frac{1}{8}\ln\left|\frac{1 + \sin 2t}{\cos 2t}\right| = \frac{1}{8}\ln\left|\frac{1}{\cos 2t} + \frac{\sin 2t}{\cos 2t}\right|$$

$$= \frac{1}{8}\ln|\sec 2t + \tan 2t|$$

so the results obtained by hand and with Mathematica are the same.

```
In[1394]:= u1p = 1/8
              Det[{{0, Cos[2t], Sin[2t]},
                    {0, -2 Sin[2t], 2 Cos[2t]},
                    {Sec[2t], -4 Cos[2t], -4 Sin[2t]}}]//
              Simplify
```

$$Out[1394]= \frac{1}{4}\ Sec[2\ t]$$

```
In[1395]:= Integrate[u1p, t]
```

$$Out[1395]= -\frac{1}{8}\ Log[Cos[t]-Sin[t]]+\frac{1}{8}\ Log[Cos[t]+Sin[t]]$$

```
In[1396]:= u2p = Simplify[1/8 Det[{{1, 0, Sin[2t]},
                    {0, 0, 2 Cos[2t]}, {0, Sec[2t],
                    -4 Sin[2t]}}]]
```

$$Out[1396]= -\frac{1}{4}$$

```
In[1397]:= Integrate[u2p, t]
```

$$Out[1397]= -\frac{t}{4}$$

```
In[1398]:= u3p = Simplify[1/8 Det[{{1, Cos[2t], 0},
                    {0, -2 Sin[2t], 0},
                    {0, -4 Cos[2t], Sec[2t]}}]]
```

$$Out[1398]= -\frac{1}{4}\ Tan[2\ t]$$

```
In[1399]:= Integrate[u3p, t]
```

$$Out[1399]= \frac{1}{8}\ Log[Cos[2\ t]]$$

Thus, a particular solution of the nonhomogeneous equation is

$$y_p = \frac{1}{8}\ln|\sec 2t + \tan 2t| - \frac{1}{4}t\cos 2t + \frac{1}{8}\ln|\cos 2t|\sin 2t$$

and a general solution is $y = y_h + y_p$. We verify that the calculations using DSolve return an equivalent solution.

```
In[1400]:= gensol =
           DSolve[y'''[t] + 4y'[t] == Sec[2t], y[t], t]//
           Simplify
Out[1400]= {{y[t] → 1/8 (8 C[3] - 2 (t - 2 C[1]) Cos[2 t]
                    - Log[Cos[t] - Sin[t]] + Log[Cos[t]
                    + Sin[t]] + (4 C[2]
                    + Log[Cos[2 t]]) Sin[2 t])}}
```

■

6.3.4 Laplace Transform Methods

The *method of Laplace transforms* can be useful when the forcing function is piecewise-defined or periodic.

Definition 4 (Laplace Transform and Inverse Laplace Transform). *Let $y = f(t)$ be a function defined on the interval $[0, \infty)$. The **Laplace transform** of is the function (of s)*

$$F(s) = \mathcal{L}\{f(t)\} = \int_0^\infty e^{-st}f(t)\,dt, \tag{6.25}$$

*provided the improper integral exists. $f(t)$ is the **inverse Laplace transform** of $F(s)$ means that $\mathcal{L}\{f(t)\} = F(s)$ and we write $\mathcal{L}^{-1}\{F(s)\} = f(t)$.*

1. LaplaceTransform[f[t],t,s] computes $\mathcal{L}\{f(t)\} = F(s)$.
2. InverseLaplaceTransform[F[s],t,s] computes $\mathcal{L}^{-1}\{F(s)\} = f(t)$.
3. UnitStep[t] returns $\mathcal{U}(t) = \begin{cases} 0, & t < 0 \\ 1, & t \geq 0. \end{cases}$

Typically, when we use Laplace transforms to solve a differential equation for a function $y(t)$, we will compute the Laplace transform of each term of the differential equation, solve the resulting algebraic equation for the Laplace transform of $y(t)$, $\mathcal{L}\{y(t)\} = Y(s)$, and finally determine $y(t)$ by computing the inverse Laplace transform of $Y(s)$, $\mathcal{L}^{-1}\{Y(s)\} = y(t)$.

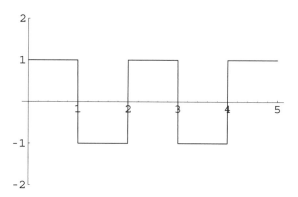

Figure 6-22   Plot of $f(t)$ for $0 \le t \le 5$

EXAMPLE 6.3.7: Let $y = f(t)$ be defined recursively by $f(t) = \begin{cases} 1, \ 0 \le t < 1 \\ -1, \ 1 \le t < 2 \end{cases}$

and $f(t) = f(t - 2)$ if $t \ge 2$. Solve $y'' + 4y' + 20y = f(t)$.

SOLUTION: We begin by defining and graphing $y = f(t)$ for $0 \le t \le 5$ in Figure 6-22.

```
In[1401]:= Clear[f, g, u, y1, y2, sol]

           f[t_] := 1/; 0 ≤ t < 1

           f[t_] := -1/; 1 ≤ t ≤ 2

           f[t_] := f[t - 2]/; t > 2

           Plot[f[t], {t, 0, 5},
             Ticks → {Automatic, {-2, -1, 0, 1, 2}},
             PlotRange → {-2, 2}]
```

We then define lhs to be the left-hand side of the equation $y'' + 4y' + 20y = f(t)$,

```
In[1402]:= Clear[y, x, lhs, stepone, steptwo]

           lhs = y''[t] + 4 y'[t] + 20 y[t];
```

and compute the Laplace transform of lhs with LaplaceTransform, naming the result stepone.

In[1403]:= **stepone = LaplaceTransform[lhs,t,s]**

Out[1403]= 20 LaplaceTransform[y[t],t,s]
   +s² LaplaceTransform[y[t],t,s]
   +4 (s LaplaceTransform[y[t],t,s] - y[0])
   -s y[0] - y'[0]

Let lr denote the Laplace transform of the right-hand side of the equation, $f(t)$. We now solve the equation $20ly + 4sly + s^2ly - 4y(0) - sy(0) - y'(0) = $ lr for ly and name the resulting output steptwo.

In[1404]:= **steptwo = Solve[stepone == lr,**
         LaplaceTransform[y[t],t,s]]

Out[1404]= {{LaplaceTransform[y[t],t,s] →
   $$\frac{lr + 4\,y[0] + s\,y[0] + y'[0]}{20 + 4\,s + s^2}\}\}$$

In[1405]:= **stepthree = ExpandNumerator[**
         steptwo[[1,1,2]],lr]

Out[1405]= $\dfrac{lr + 4\,y[0] + s\,y[0] + y'[0]}{20 + 4\,s + s^2}$

To find $y(t)$, we must compute the inverse Laplace transform of $\mathcal{L}\{y(t)\}$; the formula for which is explicitly obtained from steptwo with steptwo[[1,1,2]]. First, we rewrite : $\mathcal{L}\{y(t)\}$. Then,

$$y(t) = \mathcal{L}^{-1}\left\{\frac{\mathcal{L}\{f(t)\}}{s^2 + 4s + 20} + \frac{4y(0) + sy(0) + y'(0)}{s^2 + 4s + 20}\right\}$$

$$= \mathcal{L}^{-1}\left\{\frac{\mathcal{L}\{f(t)\}}{s^2 + 4s + 20}\right\} + \mathcal{L}^{-1}\left\{\frac{4y(0) + sy(0) + y'(0)}{s^2 + 4s + 20}\right\}.$$

Completing the square yields $s^2 + 4s + 20 = (s+2)^2 + 16$. Because

$$\mathcal{L}^{-1}\left\{\frac{b}{(s-a)^2 + b^2}\right\} = e^{at}\sin bt \quad \text{and} \quad \mathcal{L}^{-1}\left\{\frac{s-a}{(s-a)^2 + b^2}\right\} = e^{at}\cos bt,$$

the inverse Laplace transform of

$$\frac{4y(0) + sy(0) + y'(0)}{s^2 + 4s + 20} = y(0)\frac{s+2}{(s+2)^2 + 4^2} + \frac{y'(0) + 2y(0)}{4}\frac{4}{(s+2)^2 + 4^2}$$

is

$$y(0)e^{-2t}\cos 4t + \frac{y'(0) + 2y(0)}{4}e^{-2t}\sin 4t,$$

which is defined as $y_1(t)$. We perform these steps with Mathematica by first using `InverseLaplaceTransform` to calculate $\mathcal{L}^{-1}\left\{\dfrac{4y(0) + sy(0) + y'(0)}{s^2 + 4s + 20}\right\}$, naming the result `stepfour`.

> *In[1406]:=* **stepfour = InverseLaplaceTransform$\Big[$**
>
> $-\dfrac{-4\,y[0] - s\,y[0] - y'[0]}{20 + 4\,s + s^2}$ **, s, t$\Big]$**
>
> *Out[1406]=* $-\dfrac{1}{8}\,\mathrm{i}\,\mathrm{e}^{(-2-4\,\mathrm{i})\,t}\,\Big(\big((-2 + 4\,\mathrm{i}) + (2 + 4\,\mathrm{i})\,\mathrm{e}^{8\,\mathrm{i}\,t}\big)\,y[0]$
> $+\big(-1 + \mathrm{e}^{8\,\mathrm{i}\,t}\big)\,y'[0]\Big)$

To see that this is a real-valued function, we use `ComplexExpand` together with `Simplify`.

> *In[1407]:=* **stepfive = ComplexExpand[stepfour]//Simplify**
>
> *Out[1407]=* $\dfrac{1}{4}\,\mathrm{e}^{-2\,t}\,(4\,\mathrm{Cos}[4\,t]\,y[0] + \mathrm{Sin}[4\,t]\,(2\,y[0] + y'[0]))$

Because $y'(0)$ is assumed to be a real number, the imaginary part of $y'(0)$ is 0; the real part of $y'(0)$ is $y'(0)$.

> *In[1408]:=* **y1[t_] =**
> **stepfive/.{Im[y'[0]] → 0, Re[y'[0]] → y'[0]}//**
> **Simplify**
>
> *Out[1408]=* $\dfrac{1}{4}\,\mathrm{e}^{-2\,t}\,(4\,\mathrm{Cos}[4\,t]\,y[0] + \mathrm{Sin}[4\,t]\,(2\,y[0] + y'[0]))$

To compute the inverse Laplace transform of $\dfrac{\mathcal{L}\{f(t)\}}{s^2 + 4s + 20}$, we begin by computing $\mathtt{lr} = \mathcal{L}\{f(t)\}$. Let $\mathcal{U}_a(t) = \begin{cases} 1, & t \geq a \\ 0, & t < a \end{cases}$. Then, $\mathcal{U}_a(t) = \mathcal{U}(t - a) =$ `UnitStep[t - a]`.

The periodic function $f(t) = \begin{cases} 1, & 0 \leq t < 1 \\ -1, & 1 \leq t < 2 \end{cases}$ and $f(t) = f(t - 2)$ if $t \geq 2$ can be written in terms of step functions as

$$f(t) = \mathcal{U}_0(t) - 2\mathcal{U}_1(t) + 2\mathcal{U}_2(t) - 2\mathcal{U}_3(t) + 2\mathcal{U}_4(t) - \dots$$
$$= \mathcal{U}(t) - 2\mathcal{U}(t - 1) + 2\mathcal{U}(t - 2) - 2\mathcal{U}(t - 3) + 2\mathcal{U}(t - 4) - \dots$$
$$= \mathcal{U}(t) + 2\sum_{n=1}^{\infty} (-1)^n \mathcal{U}(t - n).$$

The Laplace transform of $\mathcal{U}_a(t) = \mathcal{U}(t-a)$ is $\frac{1}{s}e^{-as}$ and the Laplace transform of $f(t)\mathcal{U}_a(t) = f(t)\mathcal{U}(t-a)$ is $e^{-as}F(s)$, where $F(s)$ is the Laplace transform of $f(t)$. Then,

$$\mathtt{1r} = \frac{1}{s} - \frac{2}{s}e^{-s} + \frac{2}{s}e^{-2s} - \frac{2}{s}e^{-3s} + \ldots$$
$$= \frac{1}{s}\left(1 - 2e^{-s} + 2e^{-2s} - 2e^{-3s} + \ldots\right)$$

and

$$\frac{\mathtt{1r}}{s^2 + 4s + 20} = \frac{1}{s\left(s^2 + 4s + 20\right)}\left(1 - 2e^{-s} + 2e^{-2s} - 2e^{-3s} + \ldots\right)$$
$$= \frac{1}{s\left(s^2 + 4s + 20\right)} + 2\sum_{n=1}^{\infty}(-1)^n\frac{e^{-ns}}{s\left(s^2 + 4s + 20\right)}.$$

Because $\frac{1}{s^2 + 4s + 20} = \frac{1}{4}\frac{1}{(s+2)^2 + 4^2}$, $\mathcal{L}^{-1}\left\{\frac{1}{s\left(s^2 + 4s + 20\right)}\right\} = \int_0^t \frac{1}{4}e^{-2\alpha}$
$\sin 4\alpha\, d\alpha$, computed and defined to be the function $g(t)$.

$In[1409] := \mathbf{g[t\_]} = \int_0^t \frac{1}{4}\, \mathbf{Exp[-2\,\alpha]}\ \mathbf{Sin[4\,\alpha]}\,\mathbf{d\alpha}$

$Out[1409] = -\frac{1}{40}\, e^{-2\,t}\ (-2\,e^{2\,t} + 2\,Cos[4\,t] + Sin[4\,t])$

Alternatively, we can use `InverseLaplaceTransform` to obtain the same result.

$In[1410] := \mathbf{g[t\_]} = \mathbf{InverseLaplaceTransform}\Big[$
$$\frac{1}{\mathtt{s\ (s^2 + 4\,s + 20)}}, \mathbf{s, t}\Big]$$

$Out[1410] = \frac{1}{80}\,\left(4 - (2 + i)\ e^{(-2-4\,i)\,t} - (2 - i)\ e^{(-2+4\,i)\,t}\right)$

Then, $\mathcal{L}^{-1}\left\{2(-1)^n\frac{e^{-ns}}{s\left(s^2 + 4s + 20\right)}\right\} = 2(-1)^n g(t-n)\mathcal{U}(t-n)$ and the inverse Laplace transform of

$$\frac{1}{s\left(s^2 + 4s + 20\right)} + 2\sum_{n=1}^{\infty}(-1)^n\frac{e^{-ns}}{s\left(s^2 + 4s + 20\right)}$$

is

$$y_2(t) = g(t) + 2\sum_{n=1}^{\infty}(-1)^n g(t - n)\mathcal{U}(t - n).$$

It then follows that

$$y(t) = y_1(t) + y_2(t)$$

$$= y(0)e^{-2t}\cos 4t + \frac{y'(0) + 2y(0)}{4}e^{-2t}\sin 4t + 2\sum_{n=1}^{\infty}(-1)^n g(t-n)\mathcal{U}(t-n),$$

where $g(t) = \frac{1}{20} - \frac{1}{20}e^{-2t}\cos 4t - \frac{1}{40}e^{-2t}\sin 4t$.

To graph the solution for various initial conditions on the interval $[0, 5]$, we define $y_2(t) = g(t) + 2\sum_{n=1}^{5}(-1)^n g(t-n)\mathcal{U}(t-n)$, sol, and inits. (Note that we can graph the solution for various initial conditions on the interval $[0, m]$ by defining $y_2(t) = g(t) + 2\sum_{n=1}^{m}(-1)^n g(t-n)\mathcal{U}(t-n)$.)

$$In[1411] := \texttt{y2[t\_] := g[t] + 2} \sum_{n=1}^{5} \texttt{(-1)}^n\texttt{g[t-n] UnitStep[t-n]}$$

$$In[1412] := \texttt{sol[t\_] := y1[t] + y2[t]}$$

$$In[1413] := \texttt{inits = \{-1/2, 0, 1/2\};}$$

We then create a table of graphs of sol[t] on the interval $[0, 5]$ corresponding to replacing $y(0)$ and $y'(0)$ by the values $-1/2$, 0, and $1/2$ and then displaying the resulting graphics array in Figure 6-23.

```
In[1414] := graphs =
              Table[
                Plot[
                  sol[t] /. {y[0] → inits[[i]],
                      y'[0] → inits[[j]]}, {t, 0, 5},
                  DisplayFunction → Identity], {i, 1, 3},
                {j, 1, 3}]; Show[GraphicsArray[graphs]]
```

∎

Application: The Convolution Theorem

Sometimes we are required to determine the inverse Laplace transform of a product of two functions. Just as in differential and integral calculus when the derivative and integral of a product of two functions did not produce the product of the derivatives and integrals, respectively, neither does the inverse Laplace transform of the product yield the product of the inverse Laplace transforms. *The Convolution Theorem* tells us how to compute the inverse Laplace transform of a product of two functions.

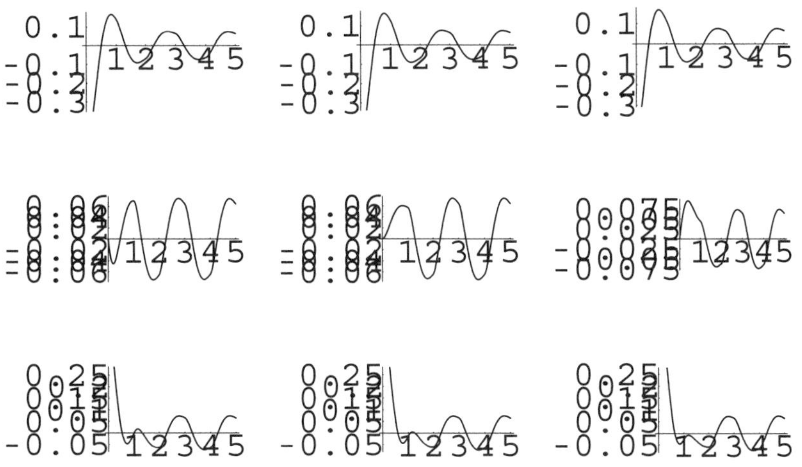

Figure 6-23 Solutions to a differential equation with a piecewise-defined periodic forcing function

Theorem 24 (The Convolution Theorem). *Suppose that $f(t)$ and $g(t)$ are piecewise continuous on $[0, \infty)$ and both are of exponential order. Further, suppose that the Laplace transform of $f(t)$ is $F(s)$ and that of $g(t)$ is $G(s)$. Then,*

$$\mathcal{L}^{-1}\{F(s)G(s)\} = \mathcal{L}^{-1}\{\mathcal{L}\{(f * g)(t)\}\} = \int_0^t f(t - v)g(v)\,dv. \qquad (6.26)$$

Note that $(f * g)(t) = \int_0^t f(t - v)g(v)\,dv$ is called the **convolution integral**.

EXAMPLE 6.3.8 (L–R–C Circuits): The initial-value problem used to determine the charge $q(t)$ on the capacitor in an L–R–C circuit is

$$\begin{cases} L\frac{d^2Q}{dt^2} + R\frac{dQ}{dt} + \frac{1}{C}Q = f(t) \\ Q(0) = 0,\ \frac{dQ}{dt}(0) = 0, \end{cases}$$

where L denotes inductance, $dQ/dt = I$, $I(t)$ current, R resistance, C capacitance, and $E(t)$ voltage supply. Because $dQ/dt = I$, this differential equation can be represented as

$$L\frac{dI}{dt} + RI + \frac{1}{C}\int_0^t I(u)\,du = E(t).$$

Note also that the initial condition $Q(0) = 0$ is satisfied because $Q(0) = \frac{1}{C}\int_0^0 I(u)\,du = 0$. The condition $dQ/dt(0) = 0$ is replaced by $I(0) = 0$. (a)

Solve this *integrodifferential equation,* an equation that involves a derivative as well as an integral of the unknown function, by using the Convolution theorem. (b) Consider this example with constant values $L = C = R = 1$ and $E(t) = \begin{cases} \sin t, & 0 \leq t < \pi/2 \\ 0, & t \geq \pi/2 \end{cases}$. Determine $I(t)$ and graph the solution.

SOLUTION: We proceed as in the case of a differential equation by taking the Laplace transform of both sides of the equation. The Convolution theorem, equation (6.26), is used in determining the Laplace transform of the integral with

$$\mathcal{L}\left\{\int_0^t I(u)\,du\right\} = \mathcal{L}\{1 * I(t)\} = \mathcal{L}\{1\}\,\mathcal{L}\{I(t)\} = \frac{1}{s}\mathcal{L}\{I(t)\}.$$

Therefore, application of the Laplace transform yields

$$Ls\mathcal{L}\{I(t)\} - LI(0) + R\mathcal{L}\{I(t)\} + \frac{1}{C}\frac{1}{s}\mathcal{L}\{I(t)\} = \mathcal{L}\{E(t)\}.$$

Because $I(0) = 0$, we have $Ls\mathcal{L}\{I(t)\} + R\mathcal{L}\{I(t)\} + \frac{1}{C}\frac{1}{s}\mathcal{L}\{I(t)\} = \mathcal{L}\{E(t)\}$.

Simplifying and solving for $\mathcal{L}\{I(t)\}$ results in $\mathcal{L}\{I(t)\} = \dfrac{Cs\mathcal{L}\{E(t)\}}{LCs^2 + RCs + 1}$

```
In[1415]:= Clear[i]

             LaplaceTransform[l i'[t] + r i[t], t, s]
Out[1415]= r LaplaceTransform[i[t], t, s]
            +l (-i[0] + s LaplaceTransform[i[t], t, s])
```

```
In[1416]:= Solve[l s lapi + r lapi + lapi/(c s) == lape, lapi]
                              c lape s
Out[1416]= {{lapi → ──────────────────}}
                          1 + c r s + c l s²
```

so that $I(t) = \mathcal{L}^{-1}\left\{\dfrac{Cs\mathcal{L}\{E(t)\}}{LCs^2 + RCs + 1}\right\}$. For (b), we note that

$E(t) = \begin{cases} \sin t, & 0 \leq t < \pi/2 \\ 0, & t \geq \pi/2 \end{cases}$ can be written as $E(t) = \sin t\,(\mathcal{U}(t) - \mathcal{U}(t - \pi/2))$.

We define and plot the forcing function $E(t)$ on the interval $[0, \pi]$ in Figure 6-24.

We use lowercase letters to avoid any possible ambiguity with built-in Mathematica functions, like E and I.

```
In[1417]:= e[t_] := Sin[t] (UnitStep[t] - UnitStep[t - π/2])

            p1 = Plot[e[t], {t, 0, π}]
```

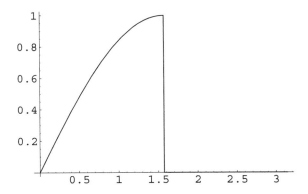

Figure 6-24   Plot of $E(t) = \sin t \, (\mathcal{U}(t) - \mathcal{U}(t - \pi/2))$

Next, we compute the Laplace transform of $\mathcal{L}\{E(t)\}$ with LaplaceTransform. We call this result lcape.

In[1418]:= **lcape = LaplaceTransform[e[t], t, s]**

Out[1418]= $\dfrac{1}{1+s^2} - \dfrac{e^{-\frac{\pi s}{2}} s}{1+s^2}$

Using the general formula obtained for the Laplace transform of $I(t)$, we note that the denominator of this expression is given by $s^2 + s + 1$ which is entered as denom. Hence, the Laplace transform of $I(t)$, called lcapi, is given by the ratio s lcape/denom.

In[1419]:= **denom = s^2 + s + 1;**

In[1420]:= **lcapi = s lcape/denom;**

   lcapi = Simplify[lcapi]

Out[1420]= $\dfrac{s - e^{-\frac{\pi s}{2}} s^2}{1 + s + 2 s^2 + s^3 + s^4}$

We determine $I(t)$ with InverseLaplaceTransform.

In[1421]:= **i[t_] = InverseLaplaceTransform[lcapi, s, t]**

Out[1421]= $\mathrm{Sin}[t] - \dfrac{2\, e^{-t/2}\, \mathrm{Sin}\left[\frac{\sqrt{3}\,t}{2}\right]}{\sqrt{3}}$

$-\left(\mathrm{Sin}[t] + \dfrac{1}{3} e^{\frac{1}{2}\left(\frac{\pi}{2}-t\right)} \left(-3\,\mathrm{Cos}\left[\frac{1}{2}\sqrt{3}\left(-\frac{\pi}{2}+t\right)\right]\right.\right.$

$\left.\left.+\sqrt{3}\,\mathrm{Sin}\left[\frac{1}{2}\sqrt{3}\left(-\frac{\pi}{2}+t\right)\right]\right)\right) \mathrm{UnitStep}\left[-\frac{\pi}{2}+t\right]$

This solution is plotted in p2 (in black) and displayed with the forcing function (in gray) in Figure 6-25. Notice the effect that the forcing function has on the solution to the differential equation.

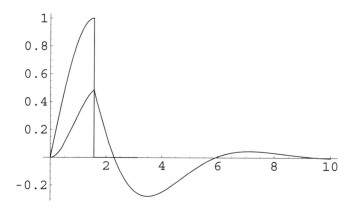

Figure 6-25   *I(t)* (in black) and *E(t)* (in gray)

In[1422]:= **p2 = Plot[i[t], {t, 0, 10},**
           DisplayFunction → Identity];
           Show[p1, p2, PlotRange → All,
           DisplayFunction → $DisplayFunction]

In this case, we see that we can use DSolve to solve the initial value problem

$$\begin{cases} Q'' + Q' + q = E(t) \\ Q(0) = 0, \ Q'(0) = 0 \end{cases}$$

as well. However, the result is very lengthy so only a portion is displayed here using Short.

In[1423]:= **sol =**
        DSolve[{q''[t] + q'[t] + q[t] == e[t],
          q[0] == 0, q'[0] == 0}, q[t], t];

In[1424]:= **Short[sol]**

Out[1424]= $\left\{\left\{q[t] \rightarrow \dfrac{8\, e^{\ll 1 \gg}\, (\ll 1 \gg)}{3 \ll 4 \gg}\right\}\right\}$

We see that this result is a real-valued function using ComplexExpand followed by Simplify.

In[1425]:= **q[t_] = ComplexExpand[sol[[1, 1, 2]]]//**
         Simplify

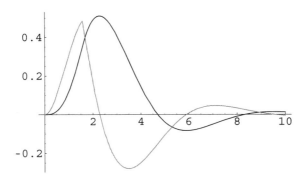

Figure 6-26 $Q(t)$ (in black) and $I(t) = Q'(t)$ (in gray)

$$Out [1425] = \frac{1}{3} \left(e^{-t/2} \right.$$

$$\left(- 3 \, e^{t/2} \, Cos[t] + 3 \, Cos \left[\frac{\sqrt{3} \, t}{2} \right] + \sqrt{3} \, Sin \left[\frac{\sqrt{3} \, t}{2} \right] \right)$$

UnitStep[t]

$$+ \left(3 \, Cos[t] - 2 \, \sqrt{3} \, e^{\frac{1}{4} \, (\pi - 2 \, t)} \, Sin \left[\frac{1}{4} \, \sqrt{3} \, (\pi - 2 \, t) \right] \right)$$

$$\left. UnitStep \left[- \frac{\pi}{2} + t \right] \right)$$

We use this result to graph $Q(t)$ and $I(t) = Q'(t)$ in Figure 6-26.

```
In[1426] := Plot[{q[t], q'[t]}, {t, 0, 10},
              PlotStyle → {GrayLevel[0], GrayLevel[0.5]}]
```

■

Application: The Dirac Delta Function

Let $\delta(t - t_0)$ denote the (generalized) function with the two properties

1. $\delta(t - t_0) = 0$ if $t \neq t_0$ and
2. $\int_{-\infty}^{\infty} \delta(t - t_0) \, dt = 1$

which is called the **Dirac delta function** and is quite useful in the definition of impulse forcing functions that arise in some differential equations. The Laplace transform of $\delta(t - t_0)$ is $\mathcal{L}\{\delta(t - t_0)\} = e^{-st_0}$. The Mathematica function DiracDelta represents the δ distribution.

```
In[1427] := LaplaceTransform[DiracDelta[t - t0], t,
              s]
```

$$Out [1427] = e^{-s \, t0}$$

EXAMPLE 6.3.9: Solve $\begin{cases} x'' + x' + x = \delta(t) + \mathcal{U}(t - 2\pi) \\ x(0) = 0, \ x'(0) = 0 \end{cases}$.

SOLUTION: We define eq to be the equation $x'' + x' + x = \delta(t) + \mathcal{U}(t - 2\pi)$ and then use LaplaceTransform to compute the Laplace transform of eq, naming the resulting output leq. The symbol LaplaceTransform [x[t],t,s] represents the Laplace transform of x[t]. We then apply the initial conditions $x(0) = 0$ and $x'(0) = 0$ to leq and name the resulting output ics.

```
In[1428]:= Clear[x, eq]

            eq = x''[t] + x'[t] + x[t] ==
                DiracDelta[t] + UnitStep[t - 2 π];
            leq = LaplaceTransform[eq, t, s]
Out[1428]= LaplaceTransform[x[t], t, s]
            + s LaplaceTransform[x[t], t, s]
            + s² LaplaceTransform[x[t], t, s]
                                                  e^(-2 π s)
            - x[0] - s x[0] - x'[0] == 1 + ─────────
                                                      s

In[1429]:= ics = leq/.{x[0] → 0, x'[0] → 0}

Out[1429]= LaplaceTransform[x[t], t, s]
            + s LaplaceTransform[x[t], t, s]
                                                                e^(-2 π s)
            + s² LaplaceTransform[x[t], t, s] == 1 + ─────────
                                                                    s
```

Next, we use Solve to solve the equation ics for the Laplace transform of $x(t)$. The expression for the Laplace transform is extracted from lapx with lapx[[1,1,2]].

```
In[1430]:= lapx =
                Solve[ics, LaplaceTransform[x[t], t, s]]
Out[1430]= {{LaplaceTransform[x[t], t, s] →
                e^(-2 π s) (1 + e^(2 π s) s)
                ───────────────────────────}}
                      s (1 + s + s²)
```

To find $x(t)$, we must compute the inverse Laplace transform of the Laplace transform of $\mathcal{L}\{x(t)\}$ obtained in lapx. We use Inverse LaplaceTransform to compute the inverse Laplace transform of lapx[[1,1,2]] and name the resulting function x[t].

```
In[1431]:= x[t_] = InverseLaplaceTransform[
                lapx[[1, 1, 2]], s, t]
```

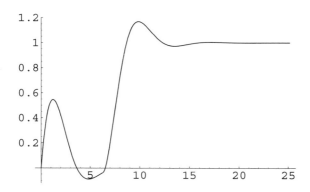

Figure 6-27   Plot of $x(t)$ on the interval $[0, 8\pi]$

$$Out[1431] = \frac{2\,e^{-t/2}\,\text{Sin}\left[\frac{\sqrt{3}\,t}{2}\right]}{\sqrt{3}}$$

$$+\frac{1}{3}\left(3 - e^{\pi - \frac{t}{2}}\left(3\,\text{Cos}\left[\frac{1}{2}\,\sqrt{3}\,(-2\,\pi + t)\right]\right.\right.$$

$$\left.\left.+\sqrt{3}\,\text{Sin}\left[\frac{1}{2}\,\sqrt{3}\,(-2\,\pi + t)\right]\right)\right)\,\text{UnitStep}[-2\,\pi + t]$$

To see that this is a real-valued function, we use `ComplexExpand` followed by `Simplify`.

$$In[1432] := \mathbf{x[t\_] = ComplexExpand[x[t]]//Simplify}$$

$$Out[1432] = \frac{1}{3}\,e^{-t}\left(2\,\sqrt{3}\,\sqrt{e^t}\,\text{Sin}\left[\frac{\sqrt{3}\,t}{2}\right]\right.$$

$$e^{t/2}\left(3\,e^{t/2} - 3\,e^{\pi}\,\text{Cos}\left[\frac{1}{2}\,\sqrt{3}\,(-2\,\pi + t)\right]\right.$$

$$\left.\left.-\sqrt{3}\,e^{\pi}\,\text{Sin}\left[\frac{1}{2}\,\sqrt{3}\,(-2\,\pi + t)\right]\right)\,\text{UnitStep}[-2\,\pi + t]\right)$$

We use `Plot` to graph the solution on the interval $[0, 8\pi]$ in Figure 6-27.

$$In[1433] := \mathbf{Plot[x[t], \{t, 0, 8\,\pi\}]}$$

Finally, we note that `DSolve` is able to solve the initial value problem directly as well. The result is very lengthy so only an abbreviated portion is displayed here using `Short`.

$$In[1434] := \mathbf{Clear[x]}$$

```
sol =
DSolve[
    {x''[t] + x'[t] + x[t] ==
        DiracDelta[t] + UnitStep[t - 2 π],
    x[0] == 0, x'[0] == 0}, x[t], t];
```

$In[1435]:= $ **Short[sol, 2]**

$Out[1435]= \left\{\left\{x[t] \to \dfrac{4\ e^{-t/2}\ (\ll 1 \gg)}{\left(-3\ \mathbb{i} + \sqrt{3}\right)\left(3\ \mathbb{i} + \sqrt{3}\right)}\right\}\right\}$

As before, we see that the result is a real-valued function using ComplexExpand followed by Simplify.

$In[1436]:= $ **ComplexExpand[sol[[1, 1, 2]]]//Simplify**

$Out[1436]= \dfrac{1}{3}\ e^{-t}\left(-2\sqrt{3}\sqrt{e^t}\ \mathrm{Sin}\left[\dfrac{\sqrt{3}\ t}{2}\right]\right.$

$+2\sqrt{3}\sqrt{e^t}\ \mathrm{Sin}\left[\dfrac{\sqrt{3}\ t}{2}\right]\mathrm{UnitStep[t]}+$

$+\left(3\ e^t - 3\ e^\pi\sqrt{e^t}\ \mathrm{Cos}\left[\dfrac{1}{2}\sqrt{3}\ (2\pi - t)\right]\right)$

$\left.+\sqrt{3}\ e^\pi\sqrt{e^t}\ \mathrm{Sin}\left[\dfrac{1}{2}\sqrt{3}\ (2\pi - t)\right]\right)\mathrm{UnitStep}[-2\pi + t]\right)$

■

6.3.5 Nonlinear Higher-Order Equations

Generally, rigorous results regarding nonlinear equations are very difficult to obtain. In some cases, analysis is best carried out numerically and/or graphically. In other situations, rewriting the equation as a system can be of benefit, which is discussed in the next section. (See Examples 6.4.5, 6.4.4, and 6.4.7.)

6.4 Systems of Equations

6.4.1 Linear Systems

We now consider first-order linear systems of differential equations:

$$\mathbf{X}' = \mathbf{A}(t)\mathbf{X} + \mathbf{F}(t), \tag{6.27}$$

where

$$\mathbf{X}(t) = \begin{pmatrix} x_1(t) \\ x_2(t) \\ \vdots \\ x_n(t) \end{pmatrix}, \quad \mathbf{A}(t) = \begin{pmatrix} a_{11}(t) & a_{12}(t) & \cdots & a_{1n}(t) \\ a_{21}(t) & a_{22}(t) & \cdots & a_{2n}(t) \\ \vdots & \vdots & \cdots & \vdots \\ a_{n1}(t) & a_{n2}(t) & \cdots & a_{nn}(t) \end{pmatrix}, \quad \text{and} \quad \mathbf{F}(t) = \begin{pmatrix} f_1(t) \\ f_2(t) \\ \vdots \\ f_n(t) \end{pmatrix}.$$

6.4.1.1 Homogeneous Linear Systems

The corresponding homogeneous system of equation (6.27) is

$$\mathbf{X}' = \mathbf{A}\mathbf{X}. \tag{6.28}$$

In the same way as with the previously discussed linear equations, a **general solution** of equation (6.27) is $\mathbf{X} = \mathbf{X}_h + \mathbf{X}_p$ where $\mathbf{X}_h$ is a *general solution* of equation (6.28) and $\mathbf{X}_p$ is a *particular solution* of the nonhomogeneous system equation (6.27).

If $\mathbf{\Phi}_1, \mathbf{\Phi}_2, \ldots, \mathbf{\Phi}_n$ are n linearly independent solutions of equation (6.28), a **general solution** of equation (6.28) is

$$\mathbf{X} = c_1\mathbf{\Phi}_1 + c_2\mathbf{\Phi}_2 + \cdots + c_n\mathbf{\Phi}_n = \begin{pmatrix} \mathbf{\Phi}_1 & \mathbf{\Phi}_2 & \cdots & \mathbf{\Phi}_n \end{pmatrix} \begin{pmatrix} c_1 \\ c_2 \\ \vdots \\ c_n \end{pmatrix} = \mathbf{\Phi}\mathbf{C},$$

A **particular solution** to a system of ordinary differential equations is a set of functions that satisfy the system but do not contain any arbitrary constants. That is, a particular solution to a system is a set of specific functions, *containing no arbitrary constants*, that satisfy the system.

where

$$\mathbf{\Phi} = \begin{pmatrix} \mathbf{\Phi}_1 & \mathbf{\Phi}_2 & \cdots & \mathbf{\Phi}_n \end{pmatrix} \quad \text{and} \quad \mathbf{C} = \begin{pmatrix} c_1 \\ c_2 \\ \vdots \\ c_n \end{pmatrix}.$$

$\mathbf{\Phi}$ is called a **fundamental matrix** for equation (6.28). If $\mathbf{\Phi}$ is a fundamental matrix for equation (6.28), $\mathbf{\Phi}' = \mathbf{A}\mathbf{\Phi}$ or $\mathbf{\Phi}' - \mathbf{A}\mathbf{\Phi} = \mathbf{0}$.

$A(t)$ constant

Suppose that $\mathbf{A}(t) = \mathbf{A}$ has constant real entries. Let λ be an eigenvalue of $\mathbf{A}$ with corresponding eigenvector $\mathbf{v}$. Then, $\mathbf{v}e^{\lambda t}$ is a solution of $\mathbf{X}' = \mathbf{A}\mathbf{X}$.

If $\lambda = \alpha + \beta i, \beta \neq 0$, is an eigenvalue of $\mathbf{A}$ and has corresponding eigenvector $\mathbf{v} = \mathbf{a} + \mathbf{b}i$, two linearly independent solutions of $\mathbf{X}' = \mathbf{A}\mathbf{X}$ are

$$e^{\alpha t}(\mathbf{a}\cos\beta t - \mathbf{b}\sin\beta t) \quad \text{and} \quad e^{\alpha t}(\mathbf{a}\sin\beta t + \mathbf{b}\cos\beta t). \tag{6.29}$$

EXAMPLE 6.4.1: Solve each of the following systems.

(a) $\mathbf{X}' = \begin{pmatrix} -1/2 & -1/3 \\ -1/3 & -1/2 \end{pmatrix}\mathbf{X}$; (b) $\begin{cases} x' = \frac{1}{2}y \\ y' = -\frac{1}{8}x \end{cases}$; (c) $\begin{cases} dx/dt = -\frac{1}{4}x + 2y \\ dy/dt = -8x - \frac{1}{4}y. \end{cases}$

SOLUTION: (a) With `Eigensystem`, we see that the eigenvalues and eigenvectors of $\mathbf{A} = \begin{pmatrix} -1/2 & -1/3 \\ -1/3 & -1/2 \end{pmatrix}$ are $\lambda_1 = -1/6$ and $\lambda_2 = -5/6$ and

$\mathbf{v}_1 = \begin{pmatrix} -1 \\ 1 \end{pmatrix}$ and $\mathbf{v}_2 = \begin{pmatrix} 1 \\ 1 \end{pmatrix}$, respectively.

```
In[1437]:= capa = {{-1/2, -1/3}, {-1/3, -1/2}};
           Eigensystem[capa]
```

$$Out[1437]= \left\{\left\{-\frac{5}{6}, -\frac{1}{6}\right\}, \{\{1, 1\}, \{-1, 1\}\}\right\}$$

Then $\mathbf{X}_1 = \begin{pmatrix} -1 \\ 1 \end{pmatrix} e^{-t/6}$ and $\mathbf{X}_2 = \begin{pmatrix} 1 \\ 1 \end{pmatrix} e^{-5t/6}$ are two linearly independent

solutions of the system so a general solution is $\mathbf{X} = \begin{pmatrix} -e^{-t/6} & e^{-5t/6} \\ e^{-t/6} & e^{-5t/6} \end{pmatrix} \begin{pmatrix} c_1 \\ c_2 \end{pmatrix}$;

a fundamental matrix is $\boldsymbol{\Phi} = \begin{pmatrix} -e^{-t/6} & e^{-5t/6} \\ e^{-t/6} & e^{-5t/6} \end{pmatrix}$.

We use DSolve to find a general solution of the system by entering

```
In[1438]:= gensol = DSolve[{x'[t] == -1/2x[t] - 1/3y[t],
             y'[t] == -1/3x[t] - 1/2y[t]}, {x[t], y[t]},
             t]
```

$$Out[1438]= \left\{\left\{x[t] \to e^{-5\ t/6}\ C[1] - e^{-t/6}\ C[2], y[t] \to e^{-5\ t/6}\ C[1] + e^{-t/6}\ C[2]\right\}\right\}$$

We graph the direction field with PlotVectorField, which is contained in the PlotField package located in the **Graphics** directory, in Figure 6-28.

Remark. After you have loaded the PlotField package,

```
PlotVectorField[{f[x,y],g[x,y]},{x,a,b},{y,c,d}]
```

generates a basic direction field for the system $\{x' = f(x, y),\ y' = g(x, y)\}$ for $a \le x \le b$ and $c \le y \le d$.

```
In[1439]:= << Graphics`PlotField`
```

```
In[1440]:= pvf = PlotVectorField[{-1/2x - 1/3y, -1/3x - 1/2y},
             {x, -1, 1}, {y, -1, 1}, Axes- > Automatic]
```

Several solutions are also graphed with ParametricPlot and shown together with the direction field in Figure 6-29.

```
In[1441]:= initsol = DSolve[{x'[t] == -1/2x[t] - 1/3y[t],
             y'[t] == -1/3x[t] - 1/2y[t], x[0] == x0, y[0] ==
             y0}, {x[t], y[t]}, t]
```

$$Out[1441]= \left\{\left\{x[t] \to -e^{-5\ t/6}\ \left(\frac{1}{2}\ (-x0 - y0) + \frac{1}{2}\ e^{2\ t/3}\ (-x0 + y0)\right),\right.\right.$$
$$\left.\left. y[t] \to e^{-5\ t/6}\ \left(\frac{1}{2}\ e^{2\ t/3}\ (-x0 + y0) + \frac{x0 + y0}{2}\right)\right\}\right\}$$

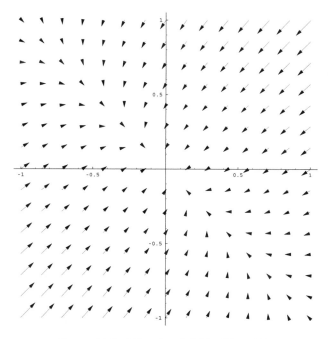

Figure 6-28   Direction field for $\mathbf{X}' = \mathbf{A}\mathbf{X}$

```
In[1442]:= t1 = Table[ParametricPlot[
              Evaluate[{x[t],y[t]}/.initsol/.{x0->1,y0->i}],
              {t,0,15},DisplayFunction->Identity,
              PlotStyle->GrayLevel[0.3]],{i,-1,1,2/8}];
           t2 = Table[ParametricPlot[
              Evaluate[{x[t],y[t]}/.initsol/.{x0->-1,y0->i}],
              {t,0,15},DisplayFunction->Identity,
              PlotStyle->GrayLevel[0.3]],{i,-1,1,2/8}];
           t3 = Table[ParametricPlot[
              Evaluate[{x[t],y[t]}/.initsol/.{x0->i,y0->1}],
              {t,0,15},DisplayFunction->Identity,
              PlotStyle->GrayLevel[0.3]],{i,-1,1,2/8}];
           t4 = Table[ParametricPlot[
              Evaluate[{x[t],y[t]}/.initsol/.{x0->i,y0->-1}],
              {t,0,15},DisplayFunction->Identity,
              PlotStyle->GrayLevel[0.3]],{i,-1,1,2/8}];

In[1443]:= Show[t1,t2,t3,t4,
              pvf,DisplayFunction->$DisplayFunction,
              AspectRatio->Automatic]
```

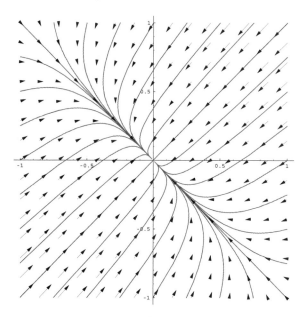

Figure 6-29   Direction field for $\mathbf{X}' = \mathbf{AX}$ along with various solution curves

(b) In matrix form the system is equivalent to the system $\mathbf{X}' = \begin{pmatrix} 0 & 1/2 \\ -1/8 & 0 \end{pmatrix} \mathbf{X}$.

As in (a), we use Eigensystem to see that the eigenvalues and eigen-

vectors of $\mathbf{A} = \begin{pmatrix} 0 & 1/2 \\ -1/8 & 0 \end{pmatrix}$ are $\lambda_{1,2} = 0 \pm \frac{1}{4}i$ and $\mathbf{v}_{1,2} = \begin{pmatrix} 1 \\ 0 \end{pmatrix} \pm \begin{pmatrix} 0 \\ 1/2 \end{pmatrix} i$.

```
In[1444]:= capa = {{0, 1/2}, {-1/8, 0}};
           Eigensystem[capa]
```

$$Out[1444]= \left\{\left\{-\frac{i}{4}, \frac{i}{4}\right\}, \{\{2\ i, 1\}, \{-2\ i, 1\}\}\right\}$$

Two linearly independent solutions are then $\mathbf{X}_1 = \begin{pmatrix} 1 \\ 0 \end{pmatrix} \cos \frac{1}{4}t - \begin{pmatrix} 0 \\ 1/2 \end{pmatrix} \sin \frac{1}{4}t =$

$\begin{pmatrix} \cos \frac{1}{4}t \\ -\frac{1}{2} \sin \frac{1}{4}t \end{pmatrix}$ and $\mathbf{X}_2 = \begin{pmatrix} 1 \\ 0 \end{pmatrix} \sin \frac{1}{4}t + \begin{pmatrix} 0 \\ 1/2 \end{pmatrix} \cos \frac{1}{4}t = \begin{pmatrix} \sin \frac{1}{4}t \\ \frac{1}{2} \cos \frac{1}{4}t \end{pmatrix}$ and a gen-

eral solution is $\mathbf{X} = c_1 \mathbf{X}_1 + c_2 \mathbf{X}_2 = \begin{pmatrix} \cos \frac{1}{4}t & \sin \frac{1}{4}t \\ -\frac{1}{2} \sin \frac{1}{4}t & \frac{1}{2} \cos \frac{1}{4}t \end{pmatrix} \begin{pmatrix} c_1 \\ c_2 \end{pmatrix}$ or $x =$

$c_1 \cos \frac{1}{4}t + c_2 \sin \frac{1}{4}t$ and $y = -c_1 \frac{1}{2} \sin \frac{1}{4}t + \frac{1}{2}c_2 \cos \frac{1}{4}t$.

As before, we use DSolve to find a general solution.

In[1445]:= **gensol = DSolve[{x'[t] == 1/2y[t], y'[t] ==**
                           -1/8x[t]}, {x[t], y[t]}, t]

Out[1445]= $\left\{\left\{x[t] \rightarrow -2 \ C[1] \ Cos\left[\frac{t}{4}\right] + 2 \ C[2] \ Sin\left[\frac{t}{4}\right],\right.\right.$

$\left.\left. y[t] \rightarrow C[2] \ Cos\left[\frac{t}{4}\right] + C[1] \ Sin\left[\frac{t}{4}\right]\right\}\right\}$

Initial-value problems for systems are solved in the same way as for other equations. For example, entering

In[1446]:= **partsol = DSolve[{x'[t] == 1/2y[t],**
                              y'[t] == -1/8x[t], x[0] == 1,
                              y[0] == -1}, {x[t], y[t]},
                              t]

Out[1446]= $\left\{\left\{x[t] \rightarrow -2 \ \left(-\frac{1}{2} \ Cos\left[\frac{t}{4}\right] + Sin\left[\frac{t}{4}\right]\right),\right.\right.$

$\left.\left. y[t] \rightarrow - Cos\left[\frac{t}{4}\right] - \frac{1}{2} \ Sin\left[\frac{t}{4}\right]\right\}\right\}$

finds the solution that satisfies $x(0) = 1$ and $y(0) = -1$.

We graph $x(t)$ and $y(t)$ together as well as parametrically with Plot and ParametricPlot, respectively, in Figure 6-30.

In[1447]:= **p1 = Plot[Evaluate[{x[t], y[t]}/.partsol], {t, 0, 8π},**
             PlotStyle- > {GrayLevel[0], GrayLevel[0.4]},
             DisplayFunction- > Identity];
          p2 = ParametricPlot[
             Evaluate[{x[t], y[t]}/.partsol], {t, 0, 8π},
             DisplayFunction- > Identity,
             AspectRatio- > Automatic];
          Show[GraphicsArray[{p1, p2}]]

We can also use PlotVectorField and ParametricPlot to graph the direction field and/or various solutions as we do next in Figure 6-31.

In[1448]:= **pvf = PlotVectorField[{1/2y, -1/8x}, {x, -2, 2},**
             {y, -1, 1}, DisplayFunction- > Identity];

In[1449]:= **initsol = DSolve[{x'[t] == 1/2y[t],**
             y'[t] == -1/8x[t], x[0] == x0, y[0] == y0},
             {x[t], y[t]}, t]

Out[1449]= $\left\{\left\{x[t] \rightarrow -2 \ \left(-\frac{1}{2} \ x0 \ Cos\left[\frac{t}{4}\right] - y0 \ Sin\left[\frac{t}{4}\right]\right),\right.\right.$

$\left.\left. y[t] \rightarrow y0 \ Cos\left[\frac{t}{4}\right] - \frac{1}{2} \ x0 \ Sin\left[\frac{t}{4}\right]\right\}\right\}$

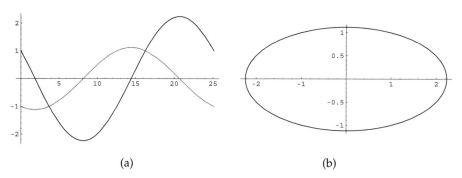

(a)   (b)

Figure 6-30   (a) Graph of $x(t)$ and $y(t)$. (b) Parametric plot of $x(t)$ versus $y(t)$

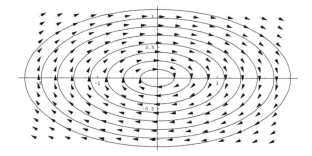

Figure 6-31   Notice that all nontrivial solutions are periodic

```
In[1450] := t1 = Table[ParametricPlot[
               Evaluate[{x[t],y[t]}/.initsol/
               .{x0->i,y0->i}],
               {t,0,8π},DisplayFunction->Identity,
               PlotStyle->GrayLevel[0.3]],
               {i,0,1,1/8}];
```

```
In[1451] := Show[t1,pvf,DisplayFunction->
               $DisplayFunction,AspectRatio->Automatic]
```

(c) In matrix form, the system is equivalent to the system $\mathbf{X}' = \begin{pmatrix} -\frac{1}{4} & 2 \\ -8 & -\frac{1}{4} \end{pmatrix} \mathbf{X}$. The eigenvalues and corresponding eigenvectors of $\mathbf{A} = \begin{pmatrix} -\frac{1}{4} & 2 \\ -8 & -\frac{1}{4} \end{pmatrix}$ are found to be $\lambda_{1,2} = -\frac{1}{4} \pm 4i$ and $\mathbf{v}_{1,2} = \begin{pmatrix} 0 \\ 2 \end{pmatrix} \pm \begin{pmatrix} 1 \\ 0 \end{pmatrix} i$ with Eigensystem.

```
In[1452]:= capa = {{-1/4, 2}, {-8, -1/4}};
           Eigensystem[capa]
```

$$Out[1452]= \left\{\left\{-\frac{1}{4}-4\ i,\ -\frac{1}{4}+4\ i\right\},\ \{\{i,\ 2\},\ \{-i,\ 2\}\}\right\}$$

A general solution is then

$$\mathbf{X} = c_1\mathbf{X}_1 + c_2\mathbf{X}_2$$

$$= c_1 e^{-t/4}\left(\begin{pmatrix}1\\0\end{pmatrix}\cos 4t - \begin{pmatrix}0\\2\end{pmatrix}\sin 4t\right) + c_2 e^{-t/4}\left(\begin{pmatrix}1\\0\end{pmatrix}\sin 4t + \begin{pmatrix}0\\2\end{pmatrix}\cos 4t\right)$$

$$= e^{-t/4}\left[c_1\begin{pmatrix}\cos 4t\\-2\sin 4t\end{pmatrix} + c_2\begin{pmatrix}\sin 4t\\2\cos 4t\end{pmatrix}\right] = e^{-t/4}\begin{pmatrix}\cos 4t & \sin 4t\\-2\sin 4t & 2\cos 4t\end{pmatrix}\begin{pmatrix}c_1\\c_2\end{pmatrix}$$

or $x = e^{-t/4}(c_1\cos 4t + c_2\sin 4t)$ and $y = e^{-t/4}(2c_2\cos 4t - 2c_1\sin 4t)$. We confirm this result using DSolve.

```
In[1453]:= gensol = DSolve[
                {x'[t] == -1/4x[t] + 2y[t],
                 y'[t] == -8x[t] - 1/4y[t]}, {x[t], y[t]}, t]
```

$$Out[1453]= \left\{\left\{x[t] \to C[2]\ \left(-\frac{1}{2}\ i\ \mathrm{Cos}\left[\left(4+\frac{i}{4}\right)\ t\right]\right.\right.\right.$$

$$+\frac{1}{2}\ i\ \mathrm{Cosh}\left[\left(\frac{1}{4}+4\ i\right)\ t\right]+\frac{1}{2}\ \mathrm{Sin}\left[\left(4+\frac{i}{4}\right)\ t\right]$$

$$\left.-\frac{1}{2}\ i\ \mathrm{Sinh}\left[\left(\frac{1}{4}+4\ i\right)\ t\right]\right)$$

$$+C[1]\ \left(-\frac{1}{2}\ \mathrm{Cos}\left[\left(4+\frac{i}{4}\right)\ t\right]\right.$$

$$-\frac{1}{2}\ \mathrm{Cosh}\left[\left(\frac{1}{4}+4\ i\right)\ t\right]-\frac{1}{2}\ i\ \mathrm{Sin}\left[\left(4+\frac{i}{4}\right)\ t\right]$$

$$\left.+\frac{1}{2}\ \mathrm{Sinh}\left[\left(\frac{1}{4}+4\ i\right)\ t\right]\right),$$

$$y[t] \to C[2]\ \left(\mathrm{Cos}\left[\left(4+\frac{i}{4}\right)\ t\right]+\mathrm{Cosh}\left[\left(\frac{1}{4}+4\ i\right)\ t\right]\right.$$

$$\left.+i\ \mathrm{Sin}\left[\left(4+\frac{i}{4}\right)\ t\right]-\mathrm{Sinh}\left[\left(\frac{1}{4}+4\ i\right)\ t\right]\right)$$

$$+C[1]\ \left(-i\ \mathrm{Cos}\left[\left(4+\frac{i}{4}\right)\ t\right]+i\ \mathrm{Cosh}\left[\left(\frac{1}{4}+4\ i\right)\ t\right]\right.$$

$$\left.\left.\left.+\mathrm{Sin}\left[\left(4+\frac{i}{4}\right)\ t\right]-i\ \mathrm{Sinh}\left[\left(\frac{1}{4}+4\ i\right)\ t\right]\right)\right\}\right\}$$

```
In[1454]:= gensol[[1, 1, 2]]
```

$$Out[1454]= C[2]\ \left(-\frac{1}{2}\ i\ \mathrm{Cos}\left[\left(4+\frac{i}{4}\right)\ t\right]+\frac{1}{2}\ i\ \mathrm{Cosh}\left[\left(\frac{1}{4}+4\ i\right)\ t\right]\right.$$

$$\left.+\frac{1}{2}\ \mathrm{Sin}\left[\left(4+\frac{i}{4}\right)\ t\right]-\frac{1}{2}\ i\ \mathrm{Sinh}\left[\left(\frac{1}{4}+4\ i\right)\ t\right]\right)$$

$$+C[1]\ \left(-\frac{1}{2}\ \mathrm{Cos}\left[\left(4+\frac{i}{4}\right)\ t\right]-\frac{1}{2}\ \mathrm{Cosh}\left[\left(\frac{1}{4}+4\ i\right)\ t\right]\right.$$

$$\left.-\frac{1}{2}\ i\ \mathrm{Sin}\left[\left(4+\frac{i}{4}\right)\ t\right]+\frac{1}{2}\ \mathrm{Sinh}\left[\left(\frac{1}{4}+4\ i\right)\ t\right]\right)$$

In[1455]:= **ComplexExpand[gensol[[1,1,2]]]//Simplify**

Out[1455]= (C[1] Cos[4 t]

$\qquad$ -C[2] Sin[4 t]) $\left(-\text{Cosh}\left[\dfrac{t}{4}\right]+\text{Sinh}\left[\dfrac{t}{4}\right]\right)$

In[1456]:= **(C[1] Cos[4 t] - C[2] Sin[4 t]) $\left(-e^{-t/4}\right)$**

In[1457]:= **ComplexExpand[gensol[[1,2,2]]]//Simplify**

Out[1457]= 2 (C[2] Cos[4 t]

$\qquad$ +C[1] Sin[4 t]) $\left(\text{Cosh}\left[\dfrac{t}{4}\right]-\text{Sinh}\left[\dfrac{t}{4}\right]\right)$

In[1458]:= **2 (C[2] Cos[4 t] + C[1] Sin[4 t]) $(e^{-t/4})$**

In this case, we obtained the real form of the solution by selecting the portion of the expression that we wanted to write in terms of exponential functions

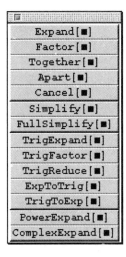

and then accessed TrigToExp from the **Algebraic Manipulation** palette to obtain the result.

```
ComplexExpand[gensol[[1, 2, 2]]] // Simplify

2 (C[2] Cos[4 t] + C[1] Sin[4 t]) (Cosh[t/4] - Sinh[t/4])

2 (C[2] Cos[4 t] + C[1] Sin[4 t]) (E^-t/4)
```

We use `PlotVectorField` and `ParametricPlot` to graph the direction field associated with the system along with various solutions in Figure 6-32.

```
In[1459]:= pvf = PlotVectorField[{1/4x + 2y, -8x - 1/4y},
                  {x, -1, 1}, {y, -1, 1}, Axes- > Automatic,
                  DisplayFunction- > Identity];

In[1460]:= initsol = DSolve[{x'[t] == -1/4x[t] + 2y[t],
                  y'[t] == -8x[t] - 1/4y[t],
                  x[0] == x0, y[0] == y0}, {x[t], y[t]}, t]
```

$$Out[1460]= \left\{\left\{x[t] \to x0\ Cos[4\ t]\ Cosh\left[\frac{t}{4}\right]\right.\right.$$
$$+\frac{1}{2}\ y0\ Cosh\left[\frac{t}{4}\right]\ Sin[4\ t] - x0\ Cos[4\ t]\ Sinh\left[\frac{t}{4}\right]$$
$$-\frac{1}{2}\ y0\ Sin[4\ t]\ Sinh\left[\frac{t}{4}\right],$$
$$y[t] \to 2\ \left(\frac{1}{2}\ y0\ Cos[4\ t]\ Cosh\left[\frac{t}{4}\right]\right.$$
$$-x0\ Cosh\left[\frac{t}{4}\right]\ Sin[4\ t] - \frac{1}{2}\ y0\ Cos[4\ t]\ Sinh\left[\frac{t}{4}\right]$$
$$\left.\left.\left.+x0\ Sin[4\ t]\ Sinh\left[\frac{t}{4}\right]\right)\right\}\right\}$$

```
In[1461]:= t1 = Table[ParametricPlot[
                  Evaluate[{x[t], y[t]}/.initsol/
                  .{x0- > 1, y0- > i}],
                  {t, 0, 15}, DisplayFunction- > Identity,
                  PlotStyle- > GrayLevel[0.3]],
                  {i, -1, 1, 2/8}];

In[1462]:= Show[t1, pvf, DisplayFunction- >
                  $DisplayFunction, PlotRange- > {{-1, 1}, {-1, 1}},
                  AspectRatio- > Automatic]
```

Last, we illustrate how to solve an initial-value problem and graph the resulting solutions by finding the solution that satisfies the initial conditions $x(0) = 100$ and $y(0) = 10$ and then graphing the results with `Plot` and `ParametricPlot` in Figure 6-33.

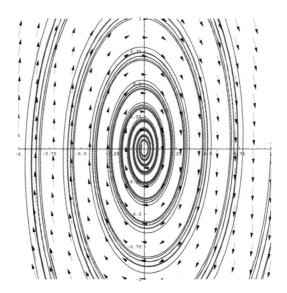

Figure 6-32   Various solutions and direction field associated with the system

$In[1463]:=$ **partsol = DSolve[{x'[t] == -1/4x[t] + 2y[t],**
y'[t] == -8x[t] - 1/4y[t], x[0] == 100,
y[0] == 10}, {x[t], y[t]}, t]

$Out[1463]=$ $\left\{\left\{x[t] \rightarrow 100 \; Cos[4 \; t] \; Cosh\left[\dfrac{t}{4}\right]\right.\right.$

$+5 \; Cosh\left[\dfrac{t}{4}\right] \; Sin[4 \; t] - 100 \; Cos[4 \; t] \; Sinh\left[\dfrac{t}{4}\right]$

$-5 \; Sin[4 \; t] \; Sinh\left[\dfrac{t}{4}\right],$

$y[t] \rightarrow 2 \; \left(5 \; Cos[4 \; t] \; Cosh\left[\dfrac{t}{4}\right]\right.$

$-100 \; Cosh\left[\dfrac{t}{4}\right] \; Sin[4 \; t] - 5 \; Cos[4 \; t] \; Sinh\left[\dfrac{t}{4}\right]$

$\left.\left.\left.+100 \; Sin[4 \; t] \; Sinh\left[\dfrac{t}{4}\right]\right)\right\}\right\}$

$In[1464]:=$ **p1 = Plot[Evaluate[{x[t], y[t]}/.partsol], {t, 0, 20},**
PlotStyle- > {GrayLevel[0], GrayLevel[0.4]},
DisplayFunction- > Identity, PlotRange- > All];
p2 = ParametricPlot[
Evaluate[{x[t], y[t]}/.partsol], {t, 0, 20},
DisplayFunction- > Identity,
AspectRatio- > Automatic];
Show[GraphicsArray[{p1, p2}]]

■

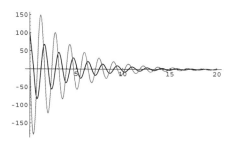

Figure 6-33   (a) Graph of $x(t)$ and $y(t)$. (b) Parametric plot of $x(t)$ versus $y(t)$. (For help with Show and GraphicsArray use the **Help Browser**)

Application: The Double Pendulum

The motion of a double pendulum is modeled by the system of differential equations

$$\begin{cases} (m_1 + m_2)l_1{}^2\dfrac{d^2\theta_1}{dt^2} + m_2l_1l_2\dfrac{d^2\theta_2}{dt^2} + (m_1 + m_2)l_1g\theta_1 = 0 \\ m_2l_2{}^2\dfrac{d^2\theta_2}{dt^2} + m_2l_1l_2\dfrac{d^2\theta_1}{dt^2} + m_2l_2g\theta_2 = 0 \end{cases}$$

using the approximation $\sin\theta \approx \theta$ for small displacements. θ_1 represents the displacement of the upper pendulum and θ_2 that of the lower pendulum. Also, m_1 and m_2 represent the mass attached to the upper and lower pendulums, respectively, while the length of each is given by l_1 and l_2.

EXAMPLE 6.4.2: Suppose that $m_1 = 3$, $m_2 = 1$, and each pendulum has length 16. If $\theta_1(0) = 1$, $\theta_1{}'(0) = 0$, $\theta_2(0) = -1$, and $\theta_2{}'(0) = 0$, solve the double pendulum problem using $g = 32$. Plot the solution.

SOLUTION: In this case, the system to be solved is

$$\begin{cases} 4 \cdot 16^2\dfrac{d^2\theta_1}{dt^2} + 16^2\dfrac{d^2\theta_2}{dt^2} + 4 \cdot 16 \cdot 32\theta_1 = 0 \\ 16^2\dfrac{d^2\theta_2}{dt^2} + 16^2\dfrac{d^2\theta_1}{dt^2} + 16 \cdot 32\theta_2 = 0, \end{cases}$$

which we simplify to obtain

$$\begin{cases} 4\dfrac{d^2\theta_1}{dt^2} + \dfrac{d^2\theta_2}{dt^2} + 8\theta_1 = 0 \\ \dfrac{d^2\theta_2}{dt^2} + \dfrac{d^2\theta_1}{dt^2} + 2\theta_2 = 0. \end{cases}$$

In the following code, we let $x(t)$ and $y(t)$ represent $\theta_1(t)$ and $\theta_2(t)$, respectively. First, we use DSolve to solve the initial-value problem.

$$In[1465] := \text{sol} =$$
$$\text{DSolve}[\{4\,x''[t] + y''[t] + 8\,x[t] == 0,$$
$$x''[t] + y''[t] + 2\,y[t] == 0, x[0] == 1,$$
$$x'[0] == 1, y[0] == 0, y'[0] == -1\},$$
$$\{x[t], y[t]\}, t]$$

$$Out[1465] = \left\{\left\{x[t] \rightarrow \frac{1}{8}\left(4\,\text{Cos}[2\,t]\right.\right.\right.$$
$$\left. +4\,\text{Cos}\left[\frac{2\,t}{\sqrt{3}}\right] + 3\,\text{Sin}[2\,t] + \sqrt{3}\,\text{Sin}\left[\frac{2\,t}{\sqrt{3}}\right]\right),$$
$$y[t] \rightarrow \frac{1}{4}\left(-4\,\text{Cos}[2\,t] + 4\,\text{Cos}\left[\frac{2\,t}{\sqrt{3}}\right]\right.$$
$$\left.\left.\left.-3\,\text{Sin}[2\,t] + \sqrt{3}\,\text{Sin}\left[\frac{2\,t}{\sqrt{3}}\right]\right)\right\}\right\}$$

The **Laplace transform** of $y = f(t)$ is $F(s) = \mathcal{L}\{f(t)\} = \int_0^\infty e^{-st} f(t)\,dt$.

To solve the initial-value problem using traditional methods, we use the *method of Laplace transforms*. To do so, we define sys to be the system of equations and use LaplaceTransform to compute the Laplace transform of each equation.

In[1466]:= **step1 = LaplaceTransform[sys, t, s]**

Out[1466]= {8 LaplaceTransform[x[t], t, s]
  +s² LaplaceTransform[y[t], t, s] - s y[0]
  +4 (s² LaplaceTransform[x[t], t, s]
  - s x[0] - x'[0]) - y'[0] == 0,
  s² LaplaceTransform[x[t], t, s]
  +2 LaplaceTransform[y[t], t, s]
  +s² LaplaceTransform[y[t], t, s]
  - s x[0] - s y[0] - x'[0] - y'[0] == 0}

Next, we apply the initial conditions and solve the resulting system of equations for $\mathcal{L}\{\theta_1(t)\} = X(s)$ and $\mathcal{L}\{\theta_2(t)\} = Y(s)$.

In[1467]:= **step2 =**
         step1 /. {x[0]- >1, x'[0]- >1, y[0]- >0,
             y'[0]- >-1}

Out[1467]= {1 + 8 LaplaceTransform[x[t], t, s]
  +4 (-1 - s + s² LaplaceTransform[x[t], t, s])
  +s² LaplaceTransform[y[t], t, s] == 0,
  - s + s² LaplaceTransform[x[t], t, s]
  +2 LaplaceTransform[y[t], t, s]
  +s² LaplaceTransform[y[t], t, s] == 0}

In[1468]:= **step3 = Solve[step2,**
            { LaplaceTransform[x[t], t, s],
              LaplaceTransform[y[t], t, s]}]

Out[1468]= {{LaplaceTransform[x[t], t, s] →
  $-\dfrac{-6 - 8\,s - 3\,s^2 - 3\,s^3}{16 + 16\,s^2 + 3\,s^4}$,
  LaplaceTransform[y[t], t, s] → $-\dfrac{-8\,s + 3\,s^2}{16 + 16\,s^2 + 3\,s^4}$}}

InverseLaplaceTransform is then used to find $\theta_1(t)$ and $\theta_2(t)$.

$f(t)$ is the **inverse Laplace transform** of $F(s)$ if $\mathcal{L}\{f(t)\} = F(s)$; we write $\mathcal{L}^{-1}\{F(s)\} = f(t)$.

In[1469]:= **x[t_] = InverseLaplaceTransform[**
         $-\dfrac{-6 - 8s - 3s^2 - 3s^3}{16 + 16s^2 + 3s^4}$ **, s, t]**

Out[1469]= $\dfrac{1}{8}$

  $\left(4\ \text{Cos}[2\ t] + 4\ \text{Cos}\left[\dfrac{2\ t}{\sqrt{3}}\right] + 3\ \text{Sin}[2\ t] + \sqrt{3}\ \text{Sin}\left[\dfrac{2\ t}{\sqrt{3}}\right]\right)$

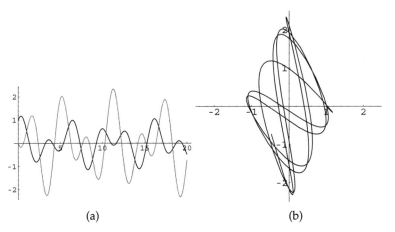

Figure 6-34   (a) $\theta_1(t)$ (in black) and $\theta_2(t)$ (in gray) as functions of t. (b) Parametric plot of $\theta_1(t)$ versus $\theta_2(t)$

$$In[1470] := \mathbf{y[t\_] = InverseLaplaceTransform}\Big[$$
$$-\frac{-8\,s + 3s^2}{16 + 16s^2 + 3s^4}, \mathbf{s, t}\Big]$$

$$Out[1470] = -\text{Cos}[2\,t] + \text{Cos}\Big[\frac{2\,t}{\sqrt{3}}\Big]$$
$$-\frac{3}{2}\,\text{Cos}[t]\,\text{Sin}[t] + \frac{1}{4}\,\sqrt{3}\,\text{Sin}\Big[\frac{2\,t}{\sqrt{3}}\Big]$$

These two functions are graphed together in Figure 6-34 (a) and parametrically in Figure 6-34 (b).

$$In[1471] := \mathbf{Plot[\{x[t], y[t]\}, \{t, 0, 20\},}$$
$$\mathbf{PlotStyle\text{-} > \{GrayLevel[0], GrayLevel[0.5]\}]}$$

$$In[1472] := \mathbf{ParametricPlot[\{x[t], y[t]\}, \{t, 0, 20\},}$$
$$\mathbf{PlotRange\text{-} > \{\{-5/2, 5/2\}, \{-5/2, 5/2\}\},}$$
$$\mathbf{AspectRatio\text{-} > 1]}$$

We can illustrate the motion of the pendulum as follows. First, we define the function pen2.

```
In[1473]:= Clear[pen2]
```

$$pen2[t\_, len1\_, len2\_] := Module\big[\{pt1, pt2\},$$

$$pt1 = \Big\{len1\ Cos\ \Big[\frac{3\pi}{2} + x[t]\Big],$$

$$len1\ Sin\ \Big[\frac{3\pi}{2} + x[t]\Big]\Big\};$$

$$pt2 =$$

$$\Big\{len1\ Cos\ \Big[\frac{3\pi}{2} + x[t]\Big] + len2\ Cos\ \Big[\frac{3\pi}{2} + y[t]\Big],$$

$$len1\ Sin\ \Big[\frac{3\pi}{2} + x[t]\Big] +$$

$$len2\ Sin\ \Big[\frac{3\pi}{2} + y[t]\Big]\Big\};$$

```
Show[
  Graphics[{Line[{{0, 0}, pt1}],
     PointSize[0.05], Point[pt1],
     Line[{pt1, pt2}], PointSize[0.05],
     Point[pt2]}], Axes → Automatic,
  Ticks → None, AxesStyle → GrayLevel[0.5],
  PlotRange → {{-32, 32}, {-34, 0}},
  DisplayFunction → Identity]]
```

Next, we define `tvals` to be a list of sixteen evenly spaced numbers between 0 and 10. `Map` is then used to apply `pen2` to the list of numbers in `tvals`. The resulting set of graphics is partitioned into four element subsets and displayed using `Show` and `GraphicsArray` in Figure 6-35.

$$In[1474]:= tvals = Table\Big[t, \Big\{t, 0, 10, \frac{10}{15}\Big\}\Big];$$

```
In[1475]:= graphs = Map[pen2[#, 16, 16]&, tvals];
```

```
In[1476]:= toshow = Partition[graphs, 4];
```

```
In[1477]:= Show[GraphicsArray[toshow]]
```

If the option `DisplayFunction->Identity` is omitted from the definition of `pen2`, we can use a `Do` loop to generate a set of graphics that can then be animated.

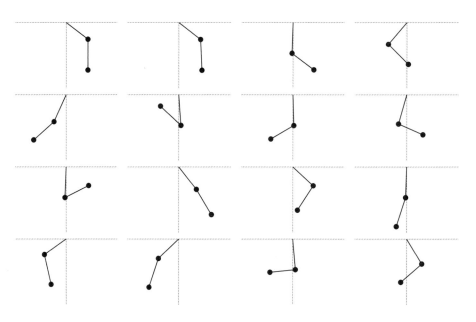

Figure 6-35   The double pendulum for 16 equally spaced values of *t* between 0 and 10

```
In[1478]:= Clear[pen2]

pen2[t_, len1_, len2_] := Module[{pt1, pt2},
    pt1 = {len1 Cos[(3π)/2 + x[t]],
        len1 Sin[(3π)/2 + x[t]]};
    pt2 =
        {len1 Cos[(3π)/2 + x[t]] + len2 Cos[(3π)/2 + y[t]],
        len1 Sin[(3π)/2 + x[t]] +
            len2 Sin[(3π)/2 + y[t]]};
    Show[
        Graphics[{Line[{{0, 0}, pt1}],
            PointSize[0.05], Point[pt1],
            Line[{pt1, pt2}], PointSize[0.05],
            Point[pt2]}], Axes → Automatic,
        Ticks → None, AxesStyle → GrayLevel[0.5],
        PlotRange → {{-32, 32}, {-34, 0}}]]
```

We show one frame from the animation that results from the Do loop

$$In[1479] := Do\left[pen2[t, 16, 16], \left\{t, 0, 10, \frac{10}{59}\right\}\right]$$

in the following screen shot.

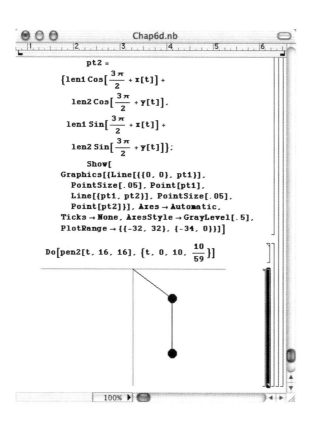

6.4.2 Nonhomogeneous Linear Systems

Generally, the method of undetermined coefficients is difficult to implement for nonhomogeneous linear systems as the choice for the particular solution must be very carefully made. Variation of parameters is implemented in much the same way as for first-order linear equations.

Let X_h be a general solution to the corresponding homogeneous system of equation (6.27), X a general solution of equation (6.27), and X_p a particular solution of equation (6.27). It then follows that $X - X_p$ is a solution to the corresponding homogeneous system so $X - X_p = X_h$ and, consequently, $X = X_h + X_p$. A particular solution of equation (6.27) is found in much the same way as with first-order linear equations. Let $\mathbf{\Phi}$ be a fundamental matrix for the corresponding homogeneous system. We assume that a particular solution has the form $\mathbf{X}_p = \mathbf{\Phi}\mathbf{U}(t)$. Differentiating $\mathbf{X}_p$ gives us

$$\mathbf{X}_p' = \mathbf{\Phi}'\mathbf{U} + \mathbf{\Phi}\mathbf{U}'.$$

Substituting into equation (6.27) results in

$$\mathbf{\Phi}'\mathbf{U} + \mathbf{\Phi}\mathbf{U}' = \mathbf{A}\mathbf{\Phi}\mathbf{U} + \mathbf{F}$$
$$\mathbf{\Phi}\mathbf{U}' = \mathbf{F}$$
$$\mathbf{U}' = \mathbf{\Phi}^{-1}\mathbf{F}$$
$$\mathbf{U} = \int \mathbf{\Phi}^{-1}\mathbf{F}\,dt,$$

where we have used the fact that $\mathbf{\Phi}'\mathbf{U} - \mathbf{A}\mathbf{\Phi}\mathbf{U} = (\mathbf{\Phi}' - \mathbf{A}\mathbf{\Phi})\,\mathbf{U} = 0$. It follows that

$$\mathbf{X}_p = \mathbf{\Phi} \int \mathbf{\Phi}^{-1}\mathbf{F}\,dt. \tag{6.30}$$

A general solution is then

$$\mathbf{X} = \mathbf{X}_h + \mathbf{X}_p$$
$$= \mathbf{\Phi}\mathbf{C} + \mathbf{\Phi} \int \mathbf{\Phi}^{-1}\mathbf{F}\,dt$$
$$= \mathbf{\Phi}\left(\mathbf{C} + \int \mathbf{\Phi}^{-1}\mathbf{F}\,dt\right) = \mathbf{\Phi} \int \mathbf{\Phi}^{-1}\mathbf{F}\,dt,$$

where we have incorporated the constant vector $\mathbf{C}$ into the indefinite integral $\int \mathbf{\Phi}^{-1}\mathbf{F}\,dt$.

EXAMPLE 6.4.3: Solve the initial-value problem

$$\mathbf{X}' = \begin{pmatrix} 1 & -1 \\ 10 & -1 \end{pmatrix}\mathbf{X} - \begin{pmatrix} t\cos 3t \\ t\sin t + t\cos 3t \end{pmatrix}, \quad \mathbf{X}(0) = \begin{pmatrix} 1 \\ -1 \end{pmatrix}.$$

Remark. In traditional form, the system is equivalent to

$$\begin{cases} x' = x - y - t\cos 3t \\ y' = 10x - y - t\sin t - t\cos 3t, \end{cases} \qquad x(0) = 1,\ y(0) = -1.$$

SOLUTION: The corresponding homogeneous system is $X'_h = \begin{pmatrix} 1 & -1 \\ 10 & -1 \end{pmatrix} X_h$.

The eigenvalues and corresponding eigenvectors of $A = \begin{pmatrix} 1 & -1 \\ 10 & -1 \end{pmatrix}$ are

$\lambda_{1,2} = \pm 3i$ and $v_{1,2} = \begin{pmatrix} 1 \\ 10 \end{pmatrix} \pm \begin{pmatrix} -3 \\ 0 \end{pmatrix} i$, respectively.

```
In[1480] := capa = {{1, -1}, {10, -1}};
            Eigensystem[capa]
Out[1480] = {{-3 i, 3 i}, {{1 - 3 i, 10}, {1 + 3 i, 10}}}
```

A fundamental matrix is $\Phi = \begin{pmatrix} \sin 3t & \cos 3t \\ \sin 3t - 3\cos 3t & \cos 3t + 3\sin 3t \end{pmatrix}$ with

inverse $\Phi^{-1} = \begin{pmatrix} \frac{1}{3}\cos 3t + \sin 3t & -\frac{1}{3}\cos 3t \\ -\frac{1}{3}\sin 3t + \cos 3t & \frac{1}{3}\sin 3t \end{pmatrix}$.

```
In[1481] := fm = {{Sin[3t], Sin[3t] - 3 Cos[3t]},
                  {Cos[3t], Cos[3t] + 3 Sin[3t]}};
            fminv = Inverse[fm]//Simplify
```

$Out[1481] = \{\{\frac{1}{3} \text{ Cos}[3\ t] + \text{Sin}[3\ t], \text{Cos}[3\ t] - \frac{1}{3} \text{ Sin}[3\ t]\},$

$\{-\frac{1}{3} \text{ Cos}[3\ t], \frac{1}{3} \text{ Sin}[3\ t]\}\}$

We now compute $\Phi^{-1}F(t)$

```
In[1482] := ft = {-t Cos[3t], -t Sin[t] - t Cos[3t]};
            step1 = fminv.ft
```

$Out[1482] = \{(-t \text{ Cos}[3\ t] - t \text{ Sin}[t]) (\text{Cos}[3\ t] - \frac{1}{3} \text{ Sin}[3\ t])$

$-t \text{ Cos}[3\ t] (\frac{1}{3} \text{ Cos}[3\ t] + \text{Sin}[3\ t]),$

$\frac{1}{3} t \text{ Cos}[3\ t]^2 + \frac{1}{3} (-t \text{ Cos}[3\ t] - t \text{ Sin}[t]) \text{ Sin}[3\ t]\}$

and $\int \Phi^{-1}F(t)\,dt$.

```
In[1483] := step2 = Integrate[step1, t]
```

$Out[1483] = \{\frac{1}{864} (-288\ t^2 + 36\ \text{Cos}[2\ t] - 216\ t\ \text{Cos}[2\ t]$

$-9\ \text{Cos}[4\ t] + 108\ t\ \text{Cos}[4\ t] - 16\ \text{Cos}[6\ t]$
$+48\ t\ \text{Cos}[6\ t] + 108\ \text{Sin}[2\ t] + 72\ t\ \text{Sin}[2\ t]$
$-27\ \text{Sin}[4\ t] - 36\ t\ \text{Sin}[4\ t] - 8\ \text{Sin}[6\ t]$
$-96\ t\ \text{Sin}[6\ t]),$

$\frac{1}{864} (72\ t^2 - 36\ \text{Cos}[2\ t] + 9\ \text{Cos}[4\ t] + 4\ \text{Cos}[6\ t]$

$+24\ t\ \text{Cos}[6\ t] - 72\ t\ \text{Sin}[2\ t] + 36\ t\ \text{Sin}[4\ t]$
$-4\ \text{Sin}[6\ t] + 24\ t\ \text{Sin}[6\ t])\}$

A general solution of the nonhomogeneous system is then $\Phi\left(\int \Phi^{-1}\mathbf{F}(t)\,dt + \mathbf{C}\right)$.

In[1484]:= **Simplify[fm.step2]**

Out[1484]= $\left\{\dfrac{1}{288}\left(27\ \text{Cos[t]} - 4\ \left(\left(1 + 6\ t + 18\ t^2\right)\ \text{Cos[3 t]}\right.\right.\right.$

$+27\ t\ \text{Sin[t]} - \text{Sin[3 t]} + 6\ t\ \text{Sin[3 t]}$

$+18\ t^2\ \text{Sin[3 t]})),$

$\dfrac{1}{288}\left(-36\ t\ \text{Cos[t]} - 4\ \left(1 - 6\ t + 18\ t^2\right)\ \text{Cos[3 t]}\right.$

$-45\ \text{Sin[t]} - 4\ \text{Sin[3 t]} - 24\ t\ \text{Sin[3 t]}$

$+72\ t^2\ \text{Sin[3 t]})\}$

It is easiest to use DSolve to solve the initial-value problem directly as we do next.

In[1485]:= **check = DSolve[{x′[t] == x[t] - y[t] - t Cos[3t],**
                             y′[t] == 10x[t] - y[t] - t Sin[t] - t Cos[3t],
                             x[0] == 1, y[0] == -1}, {x[t], y[t]}, t]

General :: "spell1" : "Possiblespellingerror :
    newsymbolnamečheckïs similar
    toexistingsymbolČheck⁚"

Out[1485]= $\left\{\left\{x[t] \rightarrow \dfrac{1}{288}\left(-9\ \text{Cos[t]} + 297\ \text{Cos[3 t]}\right.\right.\right.$

$-72\ t^2\ \text{Cos[3 t]} + 36\ t\ \text{Sin[t]}$

$+192\ \text{Sin[3 t]} - 24\ t\ \text{Sin[3 t]}),$

$y[t] \rightarrow \dfrac{1}{288}\left(-9\ \text{Cos[t]} - 36\ t\ \text{Cos[t]}\right.$

$-279\ \text{Cos[3 t]} - 72\ t\ \text{Cos[3 t]}$

$-72\ t^2\ \text{Cos[3 t]} - 45\ \text{Sin[t]}$

$+36\ t\ \text{Sin[t]} + 1107\ \text{Sin[3 t]}$

$-24\ t\ \text{Sin[3 t]} - 216\ t^2\ \text{Sin[3 t]})\}\}$

After using ?Evaluate to obtain basic information regarding the Evaluate function, the solutions are graphed with Plot and ParametricPlot in Figure 6-36.

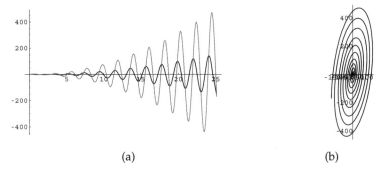

(a)   (b)

Figure 6-36   (a) Graph of $x(t)$ (in black) and $y(t)$ (in gray) . (b) Parametric plot of $x(t)$ versus $y(t)$

```
In[1486]:= ?Evaluate

"Evaluate[expr]causesexprtobeevaluatedeven
    ifitappearsastheargumentofafunction
    whoseattributesspecifythatitshouldbe
    heldunevaluated."


In[1487]:= p1 = Plot[Evaluate[{x[t],
            y[t]}/.check], {t, 0, 8π},
            PlotStyle- > {GrayLevel[0], GrayLevel[0.4]},
            DisplayFunction- > Identity];
        p2 = ParametricPlot[
            Evaluate[{x[t],y[t]}/.check], {t, 0, 8π},
            DisplayFunction- > Identity,
            AspectRatio- > Automatic];
        Show[GraphicsArray[{p1, p2}]]
```

■

6.4.3 Nonlinear Systems

Nonlinear systems of differential equations arise in numerous situations. Rigorous analysis of the behavior of solutions to nonlinear systems is usually very difficult, if not impossible.

To generate numerical solutions of equations, use NDSolve.

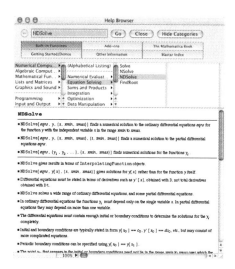

Also see Example 6.4.7.

EXAMPLE 6.4.4 (Van-der-Pol's equation): Van-der-Pol's Equation $x'' +$ $\mu\left(x^2 - 1\right)x' + x = 0$ can be written as the system

$$x' = y$$
$$y' = -x - \mu\left(x^2 - 1\right)y. \tag{6.31}$$

If $\mu = 2/3$, $x(0) = 1$, and $y(0) = 0$, (a) find $x(1)$ and $y(1)$. (b) Graph the solution that satisfies these initial conditions.

SOLUTION: We use NDSolve together to solve equation (6.31) with $\mu = 2/3$ subject to $x(0) = 1$ and $y(0) = 0$. We name the resulting numerical solution numsol.

```
In[1488]:= numsol = NDSolve[{x'[t] == y[t],
                y'[t] == -x[t] - 2/3 (x[t]^2 - 1) y[t], x[0] == 1,
                y[0] == 0}, {x[t], y[t]}, {t, 0, 30}]
Out[1488]= BoxData({{x[t] → InterpolatingFunction[{{0., 30.}},
                " <> "] [t], y[t] → InterpolatingFunction[{{0., 30.}},
                " <> "] [t]}})
```

We evaluate numsol if $t = 1$ to see that $x(1) \approx .5128$ and $y(1) \approx -.9692$.

```
In[1489]:= {x[t], y[t]}/.numsol/.t->1
Out[1489]= {{0.512849, -0.969199}}
```

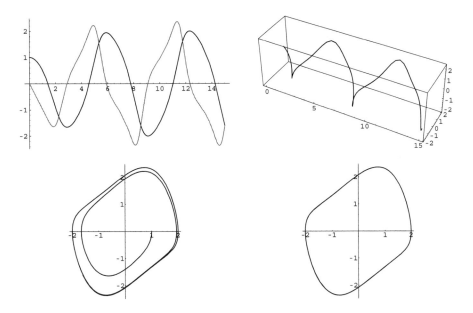

Figure 6-37   (a) $x(t)$ and $y(t)$. (b) A three-dimensional plot. (c) $x(t)$ versus $y(t)$. (d) $x(t)$ versus $y(t)$ for $20 \leq t \leq 30$

Plot, ParametricPlot, and ParametricPlot3D are used to graph $x(t)$ and $y(t)$ together in Figure 6-37 (a); a three-dimensional plot, $(t, x(t), y(t))$ is shown in Figure 6-37 (b); a parametric plot is shown in Figure 6-37 (c); and the limit cycle is shown more clearly in Figure 6-37 (d) by graphing the solution for $20 \leq t \leq 30$.

```
In[1490]:= p1 = Plot[Evaluate[{x[t],y[t]}/.numsol],{t,0,15},
                PlotStyle- > {GrayLevel[0],GrayLevel[0.4]},
                DisplayFunction- > Identity];
           p2 = ParametricPlot3D[Evaluate[{t,x[t],
                y[t]}/.numsol],{t,0,15},
                DisplayFunction- > Identity];
           p3 = ParametricPlot[Evaluate[{x[t],y[t]}/.numsol],
                {t,0,15},AspectRatio- > Automatic,
                DisplayFunction- > Identity];
           p4 = ParametricPlot[Evaluate[{x[t],y[t]}/.numsol],
                {t,20,30},AspectRatio- > Automatic,
                DisplayFunction- > Identity];
           Show[GraphicsArray[{{p1,p2},{p3,p4}}]]
```

■

Linearization

Consider the autonomous system of the form

$$\begin{aligned}
x_1' &= f_1(x_1, x_2, \ldots, x_n) \\
x_2' &= f_2(x_1, x_2, \ldots, x_n) \\
&\vdots \\
x_n' &= f_n(x_1, x_2, \ldots, x_n).
\end{aligned} \tag{6.32}$$

An **equilibrium** (or **rest**) **point**, $E = (x_1^*, x_2^*, \ldots, x_n^*)$, of equation (6.32) is a solution of the system

$$\begin{aligned}
f_1(x_1, x_2, \ldots, x_n) &= 0 \\
f_2(x_1, x_2, \ldots, x_n) &= 0 \\
&\vdots \\
f_n(x_1, x_2, \ldots, x_n) &= 0.
\end{aligned} \tag{6.33}$$

The **Jacobian** of equation (6.32) is

$$\mathbf{J}(x_1, x_2, \ldots, x_n) = \begin{pmatrix}
\frac{\partial f_1}{\partial x_1} & \frac{\partial f_1}{\partial x_2} & \cdots & \frac{\partial f_1}{\partial x_n} \\
\frac{\partial f_2}{\partial x_1} & \frac{\partial f_2}{\partial x_2} & \cdots & \frac{\partial f_2}{\partial x_n} \\
\vdots & \vdots & \cdots & \vdots \\
\frac{\partial f_n}{\partial x_1} & \frac{\partial f_n}{\partial x_2} & \cdots & \frac{\partial f_n}{\partial x_n}
\end{pmatrix}.$$

The rest point, E, is **locally stable** if and only if all the eigenvalues of $\mathbf{J}(E)$ have negative real part. If E is not locally stable, E is **unstable**.

Van-der-Pol's equationDuffing's Equation

EXAMPLE 6.4.5 (Van-der-Pol's equation): Consider the forced **pendulum equation** with damping,

$$x'' + kx' + \omega \sin x = F(t). \tag{6.34}$$

Recall the Maclaurin series for $\sin x$: $\sin x = x - \frac{1}{3!}x^3 + \frac{1}{5!}x^5 - \frac{1}{7!}x^7 + \ldots$. Using $\sin x \approx x$, equation (6.34) reduces to the linear equation $x'' + kx' + \omega x = F(t)$.

On the other hand, using the approximation $\sin x \approx x - \frac{1}{6}x^3$, we obtain $x'' + kx' + \omega\left(x - \frac{1}{6}x^3\right) = F(t)$. Adjusting the coefficients of x and x^3 and assuming that $F(t) = F \cos \omega t$ gives us **Duffing's equation**:

$$x'' + kx' + cx + \epsilon x^3 = F \cos \omega t, \tag{6.35}$$

where k and c are positive constants.

Let $y = x'$. Then, $y' = x'' = F \cos \omega t - kx' - cx - \epsilon x^3 = F \cos \omega t - ky - cx - \epsilon x^3$ and we can write equation (6.35) as the system

$$x' = y$$
$$y' = F \cos \omega t - ky - cx - \epsilon x^3. \tag{6.36}$$

Assuming that $F = 0$ results in the autonomous system

$$x' = y$$
$$y' = -cx - \epsilon x^3 - ky. \tag{6.37}$$

The rest points of system equation (6.37) are found by solving

$$x' = 0$$
$$y' = -cx - \epsilon x^3 - ky, = 0$$

resulting in $E_0 = (0, 0)$.

```
In[1491]:= Solve[{y == 0, -c x - ε x^3 - k y == 0}, {x, y}]
```
$$Out[1491] = \left\{ \{y \to 0, x \to 0\}, \left\{y \to 0, x \to -\frac{i \sqrt{c}}{\sqrt{\epsilon}}\right\}, \right.$$
$$\left\{y \to 0, x \to \frac{i \sqrt{c}}{\sqrt{\epsilon}}\right\} \right\}$$

We find the Jacobian of equation (6.37) in s1, evaluate the Jacobian at E_0,

```
In[1492]:= s1 = {{0, 1}, {-c - 3ε x^2, -k}};
           s2 = s1/.x- > 0
```
$$Out[1492] = \{\{0, 1\}, \{-c, -k\}\}$$

and then compute the eigenvalues with Eigenvalues.

```
In[1493]:= s3 = Eigenvalues[s2]
```
$$Out[1493] = \left\{ \frac{1}{2} \left(-k - \sqrt{-4 c + k^2} \right), \frac{1}{2} \left(-k + \sqrt{-4 c + k^2} \right) \right\}$$

Because k and c are positive, $k^2 - 4c < k^2$ so the real part of each eigenvalue is always negative if $k^2 - 4c \neq 0$. Thus, E_0 is locally stable.

For the autonomous system

$$x' = f(x, y)$$
$$y' = g(x, y), \tag{6.38}$$

Bendixson's theorem states that if $f_x(x, y) + g_y(x, y)$ is a continuous function that is either always positive or always negative in a particular region R of the plane, then system (6.38) has no limit cycles in R. For equation (6.37) we have

$$\frac{d}{dx}(y) + \frac{d}{dy}\left(-cx - \epsilon x^3 - ky\right) = -k,$$

which is always negative. Hence, equation (6.37) has no limit cycles and it follows that E_0 is globally, asymptotically stable.

```
In[1494]:= D[y, x] + D[-c x  - ∈ x^3 - k y, y]
Out[1494]= -k
```

We use `PlotVectorField` and `ParametricPlot` to illustrate two situations that occur. In Figure 6-38 (a), we use $c = 1$, $\epsilon = 1/2$, and $k = 3$. In this case, E_0 is a *stable node*. On the other hand, in Figure 6-38 (b), we use $c = 10$, $\epsilon = 1/2$, and $k = 3$. In this case, E_0 is a *stable spiral*.

```
In[1495]:= << Graphics`PlotField`

           pvf1 = PlotVectorField[{y, -x - 1/2x^3 - 3y},
                    {x, -2.5, 2.5}, {y, -2.5, 2.5},
                    DisplayFunction- > Identity];

In[1496]:= numgraph[init_, c_, opts__] := Module[{numsol},
           numsol = NDSolve[{x'[t] == y[t],
                    y'[t] == -c x[t] - 1/2x[t]^3 - 3y[t],
                        x[0] == init[[1]], y[0] == init[[2]]},
                    {x[t], y[t]}, {t, 0, 10}];
           ParametricPlot[Evaluate[{x[t], y[t]}/.numsol],
                    {t, 0, 10}, opts,
                    DisplayFunction- > Identity]]

In[1497]:= i1 = Table[numgraph[{2.5, i}, 1],
                    {i, -2.5, 2.5, 1/2}];
           i2 = Table[numgraph[{-2.5, i}, 1],
                    {i, -2.5, 2.5, 1/2}];
           i3 = Table[numgraph[{i, 2.5}, 1],
                    {i, -2.5, 2.5, 1/2}];
           i4 = Table[numgraph[{i, -2.5}, 1],
                    {i, -2.5, 2.5, 1/2}];

In[1498]:= c1 = Show[i1, i2, i3, i4,
                    pvf1, PlotRange- > {{-2.5, 2.5}, {-2.5, 2.5}},
                    AspectRatio- > Automatic];

In[1499]:= pvf2 = PlotVectorField[{y, -10x - 1/2x^3 - 3y},
                    {x, -2.5, 2.5}, {y, -2.5, 2.5},
                    DisplayFunction- > Identity];
```

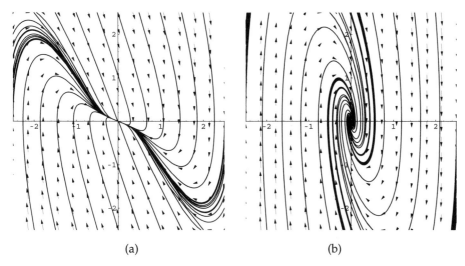

(a)   (b)

Figure 6-38   (a) The origin is a stable node . (b) The origin is a stable spiral

```
In[1500]:= i1 = Table[numgraph[{2.5,i},10],
              {i,-2.5,2.5,1/2}];
           i2 = Table[numgraph[{-2.5,i},10],
              {i,-2.5,2.5,1/2}];
           i3 = Table[numgraph[{i,2.5},10],
              {i,-2.5,2.5,1/2}];
           i4 = Table[numgraph[{i,-2.5},10],
              {i,-2.5,2.5,1/2}];

In[1501]:= c2 = Show[i1,i2,i3,i4,
              pvf2,PlotRange->{{-2.5,2.5},{-2.5,2.5}},
              AspectRatio->Automatic];

In[1502]:= Show[GraphicsArray[{c1,c2}]]
```

EXAMPLE 6.4.6 (Predator-Prey): The **predator–prey** equations take the form

$$\frac{dx}{dt} = ax - bxy$$

$$\frac{dy}{dt} = dxy - cy$$

where a, b, c, and d are positive constants. x represents the size of the prey population at time t while y represents the size of the predator population at time t. We use Solve to calculate the rest points. In this case, there is one boundary rest point, $E_0 = (0, 0)$ and one interior rest point, $E_1 = (c/d, a/b)$.

```
In[1503]:= rps = Solve[{a x - b x y == 0,
                d x y - c y == 0}, {x, y}]
```
$$Out[1503]= \left\{\{x \to 0, y \to 0\}, \left\{x \to \frac{c}{d}, y \to \frac{a}{b}\right\}\right\}$$

The Jacobian is then found using D.

```
In[1504]:= jac = {{D[a x - b x y, x], D[a x - b x y, y]},
                {D[d x y - c y, x], D[d x y - c y, y]}};
            MatrixForm[jac]
```
$$Out[1504]= \begin{pmatrix} a - b\,y & -b\,x \\ d\,y & -c + d\,x \end{pmatrix}$$

E_0 is unstable because one eigenvalue of $\mathbf{J}(E_0)$ is positive. For the linearized system, E_1 is a center because the eigenvalues of $\mathbf{J}(E_1)$ are complex conjugates.

```
In[1505]:= Eigenvalues[jac/.rps[[2]]]
```
$$Out[1505]= \left\{-\,i\ \sqrt{a}\ \sqrt{c},\ i\ \sqrt{a}\ \sqrt{c}\right\}$$

In fact, E_1 is a center for the nonlinear system as illustrated in Figure 6-39, where we have used $a = 1$, $b = 2$, $c = 2$, and $d = 1$. Notice that there are multiple limit cycles around $E_1 = (1/2, 1/2)$.

```
In[1506]:= BoxData({<< Graphics`PlotField`,
            pvf = PlotVectorField[{x - 2x y, 2x y - y},
            {x, 0, 2}, {y, 0, 2},
            DisplayFunction- > Identity]; })

In[1507]:= numgraph[init_, opts__] := Module[{numsol},
            numsol = NDSolve[{x'[t] == x[t] - 2x[t]y[t],
                    y'[t] == 2x[t]y[t] - y[t],
                        x[0] == init[[1]], y[0] == init[[2]]},
                        {x[t], y[t]}, {t, 0, 50}];
            ParametricPlot[Evaluate[{x[t], y[t]}/.numsol],
                    {t, 0, 10}, opts,
                        DisplayFunction- > Identity]]

In[1508]:= i1 = Table[numgraph[{i, i}], {i, 3/20, 1/2, 1/20}];
            Show[i1, pvf, DisplayFunction- >
            $DisplayFunction,
                PlotRange- > {{0, 2}, {0, 2}},
            AspectRatio- > Automatic]
```

In this model, a stable interior rest state is not possible.

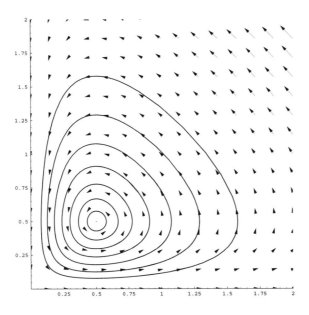

Figure 6-39   Multiple limit cycles about the interior rest point

The complexity of the behavior of solutions to the system increases based on the assumptions made. Typical assumptions include adding satiation terms for the predator (y) and/or limiting the growth of the prey (x). The **standard predator–prey equations of Kolmogorov type,**

$$x' = \alpha x\left(1 - \frac{1}{K}x\right) - \frac{mxy}{a+x}$$

$$y' = y\left(\frac{mx}{a+x} - s\right),$$

$\qquad(6.39)$

incorporate both of these assumptions.

We use Solve to find the three rest points of system (6.39). Let $E_0 = (0,0)$ and $E_1 = (k,0)$ denote the two boundary rest points, and let E_2 represent the interior rest point.

```
In[1509]:= rps = Solve[{α x (1 - 1/k x) - m x y/(a + x) == 0,
                        y (m x/(a + x) - s) == 0}, {x, y}]

Out[1509]= {{x → 0, y → 0}, {y → 0, x → k},
            {y → -(a (-k m + a s + k s) α)/(k (m - s)²), x → -(a s)/(-m + s)}}
```

The Jacobian, **J**, is calculated next in s1.

```
In[1510]:= s1 = {{D[α x (1 - 1/k x) - m x y/(a + x), x],
                  D[α x (1 - 1/k x) - m x y/(a + x), y]},
                 {D[y (m x/(a + x) - s), x],
                  D[y (m x/(a + x) - s), y]}};
           MatrixForm[s1]
```

$$Out[1510]= \begin{pmatrix} \frac{m\ x\ y}{(a+x)^2} - \frac{m\ y}{a+x} - \frac{x\ \alpha}{k} + \left(1 - \frac{x}{k}\right)\alpha & -\frac{m\ x}{a+x} \\ \left(-\frac{m\ x}{(a+x)^2} + \frac{m}{a+x}\right)y & -s + \frac{m\ x}{a+x} \end{pmatrix}$$

Because $\mathbf{J}(E_0)$ has one positive eigenvalue, E_0 is unstable.

```
In[1511]:= e0 = s1/.rps[[1]];
           MatrixForm[e0]

           eigs0 = Eigenvalues[e0]
```

$$Out[1511]= \begin{pmatrix} \alpha & 0 \\ 0 & -s \end{pmatrix}$$

$$Out[1511]= \{-s, \alpha\}$$

The stability of E_1 is determined by the sign of $m - s - am/(a + k)$.

```
In[1512]:= e1 = s1/.rps[[2]];
           MatrixForm[e1]

           eigs1 = Eigenvalues[e1]
```

$$Out[1512]= \begin{pmatrix} -\alpha & -\frac{k\ m}{a+k} \\ 0 & \frac{k\ m}{a+k} - s \end{pmatrix}$$

$$Out[1512]= \left\{\frac{k\ m}{a+k} - s,\ -\alpha\right\}$$

The eigenvalues of $\mathbf{J}(E_2)$ are quite complicated.

```
In[1513]:= e2 = s1/.rps[[3]];
           MatrixForm[e2]

           eigs2 = Eigenvalues[e2]
```

$$Out[1513] = \left(\frac{a\ s\ \alpha}{k\ (-m+s)} + \frac{a^2\ m\ s\ (-k\ m+a\ s+k\ s)\ \alpha}{k\ (m-s)^2\ (-m+s)\ \left(a - \frac{a\ s}{-m+s}\right)^2} \right.$$

$$+ \frac{a\ m\ (-k\ m+a\ s+k\ s)\ \alpha}{k\ (m-s)^2\ \left(a - \frac{a\ s}{-m+s}\right)}$$

$$+ \left(1 + \frac{a\ s}{k\ (-m+s)}\right)\alpha \qquad \frac{a\ m\ s}{(-m+s)\ \left(a - \frac{a\ s}{-m+s}\right)}$$

$$- \frac{a\ (-k\ m+a\ s+k\ s)\ \left(\frac{a\ m\ s}{(-m+s)\ \left(a - \frac{a\ s}{-m+s}\right)^2} + \frac{m}{a - \frac{a\ s}{-m+s}}\right)\alpha}{k\ (m-s)^2}$$

$$\left. -s - \frac{a\ m\ s}{(-m+s)\ \left(a - \frac{a\ s}{-m+s}\right)} \right)$$

$$Out[1513] = \left\{ \frac{1}{2\ k\ m\ (m-s)} \left(-s\ (a\ m-k\ m+a\ s+k\ s)\ \alpha - \right.\right.$$

$$\sqrt{\left(-4\ k\ m\ (m-s)\ s\right.}$$

$$\left(k\ m^2 - a\ m\ s - 2\ k\ m\ s + a\ s^2 + k\ s^2\right)\alpha +$$

$$\left.\left. s^2\ (a\ m-k\ m+a\ s+k\ s)^2\ \alpha^2\right)\right),$$

$$\frac{1}{2\ k\ m\ (m-s)} \left(-s\ (a\ m-k\ m+a\ s+k\ s)\ \alpha + \right.$$

$$\sqrt{\left(-4\ k\ m\ (m-s)\ s\right.}$$

$$\left(k\ m^2 - a\ m\ s - 2\ k\ m\ s + a\ s^2 + k\ s^2\right)\alpha +$$

$$\left.\left.\left. s^2\ (a\ m-k\ m+a\ s+k\ s)^2\ \alpha^2\right)\right)\right\}$$

Instead of using the eigenvalues, we compute the characteristic polynomial of $\mathbf{J}(E_2)$, $p(\lambda) = c_2\lambda^2 + c_1\lambda + c_0$, and examine the coefficients. Notice that c_2 is always positive.

$In[1514] := $ **cpe2 = CharacteristicPolynomial[e2, λ]//Simplify**

$$Out[1514] = \frac{a\ s\ \alpha\ (m\ (-s+\lambda) + s\ (s+\lambda))}{k\ m\ (m-s)}$$

$$+\frac{k\ (m-s)\ (-s\ \alpha\ (s+\lambda) + m\ (s\ \alpha + \lambda^2))}{k\ m\ (m-s)}$$

$In[1515] := $ **c0 = cpe2/.λ- > 0//Simplify**

$$Out[1515] = \frac{s\ (k\ (m-s) - a\ s)\ \alpha}{k\ m}$$

$In[1516] := $ **c1 = Coefficient[cpe2, λ]//Simplify**

$$Out[1516] = \frac{s\ (k\ (-m+s) + a\ (m+s))\ \alpha}{k\ m\ (m-s)}$$

$In[1517] := $ **c2 = Coefficient[cpe2, λ^2]//Simplify**

$Out[1517] = $ 1

On the other hand, c_0 and $m-s-am/(a+k)$ have the same sign because

$$In[1518]:= \texttt{c0/eigs1[[1]]//Simplify}$$
$$Out[1518]= \frac{(a+k)\ s\ \alpha}{k\ m}$$

is always positive. In particular, if $m - s - am/(a + k) < 0$, E_1 is stable. Because c_0 is negative, by Descartes' rule of signs, it follows that $p(\lambda)$ will have one positive root and hence E_2 will be unstable.

On the other hand, if $m - s - am/(a + k) > 0$ so that E_1 is unstable, E_2 may be either stable or unstable. To illustrate these two possibilities let $\alpha = K = m = 1$ and $a = 1/10$. We recalculate.

$$In[1519]:= \texttt{\alpha = 1; k = 1; m = 1; a = 1/10;}$$

```
In[1520]:= rps = Solve[{α x (1 - 1/k x) - m x y/(a + x) == 0,
              y (m x/(a + x) - s) == 0}, {x, y}]
```
$$Out[1520]= \left\{\{x \to 0, y \to 0\}, \{y \to 0, x \to 1\},\right.$$
$$\left.\left\{y \to \frac{10 - 11\ s}{100\ (-1 + s)^2}, x \to -\frac{s}{10\ (-1 + s)}\right\}\right\}$$

```
In[1521]:= s1 = {{D[α x (1 - 1/k x) - m x y/(a + x), x],
                 D[α x (1 - 1/k x) - m x y/(a + x), y]},
                {D[y (m x/(a + x) - s), x],
              D[y (m x/(a + x) - s), y]}};
            MatrixForm[s1]
```
$$Out[1521]= \begin{pmatrix} 1 - 2\ x + \dfrac{x\ y}{\left(\frac{1}{10} + x\right)^2} - \dfrac{y}{\frac{1}{10} + x} & -\dfrac{x}{\frac{1}{10} + x} \\ \left(-\dfrac{x}{\left(\frac{1}{10} + x\right)^2} + \dfrac{1}{\frac{1}{10} + x}\right)\ y & -s + \dfrac{x}{\frac{1}{10} + x} \end{pmatrix}$$

```
In[1522]:= e2 = s1/.rps[[3]];
            cpe2 = CharacteristicPolynomial[e2, λ]//Simplify
```
$$Out[1522]= \frac{-11\ s^3 + s^2\ (21 - 11\ \lambda) - 10\ \lambda^2 + s\ (-10 + 9\ \lambda + 10\ \lambda^2)}{10\ (-1 + s)}$$

$$In[1523]:= \texttt{c0 = cpe2/.\lambda- > 0//Simplify}$$
$$Out[1523]= s - \frac{11\ s^2}{10}$$

$$In[1524]:= \texttt{c1 = Coefficient[cpe2, \lambda]//Simplify}$$
$$Out[1524]= \frac{(9 - 11\ s)\ s}{10\ (-1 + s)}$$

$$In[1525]:= \texttt{c2 = Coefficient[cpe2, \lambda^2]//Simplify}$$
$$Out[1525]= 1$$

Using `InequalitySolve`, we see that

1. c_0, c_1, and c_2 are positive if $9/11 < s < 10/11$, and
2. c_0 and c_2 are positive and c_1 is negative if $0 < s < 9/11$.

In[1526] := **<< Algebra`InequalitySolve`**

InequalitySolve[c0 > 0 && c1 > 0, s]

Out[1526] = $\frac{9}{11} < s < \frac{10}{11}$

In[1527] := **InequalitySolve[c0 > 0 && c1 < 0, s]**

Out[1527] = $0 < s < \frac{9}{11}$

In the first situation, E_2 is stable; in the second E_2 is unstable.

Using $s = 19/22$, we graph the direction field associated with the system as well as various solutions in Figure 6-40. In the plot, notice that all nontrivial solutions approach $E_2 \approx (.63, .27)$; E_2 is stable – a situation that cannot occur with the standard predator–prey equations.

In[1528] := **rps/.s- > 19/22//N**

Out[1528] = {{x → 0, y → 0}, {y → 0, x → 1.},
             {y → 0.268889, x → 0.633333}}

In[1529] := **<< Graphics`PlotField`**

```
pvf = PlotVectorField[α x (1 - 1/k x) - m x y/(a + x)  ,
      y  (m x/(a + x) - 19/22)}, {x, 0, 1}, {y, 0, 1},
      DisplayFunction- > Identity];
```

In[1530] := **numgraph[init_, s_, opts___] := Module[{numsol},**
```
numsol = NDSolve[{x'[t] == α x[t]
         (1 - 1/k x[t]) - m x[t] y[t]/(a + x[t]),
           y'[t] == y[t]  (m x[t]/(a + x[t]) - s),
            x[0] == init[[1]], y[0] == init[[2]]},
          {x[t], y[t]}, {t, 0, 50}];
      ParametricPlot[Evaluate[{x[t], y[t]}/.numsol],
          {t, 0, 50}, opts, DisplayFunction- > Identity]]
```

In[1531] := **i1 = Table[numgraph[{1, i}, 19/22], {i, 0, 1, 1/10}];**
```
i2 = Table[numgraph[{i, 1}, 19/22], {i, 0, 1, 1/10}];
Show[i1, i2,  pvf,
DisplayFunction- > $DisplayFunction,
  PlotRange- > {{0, 1}, {0, 1}},
  AspectRatio- > Automatic]
```

On the other hand, using $s = 8/11$ (so that E_2 is unstable) in Figure 6-41 we see that all nontrivial solutions appear to approach a limit cycle.

In[1532] := **rps/.s- > 8/11//N**

Out[1532] = {{x → 0, y → 0}, {y → 0, x → 1.},
             {y → 0.268889, x → 0.266667}}

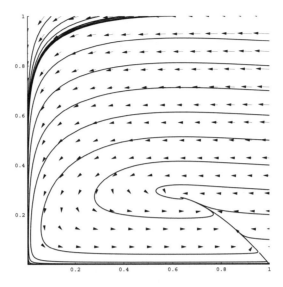

Figure 6-40 $s = 19/22$

In[1533]:= **i1 = Table[numgraph[{1, i}, 8/11], {i, 0, 1, 1/10}];**
          i2 = Table[numgraph[{i, 1}, 8/11], {i, 0, 1, 1/10}];
          p1 = Show[i1, i2, pvf,
               PlotRange- > {{0, 1}, {0, 1}},
               AspectRatio- >Automatic,
               DisplayFunction- > $DisplayFunction]

The limit cycle is shown more clearly in Figure 6-42.

In[1534]:= **numgraph[{0.759, 0.262},**
               8/11, DisplayFunction- > $DisplayFunction,
               PlotRange- > {{0, 1}, {0, 1}},
               AspectRatio- >Automatic]

Also see Example 6.4.4.

EXAMPLE 6.4.7 (Van-der-Pol's equation): In Example 6.4.4 we saw
that **Van-der-Pol's equation** $x'' + \mu(x^2 - 1)x' + x = 0$ is equivalent to
the system $\begin{cases} x' = y \\ y' = \mu(1 - x^2)y - x \end{cases}$. Classify the equilibrium points, use
NDSolve to approximate the solutions to this nonlinear system, and
plot the phase plane.

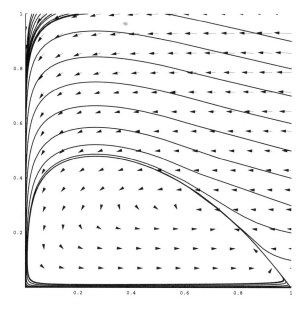

Figure 6-41 $s = 8/11$

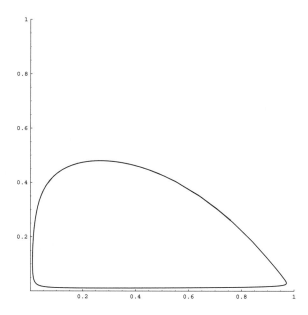

Figure 6-42   A better view of the limit cycle without the direction field

SOLUTION: We find the equilibrium points by solving $\begin{cases} y = 0 \\ \mu\left(1 - x^2\right)y - x = 0 \end{cases}$.

From the first equation, we see that $y = 0$. Then, substitution of $y = 0$ into the second equation yields $x = 0$. Therefore, the only equilibrium point is $(0, 0)$. The Jacobian matrix for this system is

$$\mathbf{J}(x, y) = \begin{pmatrix} 0 & 1 \\ -1 - 2\mu xy & -\mu\left(x^2 - 1\right) \end{pmatrix}.$$

The eigenvalues of $\mathbf{J}(0, 0)$ are $\lambda_{1,2} = \frac{1}{2}\left(\mu \pm \sqrt{\mu^2 - 4}\right)$.

```
In[1535]:= Clear[f,g]

          f[x_,y_] = y;

          g[x_,y_] = -x - μ (x² - 1) y;
```

```
In[1536]:= jac = (D[f[x,y],x]   D[f[x,y],y] );
               (D[g[x,y],x]   D[g[x,y],y] )
```

```
In[1537]:= jac /. {x- > 0, y- > 0}//Eigenvalues
```

$$Out[1537] = \left\{ \frac{1}{2}\left(\mu - \sqrt{-4 + \mu^2}\right), \frac{1}{2}\left(\mu + \sqrt{-4 + \mu^2}\right) \right\}$$

Notice that if $\mu > 2$, then both eigenvalues are positive and real. Hence, we classify $(0, 0)$ as an **unstable node**. On the other hand, if $0 < \mu < 2$, then the eigenvalues are a complex conjugate pair with a positive real part. Hence, $(0, 0)$ is an **unstable spiral**. (We omit the case $\mu = 2$ because the eigenvalues are repeated.)

We now show several curves in the phase plane that begin at various points for various values of μ. First, we define the function sol, which given μ, x_0, and y_0, generates a numerical solution to the initial-value problem

$$\begin{cases} x' = y \\ y' = \mu\left(1 - x^2\right)y - x \\ x(0) = x_0, \, y(0) = y_0 \end{cases}$$

and then parametrically graphs the result for $0 \le t \le 20$.

```
In[1538]:= Clear[sol]

        sol[μ_, {x0_, y0_}, opts___] :=
          Module[{eqone, eqtwo, solt},
            eqone = x'[t] == y[t];
        eqtwo = y'[t] == μ (1 - x[t]²) y[t] - x[t];
        solt = NDSolve[{eqone, eqtwo, x[0] == x0,
                y[0] == y0}, {x[t], y[t]}, {t, 0, 20}];
        ParametricPlot[{x[t], y[t]}/.solt,
              {t, 0, 20}, Compiled → False, opts]]
```

We then use `Table` and `Union` to generate a list of ordered pairs `initconds` that will correspond to the initial conditions in the initial-value problem.

```
In[1539]:= initconds1 = Table[{0.1 Cos[t], 0.1 Sin[t]},
              {t, 0, 2π, 2π/9}];

        initconds2 = Table[{-5, i}, {i, -5, 5, 10/9}];

        initconds3 = Table[{5, i}, {i, -5, 5, 10/9}];

        initconds4 = Table[{i, 5}, {i, -5, 5, 10/9}];

        initconds5 = Table[{i, -5}, {i, -5, 5, 10/9}];

In[1540]:= initconds = initconds1 ∪ initconds2 ∪
                initconds3 ∪ initconds4 ∪ initconds5;
```

We then use `Map` to apply `sol` to the list of ordered pairs in `initconds` for $\mu = 1/2$.

```
In[1541]:= somegraphs1 =
              Map[
                sol[1/2, #, DisplayFunction- > Identity]&,
                initconds];

In[1542]:= phase1 = Show[somegraphs1,
                PlotRange- > {{-5, 5}, {-5, 5}},
                AspectRatio- > 1, Ticks- > {{-4, 4}, {-4, 4}}];
```

Similarly, we use `Map` to apply `sol` to the list of ordered pairs in `initconds` for $\mu = 1$, $3/2$, and 3.

```
In[1543]:= somegraphs2 =
              Map[sol[1, #, DisplayFunction- > Identity]&,
                initconds];
```

```
In[1544]:= phase2 = Show[somegraphs2,
                PlotRange- > {{-5, 5}, {-5, 5}},
                AspectRatio- > 1, Ticks- > {{-4, 4}, {-4, 4}}];
```

```
In[1545]:= somegraphs3 =
              Map[
                sol[3/2, #, DisplayFunction- > Identity]&,
                initconds];
```

```
In[1546]:= phase3 = Show[somegraphs3,
                PlotRange- > {{-5, 5}, {-5, 5}},
                AspectRatio- > 1, Ticks- > {{-4, 4}, {-4, 4}}];
```

```
In[1547]:= somegraphs4 =
              Map[sol[3, #, DisplayFunction- > Identity]&,
                initconds];
```

```
In[1548]:= phase4 = Show[somegraphs3,
                PlotRange- > {{-5, 5}, {-5, 5}},
                AspectRatio- > 1, Ticks- > {{-4, 4}, {-4, 4}}];
```

We now show all four graphs together in Figure 6-43. In each figure, we see that all of the curves approach a curve called a *limit cycle*. Physically, the fact that the system has a limit cycle indicates that for all oscillations, the motion eventually becomes periodic, which is represented by a closed curve in the phase plane.

```
In[1549]:= Show[GraphicsArray[
                {{phase1, phase2}, {phase3, phase4}}]]
```

On the other hand, in Figure 6-43 we graph the solutions that satisfy the initial conditions $x(0) = 1$ and $y(0) = 0$ parametrically and individually for various values of μ. Notice that for small values of μ the system more closely approximates that of the harmonic oscillator because the damping coefficient is small. The curves are more circular than those for larger values of μ.

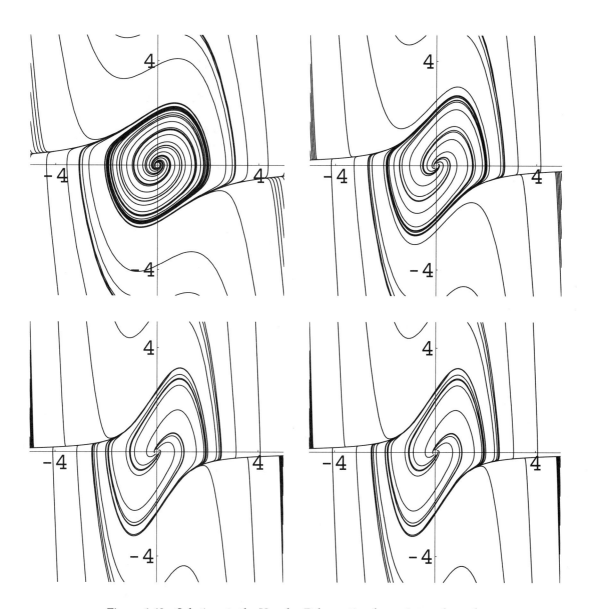

Figure 6-43   Solutions to the Van-der-Pol equation for various values of μ

```
In[1550]:= Clear[x, y, t, s]
           graph[μ_] := Module[{numsol, pp, pxy},
               numsol =
                 NDSolve[{x'[t] == y[t],
                     y'[t] == μ (1 - x[t]^2) y[t] - x[t], x[0] == 1,
                     y[0] == 0}, {x[t], y[t]}, {t, 0, 20}];
               pp = ParametricPlot[{x[t], y[t]} /. numsol,
                   {t, 0, 20}, Compiled → False,
                   PlotRange → {{-5, 5}, {-5, 5}},
                   AspectRatio → 1, Ticks → {{-4, 4}, {-4, 4}},
                   DisplayFunction → Identity];
               pxy = Plot[Evaluate[{x[t], y[t]} /. numsol],
                   {t, 0, 20},
                   PlotStyle → {GrayLevel[0],
                       GrayLevel[0.5]}, PlotRange → {-5, 5},
                   AspectRatio → 1,
                   Ticks- > {{5, 10, 15}, {-4, 4}},
                   DisplayFunction → Identity];
               GraphicsArray[{pxy, pp}]]
In[1551]:= graphs = Table[graph[i], {i, 0.25, 3, 2.75/9}];

In[1552]:= toshow = Partition[graphs, 2];

           Show[GraphicsArray[toshow]]
```

■

6.5 Some Partial Differential Equations

6.5.1 The One-Dimensional Wave Equation

Suppose that we pluck a string (like a guitar or violin string) of length p and constant mass density that is fixed at each end. A question that we might ask is: What is the position of the string at a particular instance of time? We answer this question by modeling the physical situation with a partial differential equation, namely the wave equation in one spatial variable:

$$c^2 \frac{\partial^2 u}{\partial x^2} = \frac{\partial^2 u}{\partial t^2} \qquad \text{or} \qquad c^2 u_{xx} = u_{tt}. \tag{6.40}$$

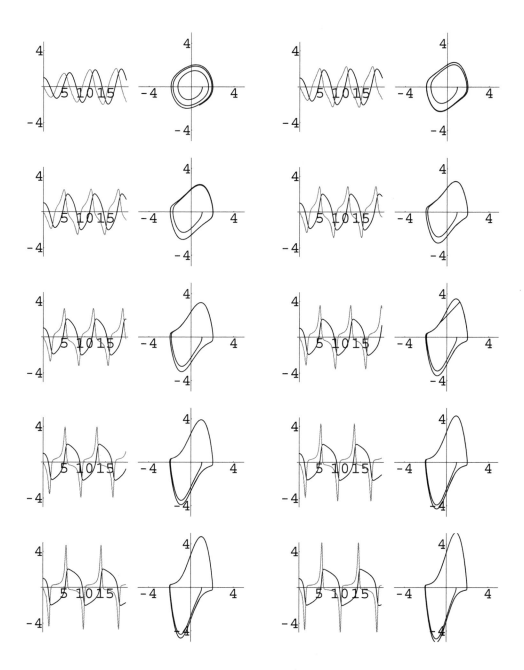

Figure 6-44 The solutions to the Van-der-Pol equation satisfying $x(0) = 1$ and $y(0) = 0$ individually (x in black and y in gray) for various values of μ

In equation (6.40), $c^2 = T/\rho$, where T is the tension of the string and ρ is the constant mass of the string per unit length. The solution $u(x, t)$ represents the displacement of the string from the x-axis at time t. To determine u we must describe the boundary and initial conditions that model the physical situation. At the ends of the string, the displacement from the x-axis is fixed at zero, so we use the homogeneous boundary conditions $u(0, t) = u(p, t) = 0$ for $t > 0$. The motion of the string also depends on the displacement and the velocity at each point of the string at $t = 0$. If the initial displacement is given by $f(x)$ and the initial velocity by $g(x)$, we have the initial conditions $u(x, 0) = f(x)$ and $u_t(x, 0) = g(x)$ for $0 \le x \le p$. Therefore, we determine the displacement of the string with the initial-boundary value problem

$$\begin{cases} c^2 \dfrac{\partial^2 u}{\partial x^2} = \dfrac{\partial^2 u}{\partial t^2}, \ 0 < x < p, \ t > 0 \\ u(0, t) = u(p, t) = 0, \ t > 0 \\ u(x, 0) = f(x), \ u_t(x, 0) = g(x), \ 0 < x < p. \end{cases} \tag{6.41}$$

λ is a constant.

This problem is solved through separation of variables by assuming that $u(x, t) = X(x)T(t)$. Substitution into equation (6.40) yields

$$c^2 X'' T = X T'' \qquad \text{or} \qquad \frac{X''}{X} = \frac{T''}{c^2 T} = -\lambda$$

so we obtain the two second-order ordinary differential equations $X'' + \lambda X = 0$ and $T'' + c^2 \lambda T = 0$. At this point, we solve the equation that involves the homogeneous boundary conditions. The boundary conditions in terms of $u(x, t) = X(x)T(t)$ are $u(0, t) = X(0)T(t) = 0$ and $u(p, t) = X(p)T(t) = 0$, so we have $X(0) = 0$ and $X(p) = 0$. Therefore, we determine $X(x)$ by solving the *eigenvalue problem*

$$\begin{cases} X'' + \lambda X = 0, \ 0 < x < p \\ X(0) = X(p) = 0. \end{cases}$$

The eigenvalues of this problem are $\lambda_n = (n\pi/p)^2$, $n = 1, 3, \ldots$ with corresponding eigenfunctions $X_n(x) = \sin(n\pi x/p)^2$, $n = 1, 3, \ldots$. Next, we solve the equation $T'' + c^2 \lambda_n T = 0$. A general solution is

$$T_n(t) = a_n \cos\left(c\sqrt{\lambda_n}t\right) + b_n \sin\left(c\sqrt{\lambda_n}t\right) = a_n \cos\frac{cn\pi t}{p} + b_n \sin\frac{cn\pi t}{p},$$

where the coefficients a_n and b_n must be determined. Putting this information together, we obtain

$$u_n(x, t) = \left(a_n \cos\frac{cn\pi t}{p} + b_n \sin\frac{cn\pi t}{p}\right)\sin\frac{n\pi x}{p},$$

so by the Principle of Superposition, we have

$$u(x, t) = \sum_{n=1}^{\infty} \left(a_n \cos \frac{c n \pi t}{p} + b_n \sin \frac{c n \pi t}{p} \right) \sin \frac{n \pi x}{p}.$$

Applying the initial displacement $u(x, 0) = f(x)$ yields

$$u(x, 0) = \sum_{n=1}^{\infty} a_n \sin \frac{n \pi x}{p} = f(x),$$

so a_n is the *Fourier sine series coefficient* for $f(x)$, which is given by

$$a_n = \frac{2}{p} \int_0^p f(x) \sin \frac{n \pi x}{p} \, dx, \quad n = 1, 2, \dots .$$

In order to determine b_n, we must use the initial velocity. Therefore, we compute

$$\frac{\partial u}{\partial t}(x, t) = \sum_{n=1}^{\infty} \left(-a_n \frac{c n \pi}{p} \sin \frac{c n \pi t}{p} + b_n \frac{c n \pi}{p} \cos \frac{c n \pi t}{p} \right) \sin \frac{n \pi x}{p}.$$

Then,

$$\frac{\partial u}{\partial t}(x, 0) = \sum_{n=1}^{\infty} b_n \frac{c n \pi}{p} \sin \frac{n \pi x}{p} = g(x)$$

so $b_n \frac{c n \pi}{p}$ represents the Fourier sine series coefficient for $g(x)$ which means that

$$b_n = \frac{p}{c n \pi} \int_0^p g(x) \sin \frac{n \pi x}{p} \, dx, \quad n = 1, 2, \dots .$$

EXAMPLE 6.5.1: Solve $\begin{cases} u_{xx} = u_{tt}, \ 0 < x < 1, \ t > 0 \\ u(0, t) = u(1, t) = 0, \ t > 0 \\ u(x, 0) = \sin \pi x, \ u_t(x, 0) = 3x + 1, \ 0 < x < 1. \end{cases}$

SOLUTION: The initial displacement and velocity functions are defined first.

> $In[1553] := $ **f[x_] = Sin[π x];**

> **g[x_] = 3x + 1;**

Next, the functions to determine the coefficients a_n and b_n in the series approximation of the solution $u(x, t)$ are defined. Here, $p = c = 1$.

> $In[1554] := $ **a$_1$ = 2 $\int_0^1$ f[x] Sin[π x] dx**

Out[1554]= 1

$In[1555]:= \mathbf{a_{n_{-}}} = 2 \int_{0}^{1} \mathbf{f[x] \ Sin[n \, \pi \, x] dx}$

$Out[1555]= \dfrac{2 \ Sin[n \, \pi]}{\pi - n^2 \, \pi}$

$In[1556]:= \mathbf{b_{n_{-}}} = \dfrac{2 \int_{0}^{1} \mathbf{g[x] \ Sin[n \, \pi \, x] dx}}{\mathbf{n \, \pi}} //\mathbf{Simplify}$

$Out[1556]= \dfrac{2 \, n \, \pi - 8 \, n \, \pi \ Cos[n \, \pi] + 6 \ Sin[n \, \pi]}{n^3 \, \pi^3}$

Because *n* represents an integer, these results indicate that $a_n = 0$ for all $n \geq 2$. We use Table to calculate the first ten values of b_n.

$In[1557]:= \mathbf{Table[\{n, b_n, b_n//N\}, \{n, 1, 10\}]//TableForm}$

Out[1557]=

1	$\dfrac{10}{\pi^2}$	1.01321
2	$-\dfrac{3}{2 \, \pi^2}$	-0.151982
3	$\dfrac{10}{9 \, \pi^2}$	0.112579
4	$-\dfrac{3}{8 \, \pi^2}$	-0.0379954
5	$\dfrac{2}{5 \, \pi^2}$	0.0405285
6	$-\dfrac{1}{6 \, \pi^2}$	-0.0168869
7	$\dfrac{10}{49 \, \pi^2}$	0.0206778
8	$-\dfrac{3}{32 \, \pi^2}$	-0.00949886
9	$\dfrac{10}{81 \, \pi^2}$	0.0125088
10	$-\dfrac{3}{50 \, \pi^2}$	-0.00607927

Notice that we define uapprox[n] so that Mathematica "remembers" the terms uapprox that are computed. That is, Mathematica does not need to recompute uapprox[n-1] to compute uapprox[n] provided that uapprox[n-1] has already been computed.

$In[1558]:= \mathbf{Clear[u, uapprox]}$

The function u defined next computes the *n*th term in the series expansion. Thus, uapprox determines the approximation of order *k* by summing the first *k* terms of the expansion, as illustrated with approx[10].

$In[1559]:= \mathbf{u[n_{-}]} = \mathbf{b_n \ Sin[n \, \pi \, t] \ Sin[n \, \pi \, x]};$

$In[1560]:= \mathbf{uapprox[k_{-}]} := \mathbf{uapprox[k]} = \mathbf{uapprox[k - 1] + u[k]};$
$\qquad\qquad \mathbf{uapprox[0]} = \mathbf{Cos[\pi \, t] \ Sin[\pi \, x]};$

$In[1561]:= \mathbf{uapprox[10]}$

Out [1561]= $\mathrm{Cos}[\pi t]\ \mathrm{Sin}[\pi x]$

$$+\frac{10\ \mathrm{Sin}[\pi t]\ \mathrm{Sin}[\pi x]}{\pi^2}-\frac{3\ \mathrm{Sin}[2\pi t]\ \mathrm{Sin}[2\pi x]}{2\pi^2}$$

$$+\frac{10\ \mathrm{Sin}[3\pi t]\ \mathrm{Sin}[3\pi x]}{9\pi^2}-\frac{3\ \mathrm{Sin}[4\pi t]\ \mathrm{Sin}[4\pi x]}{8\pi^2}$$

$$+\frac{2\ \mathrm{Sin}[5\pi t]\ \mathrm{Sin}[5\pi x]}{5\pi^2}-\frac{\mathrm{Sin}[6\pi t]\ \mathrm{Sin}[6\pi x]}{6\pi^2}$$

$$+\frac{10\ \mathrm{Sin}[7\pi t]\ \mathrm{Sin}[7\pi x]}{49\pi^2}-\frac{3\ \mathrm{Sin}[8\pi t]\ \mathrm{Sin}[8\pi x]}{32\pi^2}$$

$$+\frac{10\ \mathrm{Sin}[9\pi t]\ \mathrm{Sin}[9\pi x]}{81\pi^2}-\frac{3\ \mathrm{Sin}[10\pi t]\ \mathrm{Sin}[10\pi x]}{50\pi^2}$$

To illustrate the motion of the string, we graph uapprox[10], the tenth partial sum of the series, on the interval [0, 1] for 16 equally spaced values of *t* between 0 and 2 in Figure 6-45.

In[1562]:= **somegraphs =**
    Table$\Big[$Plot[Evaluate[uapprox[10]], {x, 0, 1},
        DisplayFunction → Identity,
        PlotRange → {-3/2, 3/2},
        Ticks → {{0, 1}, {-1, 1}}], $\Big\{$t, 0, 2, $\frac{2}{15}\Big\}\Big]$;

    toshow = Partition[somegraphs, 4];

    Show[GraphicsArray[toshow]]

If instead we wished to see the motion of the string , we can use a Do loop to generate many graphs and animate the result. We show a frame from the resulting animation.

In[1563]:= **Do$\Big[$Plot[Evaluate[uapprox[10]], {x, 0, 1},**
        PlotRange → {-3/2, 3/2},
        Ticks → {{0, 1}, {-1, 1}}], $\Big\{$t, 0, 2, $\frac{2}{59}\Big\}\Big]$;

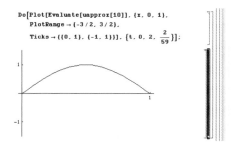

■

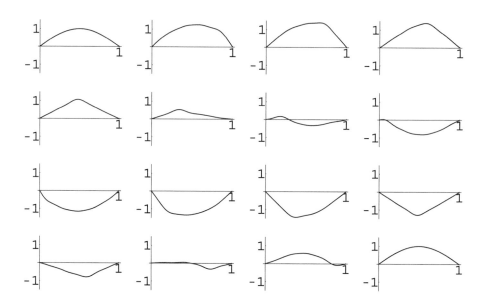

Figure 6-45   The motion of the spring for 16 equally spaced values of t between 0 and 2

6.5.2 The Two-Dimensional Wave Equation

One of the more interesting problems involving two spatial dimensions (x and y) is
the wave equation. The two-dimensional wave equation in a circular region which
is radially symmetric (not dependent on θ) with boundary and initial conditions is
expressed in polar coordinates as

$$
\begin{cases}
c^2 \left(\dfrac{\partial^2 u}{\partial r^2} + \dfrac{1}{r} \dfrac{\partial u}{\partial r} \right) = \dfrac{\partial^2 u}{\partial t^2}, \ 0 < r < \rho, \ t > 0 \\[2mm]
u(\rho, t) = 0, \ |u(0, t)| < \infty, \ t > 0 \\[2mm]
u(r, 0) = f(r), \ \dfrac{\partial u}{\partial t}(r, 0) = g(r), \ 0 < r < \rho.
\end{cases}
$$

Notice that the boundary condition $u(\rho, t) = 0$ indicates that u is fixed at zero
around the boundary; the condition $|u(0, t)| < \infty$ indicates that the solution is
bounded at the center of the circular region. Like the wave equation discussed
previously, this problem is typically solved through separation of variables by assuming a solution of the form $u(r, t) = F(r)G(t)$. Applying separation of variables
yields the solution

$$
u(r, t) = \sum_{n=1}^{\infty} (A_n \cos c k_n t + B_n \sin c k_n t) J_0(k_n r),
$$

where $\lambda_n = c\alpha_n/\rho$, and the coefficients A_n and B_n are found through application of the initial displacement and velocity functions. With

α_n represents the nth zero of the Bessel function of the first kind of order zero.

$$u(r, 0) = \sum_{n=1}^{\infty} A_n J_0(k_n r) = f(r)$$

and the orthogonality conditions of the Bessel functions, we find that

$$A_n = \frac{\int_0^\rho r f(r) J_0(k_n r) \, dr}{\int_0^\rho r [J_0(k_n r)]^2 \, dr} = \frac{2}{[J_1(\alpha_n)]^2} \int_0^\rho r f(r) J_0(k_n r) \, dr, \ n = 1, 2, \ldots.$$

Similarly, because

$$\frac{\partial u}{\partial t}(r, 0) = \sum_{n=1}^{\infty} (-ck_n A_n \sin ck_n t + ck_n B_n \cos ck_n t) J_0(k_n r)$$

we have

$$u_t(r, 0) = \sum_{n=1}^{\infty} ck_n B_n J_0(k_n r) = g(r).$$

Therefore,

$$B_n = \frac{\int_0^\rho r g(r) J_0(k_n r) \, dr}{ck_n \int_0^\rho r [J_0(k_n r)]^2 \, dr} = \frac{2}{ck_n [J_1(\alpha_n)]^2} \int_0^\rho r g(r) J_0(k_n r) \, dr, \ n = 1, 2, \ldots.$$

As a practical matter, in nearly all cases, these formulas are difficult to evaluate.

EXAMPLE 6.5.2: Solve $\begin{cases} \dfrac{\partial^2 u}{\partial r^2} + \dfrac{1}{r}\dfrac{\partial u}{\partial r} = \dfrac{\partial^2 u}{\partial t^2}, \ 0 < r < 1, \ t > 0 \\ u(1, t) = 0, \ |u(0, t)| < \infty, \ t > 0 \\ u(r, 0) = r(r - 1), \ \dfrac{\partial u}{\partial t}(r, 0) = \sin \pi r, \ 0 < r < 1. \end{cases}$

SOLUTION: In this case, $\rho = 1$, $f(r) = r(r - 1)$, and $g(r) = \sin \pi r$. To calculate the coefficients, we will need to have approximations of the zeros of the Bessel functions, so we load the **BesselZeros** package, which is contained in the **NumericalMath** folder (or directory) and define α_n to be the nth zero of $y = J_0(x)$.

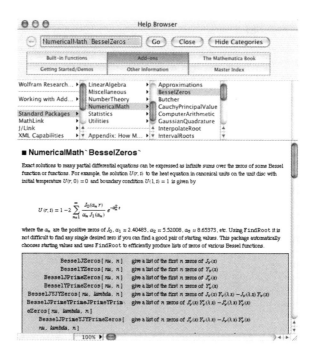

$In[1564] := $ **<< NumericalMath`BesselZeros`**

$In[1565] := $ **$\alpha_{n\_}$:= α_n = BesselJZeros[0, {n, n}][[1]]**

Next, we define the constants ρ and c and the functions $f(r) = r(r - 1)$, $g(r) = \sin \pi r$, and $k_n = \alpha_n / \rho$.

$In[1566] := $ **c = 1;**

ρ **= 1;**

f[r_] = r (r - 1);

g[r_] = Sin[π r];

$k_{n\_}$:= k_n = $\dfrac{\alpha_n}{\rho}$;

The formulas for the coefficients A_n and B_n are then defined so that an approximate solution may be determined. (We use lowercase letters to avoid any possible ambiguity with built-in *Mathematica* functions.) Note that we use NIntegrate to approximate the coefficients and avoid the difficulties in integration associated with the presence of the Bessel function of order zero.

```
In[1567] := aₙ_ :=
              aₙ = (2 NIntegrate[r f[r] BesselJ[0, kₙ r],
                    {r, 0, ρ}])/BesselJ[1, αₙ]²;

In[1568] := bₙ_ :=
              bₙ = (2 NIntegrate[r g[r] BesselJ[0, kₙ r],
                    {r, 0, ρ}])/(c kₙ BesselJ[1, αₙ]²)
```

We now compute the first ten values of A_n and B_n. Because a and b are defined using the form $a_{n\_} := a_n = \ldots$ and $b_{n\_} := b_n = \ldots$, Mathematica remembers these values for later use.

```
In[1569] := Table[{n, aₙ, bₙ}, {n, 1, 10}]//TableForm
            1    1              0.52118
            2    0.208466      -0.145776
            3    0.00763767    -0.0134216
            4    0.0383536     -0.00832269
Out[1569] = 5    0.00534454    -0.00250503
            6    0.0150378     -0.00208315
            7    0.00334937    -0.000882012
            8    0.00786698    -0.000814719
            9    0.00225748    -0.000410202
            10   0.00479521    -0.000399219
```

The nth term of the series solution is defined in u. Then, an approximate solution is obtained in uapprox by summing the first ten terms of u.

```
In[1570] := u[n_, r_, t_] := (aₙ Cos[c kₙ t] + bₙ Sin[c kₙ t])
              BesselJ[0, kₙ r];
```

$$In[1571] := \text{uapprox}[r\_, t\_] = \sum_{n=1}^{10} u[n, r, t];$$

We graph uapprox for several values of t in Figure 6-46.

```
In[1572] := somegraphs =
              Table[ParametricPlot3D[
                {r Cos[θ], r Sin[θ], uapprox[r, t]}, {r, 0, 1},
                {θ, -π, π}, Boxed → False,
                PlotRange → {-1.25, 1.25},
                BoxRatios → {1, 1, 1},
                Ticks → {{-1, 1}, {-1, 1}, {-1, 1}},
                DisplayFunction → Identity], {t, 0, 1.5, 1.5/8}];

            toshow = Partition[somegraphs, 3];

            Show[GraphicsArray[toshow]]
```

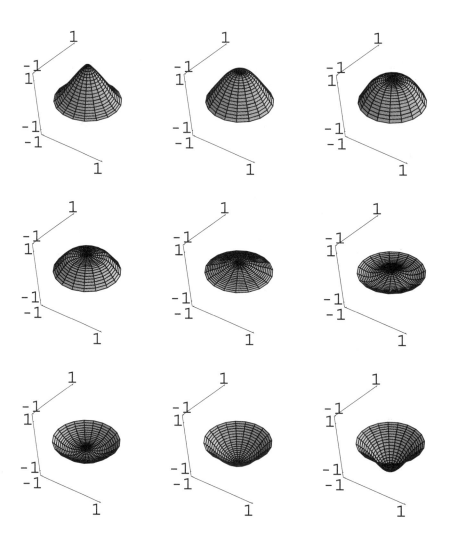

Figure 6-46   The drumhead for nine equally spaced values of t between 0 and 1.5

In order to actually watch the drumhead move, we can use a Do loop to generate several graphs and animate the result. Be aware, however, that generating many three-dimensional graphics and then animating the results uses a great deal of memory and can take considerable time, even on a relatively powerful computer. We show one frame from the animation that results from the following Do loop.

$In[1573]:=$ `Do[ParametricPlot3D[`
      `{r Cos[θ], r Sin[θ], uapprox[r, t]}, {r, 0, 1},`
      `{θ, -π, π}, Boxed → False,`
      `PlotRange → {-1.25, 1.25}, BoxRatios → {1, 1, 1},`
      `Ticks → {{-1, 1}, {-1, 1}, {-1, 1}}],`
      $\left\{t, 0, 1.5, \dfrac{1.5}{15}\right\}];$

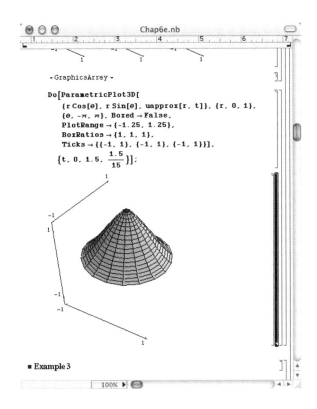

■ Example 3

■

If the displacement of the drumhead is not radially symmetric, the problem that describes the displacement of a circular membrane in its general case is

$$\begin{cases} c^2\left(\dfrac{\partial^2 u}{\partial r^2} + \dfrac{1}{r}\dfrac{\partial u}{\partial r} + \dfrac{1}{r^2}\dfrac{\partial^2 u}{\partial \theta^2}\right) = \dfrac{\partial^2 u}{\partial t^2}, \ 0 < r < \rho, \ -\pi < \theta < \pi, \ t > 0 \\ u(\rho, \theta, t) = 0, \ |u(0, \theta, t)| < \infty, \ -\pi \le \theta \le \pi, \ t > 0 \\ u(r, \pi, t) = u(r, -\pi, t), \ \dfrac{\partial u}{\partial \theta}(r, \pi, t) = \dfrac{\partial u}{\partial \theta}(r, -\pi, t), \ 0 < r < \rho, \ t > 0 \\ u(r, \theta, 0) = f(r, \theta), \ \dfrac{\partial u}{\partial t}(r, \pi, 0) = g(r, \theta), \ 0 < r < \rho, \ -\pi < \theta < \pi. \end{cases}$$

(6.42)

Using separation of variables and assuming that $u(r, \theta, t) = R(t)H(\theta)T(t)$, we obtain that a general solution is given by

$$u(r, \theta, t) = \sum_{n} a_{0n} J_0 (\lambda_{0n} r) \cos (\lambda_{0n} ct) + \sum_{m,n} a_{mn} J_m (\lambda_{mn} r) \cos (m\theta) \cos (\lambda_{mn} ct) +$$

$$\sum_{m,n} b_{mn} J_m (\lambda_{mn} r) \sin (m\theta) \cos (\lambda_{mn} ct) + \sum_{n} A_{0n} J_0 (\lambda_{0n} r) \sin (\lambda_{0n} ct) +$$

$$\sum_{m,n} A_{mn} J_m (\lambda_{mn} r) \cos (m\theta) \sin (\lambda_{mn} ct) +$$

$$\sum_{m,n} B_{mn} J_m (\lambda_{mn} r) \sin (m\theta) \sin (\lambda_{mn} ct),$$

where J_m represents the mth Bessel function of the first kind, α_{mn} denotes the nth zero of the Bessel function $y = J_m(x)$, and $\lambda_{mn} = \alpha_{mn}/\rho$. The coefficients are given by the following formulas.

$$a_{0n} = \frac{\int_0^{2\pi} \int_0^\rho f(r, \theta) J_0 (\lambda_{0n} r)\, r \, dr \, d\theta}{2\pi \int_0^\rho [J_0 (\lambda_{0n} r)]^2 \, r \, dr}$$

$$a_{mn} = \frac{\int_0^{2\pi} \int_0^\rho f(r, \theta) J_m (\lambda_{mn} r) \cos (m\theta)\, r \, dr \, d\theta}{\pi \int_0^\rho [J_m (\lambda_{mn} r)]^2 \, r \, dr}$$

$$b_{mn} = \frac{\int_0^{2\pi} \int_0^\rho f(r, \theta) J_m (\lambda_{mn} r) \sin (m\theta)\, r \, dr \, d\theta}{\pi \int_0^\rho [J_m (\lambda_{mn} r)]^2 \, r \, dr}$$

$$A_{0n} = \frac{\int_0^{2\pi} \int_0^\rho g(r, \theta) J_0 (\lambda_{0n} r)\, r \, dr \, d\theta}{2\pi \lambda_{0n} c\pi \int_0^\rho [J_0 (\lambda_{0n} r)]^2 \, r \, dr}$$

$$A_{mn} = \frac{\int_0^{2\pi} \int_0^\rho g(r, \theta) J_m (\lambda_{mn} r) \cos (m\theta)\, r \, dr \, d\theta}{\pi \lambda_{mn} c \int_0^\rho [J_m (\lambda_{mn} r)]^2 \, r \, dr}$$

$$B_{mn} = \frac{\int_0^{2\pi} \int_0^\rho g(r, \theta) J_m (\lambda_{mn} r) \sin (m\theta)\, r \, dr \, d\theta}{\pi \lambda_{mn} c \int_0^\rho [J_m (\lambda_{mn} r)]^2 \, r \, dr}$$

EXAMPLE 6.5.3: Solve $\begin{cases} 10^2 \left(\dfrac{\partial^2 u}{\partial r^2} + \dfrac{1}{r} \dfrac{\partial u}{\partial r} + \dfrac{1}{r^2} \dfrac{\partial^2 u}{\partial \theta^2} \right) = \dfrac{\partial^2 u}{\partial t^2}, \\[2mm] 0 < r < 1, \ -\pi < \theta < \pi, \ t > 0 \\[2mm] u(1, \theta, t) = 0, \ |u(0, \theta, t)| < \infty, \ -\pi \le \theta \le \pi, \ t > 0 \\[2mm] u(r, \pi, t) = u(r, -\pi, t), \ \dfrac{\partial u}{\partial \theta}(r, \pi, t) = \dfrac{\partial u}{\partial \theta}(r, -\pi, t). \\[2mm] 0 < r < 1, \ t > 0 \\[2mm] u(r, \theta, 0) = \cos (\pi r/2) \sin \theta, \\[2mm] \dfrac{\partial u}{\partial t}(r, \pi, 0) = (r - 1) \cos (\pi \theta/2), \ 0 < r < 1, \ -\pi < \theta < \pi \end{cases}$

SOLUTION: To calculate the coefficients, we will need to have approximations of the zeros of the Bessel functions, so we load the **BesselZeros** package, which is contained in the **NumericalMath** folder (or directory) and define α_{mn} to be the nth zero of $y = J_m(x)$. We illustrate the use of α_{mn} by using it to compute the first five zeros of $y = J_0(x)$.

```
In[1574] := << NumericalMath`BesselZeros`
```

```
In[1575] := α_{m_,n_} := α_{m,n} = BesselJZeros[m, {n, n}][[1]]
```

```
In[1576] := Table[α_{0,n}, {n, 1, 5}]
Out[1576] = {2.40483, 5.52008, 8.65373, 11.7915, 14.9309}
```

The appropriate parameter values as well as the initial condition functions are defined as follows. Notice that the functions describing the initial displacement and velocity are defined as the product of functions. This enables the subsequent calculations to be carried out using NIntegrate.

```
In[1577] := Clear[a, f, f1, f2, g1, g2, A, c, g, capa, capb, b]

            c = 10;

            ρ = 1;

            f1[r_] = Cos[π r / 2];

            f2[θ_] = Sin[θ];

            f[r_, θ_] := f[r, θ] = f1[r] f2[θ];

            g1[r_] = r - 1;

            g2[θ_] = Cos[π θ / 2];

            g[r_, θ_] := g[r, θ] = g1[r] g2[θ];
```

The coefficients a_{0n} are determined with the function a.

```
In[1578] := Clear[a]
```

$In[1579] := $ **a[n_] :=**
    a[n] = N[

      (NIntegrate[f1[r] BesselJ[0, $\alpha_{0,n}$ r] r,
        {r, 0, ρ}] NIntegrate[f2[t], {t, 0, 2π}])/
      (2π NIntegrate[r BesselJ[0, $\alpha_{0,n}$ r]^2,
      {r, 0, ρ}])];

Hence, as represents a table of the first five values of a_{0n}. Chop is used
to round off very small numbers to zero.

$In[1580] := $ **as = Table[a[n]//Chop, {n, 1, 5}]**
$Out[1580] = $ {0, 0, 0, 0, 0}

Because the denominator of each integral formula used to find a_{mn} and
b_{mn} is the same, the function bjmn which computes this value is defined
next. A table of nine values of this coefficient is then determined.

$In[1581] := $ **bjmn[m_, n_] :=**
    bjmn[m, n] =
    N[NIntegrate[r BesselJ[m, $\alpha_{m,n}$ r]^2, {r, 0, ρ}]]

    Table[Chop[bjmn[m, n]], {m, 1, 3}, {n, 1, 3}]
$Out[1581] = $ {{0.0811076, 0.0450347, 0.0311763},
     {0.0576874, 0.0368243, 0.0270149},
     {0.0444835, 0.0311044, 0.0238229}}

We also note that in evaluating the numerators of a_{mn} and b_{mn} we must
compute $\int_0^\rho r f_1(r) J_m (\alpha_{mn}r)\, dr$. This integral is defined in fbjmn and the
corresponding values are found for $n = 1, 2, 3$ and $m = 1, 2, 3$.

$In[1582] := $ **Clear[fbjmn]**

    fbjmn[m_, n_] :=
    fbjmn[m, n] =
    N[NIntegrate[f1[r] BesselJ[m, $\alpha_{m,n}$ r] r,
    {r, 0, ρ}]]

    Table[Chop[fbjmn[m, n]], {m, 1, 3}, {n, 1, 3}]
$Out[1582] = $ {{0.103574, 0.020514, 0.0103984},
     {0.0790948, 0.0275564, 0.0150381},
     {0.0628926, 0.0290764, 0.0171999}}

The formula to compute a_{mn} is then defined and uses the information
calculated in fbjmn and bjmn. As in the previous calculation, the coefficient values for $n = 1, 2, 3$ and $m = 1, 2, 3$ are determined.

```
In[1583]:= a[m_, n_] :=
              a[m, n] =
                N[
                    (fbjmn[m, n] NIntegrate[f2[t] Cos[m t],
                        {t, 0, 2π}])/(π bjmn[m, n])];

          Table[Chop[a[m, n]], {m, 1, 3}, {n, 1, 3}]
Out[1583]= {{0, 0, 0}, {0, 0, 0}, {0, 0, 0}}
```

A similar formula is then defined for the computation of b_{mn}.

```
In[1584]:= b[m_, n_] :=
              b[m, n] =
                N[
                    (fbjmn[m, n] NIntegrate[f2[t] Sin[m t],
                        {t, 0, 2π}])/(π bjmn[m, n])];

          Table[Chop[b[m, n]], {m, 1, 3}, {n, 1, 3}]
Out[1584]= {{1.277, 0.455514, 0.333537}, {0, 0, 0}, {0, 0, 0}}
```

Note that defining the coefficients in this manner `a[m_,n_]:=`
`a[m,n]=...` and `b[m_,n_]:=b[m,n]=...` so that Mathematica "remembers" previously computed values which reduces computation time.
The values of A_{0n} are found similarly to those of a_{0n}. After defining the
function `capa` to calculate these coefficients, a table of values is then
found.

```
In[1585]:= capa[n_] :=
              capa[n] =
                N[
                    (NIntegrate[g1[r] BesselJ[0, α_{0,n} r] r, {r, 0, ρ}]
                        NIntegrate[g2[t], {t, 0, 2π}])/
                    (2π c α_{0,n} NIntegrate[r BesselJ[0, α_{0,n} r]²,
                        {r, 0, ρ}])];

          Table[Chop[capa[n]], {n, 1, 6}]
Out[1585]= {0.00142231, 0.0000542518, 0.0000267596,
              6.41976 × 10⁻⁶, 4.95843 × 10⁻⁶, 1.88585 × 10⁻⁶}
```

The value of the integral of the component of g, `g1`, which depends on
r and the appropriate Bessel functions, is defined as `gbjmn`.

```
In[1586]:= gbjmn[m_, n_] := gbjmn[m, n] = NIntegrate[g1[r] *
              BesselJ[m, α_{m,n} r] r, {r, 0, ρ}]//N

          Table[gbjmn[m, n]//Chop, {m, 1, 3}, {n, 1, 3}]
```

Out[1586]= {{-0.0743906, -0.019491, -0.00989293},
        {-0.0554379, -0.0227976, -0.013039},
        {-0.0433614, -0.0226777, -0.0141684}}

Then, A_{mn} is found by taking the product of integrals, gbjmn depending on r and one depending on θ. A table of coefficient values is generated in this case as well.

In[1587]:= **capa[m_, n_] :=**
          capa[m, n] =
            N[
                (gbjmn[m, n] NIntegrate[g2[t] Cos[mt],
                        {t, 0, 2π}])/($\pi \alpha_{m,n}$ c bjmn[m, n])];

        Table[Chop[capa[m, n]], {m, 1, 3}, {n, 1, 3}]

Out[1587]= {{0.0035096, 0.000904517, 0.000457326},
        {-0.00262692, -0.00103252, -0.000583116},
        {-0.000503187, -0.000246002, -0.000150499}}

Similarly, the B_{mn} are determined.

In[1588]:= **capb[m_, n_] :=**
          capb[m, n] =
            N[
                (gbjmn[m, n] NIntegrate[g2[t] Sin[mt],
                        {t, 0, 2π}])/($\pi \alpha_{m,n}$ c bjmn[m, n])];

        Table[Chop[capb[m, n]], {m, 1, 3}, {n, 1, 3}]

Out[1588]= {{0.00987945, 0.00254619, 0.00128736},
        {-0.0147894, -0.00581305, -0.00328291},
        {-0.00424938, -0.00207747, -0.00127095}}

Now that the necessary coefficients have been found, we construct an approximate solution to the wave equation by using our results. In the following, term1 represents those terms of the expansion involving a_{0n}, term2 those terms involving a_{mn}, term3 those involving b_{mn}, term4 those involving A_{0n}, term5 those involving A_{mn}, and term6 those involving B_{mn}.

```
In[1589] := Clear[term1, term2, term3, term4, term5, term6]

    term1[r_, t_, n_] :=
      a[n] BesselJ[0, α0,n r] Cos[α0,n c t];

    term2[r_, t_, θ_, m_, n_] :=
      a[m, n] BesselJ[m, αm,n r] Cos[m θ] Cos[αm,n c t];

    term3[r_, t_, θ_, m_, n_] :=
      b[m, n] BesselJ[m, αm,n r] Sin[m θ] Cos[αm,n c t];

    term4[r_, t_, n_] :=
      capa[n] BesselJ[0, α0,n r] Sin[α0,n c t];

    term5[r_, t_, θ_, m_, n_] :=
      capa[m, n] BesselJ[m, αm,n r] Cos[m θ] Sin[αm,n c t];

    term6[r_, t_, θ_, m_, n_] :=
      capb[m, n] BesselJ[m, αm,n r] Sin[m θ] Sin[αm,n c t];
```

Therefore, our approximate solution is given as the sum of these terms as computed in u.

```
In[1590] := Clear[u]
```

$$u[r\_, t\_, th\_] :=$$

$$\sum_{n=1}^{5} term1[r, t, n] + \sum_{m=1}^{3} \sum_{n=1}^{3} term2[r, t, th, m, n] +$$

$$\sum_{m=1}^{3} \sum_{n=1}^{3} term3[r, t, th, m, n] + \sum_{n=1}^{5} term4[r, t, n] +$$

$$\sum_{m=1}^{3} \sum_{n=1}^{3} term5[r, t, th, m, n] +$$

$$\sum_{m=1}^{3} \sum_{n=1}^{3} term6[r, t, th, m, n];$$

```
    uc = Compile[{r, t, th}, u[r, t, th]]
Out[1590] = CompiledFunction[{r, t, th}, u[r, t, th],
              -CompiledCode-]
```

The solution is *compiled* in uc. The command Compile is used to compile functions. Compile returns a CompiledFunction which represents the compiled code. Generally, compiled functions take less time to perform computations than uncompiled functions, although compiled functions can only be evaluated for numerical arguments.

Next, we define the function `tplot` which uses `ParametricPlot3D` to produce the graph of the solution for a particular value of *t*. Note that the *x* and *y* coordinates are given in terms of polar coordinates.

```
In[1591] := Clear[tplot]

            tplot[t_] := ParametricPlot3D[
               {r Cos[θ], r Sin[θ], uc[r, t, θ]}, {r, 0, 1},
               {θ, -π, π}, PlotPoints → {20, 20},
               BoxRatios → {1, 1, 1}, Shading → False,
               Axes → False, Boxed → False,
               DisplayFunction → Identity]
```

A table of nine plots for nine equally spaced values of *t* from *t* = 0 to *t* = 1 using increments of 1/8 is then generated. This table of graphs is displayed as a graphics array in Figure 6-47.

```
In[1592] := somegraphs = Table[tplot[t], {t, 0, 1, 1/8}];

            toshow = Partition[somegraphs, 3];

            Show[GraphicsArray[toshow]]
```

Of course, we can generate many graphs with a `Do` loop and animate the result as in the previous example. Be aware, however, that generating many three-dimensional graphics and then animating the results uses a great deal of memory and can take considerable time, even on a relatively powerful computer.

∎

6.5.3 Other Partial Differential Equations

A partial differential equation of the form

$$a(x, y, u)\frac{\partial u}{\partial x} + b(x, y, u)\frac{\partial u}{\partial y} = 0c(x, y, u) \tag{6.43}$$

is called a **first-order, quasi-linear partial differential equation**. In the case when $c(x, y, u) =$, equation (6.43) is **homogeneous**; if a and b are independent of u, equation (6.43) is **almost linear**; and when $c(x, y, u)$ can be written in the form $c(x, y, u) =$

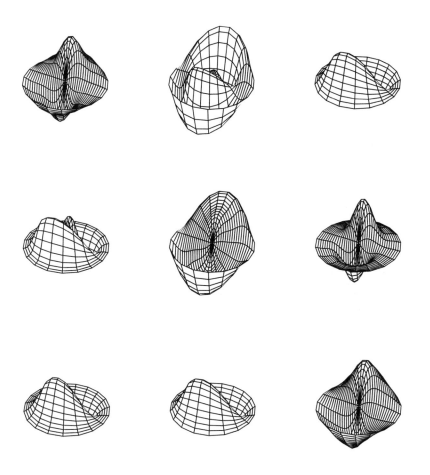

Figure 6-47   The drumhead for nine equally spaced values of t from $t = 0$ to $t = 1$

$d(x, y)u + s(x, y)$, equation (6.43) is **linear**. Quasi-linear partial differential equations can frequently be solved using the *method of characteristics*.

EXAMPLE 6.5.4:   Use the *method of characteristics* to solve the initial value problem $\begin{cases} -3xtu_x + u_t = xt \\ u(x, 0) = x. \end{cases}$

SOLUTION: For this problem, the *characteristic system* is

$$\partial x/\partial r = -3xt, \qquad\qquad x(0, s) = s$$
$$\partial t/\partial r = 1, \qquad\qquad t(0, s) = 0$$
$$\partial u/\partial r = xt, \qquad\qquad u(0, s) = s.$$

We begin by using DSolve to solve $\partial t/\partial r = 1, t(0, s) = 0$

```
In[1593]:= d1 = DSolve[{D[t[r],r] == 1, t[0] == 0}, t[r], r]
Out[1593]= {{t[r] → r}}
```

and obtain $t = r$. Thus, $\partial x/\partial r = -3xr, x(0, s) = s$ which we solve next

```
In[1594]:= d2 = DSolve[{D[x[r],r] == -3 x[r] r, x[0] == s},
              x[r], r]
Out[1594]= {{x[r] → e^{-\frac{3r^2}{2}} s}}
```

and obtain $x = se^{-3r^2/2}$. Substituting $r = t$ and $x = se^{-3r^2/2}$ into $\partial u/\partial r = xt$, $u(0, s) = s$ and using DSolve to solve the resulting equation yields the following result, named d3.

```
In[1595]:= d3 = DSolve[{D[u[r],r] == e^{-\frac{3r^2}{2}} s r, u[0] == s},
              u[r], r]
Out[1595]= {{u[r] → \frac{1}{3} e^{-\frac{3r^2}{2}} (-1 + 4 e^{\frac{3r^2}{2}}) s}}
```

To find $u(x, t)$, we must solve the system of equations

$$\begin{cases} t = r \\ x = se^{-3r^2/2} \end{cases}$$

for r and s. Substituting $r = t$ into $x = se^{-3r^2/2}$ and solving for s yields $s = xe^{3t^2}/2$. Thus, the solution is given by replacing the values obtained above in the solution obtained in d3. We do this below by using ReplaceAll (/.) to replace each occurrence of r and s in d3[[1,1,2]], the solution obtained in d3, by the values $r = t$ and $s = xe^{3t^2}/2$. The resulting output represents the solution to the initial value problem.

```
In[1596]:= d3[[1, 1, 2]] /. {r- > t, s- > x Exp[3/2 t^2]}//
              Simplify
Out[1596]= \frac{1}{3} (-1 + 4 e^{\frac{3t^2}{2}}) x
```

In this example, DSolve can also solve this first-order partial differential equation.

Next, we use DSolve to find a general solution of $-3xtu_x + u_t = xt$ and name the resulting output gensol.

> *In[1597]:=* **gensol =**
> **DSolve[-3x t D[u[x, t], x] + D[u[x, t], t] == x t,**
> **u[x, t], {x, t}]**
>
> *Out[1597]=* $\left\{\left\{u[x, t] \rightarrow \frac{1}{3}\left(-x + 3 C[1]\left[\frac{1}{6}(3 t^2 + 2 \text{Log}[x])\right]\right)\right\}\right\}$

The output

> *Out[1597]=* $C[1]\left[-\frac{3 t^2}{2} - \text{Log}[x]\right]$

represents an arbitrary function of $-\frac{3}{2}t^2 - \ln x$. The explicit solution is extracted from gensol with gensol[[1,1,2]], the same way that results are extracted from the output of DSolve commands involving ordinary differential equations.

> *In[1598]:=* **gensol[[1, 1, 2]]**
>
> *Out[1598]=* $\frac{1}{3}\left(-x + 3 C[1]\left[\frac{1}{6}(3 t^2 + 2 \text{Log}[x])\right]\right)$

To find the solution that satisfies $u(x, 0) = x$ we replace each occurrence of t in the solution by 0.

> *In[1599]:=* **gensol[[1, 1, 2]] /. t- > 0**
>
> *Out[1599]=* $\frac{1}{3}\left(-x + 3 C[1]\left[\frac{\text{Log}[x]}{3}\right]\right)$

Thus, we must find a function $f(x)$ so that

$$-\frac{1}{2}x + f(\ln x) = x$$

$$f(\ln x) = \frac{3}{2}x.$$

Certainly $f(t) = \frac{4}{3}e^{-t}$ satisfies the above criteria. We define $f(t) = \frac{4}{3}e^{-t}$ and then compute $f(\ln x)$ to verify that $f(\ln x) = \frac{3}{2}x$.

> *In[1600]:=* **Clear[f]**
>
> **f[t_] = 4 Exp[-t]/3;**
>
> **f[- Log[x]]**
>
> *Out[1600]=* $\frac{4 x}{3}$

Thus, the solution to the initial value problem is given by $-\frac{1}{3}x + f\left(-\frac{3}{2}t^2 - \ln x\right)$ which is computed and named sol. Of course, the result returned is the same as that obtained previously.

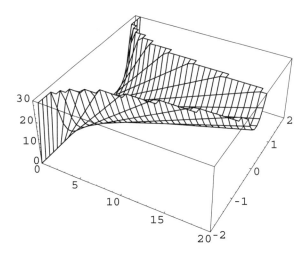

Figure 6-48   Plot of $u(x, t) = \frac{1}{3}x\left(4e^{3t^2/2} - 1\right)$

$In[1601]:= $ **sol = Simplify$\left[-\dfrac{\text{x}}{3} + \text{f}\left[-\dfrac{3\,\text{t}^2}{2} - \text{Log[x]} \right] \right]$**

$Out[1601]= $ $\dfrac{1}{3}\left(-1 + 4\,e^{\frac{3\,t^2}{2}} \right)\,x$

Last, we use Plot3D to graph sol on the rectangle $[0, 20] \times [-2, 2]$ in Figure 6-48. The option ClipFill->None is used to indicate that portions of the resulting surface which extend past the bounding box are not shown: nothing is shown where the surface is clipped.

$In[1602]:= $ **Plot3D[sol, {x, 0, 20}, {t, -2, 2},**
              PlotRange → {0, 30}, PlotPoints → 30,
              ClipFill → None, Shading → False]

■

Bibliography

[1] Abell, Martha and Braselton, James, *Differential Equations with Mathematica*, Third Edition, Academic Press, 2004.

[2] Abell, Martha and Braselton, James, *Modern Differential Equations*, Second Edition, Harcourt, 2001.

[3] Abell, Martha L., Braselton, James P., and Rafter, John A., *Statistics with Mathematica*, Academic Press, 1999.

[4] Barnsley, Michael, *Fractals Everywhere*, Second Edition, Morgan Kaufmann, 2000.

[5] Braselton, James P., Abell, Martha L., and Braselton, Lorraine M., "When is a surface *not* orientable?", *International Journal of Mathematical Education in Science and Technology*, Volume 33, Number 4, 2002, pp. 529–541.

[6] Devaney, Robert L. and Keen, Linda (eds.), *Chaos and Fractals: The Mathematics Behind the Computer Graphics*, Proceedings of Symposia in Applied Mathematics, Volume 39, American Mathematical Society, 1989.

[7] Edwards, C. Henry and Penney, David E., *Calculus with Analytic Geometry*, Fifth Edition, Prentice Hall, 1998.

[8] Edwards, C. Henry and Penney, David E., *Differential Equations and Boundary Value Problems: Computing and Modeling*, Third Edition, Pearson/Prentice Hall, 2004.

[9] Gaylord, Richard J., Kamin, Samuel N., and Wellin, Paul R., *Introduction to Programming with Mathematica*, Second Edition, TELOS/Springer-Verlag, 1996.

[10] Graff, Karl F., *Wave Motion in Elastic Solids*, Oxford University Press/Dover, 1975/1991.

[11] Gray, Alfred, *Modern Differential Geometry of Curves and Surfaces*, Second Edition, CRC Press, 1997.

[12] Gray, John W., *Mastering Mathematica: Programming Methods and Applications*, Second Edition, Academic Press, 1997.

[13] Kyreszig, Erwin, *Advanced Engineering Mathematics*, Seventh Edition, John Wiley & Sons, 1993.

[14] Larson, Roland E., Hostetler, Robert P., and Edwards, Bruce H., *Calculus with Analytic Geometry*, Sixth Edition, Houghton Mifflin, 1998.

[15] Maeder, Roman E., *The Mathematica Programmer II*, Academic Press, 1996.

[16] Maeder, Roman E., *Programming in Mathematica*, Third Edition, Addison-Wesley, 1996.

[17] Robinson, Clark, *Dynamical Systems: Stability, Symbolic Dynamics, and Chaos*, Second Edition, CRC Press, 1999.

[18] Smith, Hal L. and Waltman, P., *The Theory of the Chemostat: Dynamics of Microbial Competition*, Cambridge University Press, 1995.

[19] Stewart, James *Calculus: Concepts and Contexts*, Second Edition, Brooks/Cole, 2001.

[20] Weisstein, Eric W., *CRC Concise Encyclopedia of Mathematics*, CRC Press, 1999.

[21] Wolfram, Stephen, *A New Kind of Science*, Wolfram Media, 2002.

[22] Wolfram, Stephen, *The Mathematica Book*, Fourth Edition, Wolfram Media, 2004.

[23] Zwillinger, Daniel, *Handbook of Differential Equations*, Second Edition, Academic Press, 1992.

Index

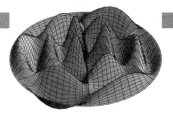